PHARMACOLOGY - RESEARCH, SAFETY TESTING AND REGULATION

ADENOSINE SIGNALING MECHANISMS

PHARMACOLOGY, FUNCTIONS AND THERAPEUTIC ASPECTS

PHARMACOLOGY - RESEARCH, SAFETY TESTING AND REGULATION

Additional books in this series can be found on Nova's website
under the Series tab.

Additional e-books in this series can be found on Nova's website
under the e-book tab.

PHARMACOLOGY - RESEARCH, SAFETY TESTING AND REGULATION

ADENOSINE SIGNALING MECHANISMS

PHARMACOLOGY, FUNCTIONS AND THERAPEUTIC ASPECTS

VICKRAM RAMKUMAR
AND
ROBERTO PAES DE CARVALHO
EDITORS

NOVA BIOMEDICAL

New York

Library of Congress Cataloging-in-Publication Data

Library of Congress Control Number: 2015942933
ISBN: 978-1-63483-186-4

Published by Nova Science Publishers, Inc. † New York

Contents

PREFACE

The field of adenosine and adenosine receptors dates back to the early work of Drury and Szent-Gyorgyi in the 1920s. However, it was only after several decades later, in the 1960s, that Berne and colleagues showed that adenosine released during ischemia increased coronary vessel dilation. Subsequent classification of purinergic receptors into P_1 and P_2 subclasses allowed researchers to focus on the adenosine preferring (P_1) receptors or the ATP preferring (P_2) receptors. An explosion of research in both of these areas ensued in the late 1980s through the 1990s. Much of these early studies characterized the physiological and pharmacological properties of these receptors in terms of second messenger signaling and regulation. These studies were aided, in large parts, by the availability of excellent receptor-specific drugs, radioligands and techniques useful for studying G protein coupling proteins and second messenger production. The subsequent cloning of four adenosine receptors further expanded research in the area of the P_1 adenosine receptors, especially in the area of receptor regulation at the level of the genes.

An interesting development in the area of adenosine receptor research is the finding that these receptors can exist and function as oligomers. A homodimer defines two adenosine receptor molecules of the same class, while heterodimer defines oligomerization of different adenosine receptor or different receptor molecules. These oligomers have different biochemical and functional properties and localization compared to the individual receptors. Furthermore, ligation of one of the receptor could dramatically alter the properties of the oligomer, such as post-receptor interactions. This has led to a diversity of interactions and functions ascribed to each adenosine receptor. Differences in subunit composition of the oligomers could produce different functions and help to fine-tune receptor function at a particular location. However, these differences could also make it more difficult to predict the exact function of adenosine receptor specific drugs.

The past decade has seen an increase in the number of studies defining the possible involvement of adenosine receptors in modulating various disease processes. Beneficial roles of adenosine receptor drugs have been described for various disease models, including cancer, Parkinson's disease and various neurological diseases, asthma, sleep, inflammation, cardiovascular disease and hearing loss. To date various preclinical and clinical trials are underway to examine the utility of selective adenosine receptor drugs to treat a number of these conditions. A potential problem with drug treatment is to provide a clinical benefit without significant side effects. This could be enhanced by using selective drugs and/or localized delivery of these drugs. An alternative treatment strategy to boost local adenosine levels in tissues is to use adenosine transport inhibitors, such as dipyridamole. Several chapters in this book examine the utility of adenosine receptor drug therapy to treat diseases.

The past decade has also seen an increase in the use of caffeine, an adenosine receptor antagonist which is a common ingredient in various beverages such as coffee, tea and carbonated beverages. High doses of caffeine are consumed by pre-teens and teenagers in "energy" drinks which could affect their behavior and brain development. Moreover, consumption of alcohol with "energy" drinks on college campuses has increased the risk of alcohol-induced accidents and deaths. These are important health issues which require immediate attention from manufacturers of "energy" drinks and governmental agencies.

This book provides a description of the current status of adenosine receptor research. Several chapters provide insights into the properties and functions of adenosine receptors and their endogenous activator, adenosine. The remaining chapters provide a description of how these receptors are regulated in disease processes and how activation or inhibition of these receptors could alter the course of these diseases. It is our hope that the information in this book would be useful for both basic science and clinical researchers and physicians and provide a framework to conduct future research in this area of adenosine receptor research.

Vickram Ramkumar,
Southern Illinois University School of Medicine, Springfield, IL, USA

Roberto Paes de Carvalho,
Program of Neurosciences and Department of Neurobiology,
Fluminense Federal University, Niteroi, Brazil

AUTHORS' LIST

José Luis Albasanz
Departamento de Química Inorgánica, Orgánica y Bioquímica, Facultad de Ciencias y Tecnologías Químicas / Facultad de Medicina de Ciudad Real, Centro Regional de Investigaciones Biomédicas, Universidad de Castilla-La Mancha, Ciudad Real, Spain.

Alex Bicket
Department of Chemistry and Biology, Faculty of Science, Ryerson University, Toronto, ON, Canada.

Carina R. Boeck
Programa de Pos-graduacao em Nanociencias, Centro Universitario Franciscano, 97010-491, Santa Maria, RS, Brazil.

Vikrant Borse
Department of Pharmacology, Southern Illinois University School of Medicine, Springfield, IL 62702, USA.

Rafael Brito
Neuroscience Program, Neurobiology Department, Fluminense Federal University, Brazil.

Felipe Cabral-Miranda
Instituto de Biofisica, Universidade Federal do Rio de Janeiro, Rio de Janeiro, Cidade Universitaria, RJ, Brazil.

Karin C. Calaza
Neuroscience Program, Neurobiology Department, Fluminense Federal University, Brazil.

Paula Campello-Costa
Programa de Neurociencias, Departamento de Neurobiologia, Instituto de Biologia, Universidade Federal Fluminense, Caixa Postal 100180, Niteroi, CEP 24001-970, RJ, Brazil

and Instituto de Biofisica, Universidade Federal do Rio de Janeiro, Rio de Janeiro, Cidade Universitaria, RJ, Brazil.

Michele B. de Carvalho
Programa de Neurociencias, Departamento de Neurobiologia, Instituto de Biologia, Universidade Federal Fluminense, Caixa Postal 100180, Niteroi, CEP 24001-970, RJ, Brazil.

Francisco Ciruela
Unitat de Farmacologia, Departament de Patologia i Terapèutica Experimental, Facultat de Medicina, IDIBELL, Universitat de Barcelona, L'Hospitalet de Llobregat, 08907 Barcelona, Spain.

Imogen R. Coe
Department of Chemistry and Biology, Faculty of Science, Ryerson University, Toronto, ON, Canada.

Leandra C. Constantino
Programa de Pos-graduacao em Bioquimica, Departamento de Bioquímica, Centro de Ciências Biológicas, Universidade Federal de Santa Catarina, UFSC, Campus Trindade, 88040-900 Florianópolis, SC, Brazil.

Tharine Dal-Cim
Departamento de Bioquímica, Centro de Ciências Biológicas, Universidade Federal de Santa Catarina, UFSC, Campus Trindade, 88040-900 Florianópolis, SC, Brazil.

Maria Rosaria Domenici
Section of Central Nervous System Pharmacology, Department of Therapeutic Research and Medicines Evaluation, Istituto Superiore di Sanità, Rome, Italy.

Víctor Fernández-Dueñas
Unitat de Farmacologia, Departament de Patologia i Terapèutica Experimental, Facultat de Medicina, IDIBELL, Universitat de Barcelona, L'Hospitalet de Llobregat, 08907 Barcelona, Spain.

Sumana Ghosh
Department of Pharmacology, Southern Illinois University School of Medicine, Springfield, IL 62702, USA.

Elizabeth Giestal-de-Araujo
Universidade Federal Fluminense – Niteroi, Brazil .

Vanessa G. Goulart
Programa de Neurociencias, Departamento de Neurobiologia, Instituto de Biologia, Universidade Federal Fluminense, Caixa Postal 100180, Niteroi, CEP 24001-970, RJ, Brazil

Natalia Grañé-Boladeras
Department of Chemistry and Biology, Faculty of Science, Ryerson University, Toronto, ON, Canada.

Elisa Guimarães-Souza
Neuroscience Program, Neurobiology Department, Fluminense Federal University, Brazil.

Tejbeer Kaur
Department of Otolaryngology, Washington University School of Medicine, St. Louis, MO 63110, USA.

Debora Lanzaster
Departamento de Bioquímica, Centro de Ciências Biológicas, Universidade Federal de Santa Catarina, UFSC, Campus Trindade, 88040-900 Florianópolis, SC, Brazil.

Mairena Martin
Departamento de Química Inorgánica, Orgánica y Bioquímica, Facultad de Ciencias y Tecnologías Químicas / Facultad de Medicina de Ciudad Real, Centro Regional de Investigaciones Biomédicas, Universidad de Castilla-La Mancha, Ciudad Real; Spain.

Alberto Martire
Section of Central Nervous System Pharmacology, Department of Therapeutic Research and Medicines Evaluation, Istituto Superiore di Sanità, Rome, Italy.

Simone Molz
Curso de Farmacia, Universidade do Contestado – UnC, 89460-000, Canoinhas, SC, Brazil.

Debashree Mukherjea
Department of Surgery (Otolaryngology), Southern Illinois University School of Medicine, Springfield, IL 62702, USA.

Mark E. Olah
Department of Pharmaceutical and Biomedical Sciences, Raabe College of Pharmacy, Ohio Northern University, Ada, OH 45810, USA.

Nádia A. de Oliveira
Program of Neurosciences and Department of Neurobiology, Fluminense Federal University, Niteroi, Brazil.

Roberto Paes-de-Carvalho
Program of Neurosciences and Department of Neurobiology, Fluminense Federal University, Niteroi, Brazil.

Timothy M. Palmer
Institute of Cardiovascular and Medical Sciences, University of Glasgow, Glasgow G12 8QQ, Scotland, United Kingdom and School of Pharmacy, University of Bradford, Bradford BD7 1DP, West Yorkshire, United Kingdom.

Mariana R. Pereira
Program of Neurosciences and Department of Neurobiology, Fluminense Federal University, Niteroi, Brazil.

Danniel Pereira-Figueiredo
Neuroscience Program, Neurobiology Department, Fluminense Federal University, Brazil.

Rafael Perígolo-Vicente
Universidade Federal do Rio de Janeiro – Rio de Janeiro, Brazil and Universidade Federal Fluminense, Niteroi, Brazil.

Patrizia Popoli
Section of Central Nervous System Pharmacology, Department of Therapeutic Research and Medicines Evaluation, Istituto Superiore di Sanità, Rome, Italy.

Camila C. Portugal
Program of Neurosciences and Department of Neurobiology, Fluminense Federal University, Niteroi, Brazil; and Instituto de Investigacao e Inovacao em Saude and Instituto de Biologia Molecular e Celular (IBMC), Universidade do Porto, Porto, Portugal.

Vickram Ramkumar
Department of Pharmacology, Southern Illinois University School of Medicine, Springfield, IL 62702, USA.

Joaquim A. Ribeiro
Instituto de Farmacologia e Neurociências, Faculdade de Medicina, Universidade de Lisboa, and Unidade de Neurociências, Instituto de Medicina Molecular, Universidade de Lisboa, Lisboa Portugal.

Henrique Rocha-Mendonça
Programa de Neurociencias, Departamento de Neurobiologia, Instituto de Biologia, Universidade Federal Fluminense, Caixa Postal 100180, Niteroi, CEP 24001-970, RJ, Brazil.

Diogo M. Rombo
Instituto de Farmacologia e Neurociências, Faculdade de Medicina, Universidade de Lisboa, and Unidade de Neurociências, Instituto de Medicina Molecular, Universidade de Lisboa, Lisboa Portugal.

Claire Rutherford
Institute of Cardiovascular and Medical Sciences, University of Glasgow, Glasgow G12 8QQ, Scotland, United Kingdom.

Leonard P. Rybak
Department of Surgery (Otolaryngology), Southern Illinois University School of Medicine, Springfield, IL 62702, USA.

Alexandre dos Santos-Rodrigues
Program of Neurosciences and Department of Neurobiology, Institute of Biology, Fluminense Federal University, Niteroi, RJ, Brazil.

Ana M. Sebastião
Instituto de Farmacologia e Neurociências, Faculdade de Medicina, Universidade de Lisboa, and Unidade de Neurociências, Instituto de Medicina Molecular, Universidade de Lisboa, Lisboa Portugal.

Kelly Sheehan
Department of Surgery (Otolaryngology), Southern Illinois University School of Medicine, Springfield, IL 62702, USA.

Sandeep Sheth
Department of Pharmacology, Southern Illinois University School of Medicine, Springfield, IL 62702, USA.

Renato Socodato
Program of Neurosciences and Department of Neurobiology, Fluminense Federal University, Niteroi, Brazil; and Instituto de Investigacao e Inovacao em Saude and Instituto de Biologia Molecular e Celular (IBMC), Universidade do Porto, Porto, Portugal.

Carla I. Tasca
Departamento de Bioquímica, Programa de Pós-graduação em Bioquímica, Programa de Pós-graduação em Neurociências, Centro de Ciências Biológicas, Universidade Federal de Santa Catarina, UFSC, Campus Trindade, 88040-900 Florianópolis, SC, Brazil.

Ana L. Tavares-Gomes
Programa de Neurociencias, Departamento de Neurobiologia, Instituto de Biologia, Universidade Federal Fluminense, Caixa Postal 100180, Niteroi, CEP 24001-970, RJ, Brazil.

Maria Teresa Tebano
Section of Central Nervous System Pharmacology, Department of Therapeutic Research and Medicines Evaluation, Istituto Superiore di Sanità, Rome, Italy.

Srinivisan Tupal
Department of Pharmacology, Southern Illinois University School of Medicine, Springfield, IL 62702, USA.

Maliha Zafar
Department of Chemistry and Biology, Faculty of Science, Ryerson University, Toronto, ON, Canada.

In: Adenosine Signaling Mechanisms
Editors: V. Ramkumar and R. Paes de Carvalho
ISBN: 978-1-63483-186-4

Chapter 1

ADENOSINE AND ADENOSINE RECEPTORS: AN OVERVIEW

Vickram Ramkumar**, *Sandeep Sheth and Srinivisan Tupal
Department of Pharmacology, Southern Illinois University School of Medicine, Springfield, IL, US

ABSTRACT

Since its discovery as a regulator of cardiovascular system, much has been learnt about the targets of action of adenosine and the various other functions it regulates. To date, this nucleoside has been shown to control various physiological processes such as sleep, metabolic activity, cardiovascular function, diuresis, pain and immune function. Adenosine is not only a significant contributor to cytoprotection and cell survival but also appears to possess antitumor properties against certain types of cancers. Receptor targets for adenosine (adenosine receptors) have been identified, which show differential tissue distribution and coupling to effectors. A number of agonist and antagonist drugs have been developed which show relative selectivity for each receptor subtype. The cloning of four adenosine receptors has greatly accelerated the pace of research concerning the nature of the ligand binding site and the coupling of these receptors to G proteins and different effectors. This chapter reviews some of the historical progression of adenosine receptor research and presents some areas in which research has led to identification of potential targets for drug treatment. An overview of the upcoming chapters is also discussed.

ADENOSINE RECEPTOR CHARACTERIZATION AND FUNCTION

Adenosine Synthesis and Metabolism

Adenosine is produced from its immediate precursor, 5'-adenosine monophosphate (5'-AMP), by the action of 5'-nucleotidase. Normal level of adenosine is controlled by

*Corresponding author: E-mail: vramkumar@siumed.edu, Phone: +1(217) 545-2171, Fax: +1(217) 545-0145.

metabolism by adenosine deaminase to inosine and hypoxanthine or by adenosine kinase to 5'-AMP and subsequently to ADP and ATP. Various adenosine transporters are also involved in shuttling adenosine across the plasma membrane. These include equilibrative nucleoside transporters (ENTs) and the concentrative nucleoside transporters (CNTs) [1]. A detailed discussion of these transporters is presented in Chapter 2 of this book by Imogen Coe and co-workers.

Characterization of Adenosine Receptors

Drury and Szent-Gyorgyi [2] were the first to describe effects of adenylyl purines on the heart and coronary vessels. These investigators showed that intravenous administration of these substances produced bradycardia and enhanced coronary blood flow. Several decades later, studies by Berne et al., [3] showed that adenosine released during hypoxia mediates coronary blood flow. These effects are mediated by adenosine receptors (ARs) that are sensitive to circulating levels of endogenous adenosine. These receptors were originally characterized by Burnstock [4] as P_1 purinergic receptors, as opposed to P_2 purinergic receptors which bind ATP and ADP. P_1 and P_2 purinergic receptors could also be differentiated by their differential sensitivity to antagonists. The P_1 receptors were inhibited by methylxanthine antagonists such as theophylline, caffeine and 3-isobutyl-1-methylxanthine while the P_2 receptors were inhibited by quinidine and high concentrations of 2-substituted imidazoline drugs. Subsequent studies identified two distinct P_1 subclasses, namely the A_1 and A_2ARs, based on their potencies for agonists. The A_1AR was classified based on the rank order of potency for responses – *R*-phenylisopropyladenosine (*R*-PIA) = cyclohexyladenosine (CHA) > 5'-N-ethylcarboxamide adenosine (NECA) > *S*-PIA. Responses mediated by the A_2AR showed the potency NECA > *R*-PIA = 2-chloroadenosine > *S*-PIA [5,6]. The A_1 and A_2AR also differed in their G protein coupling and ability to regulate adenylyl cyclase activity. While the A_1AR reduced adenylyl cyclase activity in many tissues, activation of the A_{2A}R increased enzyme activity. At the time, the second messenger resulting from enzyme activity, cyclic AMP, was believed to mediate all of the physiological roles of adenosine. Other effector systems for the A_1AR were later identified. Activation of the A_1AR was shown to inhibit voltage-dependent Ca^{2+} current [7,8], stimulate cardiac K^+ conductance [8-10], activate the low K_m cyclic AMP phosphodiesterase [11, 12], guanylate cyclase [13] and potentiate phosphatidylinositol hydrolysis [14, 15].

Early characterization of the A_1 and A_2AR was made possible by receptor binding assays using radiolabeled AR drugs. In their study, Malbon et al., [16] showed high levels and methylxanthine-sensitive binding of [^{3}H]adenosine to adipocyte membranes. The rapid degradation of the radioligand required the use of adenosine deaminase inhibitors for this study. In retrospect, the high picomole/mg protein of AR detected in adipocyte membranes suggests that the radioligand was also binding to other non-receptor targets of adenosine. The development of other radioligands which were resistant to metabolism by adenosine deaminase accelerated the pace of research in this area. Several of these were agonist radioligands ([^{3}H]PIA, [^{3}H]CHA, and [^{3}H]2-chloroadenosine), while [^{3}H]DPCPX is an antagonist for the A_1AR. Although the use of agonist radioligands improved the quality and reproducibility of radioligand binding experiments, these ligands could only detect the high affinity state of the A_1AR/G_i complex, which resulted in an underestimation of receptor density. In 1986, Jacobson et al., [17] developed [^{3}H]xanthine amine congener ([^{3}H]XAC]), a high affinity A_1AR selective antagonist radioligand. This radioligand, along with [^{3}H]8-cyclopentyl-1,3-dipropylxanthine (DPCPX) [18], were widely used in the characterization and purification of the A_1AR. Photoaffinity labelling of the A_1AR was subsequently shown using agonist [19, 20] and antagonist radioligands [21, 22]. From these studies, it appeared that the photoaffinity labelled A_1AR migrated as a 34-38 kDa protein on SDS PAGE [19-22]. Early characterization of the A_2AR was made possible using [^{3}H]NECA, even though this drug has the limitation of binding to both A_1 and A_2AR [23]. The synthesis of 2-[4-(2-[(4-aminophenyl]methylcarbonyl)ethyl)phenyl]ethylamino-5'-N-ethylcarboxamidoadenosine (PAPA-APEC) provided an important tool for studying the A_2AR. The iodinated form of PAPA-APEC was useful for studying A_{2A}R pharmacology, structure [20] and regulation [24,25].

Using [^{3}H]XAC to label the A_1AR, it was shown that the interaction of agonists with this receptor was sensitive to GTP analogs. Accordingly, GTP analogs decreased the high affinity binding of agonists, as would be expected for a G protein-coupled receptor [26]. Surprisingly, it was shown that GTP analogs also affect the binding of the antagonist, albeit in the opposite direction. It was shown that GTP analogs increase the maximum binding capacity of the A_1AR defined by [^{3}H]DPX [27] or binding affinity using [^{3}H]XAC [28]. These data suggest that guanine nucleotides can reciprocally modulate agonist and antagonist binding to the A_1AR and that uncoupling of the A_1AR from its G proteins (using guanine nucleotides) facilitates antagonist binding. An earlier study had shown that the coupling of the A_1AR to the G_i protein was extremely "tight" even in the absence of agonist [29]. This tight coupling might put some constraints on the A_1AR confirmation which is relieved when the receptor is dissociated from its G_i proteins (i.e., with guanine nucleotides). Theoretically, the uncoupled receptor conformation possesses a higher affinity for antagonists. One argument against this explanation is that guanine nucleotides reduce the binding of endogenous adenosine to the A_1AR (even in presence of adenosine deaminase) and thereby "appear" to stimulate antagonist binding [30].

Purification and Cloning of the Adenosine Receptors

In the late 1980s, a number of laboratories devoted considerable effort and resources to purify the A_1AR. It was reasoned that the availability of a purified receptor protein would allow for partial amino acid sequencing and cloning of this receptor. While partial purification of the A_1AR was shown by a number of laboratories the level of purity and the limited quantity of the protein obtained did not allow for sequencing studies [31]. However, in 1990, Hiro Nakata [32] was able to obtain more than 25,000-fold purification of the A_1AR, using a sequence of antagonist affinity columns, hydroxyapatite and gel permeation columns. Interestingly, the A_1AR was originally cloned as an orphan receptor (termed RDC7) from dog thyroid cDNA library based on degenerate primers of conserved sequences among G protein coupled receptors [33] and not on protein sequencing data. These investigators demonstrate high levels of A_1AR transcripts in the brain, thyroid, and testis and lesser amounts in the heart and kidney [33]. When expressed in Chinese hamster ovary (CHO) cells, the receptor showed the appropriate pharmacology and inhibition of forskolin stimulated adenylyl cyclase activity. A subsequent study cloned the A_1AR and characterized the bovine brain cDNA library and identified histidine-258 and -271 in the receptor sequence as being important drug binding [34]. The cloning and expression of the A_1AR and the A_2AR [33] and subsequently of the $A_{2B}AR$ [35, 36] has stimulated a great deal of research into the interaction of these receptors with G proteins and various signalling pathways.

Cloning and identifying the A_3AR also occurred through a circuitous route. The receptor clone (R226) was identified from a rat striatal cDNA library. When expressed in COS-7 cells, clone R226 bound to two AR agonist radioligand N^6-2-(4-amino-3-iodophenyl)-ethyladenosine [^{125}I]APNEA and [^{3}H]NECA, but with pharmacology distinct from the A_1 and A_2AR, thus the designation of A_3AR [37]. The identification of this new class of ARs helped to explain previous observations of a different class of AR which did not fit the traditional A_1 and A_2AR class [38-41]. A subsequent study in RBL-2H3 cells showed that the A_3AR was the putative AR which mediates adenosine-enhanced degranulation of these rat basophilic cells [42].

Adenosine Receptors Regulation in Disease States

The availability of ligands for studying AR levels and their G coupling proteins enabled a more detailed examination of the function and expression of these receptors in different disease conditions. The rat adipocyte model proved excellent for these studies since the A_1AR produced significant suppression of adenylyl cyclase activity in these cells. A few examples of changes in AR functions and regulation in different conditions are listed below.

Adenosine inhibits lipolysis by acting on A_1AR on adipocytes. In the hypothyroid state, the inhibitory effect of adenosine on lipolysis is enhanced. This effect was not linked to changes in A_1AR levels, but could result from increased levels of G_i coupling proteins [43] which transduce the receptor signal to inhibition of adenylyl cyclase. Since increase in the inhibitory G_i protein signal would be expected to negatively influence the G_s pathway, this finding could also explain why the stimulation of lipolysis by catecholamines is reduced in hypothyroidism. In contrast to the hypothyroid state, hyperthyroidism is associated with increased lipolysis [44] which likely results from a reduction in adipocyte A_1AR levels and reduction in tonic inhibition of lipolysis via this receptor [45]. A subsequent study, using a hyperthyroid model in rats, demonstrated reduced basal and stimulated adenyl cyclase activity in adipocytes obtained from hyperthyroid rats. No change in A_1AR or β-adrenergic receptors was observed [46]. The authors concluded that the deficit in stimulated adenylyl cyclase likely resulted from a decrease in the catalytic moiety of adenylyl cyclase.

Early studies suggest a reduction in the antilipolytic action of adenosine, mediated via the A_1AR, is reduced in obese subjects [47]. These investigators observed a negative correlation between body weight and the effectiveness of the agonist, *R*-PIA. This reduction in agonist efficacy was accompanied by a decrease in A_1AR-mediated inhibition of cyclic AMP accumulation in adipocytes. One explanation for these findings is that the A_1AR becomes partially desensitized during obesity, probably as a result of increased release of adenosine into the extracellular space. Support for this was provided in a study by Hoffman et al., [48] which showed that a 6 day infusion of *R*-PIA in rats decreased the antilipolytic action of this drug tested in isolated adipocytes along with the a reduction in inhibition of adenylyl cyclase activity. An interesting finding of this study was the potentiation of β-adrenergic receptor and forskolin stimulation of cyclic AMP accumulation in adipocytes desensitized to *R*-PIA. In contrast to obesity, the antilipolytic action of adenosine is enhanced in aging, which could account for the decreased lipolytic action of β-adrenergic receptor agonists [49].

Starvation can also regulate the function of adenosine on lipolysis in human. It appears that in the fed state which reflects normal levels of adenosine, the preferred target of catecholamines is the β-adrenergic receptor which promotes lipolysis. However, starvation reduces the levels of adenosine and the modulatory role of this nucleoside and shifts the preferred target of catecholamines to the α_2-adrenergic receptor, which mediate inhibition of lipolysis [50]. Thus, adenosine levels determine the function of catecholamines on lipolysis.

Desensitization and Sensitization of Adenosine Receptors

The use of radioligands for the A_1 and A_2ARs led to a significant amount of research into defining molecular pathways by which these receptors become tolerant (desensitized) or sensitized upon prolonged exposure to the ligands. Early studies of A_1AR receptor desensitization used acutely prepared or cultured adipocytes obtained from rat epididymal fat pads. In an elegant study, Hoffman et al., [48] showed that infusion of *R*-PIA to rats for 6 days led to tolerance development to the normal antilipolytic actions of the drug. This was associated with reduced *R*-PIA dependent inhibition of isoproterenol-induced cyclic AMP accumulation. An interesting observation was that the isoproterenol-stimulated cyclic AMP accumulation in the *R*-PIA tolerant adipocytes was higher that the control adipocytes. This suggests that the β-adrenergic receptor signalling was enhanced in the *R*-PIA tolerant adipocytes, probably due to a loss of tonic inhibition by the A_1AR. In a subsequent study in rats, it was shown that a similar *R*-PIA treatment protocol led to desensitization and decreased surface labelling of the A_1AR along with reductions in the G_i proteins. Interestingly, adipocytes which were desensitized to *R*-PIA showed increased G_s proteins but no changes in [^{125}I]iodocyanopindolol binding to β receptors [51]. An extensive discussion on AR desensitization could be found in the chapter by Tim Palmer (Chapter 3). In contrast to agonist-mediated desensitization, treatment of rats with caffeine increased A_1AR high affinity state and the levels of G_i proteins [52]. These changes could sensitize the A_1AR to the endogenous agonist adenosine.

CAFFEINE

Caffeine is the most widely consumed behaviourally active compound in the world. The drug is usually consumed in coffee, tea or carbonated beverages. Average intake can range between 80-400 mg daily resulting in blood levels of 5-20 μM [53]. Caffeine can affect the normal physiological processes leading to arousal, cardiac stimulant, changes in respiration rate and diuresis. The drug also has limited clinical utility in the treatment of narcolepsy, asthma and apnea. CNS stimulant actions of caffeine and various methylxanthines were originally attributed to their ability to inhibit binding to brain ARs [54]. Low dose caffeine stimulates motor activity in human while high dose caffeine has a depressant effect. Knockout of the A_{2A}AR led to a depressant effect of caffeine [55].

Caffeine dependence is observed in some individuals in whom caffeine is abruptly terminated after a prolonged period of use. Caffeine withdrawal was implicated in the development of the headaches. Increased incidence of headaches was observed in individuals who fast during the observance of Ramadan (Moslems' month of fasting) [56] or Yom Kippur (Jewish Day of Atonement) [57] compared to non-fasters. In addition to headaches, withdrawal symptoms include drowsiness, fatigue, difficulty concentration, and irritability and are observed during 12-48 h following withdrawal. Caffeine does not produce the same amount of dependency compared to classical drugs of abuse, such as opiates and cocaine [53]. This is due, in part, to a lower level of release of dopamine from nucleus accumbens with caffeine as compared to classical drugs of abuse [58]. One idea which has been pursued to explain the caffeine withdrawal symptoms is the targets of therapeutic levels of caffeine, namely the ARs, are upregulated and inhibited during the period of caffeine consumption. This inhibition is abruptly relieved upon withdrawal of caffeine, allowing the endogenous ligand (adenosine) full access to these receptors. Sensitization of the A_1AR signalling following chronic caffeine administration has been observed in experimental animals [52,59].

A growing trend in recent years, especially among pre-teens, teenagers and young adults, is the consumption of energy drinks which contain high levels of caffeine. These individuals account for about 50% of the energy drink market of ~$9 billion in 2011, which is expected to increase to $13 billion by 2015 (www.statista.com/topics/1687/energy-drinks). The total amount of caffeine present in a can of energy drink could range from 80 mg to 500 mg [60]. The consumption of energy drinks has significantly increased the number of Emergency Department visits in the United States from caffeine overdoses. The affected individuals present with a range of symptoms including tachycardia, seizures, cardiac dysrhythmias, hypertension and kidney failure [61]. Future studies are needed to determine what effects chronic consumption of caffeine would have on normal development and the future health of individuals at these young ages. The consumption of energy drink along with alcohol, observed on many college campuses, is even more problematic. This combination appears to increase total alcohol consumption, increase risk taking behavior, and greatly impair driving ability [62].

THERAPEUTIC POTENTIALS OF THE ADENOSINE RECEPTORS

A major role of the AR under normal physiological conditions is to provide cytoprotection. This is observed in the central nervous system, where adenosine suppresses neuronal firing rate and decrease the release of excitatory neurotransmitters. During ischemia, the levels of endogenous adenosine can increase to micromolar concentrations, which confer neuroprotection. More details concerning roles of adenosine and adenosine receptors in neuroprotection are presented in chapters by Ana Sebastião and colleagues (Chapter 10), Popoli and co-workers (Chapter 13) and Rocha-Mendonça and co-workers (Chapter 14) in this book. Adenosine released during transient applications of low levels of ischemic stress could provide protection against a more pronounced episode of ischemia (preconditioning). A more extensive discussion of adenosine and adenosine receptors in preconditioning is presented in the chapter by Carla Tasca and co-workers (Chapter 12) in this book. Preserving extracellular adenosine by blocking its elimination pathways could serve as a useful neuroprotective strategy against ischemia [63]. A major mechanism which contributes to neuroprotection appears to involve the ability of adenosine to suppress inflammation [64]. A role of ARs in the regulation of various cytokines is discussed in a chapter by Perígolo-Vicente and Giestal-de-Araujo (Chapter 7).

Adenosine is a major inducer of sleep and is likely involved in the rebound sleep following sleep deprivation [65], presumably by regulating basal forebrain A_1AR [66]. Adenosine can also influence locomotion by activating A_1AR present on medium spiny GABAergic neurons of the direct pathway of the striatum, leading to suppression of dopamine D_1 receptor function. In addition, adenosine can also activate $A_{2A}AR$ present on medium spiny GABAergic neurons present in indirect pathway of the striatum, leading to suppression of D_2 dopamine receptor function. The functional interplay between the A_{2A} and D_2 dopamine receptor could be exploited for the treatment of Parkinson's disease [67]. The functions of the AR are mediated by homo- and heterodimerization of these receptors. A role of AR dimerization in the treatment of Parkinson's disease is presented in a chapter by Ciruela and co-workers (Chapter 11) in this book. A number of laboratories, including mine, have shown otoprotective role of the A_1AR in the cochlea. Studies to date indicate that this action is mediated by the A_1AR expressed in the cochlea and that A_1AR agonist could be useful otoprotective agents [68]. A more detailed discussion of adenosine and the peripheral auditory system is presented by Ramkumar and co-workers (Chapter 9) in this book.

Since its discovery as a modulator of cardiovascular function by Drury and Szent-Gyorgyi [2], adenosine and ARs have played a prominent role in cardiovascular research. The nucleoside is termed a "retaliatory metabolite" based on its negative feedback role to dampen cellular activities which stimulate its production. As in the central nervous system, adenosine provides protection in the cardiovascular system by decreasing heart rate and increasing vasodilation in areas of ischemia. A more detailed discussion of the role of adenosine and ARs in the cardiovascular system is presented in the chapter by Mark Olah (Chapter 4). A growing body of literature demonstrates a role of ARs, especially the A_3AR subtypes, in cancers. In a number of cancers, activation of the A_3AR confers anti-proliferative and anti-metastatic properties [69]. A detailed discussion of the roles of ARs in various cancers is presented by Sheth and co-workers (Chapter 8) in this book. Early studies demonstrated a role of adenosine in the regulation of cyclic AMP in embryonic chick retina [70] and the expression of the A_1AR during chick retinal development [71]. Expression of ARs in the retina has also been described previously [72]. A more detailed characterization of the expression and roles of ARs in the chick retina are presented in a chapter by Calaza and co-workers (Chapter 6). Developmental regulation of ARs in retinal development and their roles in cytoprotection are presented in a chapter by Paes de Carvalho and co-workers (Chapter 5). Effective drug delivery of adenosine-based drugs would be essential for treating various diseases in which ARs are implicated. The use of nanotechnology for delivery of adenosine-based drugs is described in chapter by Albasanz and Martin (Chapter 15).

Conclusion

This review highlights the progress of AR research over the past 40 years and identifies significant contributors to our current understanding of this area. This has led to a better understanding of the diversity of receptor targets and signaling pathways regulated by adenosine. In the process, a number of diseases were identified in which adenosine receptors have been implicated. The use of selective agonist or antagonist drugs for these receptors could provide effective treatment options for these conditions. An important hurdle to overcome is how to effectively target the drugs to their potential sites of action in order to decrease potential systemic side effects. This could be easily accomplished by direct drug delivery to the inner ear (by transtympanic injections), eye or into localized tumors. However, localizing the drug could be more difficult when the affected organ is the brain, heart or kidneys. Several areas of research which appear promising include using AR drugs to treat ischemia, Parkinson's disease, inflammation, cancer and hearing loss. With continued funding and the attraction of dedicated young investigators to this area, the future of adenosine receptor research looks bright.

REFERENCES

[1] Sheth, S; Brito, R; Mukherjea, D; Rybak, LP; Ramkumar, V. Adenosine receptors: expression, function and regulation. *Int. J. Mol. Sci.*, 2014, 15, 2024-2052.

[2] Drury, AN; Szent-Gyorgyi, A. The physiological activity of adenine compounds with especial reference to their action upon the mammalian heart. *J. Physiol.*, 1929, 68, 213-237.

[3] Katori, M; Berne, RM. Release of adenosine from anoxic hearts. Relationship to coronary flow. *Circ. Res.* 1966, 19, 420-425.

[4] Burnstock, G. Cell membrane receptors for drugs and hormones: a multidisciplinary approach, p. 107-118. Eds. Bolis, L., Straub, RW., Raven Press, New York, 1978.

[5] van Calker, D; Müller, M; Hamprecht, B. Adenosine regulates via two different types of receptors, the accumulation of cyclic AMP in cultured brain cells. *J. Neurochem.*, 1979, 33, 999-1005.

[6] Londos, C; Cooper, DM; Wolff, J. Subclasses of external adenosine receptors. *Proc. Natl. Acad. Sci. U S A.*, 1980, 77, 2551-2554.

[7] Dolphin, AC; Prestwich, SA; Forda, SR. Adenosine: Receptors and modulation of cell function, p. 107. Alan R. Liss, New York, New York, NY., 1985.

[8] Schubert, P. Adenosine: Receptor and modulation of cell function. pp. 233-239, Eds. Stafanovich, V., Rudolphi, K., Schubert, P. IRL Press Limited, Oxford, England, 1985.

[9] Böhm, M; Brückner, R; Neumann, J; Schmitz, W; Scholz, H; Starbatty, J. Role of guanine nucleotide-binding protein in the regulation by adenosine of cardiac potassium conductance and force of contraction. Evaluation with pertussis toxin. *Naunyn-Schmiedebergs Arch. Pharmacol.*, 1986, 332, 403-405.

[10] Kurachi, Y; Nakajima, T; Sugimoto, T. On the mechanism of activation of muscarinic K+ channels by adenosine in isolated atrial cells: involvement of GTP-binding proteins. *Pflugers Arch.* 1986, 407, 264-274.

[11] Teo, TS; Ooi, SO; Wong, EH. Stimulation of rat fat cell phosphodiesterase by adenosine. *FEBS Lett.*, 1981, 128, 75-78.

[12] de Mazancourt, P; Giudicelli, Y. Guanine nucleotides and adenosine "R(i)'-site analogues stimulate the membrane-bound low-K(m) cyclic AMP phosphodiesterase of rat adipocytes. *FEBS Lett.*, 173, 385-388.
[13] Kurtz A. Adenosine stimulates guanylate cyclase activity in vascular smooth muscle cells. *J. Biol. Chem.*, 1987, 262, 6296-6300.
[14] Hollingsworth, EB; De La Cruz, A; Daly, JW. Accumulations of inositol phosphates and cyclic AMP in brain slices: Synergistic interactions of histamine and 2-chloroadenosine. *Eur. J. Pharmacol.*, 122, 45-50.
[15] Hill, SJ; Kendall, DA. Studies on the adenosine-receptor mediating the augmentation of histamine-induced inositol phospholipid hydrolysis in guinea-pig cerebral cortex. *Br. J. Pharmacol.*, 1987, 91, 661-669.
[16] Malbon, CC; Hert, RC; Fain, JN. Characterization of [3H]adenosine binding to fat cell membranes. *J. Biol. Chem.*, 1978, 253, 3114-3122.
[17] Jacobson, KA; Ukena, D; Kirk, KL; Daly, JW. [^{3}H]xanthine amine congener of 1,3-dipropyl-8-phenylxanthine: an antagonist radioligand for adenosine receptors. *Proc. Natl. Acad. Sci. U. S. A.*, 1986, 83, 4089-4093.
[18] Bruns, RF; Fergus, JH; Badger, EW; Bristol, JA; Santay, LA; Hartman, JD; Hays, SJ; Huang, CC. Binding of the A1-selective adenosine antagonist 8-cyclopentyl-1,3-dipropylxanthine to rat brain membranes. *Naunyn Schmiedebergs Arch. Pharmacol.*, 1987, 335, 59-63.
[19] Klotz, KN; Cristalli, G; Grifantini, M; Vittori, S; Lohse, MJ. Photoaffinity labeling of A_1-adenosine receptors. *J Biol Chem.*, 1985, 260, 14659-14664.
[20] Ramkumar, V; Barrington, WW; Jacobson, KA; Stiles, GL. Demonstration of both A_1 and A_2 adenosine receptors in DDT_1 MF-2 smooth muscle cells. *Mol. Pharmacol.*, 1990, 37, 149-156.
[21] Earl, CQ; Patel, A; Craig, RH; Daluge, SM; Linden, J. Photoaffinity labeling adenosine A_1 receptors with an antagonist ^{125}I-labeled aryl azide derivative of 8-phenylxanthine. *J. Med. Chem.*, 1988, 31, 752-756.
[22] Stiles, GL; Jacobson, KA. A new high affinity, iodinated adenosine receptor antagonist as a radioligand/photoaffinity crosslinking probe. *Mol. Pharmacol.*, 32, 184-188.
[23] Green, RM; Stiles, GL. Chronic caffeine ingestion sensitizes the A_1 adenosine receptor-adenylate cyclase system in rat cerebral cortex. *J. Clin. Invest.*, 1986, 77, 222-227.
[24] Barrington, WW; Jacobson, KA; Hutchison, AJ; Williams, M; Stiles, GL. Identification of the A_2 adenosine receptor binding subunit by photoaffinity crosslinking. *Proc. Natl. Acad. Sci. USA.*, 1989, 86, 6572-6576.
[25] Ramkumar, V; Olah, ME; Jacobson, KA; Stiles, GL. Distinct pathways of desensitization of A_1- and A_2-adenosine receptors in DDT1 MF-2 cells. *Mol. Pharmacol.*, 1991, 40, 639-647.
[26] Ramkumar, V; Pierson, G; Stiles, GL. Adenosine receptors: clinical implications and biochemical mechanisms. *Prog. Drug Res.*, 1988. 32, 195-247.
[27] Yeung, SM; Green, RD. Agonist and antagonist affinities for inhibitory adenosine receptors are reciprocally affected by 5'-guanylylimidodiphosphate or N-ethylmaleimide. *J. Biol. Chem.*, 258, 2334-2339.
[28] Ramkumar, V; Stiles, GL. Reciprocal modulation of agonist and antagonist binding to A_1 adenosine receptors by guanine nucleotides is mediated via a pertussis toxin-sensitive G protein. *J. Pharmacol. Exp. Ther.*, 1988, 246, 1194-1200.

[29] Stiles, GL. The A_1 adenosine receptor. Solubilization and characterization of a guanine nucleotide-sensitive form of the receptor. *J. Biol. Chem.*, 1985, 260, 6728-6732.

[30] Prater, MR; Taylor, H; Munshi, R; Linden, J. Indirect effect of guanine nucleotides on antagonist binding to A1 adenosine receptors: occupation of cryptic binding sites by endogenous vesicular adenosine. *Mol. Pharmacol.*, 1992, 42, 765-772.

[31] Olah, ME; Jacobson, KA; Stiles, GL. Purification and characterization of bovine cerebral cortex A1 adenosine receptor. *Arch. Biochem. Biophys.*, 1990, 283, 440-446.

[32] Nakata, H. A_1 adenosine receptor of rat testis membranes. Purification and partial characterization. *J. Biol. Chem.*, 1990, 265, 671-677.

[33] Libert, F; Schiffman, S. N; Lefort, A; Parmentier, M; Gerard, C; Dumont, JE; Vanderhoeghen, J.-J; Vassart, G. The orphan receptor cDNA RDC7 encodes an A1 adenosine receptor. *EMBO J.*, 1990, 10, 1677-1682.

[34] Olah, ME., Ren, H., Ostrowski, J., Jacobson, KA., Stiles, GL. Cloning, expression, and characterization of the unique bovine A_1 adenosine receptor. Studies on the ligand binding site by site-directed mutagenesis. *J. Biol. Chem.*, 1992, 267, 10764-10770.

[35] Pierce, KD; Furlong, TJ; Selbie, LA; Shine, J. Molecular cloning and expression of an adenosine A2b receptor from human brain. *Biochem. Biophys. Res. Commun.*, 1992, 187, 86-93.

[36] Rivkees, SA; Reppert, SM. RFL9 encodes an A2b-adenosine receptor. *Mol. Endocrinol.*, 1992, 6, 1598-1604.

[37] Zhou, QY; Li, C; Olah, ME; Johnson, RA; Stiles, GL; Civelli, O. Molecular cloning and characterization of an adenosine receptor: the A_3 adenosine receptor. *Proc. Natl. Acad. Sci. USA.*, 1992, 89, 7432-7436.

[38] Lin, W; Marcucci, KA; Rabin, RA; Roth, JA. 2-Chloroadenosine decreases tyrosylprotein sulfotransferase activity in the Golgi apparatus in PC12 cells. Evidence for a novel receptor. *J. Biol. Chem.*, 1991, 266, 14457-14463.

[39] Ribeiro, JA; Sebastiao, AM. (1986) Adenosine receptors and calcium: basis for proposing a third (A3) adenosine receptor. *Prog. Neurobiol.*, 26, 179-209.

[40] Oliveira, JC; Sebastiao, AM; Riberio, JA. Solubilized rat brain adenosine receptors have two high-affinity binding sites for 1,3-dipropyl-8-cyclopentylxanthine. *J. Neurochem.*, 1991, 57, 1165-1171.

[41] Ali, H; Cunha-Melo, JR; Saul, WF; Beaven, MA. Activation of phospholipase C via adenosine receptors provides synergistic signals for secretion in antigen-stimulated RBL-2H3 cells. Evidence for a novel adenosine receptor. *J. Biol. Chem.*, 1990, 265, 745-753.

[42] Ramkumar, V; Stiles, GL; Beaven, MA; Ali, H. The A_3 adenosine receptor is the unique adenosine receptor which facilitates release of allergic mediators in mast cells. *J. Biol. Chem.*, 1993, 268, 16887-16890.

[43] Malbon, CC; Rapiejko, PJ; Marigano, TJ. Fat cell adenylate cyclase system. Enhanced inhibition by adenosine and GTP in the hypothyroid rat. *J. Biol. Chem.*, 260, 2558-2564.

[44] Malbon, CC; Moreno, FJ; Cabelli, RJ; Fain, JN. Fat cell adenylate cyclase and beta-adrenergic receptors in altered thyroid state. *J. Biol. Chem.*, 253, 671-678.

[45] Rapiejko, PJ; Malbon, CC. Short-term hyperthyroidism modulates adenosine receptors and catalytic activity of adenylate cyclase in adipocytes. *Biochem. J.*, 241, 765-771.

[46] Bumgarner, JR; Ramkumar, V; Stiles, GL. Altered thyroid status regulates the adipocyte A_1 adenosine receptor-adenylate cyclase system. *Life Sci.*, 1989, 44, 1705-1712.

[47] Ohisalo, JJ; Ranta, S; Huhtaniemi, IT. Attenuated adenosine R-site effect in adipocytes in obesity . *Metabolism.*, 1986, 35, 143-146.

[48] Hoffman, BB; Chang, H; Dall'Aglio, E; Reaven, GM. Desensitization of adenosine receptor-mediated inhibition of lipolysis. The mechanism involves the development of enhanced cyclic adenosine monophosphate accumulation in tolerant adipocytes. *J. Clin. Invest.*, 1986, 78, 185-190.

[49] Hoffman, BB; Chang, H; Farahbakhsh, Z; Reaven, GM. Inhibition of lipolysis by adenosine is potentiated with age. *J. Clin. Invest.*, 1984, 74, 1750-1755.

[50] Kather, H; Wieland, E; Fischer, B; Wirth, A; Schlierf, G. *Eur. J. Clin. Invest.*, 1985, 15, 30-37.

[51] Parsons, WJ; Stiles, GL. Heterologous desensitization of the inhibitory A1 adenosine receptor-adenylate cyclase system in rat adipocytes. Regulation of both Ns and Ni. *J. Biol. Chem.*, 1987, 262, 841-847.

[52] Ramkumar, V; Bumgarner, JR; Jacobson, KA; Stiles, GL. Multiple components of the A_1 adenosine receptor-adenylate cyclase system are regulated in rat cerebral cortex by chronic caffeine ingestion. *J. Clin. Invest.*, 1988, 82, 242-247.

[53] Daly, JW; Fredholm, BB. Caffeine--an atypical drug of dependence. *Drug Alcohol Depend.*, 1998, 51, 199-206.

[54] Snyder, SH; Katims, JJ; Annau, Z; Bruns, RF; Daly, JW. Adenosine receptors and behavioral actions of methylxanthines. *Proc. Natl. Acad. Sci. USA.*, 1981, 78, 3260-3264.

[55] Ledent, C; Vaugeois, JM; Schiffmann, SN; Pedrazzini, T; El Yacoubi, M; Vanderhaeghen, JJ; Costentin, J; Heath, JK; Vassart, G; Parmentier, M. Aggressiveness, hypoalgesia and high blood pressure in mice lacking the adenosine A2a receptor. *Nature*. 1997, 388, 674-678.

[56] Awada, A; al Jumah, M. The first-of-Ramadan headache. *Headache* 1998, 39, 490-493.

[57] Mosek, A; Korczyn, AD. Yom Kippur headache. *Neurology*. 1995, 45, 1953-1955.

[58] Nehlig, A. Are we dependent upon coffee and caffeine? A review on human and animal data. *Neurosci. Biobehav. Rev.*, 1999, 23, 563-576.

[59] Green, RM; Stiles, GL. Chronic caffeine ingestion sensitizes the A1 adenosine receptor-adenylate cyclase system in rat cerebral cortex. *J. Clin. Invest.*, 1986, 77, 222-227.

[60] Food and Drug Administration. (2007). Medicines in my home: Caffeine and your body. http://wwwfdagov/downloads/Drugs/ResourcesForYou/Consumers/BuyingUsing Medicine Safely/ UnderstandingOver-the-CounterMedicines/ UCM205286pdf.

[61] Seifert, SM; Schaechter, JL; Hershorin, ER; Lipshultz, SE. Health effects of energy drinks on children, adolescents, and young adults. *Pediatrics* 2011, 127, 511-528.

[62] Wolk, BJ; Ganetsky, M; Babu, KM. Toxicity of energy drinks. *Curr. Opin. Pediatr.*, 2012, 24, 243-251.

[63] Noji, T; Karasawa, A; Kusaka, H. Adenosine uptake inhibitors. *Eur. J. Pharmacol.*, 2004, 495, 1-16.

[64] Fredholm, BB. Adenosine, an endogenous distress signal, modulates tissue damage and repair. *Cell Death Differ.*, 14,1315-1323.

[65] Porkka-Heiskanen, T; Strecker, RE; Thakkar, M; Bjorkum, AA; Greene, RW; McCarley, RW. Adenosine: a mediator of the sleep-inducing effects of prolonged wakefulness. *Science.*, 1997, 276, 1265-1268.

[66] Basheer, R; Strecker, RE; Thakkar, MM; McCarley, RW. Adenosine and sleep-wake regulation. *Prog. Neurobiol.*, 2004, 73, 379-396.

[67] Jenner, P. A_{2A} antagonists as novel non-dopaminergic therapy for motor dysfunction in PD. *Neurology.*, 2003, 61(11 Suppl 6), S32-8.

[68] Vlajkovic, SM; Housley, GD; Thorne, PR. Adenosine and the auditory system. *Curr. Neuropharmacol.*, 2009, 7, 246-756.

[69] Jajoo, S; Sheth, S; Mukherjea, D; Kaur, T; Rybak, LP; Ramkumar, V. Targeting adenosine receptors for the treatment of cancers. *Curr. Topics Biochem. Res.*, 2011, 13, 1.

[70] Paes-de-Carvalho, R; de Mello, FG. Adenosine-elicited accumulation of adenosine 3', 5'-cyclic monophosphate in the chick embryo retina. *J. Neurochem.*, 1982, 38(2), 493-500.

[71] Paes-de-Carvalho, R. Development of A1 adenosine receptors in the chick embryo retina. *J. Neurosci. Res.*, 1990, 25(2), 236-242.

[72] Blazynski, C; Perez, MT. Adenosine in vertebrate retina: localization, receptor characterization, and function. *Cell. Mol. Neurobiol.*, 1991, 11, 463-484.

In: Adenosine Signaling Mechanisms
Editors: V. Ramkumar and R. Paes de Carvalho
ISBN: 978-1-63483-186-4

Chapter 2

ADENOSINE MEMBRANE TRANSPORTERS

Natalia Grañé-Boladeras[1], Alex Bicket[1], Maliha Zafar[1], Alexandre dos Santos-Rodrigues[2] and Imogen R. Coe[1,*]

[1] Department of Chemistry and Biology, Faculty of Science, Ryerson University, Toronto, ON, Canada

[2] Program of Neurosciences and Department of Neurobiology, Institute of Biology, Fluminense Federal University, Niterói, RJ, Brazil

ABSTRACT

Adenosine signalling requires the participation of a complex but tightly connected network, the purinome, which includes purinergic ligands, receptors, enzymes and transporters. Nucleoside transporters (NTs) are key elements of the purinome since translocation of adenosine across the plasma membrane modulates the concentration of extracellular adenosine and thus the triggering of purinergic signalling. There are two families of nucleoside transporters, the SLC28 or concentrative nucleoside transporters (CNTs), and the SLC29 or equilibrative nucleoside transporters, of which 4 members (ENT1, ENT2, CNT2 and CNT3) are the primary transporters responsible for adenosine uptake. An increasing body of evidence supports the crosstalk and tight interaction between NTs and adenosine receptors (ARs) in terms of functionality at the plasma membrane but also potentially via transcriptional regulation. In this chapter we describe data supporting a role for NTs in the purinome, and discuss the potential crosstalk between ARs and NTs, which, although still largely unknown, provides a new perspective on the role of NTs in the purinome.

INTRODUCTION

Adenosine is a major player in cellular purinergic signalling and requires a molecular network of purinergic ligands (i.e., adenosine and other purine nucleotides), receptors,

[*] Corresponding Author: Dr. Imogen R. Coe, Department of Chemistry and Biology, Ryerson University, VIC-724, 350 Victoria St., Toronto, ON, Canada, E-mail: imogen.coe@ryerson.ca.

enzymes and transporters, which collectively are known as the purinome [1]. Various studies suggest that these purinergic components are closely connected to ensure coordinated responses to the physiological needs of the cell [2-4]. Therefore, the purinome is a complex and dynamic molecular network that exhibits crosstalk and cooperative interaction, and alterations within this dynamic consortium of players may lead to pathologies or dysfunctions within the cell [5-7].

Nucleoside transporters (NTs) play an important role in the purinome since they are responsible for adenosine flux and, consequently, contribute to the modulation of extracellular adenosine concentrations thereby influencing the activation of adenosine receptors (AR) and adenosine signalling [1]. Adenosine, and other nucleosides, are hydrophilic molecules that require specialized transmembrane proteins, NTs, for translocation across bi-lipid membranes. In contrast to our knowledge and understanding of the function and regulation of ARs and other components of the purinome, relatively little is known about mode of action and regulation of NTs, and their role in purinergic signalling in either normal physiology or in pathologies that result from dysfunction in purinergic signalling. An understanding of the role of NTs in the purinome, particularly in relation to other components such as receptors and metabolic enzymes, will lead to enhanced comprehension of the purinergic signalling network and to potential advances in drug discovery and development [8, 9].

In this chapter, we describe the different members of the NT families involved in adenosine uptake, and discuss several mechanisms of their regulation based on current studies. We also present evidence that suggests a finely regulated crosstalk between NTs and ARs may exist, and briefly describe the participation of NT in pathological states and role as routes of entry of therapeutic agents.

NUCLEOSIDE TRANSPORTERS: STRUCTURE, FEATURES AND TISSUE DISTRIBUTION

Nucleoside translocation across bi-lipid membranes is mediated by NTs [10] of which there are currently seven human molecular species identified, divided into two families, SLC28, the concentrative nucleoside transporters (CNTs) and SLC29, the equilibrative nucleoside transporters (ENTs). There are three isoforms of SLC28, CNT1, CNT2 and CNT3, and four isoforms of SLC29, ENT1, ENT2, ENT3 and ENT4. NT members display differential selectivity for substrates (Table 1) and ENT1, ENT2, CNT2 and CNT3 are considered to be the transporters predominantly responsible for adenosine uptake in human systems [11].

ENTs mediate facilitated diffusion of their substrate in a bi-directional manner down the endogenous substrate concentration gradient and do not require energy in the form of ATP or an ion gradient [12]. CNTs transport nucleosides using electrochemical energy present in an inherent ion gradient and typically co-transport sodium ions. CNTs also typically have a higher affinity for their respective substrates compared to ENTs [13].

Table 1. A brief summary of NT substrate specificity and tissue distribution

NT	Substrate	Affinity for Adenosine	Tissue distribution	Distinguishing/ unique feature
CNT1	Pyrimidine nucleosides	NA	Kidney, liver, small intestine.	Substrate specificity
CNT2	Purine nucleosides and uridine	High affinity Km = 8 μM	Heart, skeletal muscle, liver, kidney, intestine, pancreas, placenta and brain	Substrate specificity
CNT3	Pyrimidine and purine nucleosides	High affinity Km = 15 μM	Widely expressed but most abundant in mammary gland, pancreas, bone marrow, trachea and intestine	Can use the Na^+ or H^+ gradient for nucleoside co-transport
ENT1	Pyrimidine and purine nucleosides	Moderate affinity Km = 40μM	Ubiquitous	Very sensitive to inhibition by NBTI
ENT2	Pyrimidine and purine nucleosides and nucleobases	Moderate affinity Km = 100μM	Ubiquitous	Nucleobase transporter, role in nuclear membrane
ENT3	Pyrimidine and purine nucleosides and adenine (at pH 5.5)	Low affinity Km = 2mM	Ubiquitous	Intracellular localization, activity is pH dependent
ENT4 (PMAT)	Adenosine (at pH 5.5) and organic cations, including serotonin	Low affinity Km = 0.78mM	Ubiquitous	pH dependent adenosine transport, transports organic cations

ENTs and CNTs also transport many nucleoside analog drugs used in cancer and anti-viral therapy [14, 15] and thus NTs have been widely studied for their role in and contribution to drug efficacy. Both CNTs and ENTs are important in determining the therapeutic efficacy of nucleoside analog drugs because isoforms differ in terms of tissue expression and cellular location (i.e basal vs. apical membrane). Both families are poorly understood in terms of three-dimensional structure and mode of action and, to date, there are no reported crystal structures of mammalian NTs although a *Vibrio cholerae* CNT structure was recently reported, which shares ~40% amino acid sequence identity with human CNTs, and which suggests, intriguingly, that CNTs may form trimers [16].

The SLC29 Family (ENTs)

The putative structure of the ENTs consists of 11 transmembrane domains (TM), with an intracellular N-terminus and extracellular C-terminus. There is a large extracellular loop between TM domains 1 and 2 and a large intracellular loop between TM domains 6 and 7.

Intracellular loops contain putative phosphorylation target motifs that suggest regulation of the protein by kinases and phosphatases. The large intracellular loop also contains motifs that suggest putative intracellular protein-protein interactions. The first extracellular loop contains an N-linked glycosylation site [17] and TM domains 2, 4 and 5 of ENT1 are reported to be critical for nucleoside transport [18, 19].

ENT1

ENT1 is the first cloned and most studied member of the SLC29 family. It transports both purine and pyrimidine nucleosides, but has a higher affinity for adenosine compared to other nucleosides [10]. The human ortholog of ENT1, hENT1, is ubiquitously distributed in human tissues although expression levels vary considerably from individual to individual [20,21] and across tissue types. In the human brain, hENT1 is most abundant in the frontal and parietal lobes of the cortex [22] compared to other isoforms. hENT1 and hENT2 are primarily localized at the plasma membrane although there is evidence of their presence at the endoplasmic reticulum and nuclear membranes [23]. Therefore, hENT1 and hENT2 may be involved in translocating nucleosides from the cytosol into these membrane-bound organelles although physiological relevance remain unclear. Initially, ENTs were defined based on their ability to be inhibited by the nucleoside analog 4-nitrobenzylthioinosine (NBTI) and inhibition by NBTI has been a commonly used parameter in the characterization of ENTs. ENT1 is strongly inhibited by NBTI (in the nanomolar range) and the vasodilators dipyridamole and dilazep [24].

ENT2

ENT2 is also localized at the plasma membrane. ENT2 transports nucleosides, although at lower affinities compared to ENT1, except for inosine and, unlike ENT1, ENT2 transports nucleobases. ENT2 is inhibited by NBTI in the micromolar or above range [25, 26]. ENT2 splice variants exist which localize at the nuclear membrane and which likely play a role in maintaining nuclear pools of nucleosides/nucleobases which are needed during DNA replication [27]. ENT2 is most abundantly expressed in skeletal muscle but is also found in many other tissues including brain, heart, placenta, thymus, pancreas, prostate and kidney [20].

ENT3

ENT3 is primarily localized intracellularly and has an endosomal/lysosomal targeting sequence within its N-terminus, which is longer and distinct compared to ENT1 and ENT2. The activity of ENT3 is pH dependent (with an optimum pH=5.5) [28], which reflects its endosomal/lysosomal localization, where it is proposed to function in the release of nucleosides produced from the metabolism of nucleic acid [29]. ENT3 has a lower affinity for adenosine than ENT1 and ENT2. ENT3 is expressed in a variety of human tissues and is most abundantly found in the placenta. Also, compared to ENT1, ENT3 is much less sensitive to inhibition from NBTI and vasodilators.

ENT4

ENT4 is relatively poorly understood compared to the other ENTs. It is localized at the plasma membrane and was initially reported to be a low affinity adenosine transporter by Beal and colleagues (2004) [12]. However Engel and Wang (2005) reported no interaction with nucleosides and demonstrated that it can function as a plasma membrane monoamine transporter (PMAT), which efficiently transports serotonin [30]. Further investigation confirmed that adenosine transport ability is strictly dependent on acidic pH, while serotonin transport is pH independent [31]. ENT4 is possibly involved in the regulation of extracellular adenosine under acidic conditions such as those associated with ischemia. ENT4 is ubiquitously expressed in human tissues, but most abundantly expressed in the heart and brain [32] and is only weakly inhibited by NBTI and coronary vasodilators [32].

The SLC28 Family (CNTs)

The concentrative nucleoside transporters are proposed to have 13 TM domains, a cytosolic N-terminus and extracellular C-terminus. The extracellular C-terminus contains N-linked glycosylation sites. CNTs are nucleoside/Na^+ co-transporters and mediate a secondary active nucleoside transport coupled to a cation symport (Na^+ or H^+) in an unidirectional manner [11]. Although there are no crystal structures reported for mammalian CNTs, a *Vibrio cholerae* CNT (vcCNT) crystal structure reported in 2012 shows that this transporter functions as a trimer [16] suggesting that a similar model may exist for human CNTs.

CNT1

CNTs are not as ubiquitously expressed in as many tissues as ENTs but, like ENTs, are primarily localized at the plasma membrane. CNT1 is found most abundantly in liver, small intestine and kidney. It selectively transports pyrimidine nucleosides with a high affinity in a 1:1 ratio of Na^+ to nucleoside [11].

CNT2

CNT2 is a high affinity transporter of purine nucleosides and uridine and like CNT1, transports Na^+ and nucleoside in a 1:1 ratio. It is primarily expressed in the heart, skeletal muscles, liver, kidneys, intestine, pancreas, placenta and brain [11]. CNT2 is the highest affinity transporter within the SLC28 and SLC29 families [10].

CNT3

CNT3 is broadly selective for transport of purine and pyrimidine nucleosides and has a few unique features compared to the other family members. It transports Na^+ and nucleosides in a 2:1 ratio and it also differs from CNT1 and CNT2 in that it can use a proton (H^+) gradient, co-transporting H^+ and nucleoside in a 1:1 ratio. In the presence of both Na^+ and H^+, the coupling ratio changes to 1Na^+:1H^+:1 nucleoside [33, 34]. Although CNT3 is primarily localized at the plasma membrane, one splice variant has been reported to localize to intracellular membranes [35].

Moreover, CNT3 has a wider expression profile than CNT1 and CNT2, including mammary glands, pancreas, bone marrow, trachea and intestine [11].

Mechanisms of Regulation of Nucleoside Transporters

NTs can be regulated at various points in their life cycle, either at the level of transcription, during protein translation or after the protein has been synthesized (Figure 1). Since NTs are responsible for flux of adenosine into or out of cells, regulation of NT activity or presence at the plasma membrane may affect levels of adenosine in the extracellular space and therefore may modulate or impact the activation of ARs and subsequent signalling pathways.

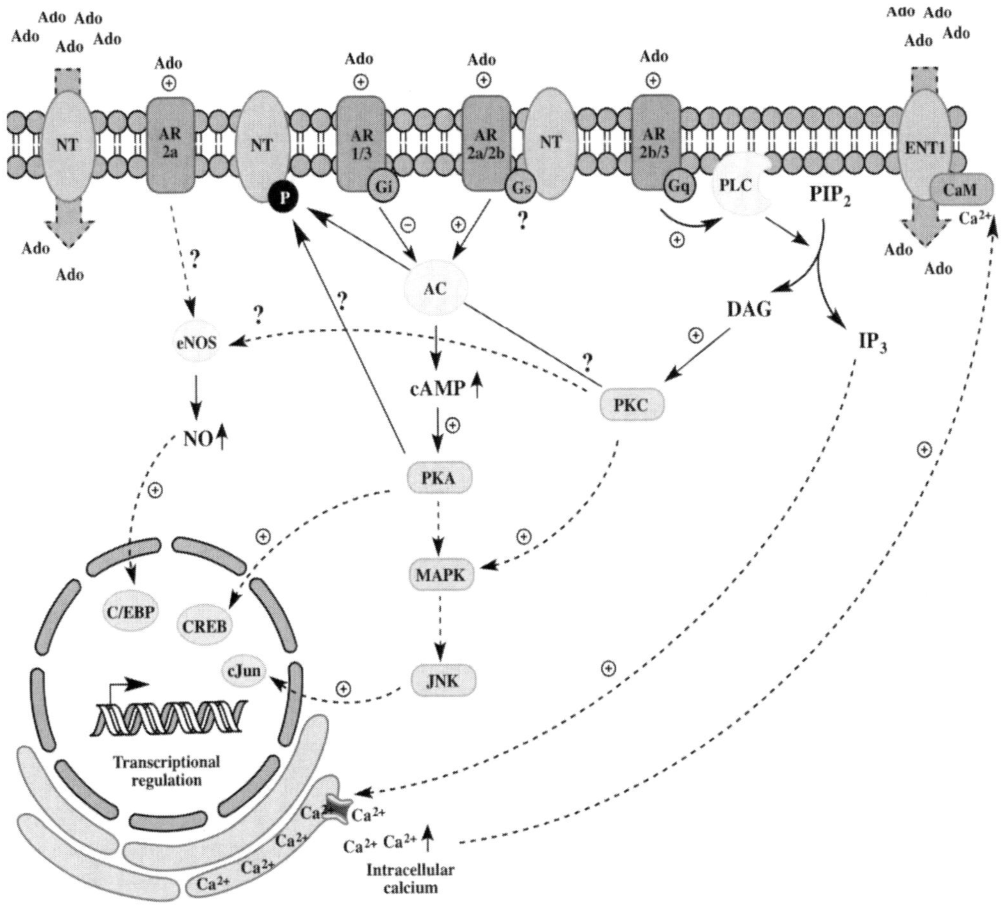

Figure 1. *Cartoon representing the crosstalk between NTs and ARs.* NTs translocate adenosine (Ado) inside the cell regulating the concentration of extracellular adenosine that activates AR. Activation of AR may activate / inhibit adenilate cyclase (AC), which produces cAMP and activates PKA. PKA can directly phosphorylate NTs and, at the same time, trigger the MAPK pathway to ultimately regulate NTs expression via transcription factors (TF).

AR can also activate phospholipase C (PLC) that generates DAG and activates PKC, which could phosphorylate NTs and also regulate expression of NTs via TF. PLC also produces IP_3, which activates calcium channels at the endoplasmic reticulum and increases intracellular calcium concentration. Calcium triggers the protein-protein interaction (PPI) between calmodulin (CaM) and ENT1 increasing the activity of the transporter at the plasma membrane and promoting adenosine uptake. In placenta, activation of PKC and AR_{2a} appears to activate eNOS, which generates NO and regulates NT's transcription via C/EBP.

Regulation of NTs at the Level of the Gene

The level of expression of a gene is primarily regulated by transcription factors (TF), which bind to specific DNA sequences including those within the promoter region [36]. Little is known about the transcriptional regulation of members of CNT family. CNT1 and CNT2 are highly present in liver and several Liver Enriched Transcription Factors (LETFs) are involved in their transcriptional regulation [37]. It has been reported that HNF-4α and bile acids are major determinants of CNT1 expression [38], whereas C/EBPα and HNF-3γ modulate CNT2 [39]. CNT1 is also a target of multifunctional cytokines such as TNF-α or IL-6, both implicated in the priming of hepatocytes before proliferation [40]. CNT2, in contrast, is highly regulated by the pro-apoptotic cytokine TGF-β1 via JNK-dependent transcriptional activation of the CNT2 promoter [41]. These data suggest that CNT genes may be subject to regulation by signalling pathways coupled to a variety of potential purinergic inputs although details remain unknown.

Transcriptional regulation of ENT1 is marginally better understood. Based on consensus sequence analyses, several transcription factors have been proposed to regulate the ENT1 promoter including, but not limited to, CREB, GATA-1 and Sp-1 for mENT1 or MAZ, Sp-1 and IL-4 for hENT1 [42-44]. More recent studies reported that activation of JNK-cJun pathway negatively regulates mENT1 [45] and that overexpression and activation of PPARα and γ result in an increase in hENT1-mediated transport activity [46]. Other studies demonstrate that ENT expression is differentially regulated in placenta such that high glucose levels, along with subsequent increased NO production, regulate ENT1 promoter activity via a MAP kinase-dependent pathway, whereas regulation of ENT2 expression is via an NO-independent mechanism [47-51]. The physiological relevance of these observations suggests that the expression of NTs is finely regulated by several independent pathways and that the presence of NTs at the plasma membrane can be altered as a mechanism of adaptation of the cell to physiological conditions. The role of purinergic inputs into these regulatory pathways is not known.

As our understanding of the role of the NTs in response to AR-dependent pathways has expanded, it has become clear that there is likely to be feedback regulation of NTs via AR-dependent pathways such that a coordinated physiological response exists to couple AR-activation with levels of NTs in response to prevailing levels of extracellular adenosine. For instance, a decrease in oxygen in the brain and/or heart leads to an activation of the transcription factor HIF-1, which in turn regulates ENT1 and ENT2 promoters to down-regulate the expression of both of these NTs leading to fewer proteins at the membrane and a decreased uptake of adenosine [52-54]. This results in an increase in the concentration of extracellular adenosine and an activation of ARs coupled to pathways that promote cellular responses to combat the ischemic insult [52-54]. We know from previous work that reduction or absence of ENT1 at the plasma membrane leads to an inherently cardio-protected phenotype by optimizing AR-dependent activation of signalling pathways [53, 55-58] so regulation of ENT transcription is an example of the coordination of various components of the purinome linking transcriptional regulation to NT expression and resulting function to the optimization of AR-coupled signalling pathways that promote cellular well-being. Direct connection of other transcriptional regulation of NTs to purinergic signalling pathways has not been confirmed but we do know that ARs are coupled to signalling pathways that act through PKA and PKC, and which can activate the MAPKs pathway [59], leading to activation of JNK and ultimately the transcriptional factor c-Jun (a putative TF for CNT2 and ENT1). Adenosine-dependent activation of PKA could lead to regulation of the transcription factor CREB which can regulate ENT1 promoter [43, 60, 61]. Studies suggest a connection between adenosine-dependent activation of $A_{2a}R$ and synthesis of NO in placenta [62] when PKC pathway is activated [63, 64]. Taken together, these data suggest that NTs may be regulated at the level of the gene in response to AR-dependent activation of signalling pathways leading to the precise regulation of the components of the purinome that are involved in adenosine flux across membranes.

In addition to studies of the transcriptional regulation of NTs, there are a number of studies of the genetic sequences of CNTs and ENTs. For both families, there are comparatively few genetic variations compared to many other transporter gene families [11, 65, 66]. Other than a frameshift mutation that prematurely truncates hENT2 [66], single nucleoside polymorphisms (SNPs) of ENT1 and ENT2 that have been characterized show proteins or cells with normal nucleoside and nucleobase uptake characteristics. In contrast, CNT genes with identified SNPs results in altered functional transporter properties [11]. Very few splice variants have been described for either CNTs or ENTs and the only known human CNT splice isoform is CNT3ins, which lacks 69 amino acids at the N-terminus and which is retained in the endoplasmic reticulum where it shows characteristic CNT3-type transport activity [35]. Within the ENT family, hENT2 is the only member possessing splice variants reported to date [67, 68]. A splice variant lacking part of the exon 4 has been described as the product of a delayed early-response gene, which is expressed in response to mitogenic stimuli. This variant encodes a protein comprising 326 amino acids located at the nuclear membrane [67]. A second splice variant with a 40 base pair deletion in exon 9 (hENT2A) was also identified. This variant generates a truncated non-functional protein of 304 amino acids that shows an intracellular vesicular distribution [68]. The physiological relevance of these isoforms remains unclear, and their role in adenosinergic processes is difficult to predict.

NTs are both necessary and sufficient for drug efficacy for most nucleoside analog drugs. Although we know relatively little about transcriptional regulation of NTs, expression profiles have been widely investigated as potential predictive indicators of drug response. However the relationship between NT expression profiles on patient outcomes is far from clear. Gemcitabine, a pyrimidine analog, is transported into cells by hENT1 and hCNT1 [69] with a minor contribution from hCNT3 [70] but there is no significant correlation between hENT1 or hCNT3 profiles and patient survival with lung cancer. The relationship between relative levels of NT expression and drug efficacy is further complicated by the observation that some drugs, such as LY2334737, a prodrug of gemcitabine, which has significant antitumor activities in human colon and lung tumor xenograft models, with higher efficacy than traditional treatment with gemcitabine [71] also increases hCNT3 transcript levels suggesting an unexpected positive feedback loop linking drug uptake to enhanced NT expression. ENT1, ENT2, ENT4 and CNT3 proteins are observed in ovarian carcinoma effusions, and patients with higher levels of ENT2 levels tended to show longer survival times [73] suggesting that in some disease contexts, the relative levels of certain isoforms or family members of NTs might be informative. For instance, a positive correlation between CNT3 expression and longer survival times has been noted [74] while CNT1 has also been suggested as a possible tumor suppressor in gemcitabine-resistant pancreatic cancer cells [75].

In addition, to the expression profiles of NTs, the profiles of drug metabolizing proteins and patient outcomes is being assessed. Analysis of the profile of expression of NTs (hENT1, hCNT1), the pro-drug activating enzyme (dCK), the ribonucleotide reductase regulatory subunits (RRM1, RRM2), and GEM catabolic enzymes (CDA, 5'NT) in primary ovarian carcinomas suggests heterogeneous pattern in the expression levels for most of the proteins, but importantly, a correlation between increasing expression of RRM2 and poorer outcomes [72]. Research linking presence and activity of NTs with drug efficacy and patient outcome is an area of high priority and is key in improving chemotherapeutic approaches using nucleoside analog drugs in the treatment of disease.

Regulation of NTs at the Level of the Protein

Localization

NTs must be correctly localized in order to be functional. Mechanisms of regulation involved in trafficking of NTs are yet not well known although trafficking of a transmembrane protein to the plasma membrane, or other intracellular organelles, usually requires the presence of a targeting sequence in the protein that leads to sorting to a specific destination and insertion of the protein into the bi-lipid membrane [36]. Recent studies have identified a specific region of the intracellular N-terminus of rCNT2 that is implicated in the sorting of the protein to the plasma membrane [76, 77]. However, very little is known about potential targeting motifs within ENTs. The RXXV motif of hENT1 and di-leucine residues in hENT2 at the C-terminus of the transporters are suspected to be necessary for the presence of ENTs at the plasma membrane and also to be involved in their sorting to the baso-lateral membrane of renal epithelial cells [68].

The two dimensional model proposed for ENTs predicts N-glycosylation sites (Asn_{48}:ENT1, Asn_{48},Asn_{57}:ENT2) in the extracellular loop between transmembrane domains 1 and 2. Mutational analyses demonstrate that loss of glycosylation at these sites leads to disrupted localization of ENT2 at the plasma membrane, suggesting that these modifications are needed for proper targeting and trafficking to plasma membrane [78], as occurs for ARs [79]. Mutational analyses of ENT1 suggest that glycosylation of the protein is required for effective nucleoside transport and/or inhibitor binding affinity, as well as maintenance of the correct ENT1 protein structure [17, 78, 80-82]. However, we have recently confirmed that glycosylation is also critical for localization and presence of ENT1 at the plasma membrane, as well as for its functionality (Bicket et al., in prep.).

The cellular machinery that is involved in the localization of NTs is yet not fully understood although microtubules have been implicated in ENT1 trafficking to the plasma membrane [83]. The life cycle of NTs likely involves regulated endocytosis and recycling [84] and our current research suggests that substrate binding and translocation triggers clathrin-dependent internalization (Zafar et al., in prep.). Regulated trafficking of CNTs to the plasma membrane involves RS1, which was initially identified as regulator of the glucose transporter SGLT1 [85]. RS1 down-regulates the localization and activity of the three members of the CNT family by acting as an inhibitor of their trafficking to the plasma membrane [86]. Another mechanism which appears to regulate trafficking of CNT3 is the TGF-β1-dependent activation of a pathway involving ERK1/2 and the small GTPase RhoA, which promotes the plasma membrane trafficking of the intracellular pools of CNT3 protein in Chronic Lymphocytic Leukaemia (CLL) [87].

Phosphorylation

Reversible phosphorylation of proteins is a ubiquitous regulatory mechanism within the cell that consists of the balance between phosphorylation by kinases, and dephosphorylation by phosphatases of target proteins resulting in the activation and repression of multiple signalling and regulatory pathways within the cell [88]. There is currently no evidence of direct phosphorylation of any member of the CNT family or of any kinase-dependent mechanism of regulation. In contrast, several studies reported a regulatory connection between ENT1 and ENT2 and the kinases, PKA, PKC and CKII, either by direct phosphorylation or signalling pathways [89]. PKA and PKC signalling pathways may be involved in both transcriptional regulation of NTs and regulation at a post-transcriptional level. Sensitivity of ENT1-dependent adenosine uptake to inhibition by ethanol is PKA- and PKC-dependent [90-93], while PKC and PKA activation induce a decrease of ENT1-dependent activity in cultured chromaffin cells and neuroblastoma cells, resulting in a reduction of adenosine uptake [94-96]. Putative CKII phosphorylation sites have been predicted in both ENT1 and ENT2 sequences [97-101], and appear to be correlated to an increased presence of the ENTs at the membrane with a consequent increase in substrate translocation and altered inhibitor sensitivity [102].

While the mechanisms involved in the kinase-dependent regulation of ENTs are still not fully understood, it has been recently demonstrated that ENT1 can be directly phosphorylated, *in vitro*, by both PKA and PKC within the large intracellular loop between transmembrane domains 6 and 7 [103]. In addition, we have confirmed direct *in vitro* phosphorylation of ENT2 by CKII and PKC (Grañé-Boladeras et al., in prep.), suggesting the ENT2 may be regulated through direct phosphorylation by PKC and CKII. Supporting this hypothesis, Lu and colleagues (2010) showed that chronic morphine treatment caused adenosine accumulation due to an alteration of adenosine uptake via ENT2 and that this attenuation of ENT2-mediated uptake was correlated with a decrease in PKC activity [104], strengthening the case for a regulatory role of PKC on ENT2, although whether this regulation is via direct phosphorylation or not remains to be determined.

Future Perspectives in the Regulation of NT: Protein-Protein Interactions

Regulation via protein-protein interactions (PPIs) is an important form of regulation within the cell. PPI may occur between proteins of different nature (i.e transmembrane proteins interacting with enzymes) or among proteins of the same family forming homo- or heteromers with different functional or localization features depending on the configuration of the oligomers, as it occurs in the case of adenosine receptors [60]. PPIs between transporters and receptors and the formation of heteromers has been previously reported in the CNS [105], thus we can speculate that similar complex formations between ARs and NTs may occur. Although there is no evidence of oligomerization of human NTs, our work confirms the existence of PPIs between ENT1 and ENT2 with both forming homo- and heteromers (Grañé-Boladeras et al., in prep). The physiological relevance of these interactions remains unclear but the existence of oligomerization suggests that this is a putative novel mechanism of regulation and/or modulation of NTs at the plasma membrane.

In addition to oligomerization of NTs, we have also discovered putative PPIs between NTs and other proteins using the innovative MYTH technique [106]. ENT2 was determined to interact with the Gs_{α} subunit, which is coupled to A_{2a} and A_{2b} receptors [27], suggesting that crosstalk between ENT2 and A_{2b}R may exist [113]. A recent study using bacterial two-hybrid screening and GST pull-down techniques confirmed PPI between the N-terminal segment of CNT2, aldolase B and glucose-regulated protein 58 (GRP58), which appear to be regulators of CNT2 activity and which thus link the high-affinity adenosine transporter with energy metabolism (involving ATP) in intestinal and liver parenchymal cells [125]. A recent MYTH screen of ENT1 suggested that this NT associates with calmodulin in a calcium-dependent manner (Mehrabi et al., in prep). Functional relevance of this interaction was confirmed via activation of NMDA receptors in neuronal cells, which allows extracellular calcium to enter cells, leading to a significant increase in ENT1-dependent adenosine uptake (Bicket et al., in prep.). These data suggest that ligand gated ion channels play a role in the regulation of ENT1-dependent adenosine uptake and thus extracellular levels of adenosine, thereby modulating AR-dependent activation. These data thus link glutamate-dependent calcium-regulated activity of NTs leading to modulation of AR-dependent signalling.

Nucleoside Transporters in the Purinome: Crosstalk with Adenosine Receptors

NTs modulate extracellular adenosine levels, which affects AR-dependent signalling suggesting that crosstalk via various mechanisms is likely to occur between ARs and NTs. Adenosine can also be released from the cell by ENTs [107] suggesting that the ENTs contribute to nucleoside uptake and release depending on the physiological context. In the central nervous system (CNS), ENTs can facilitate adenosine release [108], while in cardiovascular tissue the flux appears to be primarily uptake [57, 109]. This highlights the importance of understanding the cellular context and system under investigation when determining the role and contribution of NTs to purinergic pathways. Moreover, the diversity of locations in which NTs function introduces a level of complexity, which makes dissecting the role and contribution of the ENTs and CNTs to purinergic signalling in an entire organism even more challenging.

Although a clear and/or unifying mechanistic explanation for regulation and/or feedback between NTs and ARs has not been uncovered, there is compelling evidence for a link between NTs and AR-dependent signalling from research using the ENT1 knockout mouse. Hypoxic challenge is well established to lead to adenosine/AR-dependent signalling and a role for ENT1 in the clearance of adenosine from the extracellular space was confirmed in the ENT1 knockout mice [58]. Adenosine levels in the blood of ENT1 knock out mice are significantly higher compared to circulating adenosine levels in wild type mice and the knock-out mice possess a phenotype that reflects enhanced AR-dependent signalling in the heart [58]. The absence of ENT1 results in an amplification of cardioprotective pathways, including those that are AR-dependent, confirming the interconnection between ENT1 and AR-dependent signalling [110]. Overall, ENT1 knock out mice express a cardioprotective phenotype and the underlying mechanisms responsible for this effect appear to involve, at least in part, AR-activation supporting the concept that ENT1 modulates AR dependent signalling [56] possibly via A_1 receptor and A_3 receptor dependent pathways and cAMP signalling [58]. ENT1 knock out mice also show increase in adenosine deaminase and A_{2a} receptor mRNA and protein levels [111] supporting a role of the NTs in the overall purinome.

ENT2 knock out mice show reduced levels of pulmonary edema and improved gas exchange following induced acute lung injury compared to controls. ENT2 knock out mice have elevated adenosine levels in the bronchoalveolar fluid, and the protective effects in dipyridamole-dependent attenuation of lung inflammation is lost with *Adora2b* gene deletion supporting a link between modulating of adenosine levels by ENT2 and $A_{2B}R$ dependent signalling. AR knock out mice have also been useful in demonstrating relationships between

NTs and ARs. For instance, AR-knockout mice with acute kidney injury show renal protection by ENT inhibitors [112] and this protection is coupled to A_{2b} receptor activation. Apparent crosstalk between ENT2 and A_{2b} receptors may also exist in alveolar epithelia resulting in adenosine-stimulated lung protection during acute lung injury [113].

The data derived from studies on NT and AR knockout animals supports previous findings that suggest an association between ENT1 and A_1 receptors based on pharmacological data. For instance, [^{3}H]-NBTI binding sites (an indication of the presence of ENT1 proteins) are found in similar locations to A_1 receptors in plexiform layers of chick retina [114] and co-localization of ENT1 and A_1 receptors is observed in brain structures in humans and rats [22, 115] suggesting that ENT1 could influence neuromodulatory actions of A_1 receptors. Inhibition of ENT1-dependent adenosine uptake leads to increased extracellular adenosine concentrations resulting in an A_1R-dependent decrease in glutamatergic synaptic transmission [115]. In rat hippocampus, A_{2a} receptor activation appears to regulate NT activity [116] via a feedback mechanism that is not well understood. For instance, removal of adenosine from the apical surface of airway epithelia is mediated in human cells by A_1 receptor, CNT2, and CNT3 but not by the ENTs [117]. Thus, similar, but not identical, relationships between ARs and NTs appear to be present in a variety of cell types and taken together, it is clear that NTs and AR-dependent signalling are closely connected although the context and the outcomes can vary.

Much work has focused on the role the ENTs but previous studies suggest an interaction between CNTs and ARs [118, 119] and recent data suggest that there is direct crosstalk between CNT3 and purinergic receptors in biliary epithelia [4]. This study defines the A_{2a} receptor dependent regulation of CNT3, via pathways involving cAMP, PKA, ERK, CREB, intracellular trafficking, and the phosphorylation of AMPK. Other studies have identified CNT2 as a potential modulator of adenosine levels, and thus AR-dependent signalling, in the central nervous system of rats [118, 119]. Intriguingly, CNT2 mRNA levels (but not ENT1 mRNA levels) may also be regulated by changes in extracellular adenosine concentration or by sleep deprivation [119, 120].

There is a growing body of literature, which suggests that NTs are closely associated with the ARs and that the crosstalk between these proteins is likely to be extensive. While there is a close association with the NT-dependent modulation of adenosine levels and AR-dependent effects, there are currently no data to support a close physical proximity between NT proteins and AR proteins. However, a complex between A_3 receptors and serotonin transporters (SERT) has been observed [121] and activation of A_3 receptors affects the levels of SERT-A_3 complexes as well as SERT activity [122, 123] suggesting that a similar mechanism may exist for NTs and ARs. An even more intriguing form of action of NTs, is the concept of NTs playing roles in disease or dysfunction that transcend their roles as drug transporters as was recently proposed for hCNT1 (which does not transport adenosine). hCNT1 expression in pancreatic cancer cells altered cell cycle progression, basal phosphorylation of intracellular signaling pathways and induced more cell death in a manner that was independent of transport activity suggesting that, interestingly, CNT1 could be acting as a transceptor, a transporter which transduces signals to the intracellular milieu [124]. Whether this occurs for purine-selective or preferring NTs remains to be seen.

Conclusion

NTs play critical roles in the movement of adenosine into and out of cells and probably, within cells. By altering and modulating adenosine levels in the extracellular space, they affect adenosine-receptor dependent signalling that can have a wide variety of effects and cellular responses. Despite their central role within the purinome and particularly their close relationship with ARs, NTs are still not well understood in terms of their structure, mechanism of action or regulation. Future studies focussed on a deeper understanding of the nature of the NT-AR crosstalk will provide novel information that will be relevant to normal physiological conditions, pathophysiologies and various types of disease treatments. Our future research is aimed at enhancing our understanding of these important contributors to the purinome and fully uncovering the nature of their intriguing relationship with ARs.

References

[1] Volonté, C; D'Ambrosi, N. Membrane compartments and purinergic signalling: the purinome, a complex interplay among ligands, degrading enzymes, receptors and transporters. *FEBS J.* 2009, 276(2), 318–29.

[2] Burnstock, G; Fredholm, BB; Verkhratsky, A. Adenosine and ATP receptors in the brain. *Curr. Top. Med. Chem.* 2011, 11(8), 973–1011.

[3] Coddou, C; Yan, Z; Obsil, T; Huidobro-Toro, JP; Stojilkovic, SS. Activation and regulation of purinergic P2X receptor channels. *Pharmacol. Rev.* 2011, 63(3), 641–83.

[4] Godoy, V; Banales, JM; Medina, JF; Pastor-Anglada, M. Functional crosstalk between the adenosine transporter CNT3 and purinergic receptors in the biliary epithelia. *J. Hepatol.* 2014. [Epub ahead of print]

[5] Schetinger, MRC; Morsch, VM; Bonan, CD; Wyse, ATS. NTPDase and 5'-nucleotidase activities in physiological and disease conditions: new perspectives for human health. *Biofactors.* 2007, 31(2), 77–98.

[6] Samsel, M; Dzierzbicka, K. Therapeutic potential of adenosine analogues and conjugates. *Pharmacol. Rep.* 2011, 63(3), 601–17.

[7] Ferrero, ME. Purinoceptors in inflammation: potential as anti-inflammatory therapeutic targets. *Front. Biosci, (Landmark Ed).* 2011, 16, 2172–86.

[8] Murray, JM; Bussiere, DE. Targeting the purinome. *Methods Mol. Biol.* 2009, 575, 47–92.

[9] Knapp, M; Bellamacina, C; Murray, JM; Bussiere, DE. Targeting cancer: the challenges and successes of structure-based drug design against the human purinome. *Curr. Top. Med. Chem.* 2006, 6(11), 1129–59.

[10] Molina-Arcas, M; Casado, FJ; Pastor-Anglada, M. Nucleoside transporter proteins. *Curr. Vasc. Pharmacol.* 2009, 7(4), 426–34.

[11] Young, JD; Yao, SYM; Baldwin, JM; Cass, CE; Baldwin, SA. The human concentrative and equilibrative nucleoside transporter families, SLC28 and SLC29. *Mol. Aspects Med.* 2013, 34(2-3), 529–47.

[12] Beal, PR; Yao, SYM; Baldwin, SA; Young, JD; King, AE; Cass, CE. The equilibrative nucleoside transporter family, SLC29. *Pflügers Arch.* 2004, 447(5), 735–43.

[13] Gray, JH; Owen, RP; Giacomini, KM. The concentrative nucleoside transporter family, SLC28. *Pflügers Arch.* 2004, 447(5), 728–34.

[14] Damaraju, VL; Damaraju, S; Young, JD; Baldwin, SA; Mackey, J; Sawyer, MB; et al. Nucleoside anticancer drugs: the role of nucleoside transporters in resistance to cancer chemotherapy. *Oncogene.* 2003, 22(47), 7524–36.

[15] Pastor-Anglada, M; Cano-Soldado, P; Molina-Arcas, M; Lostao, MP; Larráyoz, I; Martínez-Picado, J; et al. Cell entry and export of nucleoside analogues. *Virus Res.* 2005, 107(2), 151–64.

[16] Johnson, ZL; Cheong, C-G; Lee, S-Y. Crystal structure of a concentrative nucleoside transporter from Vibrio cholerae at 2.4 Å. *Nature.* 2012, 11 (7390):489-93.

[17] Vickers, MF; Mani, RS; Sundaram, M; Hogue, DL; Young, JD; Baldwin, SA; et al. Functional production and reconstitution of the human equilibrative nucleoside transporter (hENT1) in Saccharomyces cerevisiae. Interaction of inhibitors of nucleoside transport with recombinant hENT1 and a glycosylation-defective derivative (hENT1/N48Q). *Biochem J.* 1999, 339(Pt1), 21–32.

[18] Endres, CJ; Unadkat, JD. Residues Met89 and Ser160 in the human equilibrative nucleoside transporter 1 affect its affinity for adenosine, guanosine, S6-(4-nitrobenzyl)-mercaptopurine riboside, and dipyridamole. *Mol Pharmacol.* 2005, 67(3), 837–44.

[19] Sengupta, DJ; Lum, PY; Lai, Y; Shubochkina, E; Bakken, AH; Schneider, G; et al. A single glycine mutation in the equilibrative nucleoside transporter gene, hENT1, alters nucleoside transport activity and sensitivity to nitrobenzylthioinosine. *Biochemistry.* 2002, 41(5), 1512–9.

[20] Pennycooke, M; Chaudary, N; Shuralyova, I; Zhang, Y; Coe, IR. Differential expression of human nucleoside transporters in normal and tumor tissue. *Biochem. Biophys. Res. Commun.* 2001, 280(3), 951–9.

[21] Marvi, M; Rose, JB; Bang, A; Moon, BC; Pozeg, Z; Ibrahim, M; et al. Nucleoside transporter expression profiles in human cardiac tissue show striking individual variability with overall predominance of hENT1. *Eur. J. Pharm. Sci.* 2010, 41(5), 685–91.

[22] Jennings, LL; Hao, C; Cabrita, MA; Vickers, MF; Baldwin, SA; Young, JD; et al. Distinct regional distribution of human equilibrative nucleoside transporter proteins 1 and 2 (hENT1 and hENT2) in the central nervous system. *Neuropharmacology.* 2001, 40(5), 722–31.

[23] Mani, RS; Hammond, JR; Marjan, JM; Graham, KA; Young, JD; Baldwin, SA; et al. Demonstration of equilibrative nucleoside transporters (hENT1 and hENT2) in nuclear envelopes of cultured human choriocarcinoma (BeWo) cells by functional reconstitution in proteoliposomes. *J. Biol. Chem.* 1998, 273(46), 30818–25.

[24] Griffiths, M; Beaumont, N; Yao, SYM; Sundaram, M; Boumah, CE; Davies, A; et al. Cloning of a human nucleoside transporter implicated in the cellular uptake of adenosine and chemotherapeutic drugs. *Nat. Med.* 1997, 3(1), 89–93.

[25] Griffiths, M; Yao, SYM; Abidi, F; Phillips, SE; Cass, CE; Young, JD; et al. Molecular cloning and characterization of a nitrobenzylthioinosine-insensitive (ei) equilibrative nucleoside transporter from human placenta. *Biochem. J.* 1997, 328(Pt3), 739–43.

[26] Crawford, CR; Patel, DH; Naeve, C; Belt, JA; Cloning of the human equilibrative, nitrobenzylmercaptopurine riboside (NBMPR)-insensitive nucleoside transporter ei by

functional expression in a transport-deficient cell line. *J. Biol. Chem.* 1998, 273(9), 5288–93.

[27] Grañé-Boladeras, N. Novel approaches to understanding hENT2 and hENT2-related proteins: from novel nuclear variants to global networks [Online]. 2012. Available from: http://diposit.ub.edu/dspace /handle/2445/36229

[28] Baldwin, SA; Yao, SYM; Hyde, RJ; Ng, AML; Foppolo, S; Barnes, K; et al. Functional characterization of novel human and mouse equilibrative nucleoside transporters (hENT3 and mENT3) located in intracellular membranes. *J. Biol. Chem.* 2005, 280(16), 15880–7.

[29] Endo, Y; Obata, T; Murata, D; Ito, M; Sakamoto, K; Fukushima, M; et al. Cellular localization and functional characterization of the equilibrative nucleoside transporters of antitumor nucleosides. *Cancer Science.* 2007, 98(10), 1633–7.

[30] Engel, K; Wang, J. Interaction of organic cations with a newly identified plasma membrane monoamine transporter. *Mol. Pharmacol.* 2005, 68(5), 1397–407.

[31] Barnes, K; Dobrzynski, H; Foppolo, S; Beal, PR; Ismat, F; Scullion, ER; et al. Distribution and functional characterization of equilibrative nucleoside transporter-4, a novel cardiac adenosine transporter activated at acidic pH. *Circ.Res.* 2006, 99(5), 510–9.

[32] Young, JD; Yao, SYM; Sun, L; Cass, CE; Baldwin, SA. Human equilibrative nucleoside transporter (ENT) family of nucleoside and nucleobase transporter proteins. *Xenobiotica.* 2008, 38(7-8), 995–1021.

[33] Slugoski, MD; Smith, KM; Mulinta, R; Ng, AML; Yao, SYM; Morrison, EL; et al. A conformationally mobile cysteine residue (Cys-561) modulates Na^+ and H^+ activation of human CNT3. *J. Biol. Chem.* 2008, 283(36), 24922–34.

[34] Slugoski, MD; Ng, AML; Yao, SYM; Smith, KM; Lin, CC; Zhang, J; et al. A proton-mediated conformational shift identifies a mobile pore-lining cysteine residue (Cys-561) in human concentrative nucleoside transporter 3. *J, Biol. Chem.* 2008, 283(13), 8496–507.

[35] Errasti-Murugarren, E; Molina-Arcas, M; Casado, FJ; Pastor-Anglada, M. A splice variant of the SLC28A3 gene encodes a novel human concentrative nucleoside transporter-3 (hCNT3) protein localized in the endoplasmic reticulum. *FASEB J.* 2009, 23(1), 172–82.

[36] Lodish, H; Berk, A; Kaiser, CA; Krieger, M; Scott, MP; Bretscher, A; et al. *Mol.Cell Biology*. W. H. Freeman; 2007.

[37] Pastor-Anglada, M; Cano-Soldado, P; Errasti-Murugarren, E; Casado, FJ. SLC28 genes and concentrative nucleoside transporter (CNT) proteins. *Xenobiotica.* 2008, 38(7-8), 972–94.

[38] Klein, K; Kullak-Ublick, GA; Wagner, M; Trauner, M; Eloranta, JJ. Hepatocyte nuclear factor-4 and bile acids regulate human concentrative nucleoside transporter-1 gene expression. *Am. J. Physiol. Gastrointest. Liver Physiol.* 2009, 296(4), G936–47.

[39] Fernández-Veledo, S; Jover, R; Casado, FJ; Gómez-Lechón, MJ; Pastor-Anglada, M. Transcription factors involved in the expression of SLC28 genes in human liver parenchymal cells. *Biochem. Biophys. Res. Commun.* 2007, 353(2), 381–8.

[40] Fernández-Veledo, S; Valdés, R; Wallenius, V; Casado, FJ; Pastor-Anglada, M. Up regulation of the high-affinity pyrimidine-preferring nucleoside transporter

concentrative nucleoside transporter 1 by tumor necrosis factor-alpha and interleukin-6 in liver parenchymal cells. *J. Hepatol.* 2004, 41(4), 538–44.

[41] Valdés, R; Fernández-Veledo, S; Aymerich, I; Casado, FJ; Pastor-Anglada, M. TGF-β transcriptionally activates the gene encoding the high-affinity adenosine transporter CNT2 in rat liver parenchymal cells. *Cell Mol, Life Sci.* 2006, 63(21), 2527–37.

[42] Choi, D-SS; Handa, M; Young, H; Gordon, AS; Diamond, I; Messing, RO. Genomic organization and expression of the mouse equilibrative, nitrobenzylthioinosine-sensitive nucleoside transporter 1 (ENT1) gene. *Biochem. Biophys. Res. Commun.* 2000, 277(1), 200–8.

[43] Abdulla, P; Coe, IR. Characterization and functional analysis of the promoter for the human equilibrative nucleoside transporter gene, hENT1. *Nucleosides Nucleotides Nucleic Acids.* 2007, 26(1), 99–110.

[44] Fernandez-Calotti, P; Galmarini, CM; Cañones, C; Gamberale, R; Saénz, D; Avalos, JS; et al. Modulation of the human equilibrative nucleoside transporter1 (hENT1) activity by IL-4 and PMA in B cells from chronic lymphocytic leukemia. *Biochem. Pharmacol.* 2008, 75(4), 857–65.

[45] Leisewitz, AV; Zimmerman, EI; Huang, M; Jones, SZ; Yang, J; Graves, LM. Regulation of ENT1 expression and ENT1-dependent nucleoside transport by c-Jun N-terminal kinase. *Biochem. Biophys. Res. Commun.* 2011, 404(1), 370–5.

[46] Montero, TD; Racordon, D; Bravo, L; Owen, GI; Bronfman, ML; Leisewitz, AV. PPARa and PPARg regulate the nucleoside transporter hENT1. *Biochem. Biophys. Res. Commun.* 2012, 419(2), 405–11.

[47] Muñoz, G; Martín, RS; Farías, M; Cea, L; Vecchiola, A; Casanello, P; et al. Insulin restores glucose inhibition of adenosine transport by increasing the expression and activity of the equilibrative nucleoside transporter 2 in human umbilical vein endothelium. *J. Cell Physiol.* 2006, 209(3), 826–35.

[48] Puebla, C; Farías, M; González, M; Vecchiola, A; Aguayo, C; Krause, B; et al. High D-glucose reduces SLC29A1 promoter activity and adenosine transport involving specific protein 1 in human umbilical vein endothelium. *J. Cell Physiol.* 2008, 215(3), 645–56.

[49] Farías, M; Puebla, C; Westermeier, F; Jo, MJ; Pastor-Anglada, M; Casanello, P; et al. Nitric oxide reduces SLC29A1 promoter activity and adenosine transport involving transcription factor complex hCHOP-C/EBP in human umbilical vein endothelial cells from gestational diabetes. *Cardiovasc, Res.* 2010, 86(1), 45–54.

[50] Westermeier, F; Salomón, C; González, M; Puebla, C; Guzmán-Gutiérrez, E; Cifuentes, F; et al. Insulin Restores Gestational Diabetes Mellitus-Reduced Adenosine Transport Involving Differential Expression of Insulin Receptor Isoforms in Human Umbilical Vein Endothelium. *Diabetes.* 2011, 60(6), 1677–87.

[51] Salomón, C; Westermeier, F; Puebla, C; Arroyo, P; Guzmán-Gutiérrez, E; Pardo, F; et al. Gestational Diabetes Reduces Adenosine Transport in Human Placental Microvascular Endothelium, an Effect Reversed by Insulin. *PLoS ONE.* 2012, 7(7), e40578.

[52] Chu, S; Xiong, W; Zhang, D; Soylu, H; Sun, C; Albensi, BC; et al. Regulation of adenosine levels during cerebral ischemia. *Nature.* 2012, 1–7.

[53] Eltzschig, HK; Abdulla, P; Hoffman, E; Hamilton, KE; Daniels, D; Schönfeld, C; et al. HIF-1-dependent repression of equilibrative nucleoside transporter (ENT) in hypoxia. *J. Exp. Med.* 2005, 202(11), 1493–505.

[54] Morote-Garcia, JC; Rosenberger, P; Nivillac, NMI; Coe, IR; Eltzschig, HK. Hypoxia-inducible factor-dependent repression of equilibrative nucleoside transporter 2 attenuates mucosal inflammation during intestinal hypoxia. *Gastroenterology*. 2009, 136(2), 607–18.

[55] Chaudary, N; Naydenova, Z; Shuralyova, I; Coe, IR. Hypoxia regulates the adenosine transporter, mENT1, in the murine cardiomyocyte cell line, HL-1. *Cardiovasc. Res.* 2004, 61(4), 780–8.

[56] Chaudary, N; Naydenova, Z; Shuralyova, I; Coe, IR. The adenosine transporter, mENT1, is a target for adenosine receptor signaling and protein kinase Cepsilon in hypoxic and pharmacological preconditioning in the mouse cardiomyocyte cell line, HL-1. *J. Pharmacol. Exp. Ther.* 2004, 310(3), 1190–8.

[57] Löffler, M; Morote-Garcia, JC; Eltzschig, SA; Coe, IR; Eltzschig, HK. Physiological roles of vascular nucleoside transporters. *Arterioscler, Thromb, Vasc, Biol.* 2007, 27(5), 1004–13.

[58] Rose, JB; Naydenova, Z; Bang, A; Eguchi, M; Sweeney, G; Choi, D-SS; et al. Equilibrative nucleoside transporter 1 plays an essential role in cardioprotection. *Am. J. Physiol. Heart. Circ. Physiol.* 2010, 298(3), H771–7.

[59] Cargnello, M; Roux, PP. Activation and function of the MAPKs and their substrates, the MAPK-activated protein kinases. *Microbiol. Mol. Biol. Rev.* 2011, 75(1), 50–83.

[60] Sheth, S; Brito, R; Mukherjea, D; Rybak, LP; Ramkumar, V. Adenosine receptors: expression, function and regulation. *Int. J. Mol. Sci.* 2014, 15(2), 2024–52.

[61] Choi, D-SS; Cascini, M-G; Mailliard, W; Young, H; Paredes, P; McMahon, T; et al. The type 1 equilibrative nucleoside transporter regulates ethanol intoxication and preference. *Nat. Neurosci.* 2004, 7(8), 855–61.

[62] San Martín, R; Sobrevia, L. Gestational diabetes and the adenosine/l-Arginine/nitric oxide (ALANO) pathway in human umbilical vein endothelium. *Placenta*. 2006, 27(1), 1–10.

[63] Wyatt, AW; Steinert, JR; Wheeler-Jones, CPD; Morgan, AJ; Sugden, D; Pearson, JD; et al. Early activation of the p42/p44MAPK pathway mediates adenosine-induced nitric oxide production in human endothelial cells: a novel calcium-insensitive mechanism. *FASEB J.* 2002, 16(12), 1584–94.

[64] Guzmn-Gutirrez, E; Arroyo, P; Pardo, F; Leiva, A; Sobrevi, L. Gestational Diabetes - Causes, Diagnosis and Treatment: The Adenosine–Insulin Signaling Axis in the Fetoplacental Endothelial Dysfunction in Gestational Diabetes. *InTech*, 2013.

[65] Gray, JH; Mangravite, LM; Owen, RP; Urban, TJ; Chan, W; Carlson, EJ; et al. Functional and genetic diversity in the concentrative nucleoside transporter, CNT1, in human populations. *Mol. Pharmacol.* 2004, 65(3), 512–9.

[66] Owen, RP; Lagpacan, LL; Taylor, TR; La Cruz De, M; Huang, CC; Kawamoto, M; et al. Functional characterization and haplotype analysis of polymorphisms in the human equilibrative nucleoside transporter, ENT2. *Drug Metab. Dispos.* 2006, 34(1), 12–5.

[67] Williams, JB; Lanahan, AA. A mammalian delayed-early response gene encodes HNP36, a novel, conserved nucleolar protein. *Biochem. Biophys. Res. Commun.* 1995, 213(1), 325–33.

[68] Mangravite, LM; Xiao, G; Giacomini, KM. Localization of human equilibrative nucleoside transporters, hENT1 and hENT2, in renal epithelial cells. *Am. J. Physiol. Renal Physiol.* 2003, 284(5), F902–10.

[69] Galmarini, CM; Mackey, JR; Dumontet, C. Nucleoside analogues: mechanisms of drug resistance and reversal strategies. *Leukemia*. 2001, 15(6), 875–90.

[70] Sève, P; Mackey, JR; Isaac, S; Trédan, O; Souquet, PJ; Pérol, M; et al. cN-II expression predicts survival in patients receiving gemcitabine for advanced non-small cell lung cancer. *Lung Cancer*. 2005, 49(3), 363–70.

[71] Pratt, SE; Durland-Busbice, S; Shepard, RL; Donoho, GP; Starling, JJ; Wickremsinhe, ER; et al. Efficacy of low-dose oral metronomic dosing of the prodrug of gemcitabine, LY2334737, in human tumor xenografts. *Mol. Cancer Ther*. 2013, 12(4), 481–90.

[72] Ferrandina, G; Mey, V; Nannizzi, S; Ricciardi, S; Petrillo, M; Ferlini, C; et al. Expression of nucleoside transporters, deoxycitidine kinase, ribonucleotide reductase regulatory subunits, and gemcitabine catabolic enzymes in primary ovarian cancer. *Cancer Chemother. Pharmacol*. 2010, 65(4), 679–86.

[73] Bock, AJ; Dong, HP; Tropé, CG; Staff, AC; Risberg, B; Davidson, B; Nucleoside transporters are widely expressed in ovarian carcinoma effusions. *Cancer Chemother. Pharmacol*. 2011, 69(2), 467–75.

[74] Maréchal, R; Mackey, JR; Lai, R; Demetter, P; Peeters, M; Polus, M; et al. Human equilibrative nucleoside transporter 1 and human concentrative nucleoside transporter 3 predict survival after adjuvant gemcitabine therapy in resected pancreatic adenocarcinoma. *Clin. Cancer Res*. 2009, 15(8), 2913–9.

[75] Bhutia, YD; Hung, SW; Patel, B; Lovin, D; Govindarajan, R. CNT1 expression influences proliferation and chemosensitivity in drug-resistant pancreatic cancer cells. *Cancer Res*. 2011, 71(5), 1825–35.

[76] Pinilla-Macua, I; Casado, FJ; Pastor-Anglada, M. Structural determinants for rCNT2 sorting to the plasma membrane of polarized and non-polarized cells. *Biochem. J*. 2012, 442(3), 517–25.

[77] Pinilla-Macua, I; Claudio-Montero, A; Pastor-Anglada, M. rCNT2 extracellular cysteines, Cys(615) and Cys(649), are important for maturation and sorting to the plasma membrane. *FEBS Lett*. 2014. [Epub ahead of print]

[78] Ward, JL; Leung, GPH; Toan, S-V; Tse, C-M. Functional analysis of site-directed glycosylation mutants of the human equilibrative nucleoside transporter-2. *Arch. Bioch. Biophys*. 2003, 411(1), 19–26.

[79] Piirainen, H; Ashok, Y; Nanekar, RT; Jaakola, V-P. Structural features of adenosine receptors: From crystal to function. *BBA - Biomembranes*. 2011, 1808(5), 1233–44.

[80] Sundaram, M; Yao, SYM; Ingram, JC; Berry, ZA; Abidi, F; Cass, CE; et al. Topology of a human equilibrative, nitrobenzylthioinosine (NBMPR)-sensitive nucleoside transporter (hENT1) implicated in the cellular uptake of adenosine and anti-cancer drugs. *J. Biol. Chem*. 2001, 276(48), 45270–5.

[81] Reyes, G; Coe, IR. Genomics and proteomics of nucleoside transporters. *Current Pharmacogenomics*. 2005.

[82] Amanchy, R; Periaswamy, B; Mathivanan, S; Reddy, R; Tattikota, SG; Pandey, A. A curated compendium of phosphorylation motifs. *Nat. Biotechnol*. 2007, 25(3), 285–6.

[83] Nivillac, NMI; Bacani, J; Coe, IR. The life cycle of human equilibrative nucleoside transporter 1: From ER export to degradation. *Exp. Cell Res*. 2011, 317(11), 1567–79.

[84] Panigrahi, R; Chandra, PK; Ferraris, P; Kurt, R; Song, K; Garry, RF; et al. Persistent HCV infection impairs ribavirin antiviral activity through clathrin-mediated trafficking of equilibrative nucleoside transporter 1. *J. Virol*. 2014. [Epub ahead of print]

[85] Korn, T. The Plasma Membrane-associated Protein RS1 Decreases Transcription of the Transporter SGLT1 in Confluent LLC-PK1 Cells. *J. Biol. Chem.* 2001, 276(48), 45330–40.

[86] Errasti-Murugarren, E; Fernandez-Calotti, P; Veyhl-Wichmann, M; Diepold, M; Pinilla-Macua, I; Pérez-Torras, S; et al. Role of the Transporter Regulator Protein (RS1) in the Modulation of Concentrative Nucleoside Transporters (CNTs) in Epithelia. *Mol. Pharmacol.* 2012, 82(1), 59–67.

[87] Fernandez-Calotti, P; Pastor-Anglada, M. All-trans-retinoic acid promotes trafficking of human concentrative nucleoside transporter-3 (hCNT3) to the plasma membrane by a TGF-beta1-mediated mechanism. *J. Biol. Chem.* 2010, 285(18), 13589–98.

[88] Newton, AC. Regulation of the ABC kinases by phosphorylation: protein kinase C as a paradigm. *Biochem J.* 2003, 370(Pt2), 361–71.

[89] Santos-Rodrigues dos, A; Grañe-Boladeras, N; Bicket, A; Coe, IR. Nucleoside transporters in the purinome. *Neurochemistry International.* 2014, 73(C), 229–37.

[90] Nagy, LE; Diamond, I; Gordon, AS. cAMP-dependent protein kinase regulates inhibition of adenosine transport by ethanol. *Mol. Pharmacol.* 1991, 40(5), 812-7.

[91] Coe, IR; Dohrman, DP; Constantinescu, A. Activation of cyclic AMP-dependent protein kinase reverses tolerance of a nucleoside transporter to ethanol. *J. Pharmacol. Exp. Ther.* 1996, 276, 365-369.

[92] Coe, IR; Yao, L; Diamond, I; Gordon, AS. The role of protein kinase C in cellular tolerance to ethanol. *J. Biol. Chem.* 1996, 271(46), 29468–72.

[93] Ramadan, A; Naydenova, Z; Stevanovic, K; Rose, JB; Coe, IR. The adenosine transporter, ENT1, in cardiomyocytes is sensitive to inhibition by ethanol in a kinase-dependent manner: implications for ethanol-dependent cardioprotection and nucleoside analog drug cytotoxicity. *Purinergic Signal.* 2014, 10(2), 305–12.

[94] Delicado, EG; Sen, RP; Miras-Portugal, MT. Effects of phorbol esters and secretagogues on nitrobenzylthioinosine binding to nucleoside transporters and nucleoside uptake in cultured chromaffin cells. *Biochem. J.* 1991, 279(Pt3), 651-5.

[95] Sen, RP; Delicado, EG; Alvarez, A; Brocklebank, AM; Wiley, JS; Miras-Portugal, MT. Flow cytometric studies of nucleoside transport regulation in single chromaffin cells. *FEBS Lett.* 1998, 422(3), 368–72.

[96] Sen, RP; Delicado, EG; Miras-Portugal, MT. Differential modulation of nucleoside transport types in neuroblastoma cells by protein kinase activation. *Neuropharmacology.* 1999, 38(7), 1009-15.

[97] Kiss, A; Farah, K; Kim, J; Garriock, RJ; Drysdale, TA; Hammond, JR. Molecular cloning and functional characterization of inhibitor-sensitive (mENT1) and inhibitor-resistant (mENT2) equilibrative nucleoside transporters from mouse brain. *Biochem. J.* 2000, 352(Pt 2), 363–72.

[98] Stolk, M; Cooper, E; Vilk, G; Litchfield, DW; Hammond, JR. Subtype-specific regulation of equilibrative nucleoside transporters by protein kinase CK2. *Biochem. J.* 2005, 386(Pt2), 281–9.

[99] Robillard, KR; Bone, DBJ; Park, JS; Hammond, JR. Characterization of mENT1 11, a Novel Alternative Splice Variant of the Mouse Equilibrative Nucleoside Transporter 1. *Mol. Pharmacol.* 2008, 74(1), 264–73.

[100] Hammond, JR; Stolk, M; Archer, RGE; McConnell, K. Pharmacological analysis and molecular cloning of the canine equilibrative nucleoside transporter 1. *Eur. J. Pharmacol.* 2004, 491(1), 9–19.

[101] Handa, M; Choi, D-SS; Caldeiro, RM; Messing, RO; Gordon, AS; Diamond, I. Cloning of a novel isoform of the mouse NBMPR-sensitive equilibrative nucleoside transporter (ENT1) lacking a putative phosphorylation site. *Gene.* 2001, 262(1-2), 301–7.

[102] Bone, DBJ; Robillard, KR; Stolk, M; Hammond, JR. Differential regulation of mouse equilibrative nucleoside transporter 1 (mENT1) splice variants by protein kinase CK2. *Mol. Membr. Biol.* 2007, 24(4), 294–303.

[103] Reyes, G; Nivillac, NMI; Karim, MZ; Desouza, L; Siu, KWM; Coe, IR. The Equilibrative Nucleoside Transporter (ENT1) can be phosphorylated at multiple sites by PKC and PKA. *Mol. Membr. Biol.* 2011, 28(6), 412–26.

[104] Lu, G; Zhou, Q-X; Kang, S; Li, Q-L; Zhao, L-C; Chen, J-D; et al. Chronic morphine treatment impaired hippocampal long-term potentiation and spatial memory via accumulation of extracellular adenosine acting on adenosine A1 receptors. *J. Neuroscience.* 2010, 30(14), 5058–70.

[105] Genedani, S; Guidolin, D; Leo, G; Filaferro, M; Torvinen, M; Woods, AS; et al. Computer-assisted image analysis of caveolin-1 involvement in the internalization process of adenosine A2A-dopamine D2 receptor heterodimers. *J. Mol. Neurosci.* 2005, 26(2-3), 177–84.

[106] Snider, J; Kittanakom, S; Damjanovic, D; Curak, J; Wong, V; Stagljar, I. Detecting interactions with membrane proteins using a membrane two-hybrid assay in yeast. *Nature Protocols.* 2010, 5(7), 1281–93.

[107] Wall, MJ; Dale, N. Neuronal transporter and astrocytic ATP exocytosis underlie activity-dependent adenosine release in the hippocampus. *J. Physiol.* 2013, 591(Pt16), 3853–71.

[108] Latini, S; Pedata, F. Adenosine in the central nervous system: release mechanisms and extracellular concentrations. *J. Neurochem.* 2001, 79(3), 463–84.

[109] Rose, JB; Coe, IR. Physiology of nucleoside transporters: back to the future. *Am. J. Physiol.* 2008, 23, 41–8.

[110] Rose, JB; Naydenova, Z; Bang, A; Ramadan, A; Klawitter, J; Schram, K; et al. Absence of equilibrative nucleoside transporter 1 in ENT1 knockout mice leads to altered nucleoside levels following hypoxic challenge. *Life Sci.* 2011, 89(17-18), 621–30.

[111] Bone, DBJ; Choi, D-SS; Coe, IR; Hammond, JR. Nucleoside/nucleobase transport and metabolism by microvascular endothelial cells isolated from ENT1-/- mice. *Am. J. Physiol. Heart Circ. Physiol.* 2010, 299(3), H847–56.

[112] Grenz, A; Bauerle, JD; Dalton, JH; Ridyard, D; Badulak, A; Tak, E; et al. Equilibrative nucleoside transporter 1 (ENT1) regulates postischemic blood flow during acute kidney injury in mice. *J. Clin. Invest.* 2012, 122(2),693–710.

[113] Eckle, T; Hughes, K; Ehrentraut, H; Brodsky, KS; Rosenberger, P; Choi, D-SS; et al. Crosstalk between the equilibrative nucleoside transporter ENT2 and alveolar Adora2b adenosine receptors dampens acute lung injury. *FASEB J.* 2013, 27(8), 3078–89.

[114] Paes de Carvalho, R; Braas, KM; Adler, R; Snyder, SH. Developmental regulation of adenosine A1 receptors, uptake sites and endogenous adenosine in the chick retina. *Brain Res. Dev. Brain Res.* 1992, 70(1), 87–95.

[115] Ackley, MA; Governo, RJM; Cass, CE; Young, JD; Baldwin, SA; King, AE. Control of glutamatergic neurotransmission in the rat spinal dorsal horn by the nucleoside transporter ENT1. *J. Physiol.* 2003, 548(2), 507–17.

[116] Pinto-Duarte, A; Coelho, JE; Cunha, RA; Ribeiro, JA; Sebastião, AM. Adenosine A2A receptors control the extracellular levels of adenosine through modulation of nucleoside transporters activity in the rat hippocampus. *J. Neurochem.* 2005, 93(3), 595–604.

[117] Hirsh, AJ; Stonebraker, JR; van Heusden, CA; Lazarowski, ER; Boucher, RC; Picher, M. Adenosine deaminase 1 and concentrative nucleoside transporters 2 and 3 regulate adenosine on the apical surface of human airway epithelia: implications for inflammatory lung diseases. *Biochemistry.* 2007, 46(36), 10373–83.

[118] Duflot, S; Riera, B; Fernández-Veledo, S; Casadó, V; Norman, RI; Casado, FJ; et al. ATP-sensitive K(+) channels regulate the concentrative adenosine transporter CNT2 following activation by A(1) adenosine receptors. *Mol. Cell Biol.* 2004, 24(7), 2710–9.

[119] Medina-Pulido, L; Molina-Arcas, M; Justicia, C; Soriano, E; Burgaya, F; Planas, AM; et al. Hypoxia and P1 receptor activation regulate the high-affinity concentrative adenosine transporter CNT2 in differentiated neuronal PC12 cells. *Biochem. J.* 2013, 454(3), 437–45.

[120] . Guillén-Gómez, E; Calbet, M; Casado, J; de Lecea, L; Soriano, E; Pastor-Anglada, M; et al. Distribution of CNT2 and ENT1 transcripts in rat brain: selective decrease of CNT2 mRNA in the cerebral cortex of sleep-deprived rats. *J. Neurochem.* 2004, 90(4), 883–93.

[121] Zhu, C-B; Lindler, KM; Campbell, NG; Sutcliffe, JS; Hewlett, WA; Blakely, RD. Colocalization and regulated physical association of presynaptic serotonin transporters with A_3 adenosine receptors. *Mol. Pharmacol.* 2011, 80(3), 458–65.

[122] Zhu, C-B; Hewlett, WA. Feoktistov, I; Biaggioni, I; Blakely, RD. Adenosine receptor, protein kinase G, and p38 mitogen-activated protein kinase-dependent up-regulation of serotonin transporters involves both transporter trafficking and activation. *Mol. Pharmacol.* 2004, 65(6), 1462–74.

[123] Zhu, C-B; Steiner, JA; Munn, JL; Daws, LC; Hewlett, WA; Blakely, RD. Rapid stimulation of presynaptic serotonin transport by A(3) adenosine receptors. *J. Pharmacol. Exp. Ther.* 2007, 322(1), 332–40.

[124] Pérez-Torras, S; Vidal-Pla, A; Cano-Soldado, P; Huber-Ruano, I; Mazo, A; Pastor-Anglada, M. Concentrative nucleoside transporter 1 (hCNT1) promotes phenotypic changes relevant to tumor biology in a translocation-independent manner. *Cell Death Dis.* 2013, 4(5), e648–9.

[125] Huber-Ruano, I; Pinilla-Macua, I; Torres, G; Casado, FJ; Pastor-Anglada, M. Link between high-affinity adenosine concentrative nucleoside transporter-2 (CNT2) and energy metabolism in intestinal and liver parenchymal cells. *J. Cell. Physiol.* 2010, 225(2), 620-30.

In: Adenosine Signaling Mechanisms
Editors: V. Ramkumar and R. Paes de Carvalho
ISBN: 978-1-63483-186-4

Chapter 3

REGULATION OF ADENOSINE RECEPTOR DESENSITISATION AND FUNCTION

Claire Rutherford[1] and Timothy M. Palmer[1,2*]
[1] Institute of Cardiovascular and Medical Sciences,
University of Glasgow, Glasgow, Scotland, United Kingdom
[2] School of Pharmacy, University of Bradford,
Bradford, West Yorkshire, United Kingdom

ABSTRACT

The ability of cells to respond to changes in extracellular concentrations of the nucleoside adenosine is determined by their complement of cell surface adenosine receptors. Four receptor subtypes, termed A_1, A_{2A}, A_{2B} and A_3, have been identified that, as well as displaying distinct pharmacological characteristics, mediate their intracellular effects by coupling to defined G-protein families and other G-protein-independent pathways. A large body of research has also identified the molecular mechanisms by which adenosine receptor-derived signalling events are regulated. This has revealed that the desensitisation of closely related adenosine receptors, which bind the same agonist in vivo (i.e., adenosine) and activate the same family of G-proteins, can be regulated by unique receptor-specific mechanisms. In this chapter, we will describe the nature of these differences, and how they could potentially be exploited to enhance the therapeutically beneficial effects of adenosine receptor modulation in vivo.

INTRODUCTION

The multiple effects of extracellular adenosine observed in many tissues are dependent on its ability to bind and activate four heterotrimeric G-protein-coupled, or "seven transmembrane-spanning," adenosine receptor (AR) subtypes, termed A_1, A_{2A}, A_{2B} and A_3 [1]. The unique effects of adenosine in defined cell types are determined by several factors,

* Address reprint requests to: Professor Tim Palmer, E-mail: T.Palmer1@bradford.ac.uk, Phone: +44(0)1274 234662

including AR expression profiles, their expression levels, and the signalling modules activated by agonist-occupied receptor proteins within a given cell type. Substantial progress has also been made in the identification of molecular processes that control AR signal output and it is now clear that for many seven transmembrane-spanning receptors (7TMRs), stimulus-induced changes in post-translational modification profiles of both ARs and their adaptor proteins can co-ordinate multiple changes in receptor function in a complex and subtype-specific manner [2, 3].

In this chapter, we will focus on several laboratories' efforts to identify the processes that control AR function and describe how this information might be applied to determine whether manipulation of such events would be useful approaches with which to control AR functionality in disease states. However to put the findings on AR regulation in context, we will first describe key mechanisms known to be important for controlling signalling from related 7TMRs.

7TMR Desensitisation Is Controlled by Multiple Post-Translational Modifications and Protein Interactions

7TMR functionality has typically been determined by interaction with heterotrimeric G-proteins, which control the activity of effector enzymes such as adenylyl cyclases to modulate levels of intracellular second messengers such as cyclic AMP [4, 5]. However studies examining the molecular basis of 7TMR desensitisation, defined as the ability of a receptor-mediated response to plateau and then diminish despite the sustained presence of agonist, have revealed that multiple intracellular proteins not only fine tune G-protein-mediated signalling at the plasma membrane but also initiate G-protein-independent signals from intracellular compartments. These details have been best characterised for the human β_2-adrenergic receptor but similar studies on many other 7TMRs have shown that the same basic principles apply, even if detailed mechanisms and functional consequences may vary [6, 7].

Desensitisation responses can be classified as either homologous or heterologous. Heterologous desensitisation is defined as a situation whereby exposure to an agonist at a given receptor also results in the desensitisation of agonist responses from multiple receptors. This can arise due to a variety of mechanisms, including phosphorylation or down-regulation of common post-receptor intermediates such as G-proteins or effector enzymes, as well as receptor modification by second messenger activated kinases such as cyclic AMP-dependent protein kinase (PKA) or protein kinase C (PKC) [8-10].

In contrast, homologues desensitisation is defined as a loss of responsiveness specifically at the receptor activated by the relevant agonist. For almost all 7TMRs, homologous desensitisation is the result of a series of temporally distinct but mechanistically inter-related processes that act at the agonist-occupied receptor. Together, these inhibit productive interaction of the agonist-occupied receptor with G-proteins, and trigger a series of protein-protein interactions that promote receptor sequestration, defined as a loss of cell surface receptor, and down-regulation, defined as a reduction in total receptor number [6, 7]. A general scheme describing mechanisms involved in homologous desensitisation for many 7TMRs is presented in Figure 1.

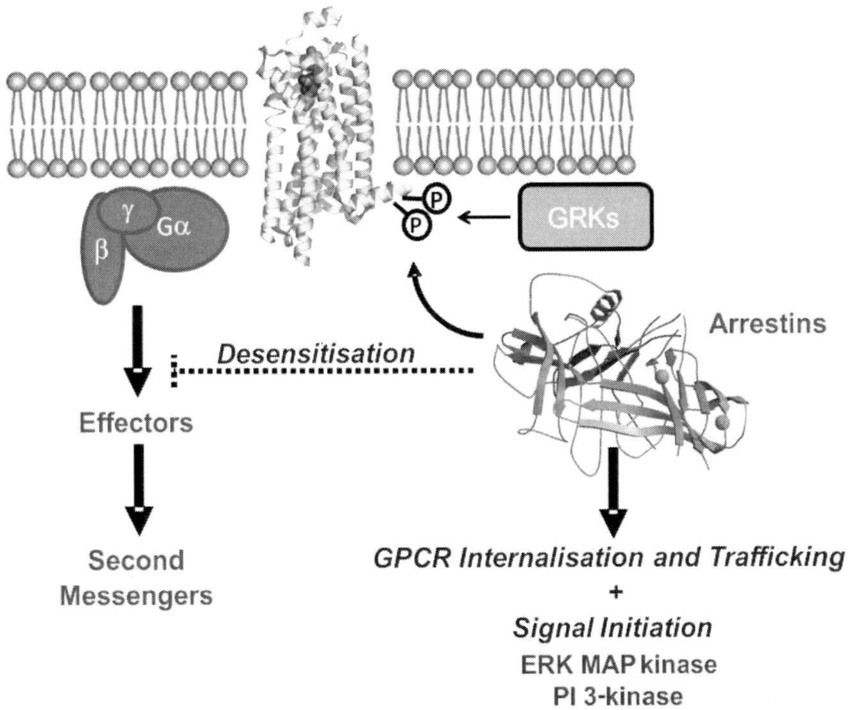

Figure 1. Model in which the capacity of 7TMRs to activate heterotrimeric G-proteins is attenuated upon phosphorylation of agonist-occupied receptors by one or more G-protein-coupled receptor kinases (GRKs) followed by the recruitment of arrestins (structure taken from PDB 2WTR). As well as triggering receptor internalisation, arrestins act as adaptor proteins to link phosphorylated 7TMRs to activation of G-protein-independent signalling pathways, including the phosphatidylinositol 3-kinase (PI 3-kinase) and extracelllar signal-regulated kinase (ERK)/MAP kinase pathways amongst others [2, 3, 6, 7]. The 7TMR structure is agonist-bound human A_{2A}AR (PDB 3QAK).

The key features of this model are as follows:-

1. Agonist binding triggers a conformational change in the receptor that initiates interaction with and activation of heterotrimeric G-proteins.
2. Agonist-bound receptors recruit one or more G-protein-coupled receptor kinases (GRKs). These constitute a family of seven related protein kinases (GRKs 1-7) that typically phosphorylate Ser and/or Thr residues within either the C-terminal cytoplasmic tail or third intracellular loops of the activated receptor [6].
3. GRK-phosphorylated receptors recruit "arrestin" proteins from the cytoplasm. While arrestin-1 and -4 are localised to retinal rods and cones respectively, arrestin-2 (also known as β-arrestin-1) and arrestin-3 (β-arrestin-2) are ubiquitously expressed and have been shown to interact with many 7TMRs [2, 3, 6, 7].
4. Arrestin binding to phosphorylated 7TMRs typically has at least three consequences:-
    I. A reduced capacity of the agonist-bound receptor to productively interact with G-proteins, thus producing a homologous desensitisation of G-protein-mediated signalling [6, 7].
    II. Accumulation of arrestin-bound receptors in cell surface clathrin-coated pits via direct interaction of arrestins with clathrin heavy chain, the β 2-adaptin

component of the AP2 adaptor protein complex and phosphoinositides amongst others. This precedes internalisation into early endosomes and either trafficking to lysosomes for receptor degradation or recycling back to the plasma membrane [2, 3]. Ubiquitylation of 7TMRs plays an important role in these trafficking events and arrestins appear to co-ordinate 7TMR ubiquitylation by simultaneously binding some of the E3 ubiquitin ligases and deubiquitinases responsible [11, 12].

III. Arrestin-bound receptors can also initiate novel G-protein-independent signalling pathways via the same receptors for which they desensitise G-protein coupling. This has been demonstrated for an expanding list of 7TMRs and shown to have important physiological roles in many cases. Several studies examining extracellular signal-regulated kinase 1/2 (ERK1/2) pathway activation by the □activation by the tor have demonstrated that G-protein- and arrestin-mediated routes to ERK1/2 activation display distinct kinetics and can direct ERK1/2 to different subcellular localisations [13].

In addition to homologous and heterologous desensitisation mechanisms, 7TMR function can additionally be regulated at transcriptional and post-transcriptional levels, as well as via receptor-specific protein-protein interactions. In the rest of this chapter, we will discuss these processes as they are currently known to apply to each of the four AR subtypes.

A_1 Adenosine Receptor

A_1ARs Are Subject to Dynamic Cell Type-Dependent Regulation

The A_1 and A_3ARs are still thought to mediate the majority of their effects through interaction with the G_i family of G-proteins. These trigger multiple signalling events, including inhibition of cAMP generation, and activation of phospholipase Cβ, inwardly rectifying potassium channels, and ERK1/2. What has also emerged is that each receptor is subject to specific regulatory mechanisms which fine-tune receptor output. The most revealing aspect of this is the distinct mechanisms by which their responses desensitise following agonist exposure. Early experiments performed by several laboratories studying A_1AR function in rat adipocytes demonstrated that cellular exposure to the AR agonist *(R)*-N^6-(phenylisopropyl)adenosine (R-PIA) over a time course of several days in vivo resulted in a desensitisation of A_1AR-mediated inhibition of lipolysis and cAMP formation [14-16]. Desensitisation was accompanied by elevated basal and isoproteronol-stimulated adenylyl cyclase activities in membranes from adipocytes isolated from treated animals [15, 16]. Detailed characterisation demonstrated that desensitisation was associated with a parallel down-regulation of A_1AR expression and a reduced proportion of A_1ARs in a high-affinity G-protein-coupled state. In addition, R-PIA treatment reduced levels of the α-subunits of G_i1 and G_i2 while increasing levels of $G_s\alpha$, which is responsible for 7TMR-mediated stimulation of adenylyl cyclase [16]. Interestingly, changes in $G_i\alpha$-subunit expression occurred in the absence of any changes in mRNA levels, suggesting that chronic agonist exposure accelerated the degradation of the activated $G_i\alpha$-subunits, although this has not been tested directly [16].

Regardless, these changes in G-protein expression had two consequences; first, the desensitisation induced by R-PIA treatment was heterologous in nature, i.e., inhibition of cAMP formation by the G_i-linked prostaglandin E_1 receptor was also desensitised presumably due to down-regulation of G_i subtypes utilised by both receptors [14-16]. Second, the correlation of increased $G_s\alpha$ with elevated basal and isoproterenol-stimulated adenylyl cyclase activity suggested that increased levels of G_s were responsible for the enhanced lipolytic effects of β-adrenergic receptor agonists in adipocytes from R-PIA-treated animals [14-16]. Similar studies by Green and co-workers examining primary rat adipocyte cultures treated in vitro with or without R-PIA have produced similar findings [17, 18].

On the back of these findings, several groups have examined A_1AR regulation in established cell lines. These include DDT_1 MF-2, a smooth muscle cell line derived from hamster vas deferens [19, 20]. Desensitisation of endogenous DDT_1 MF-2 A_1ARs in vitro requires a much shorter agonist incubation time (24 hour exposure versus 6 days required in rat adipocytes) [19]. While this involved both an uncoupling of A_1ARs from G_i-proteins and a down-regulation in the total A_1AR number, expression levels of inhibitory G-proteins were unaffected suggesting that such G_i-protein down-regulation is not required for desensitisation of A_1AR function [19]. One of the most significant findings from DDT_1 MF-2 cells is that the A_1AR is phosphorylated following exposure not only to agonist [19, 20] but activation of PKC and cAMP-dependent protein kinase (PKA) [20]. In response to agonist, the receptor appeared to be transiently phosphorylated predominantly on tyrosine residues, a novel mechanism distinct from other 7TMRs [20]. Phosphorylation also precedes a rapid clustering of A_1ARs on the cell surface and uncoupling of receptor from G-protein, a state which is maintained prior to receptor internalisation, possibly via caveolae rather than clathrin-mediated endocytosis, and subsequent down-regulation [19,20]. Another major finding is that A_1AR activation can trigger the translocation of GRK2 from the cytosol to the plasma membrane, an event which precedes the uncoupling of the receptor from G-proteins as determined by reduced binding of agonist radioligand $[^{125}I]N^6$-2-(4-amino-3-iodophenyl)ethyladenosine ($[^{125}I]$APNEA) but not antagonist radioligand $[^3H]$8-cyclopentyl-1,3-dipropylxanthine ($[^3H]$DPCPX). These data are not only consistent with some studies demonstrating an agonist-dependent phosphorylation of the A_1AR in intact cells [19, 20] but also findings demonstrating that agonist-bound A_1AR can be phosphorylated by purified GRK2 in vitro [21], although the phosphoacceptor sites have yet to be identified.

For many 7TMRs, GRK phosphorylation promotes their interaction with arrestins [3, 6]. In contrast, studies examining physical and functional interactions of the human A_1AR with arrestins have produced different findings. Analysis of A_1AR-mediated interactions with β-arrestin-2/arrestin-3 in transfected Chinese hamster ovary (CHO) cells have revealed that agonist activation triggers the accumulation of β-arrestin-2 from the cytoplasm to discrete puncta at the plasma membrane that, unusually, do not co-localise with the receptor [22]. Interestingly, redistribution of β-arrestin-2 occurs in the absence of any significant A_1AR internalisation in CHO cells, suggesting that additional factors are present in other cell types in which more rapid A_1AR internalisation is observed [19, 20, 23]. However the ability of the A_1AR to re-distribute β-arrestins [22, 24] could be important in mediating ERK1/2 phosphorylation and activation of AP-1 transcription factor complexes [24]. β-arrestin-1/arrestin-2-mediated activation of ERK1/2 also appears to be required for induction of A_1AR mRNA, thus priming cells for re-synthesis of a new cellular pool of A_1ARs following agonist withdrawal [24].

A_3 ADENOSINE RECEPTOR

Signalling from A_3ARs Is Subject to Rapid Regulation

Identification of the orphan rat clone R226 as a novel AR subtype subsequently termed the A_3AR [25] initiated new lines of investigation aimed at defining physiological roles for this receptor in both normal physiology and disease states [26, 27]. As well as being pharmacologically distinct, a wealth of studies has also demonstrated that the desensitisation of A_3AR function also differs from that observed for the A_1AR.

The isolation of an A_3AR provided a molecular basis for the effects of adenosine and other AR drugs on enhancing secretory responses in rat RBL-2H3 cells. In this system, FcεRI-activated secretion of histamine and β-hexosaminidase are synergistically enhanced by adenosine and AR agonist 5' N-ethylcarboxamidoadenosine (NECA) [28]. Like many secretagogues, A_3AR activation triggers a transient increase in intracellular calcium that is rapidly desensitised within seconds [28, 29]. Importantly, the responses of NECA-treated cells to FcεRI activation are unaffected, suggesting the phenomenon reflects a homologous desensitisation of A_3AR signalling [28]. A similar rapid desensitisation of A_3AR-mediated actin polymerisation has also been observed [30]. To examine these phenomena in greater molecular detail, transfected cell systems expressing recombinant ARs have been used extensively. This approach allows for the use of epitope- and green fluorescent protein (GFP)-tagged ARs, thereby facilitating comparison between otherwise different receptors. In addition, by comparing receptor regulation following expression in the same cell type, the possibility of cell type-specific differences is eliminated. Finally, the generation and expression of mutated and chimaeric inhibitory ARs has permitted assignment of specific functions to defined residues within each receptor.

Many groups have employed CHO cells stably expressing epitope-tagged A_1 and A_3ARs to study inhibitory AR regulation. Addition of tag sequences to the amino and carboxyl termini of each receptor alters neither their pharmacology nor signalling capacity. Moreover, the cell lines used for these studies have expressed ARs at levels typical of those found in cells types that express them endogenously, i.e., 1.0-1.5 pmol receptor/mg membrane protein, as determined by radioligand binding analyses [31-33]. Most importantly, recombinant ARs display similar time-courses for agonist-mediated desensitisation characteristic of A_1ARs endogenously expressed in smooth muscle cells [19,20] and A_3ARs expressed in RBL-2H3 cells and ADF human astrocytoma cells [29, 34]. Thus, a 10 min agonist exposure of CHO cells expressing the rat A_3AR produces a functional desensitisation of A_3AR function, as manifested by an eight-fold increase in the concentration of agonist required to half-maximally inhibit forskolin-stimulated adenylyl cyclase activity in isolated membranes, while A_1AR function is unaltered at this time point [32, 33]. Functional desensitisation of rat A_3ARs is associated with a 30-40% reduction in the number of high-affinity agonist binding sites detectable in ^{125}I ABMECA agonist radioligand binding assays, suggesting an agonist induced uncoupling of the A_3AR from its associated G-proteins that produces a desensitisation of adenylyl cyclase inhibition [32].

Initial lines of investigation using recombinant receptors expressed in CHO cells focused on assessing the phosphorylation status of the A_1 and A_3ARs following short-term agonist exposure. From these, it was found that conditions which produced a rapid functional desensitisation of A_3ARs triggered the phosphorylation of the receptor protein predominantly on threonine residues. Time-course experiments revealed that the response was sufficiently rapid to suggest a causal role in mediating rapid A_3AR desensitisation [32], while two observations implicated the GRK family as being responsible for A_3AR phosphorylation. First, under conditions in which agonist-stimulated A_3AR phosphorylation was maximal, various activators of second messenger-activated kinases had no effect [32]. Second, concentration-effect curves for NECA-stimulated A_3AR phosphorylation produced EC_{50} values equivalent to the K_i for this ligand in agonist radioligand binding competition assays [35], suggesting a link between extents of receptor occupancy and receptor phosphorylation. This was confirmed in kinase assays which demonstrated the ability of the A_3AR to serve as a substrate for agonist-dependent phosphorylation by purified recombinant GRK2, GRK3 and GRK5. Importantly, pretreatment of cells with agonist under conditions that resulted in functional desensitisation and A_3AR phosphorylation in situ significantly reduced the extent to which GRK2 could mediate agonist-stimulated A_3AR phosphorylation in vitro [32, 33]. This strongly implicates GRK2 or a kinase of similar substrate specificity as being responsible for A_3AR phosphorylation in intact cells.

The C-Terminal Domain of the A_3AR Dictates Its Rapid Desensitisation

In contrast with observations made in smooth muscle cells [19,20], no phosphorylation of the A_1AR was detectable either by agonist or various activators of second messenger-regulated kinases in CHO cells and this is associated with no functional desensitisation at these time points [33, 36, 37]. Given their distinct behaviour in CHO cells, we therefore reasoned that the A_1AR might prove a useful "acceptor" molecule into which putative regulatory domains from the A_3AR could be grafted to determine whether they could reconstitute rapid agonist-stimulated phosphorylation and desensitisation within the context of the resulting chimaera. We focused initially on the carboxyl-terminal tail distal to each receptor's putative sites of palmitate attachment, which is the region least conserved between the two receptors. The A_3AR sequence contains clusters of Ser and Thr residues, flanked in several instances on their amino-terminal side by acidic amino acids, a structural context known to favour GRK2 phosphorylation of target substrates in vitro [38]. In contrast, the A_1AR contains no potential phosphorylation sites within this region. Therefore, a chimaeric A_1CT3AR was generated in which the carboxyl-terminal nine residues of the human A_1AR distal to its site of palmitate attachment (Cys309) [37] were replaced with the corresponding fourteen amino acid region from the rat A_3AR. Following expression in CHO cells, A_1CT3AR was found to undergo rapid functional desensitisation and agonist-stimulated phosphorylation in intact cells, and serve as a substrate for GRKs 2, 3 and 5 in vitro similar to the A_3AR [33]. More restricted mutational analysis revealed that the presence of three Thr residues (Thr's 307, 318 and 319) within the A_3AR C-terminal regulatory domain was crucial in determining receptor sensitivity to agonist-stimulated phosphorylation and desensitisation, with Thr's 318 and 319 being most important and prior phosphorylation at Thr318 required in order for phosphorylation at Thr319 to proceed [39]. This "sequential" mechanism, whereby a "priming" phosphorylation event on one or more key residues is required for additional phosphorylation to occur, has also been noted for some GRK2 substrates [40] but is not a universal finding for all GRK2-phosphorylated 7TMRs [41].

A_3AR Internalisation and Interactions with Arrestins

The distinct kinetics and sensitivities of the A_1 and A_3ARs to rapid agonist-mediated desensitisation suggested that their abilities to recruit and utilise arestins to mediate receptor internalisation might also differ. We have examined this in some detail in transfected CHO cells [22,42] making the following observations. First, using a sequential cell surface biotinylation and immunoprecipitation assay, we found that while the A_1 and A_3ARs both undergo agonist-mediated sequestration away from the cell surface, these occur at distinct rates. Thus, while cell surface A_1ARs internalise with a $t_{1/2}$ of approximately 60 min, cell surface A_3ARs are lost with a $t_{1/2}$ of 15-20 min [42]. Second, A_3AR internalisation temporally follows the onset of receptor phosphorylation, suggesting that phosphorylation might be an important trigger for the onset of internalisation [42]. To test this hypothesis, loss- and gain-of-function phosphorylation site-mutated ARs were screened for their capacity to undergo agonist-mediated sequestration. An A_3AR mutant in which Thr 307,318 and 319 were mutated to Ala to render it resistant to phosphorylation failed to undergo significant agonist-stimulated internalisation over a time course in which substantial sequestration of the WT receptor was observed. This was not due to non-specific changes in conformation that resulted in an inability of the mutant receptor to interact with the cellular endocytotic machinery, as the rate of agonist-mediated internalisation of the chimaeric A_1CT3AR was identical to that of WT A_3AR [42]. The ability of the carboxyl-terminal fourteen residues of the A_3AR to confer rapid internalisation kinetics that temporally followed the onset of receptor phosphorylation suggested that A_3AR phosphorylation is necessary and sufficient to trigger rapid agonist-sequestration away from the cell surface. Moreover, internalisation is only detectable once receptor phosphorylation is maximal, implying that phosphorylation of each of Thr 307, 318 and 319 is necessary before A_3AR sequestration is initiated. Finally, agonist removal led to a recovery of cell surface A_3AR expression, although the kinetics of this process were significantly slower than those for receptor internalisation [42].

The development of GFP-tagged ARs has proved invaluable for examination of the subtype-specific dynamics of AR internalisation and identification of the subcellular compartments occupied by the internalised receptors. Consistent with data obtained from biochemical assays [42], confocal laser-scanning microscopy of cells expressing GFP-tagged A_3ARs revealed that agonist-occupied receptors were rapidly sequestered from the plasma membrane and accumulated in punctate transferrin receptor-positive endosomal vesicles. GFP-tagged A_1ARs and phosphorylation-resistant Thr307,318,319Ala-mutated A_3ARs both remained at the cell surface even after prolonged agonist treatment, while a GFP-tagged A_1CT3AR rapidly internalised and accumulated in a transferrin receptor-positive intracellular membrane pool in a manner indistinguishable from WT A_3ARs. Therefore, as well as conferring the ability to internalise, the carboxyl-terminal A_3AR regulatory domain also directs internalised receptor accumulation in endosomes [22]. Similar kinetics of internalisation and recycling have been reported for recombinant human A_3ARs expressed in CHO cells and A_3ARs expressed endogenously in ADF human astrocytoma cells [34, 43]. Given the sequence similarities of the putative C-terminal regulatory domains of the rat and human A_3ARs, these observations would suggest that very similar mechanisms control their agonist-stimulated phosphorylation and internalisation.

As interactions with arrestins regulate the internalisation of many 7TMRs, we examined the ability of the A_3AR to interact with arrestins in transfected CHO cells [22]. Treatment of A_3AR-expressing cells with agonist triggered a re-distribution of co-expressed β-arrestin-2/arrestin-3-GFP from the cytosol to small punctate spots both in the cytosol and at the plasma membrane. Interestingly these spots did not co-localise with immuno-stained A_3ARs in the same cells over a range of agonist treatment times, suggesting that the A_3AR and β-arrestin-2 either did not interact or, if they did so, the event was sufficiently short-lived to evade detection [22]. Indeed, others have employed an enzyme complementation assay to demonstrate an agonist-mediated interaction of the human A_3AR with β-arrestin-2 in CHO cells, increasing the likelihood that a very rapid and transient interaction of the rat A_3AR with β-arrestin-2 must occur [44]. It should also be noted that similar molecular mechanisms likely regulate endogenous rat A_3AR function in RBL-2H3 cells, since treatment of cells with an AR agonist over a time period sufficient to cause functional desensitisation of A_3AR-mediated calcium mobilisation and a decrease in ^{125}I-ABMECA binding similar to that observed in transfected CHO cells (T.M. Palmer, unpublished observations). However treatment of RBL-2H3 cells with AR agonist NECA has been reported to have no effect on the distribution of endogenous β-arrestin-1 or β-arrestin-2 although, as described for the recombinant A_3AR in CHO cells, any effect may have been sufficiently rapid and transient to escape detection [45].

A_3AR Responses to Chronic Agonist Treatment

Like the A_1AR, chronic agonist treatment over several hours triggers a profound functional desensitisation paralleled by a down-regulation of A_3AR protein [34] and a down-regulation of $G_i\alpha$ and β-subunits [31, 46]. Interestingly, similar to observations also made for the A_1AR, chronic agonist exposure of A_3AR-expressing cells also results in elevated adenylyl cyclase activity, leading to enhanced Ser133 phosphorylation of cyclic AMP response element binding protein (CREB), a transcription factor substrate of PKA [46]. However while adenylyl cyclase sensitisation in response to chronic A_1AR activation is thought to be a result of increased levels of $G_s\alpha$ [15, 16], this does not occur in A_3AR-expressing cells [31, 46]. Specifically, while the effect occurred over a time course of several hours, it did not require new protein synthesis as it was unaffected by protein synthesis inhibitor cycloheximide [46]. In addition, sensitisation manifested itself as an increased capacity of adenylyl cyclase to be stimulated by Gs despite no apparent change in the levels of either G_s or adenylyl cycase catalytic subunits [46]. While the mechanisms for this effect remain obscure, similar findings have been described for many other G_i-coupled 7TMRs suggesting that the key mechanisms involved are largely conserved [47]. However in terms of the A_3AR, it may explain why for certain A_3AR-mediated effects, including neuroprotection from ischaemic injury [48], acute stimulation with A_3AR agonists can produce the opposite effect of chronic agonist treatment [27].

A_{2A} ADENOSINE RECEPTOR

Regulation of A_{2A}AR Desensitisation and Function

The A_{2A}AR activates adenylyl cyclase through Gs and can also trigger activation of ERK1,2 via Gs-independent mechanisms in several cell systems [49-51]. It is distinguished from the other ARs by the presence of a 122 residue cytoplasmic C-terminal domain that is abundant in potential Ser/Thr phosphoacceptor sites but which bears no significant homology to any other protein [reviewed in 52].

Endogenous A_{2A}ARs in DDT1-MF2 hamster smooth muscle, NG108-15 rat neuroblastoma and CD4^{+} murine T cells [19, 53, 54], as well as recombinant A_{2A}ARs expressed in CHO cells [55, 56], each undergo a rapid and substantial homologous desensitisation following acute agonist exposure. In CHO cells, desensitisation is accompanied by rapid receptor phosphorylation and an accumulation of A_{2A}ARs in a so-called "light vesicle fraction" consistent with a rapid internalisation into endosomal compartments [55]. This is consistent with subsequent reports of rapid agonist-stimulated internalisation of A_{2A}ARs bound to fluorescent agonist probe Alexa488-APEC (N6-(4-Amino-3-iodobenzyl)adenosine-5′-N-ethylcarboxamidoadenosine) into Rab5-positive endosomes [57]. Detailed analysis of the mechanisms involved in A_{2A}AR desensitisation has shown that while agonist-occupied A_{2A}ARs are rapidly phosphorylated, removal of 95 residues from the C-terminal tail produced a truncated A_{2A}AR which desensitised as rapidly as the wild type receptor, indicating that this region is not required for rapid desensitisation and suggesting it may have an alternative functional role. However, mutation of Thr298 to Ala in the region of the C-terminal domain proximal to the seventh trans-membrane domain was sufficient to abolish agonist-stimulated A_{2A}AR phosphorylation and rapid desensitisation, although desensitisation in response to chronic agonist exposure was maintained suggesting that mechanisms responsible for short- and long-term adaptation are distinct [55, 56]. Several studies have proposed that GRK2 or a related kinase may be involved in controlling A_{2A}AR desensitisation but direct evidence linking GRK2 with changes in A_{2A}AR phosphorylation are lacking [53, 58, 59]. However, consistent with a possible role for GRKs, agonist treatment has been shown to stimulate A_{2A}AR interaction with β-arrestin-1/arrestin-2, which appears to be important for receptor clustering via interaction of the receptor's C-terminal domain with α-actinin prior to internalisation [60]. Importantly, inhibition of A_{2A}AR internalisation or accumulation in light vesicle membranes by pre-treatment with either sucrose or concanavalin A fails to inhibit short-term desensitisation and instead blocks the rapid recovery of responsiveness (resensitisation) following agonist removal, suggesting a role in the restoration of cell surface receptors [55, 61].

Roles for the A_{2A}AR C-Terminal Domain Independent of Receptor Desensitisation

The lack of any requirement for the majority of the A_{2A}AR's C-terminal domain for desensitisation begs the question of its functional significance. In this regard, several groups have shown that the C-terminal tail can interact with a variety of intracellular proteins with roles in intracellular signalling and trafficking. As well as α-actinin [see above], these include ARNO/cytohesin-2, the deubiquitinase USP4, TRAX (translin-X-associated protein), neuronal calcium binding protein NECAB2, synapse associated protein SAP102 and phospho-Ser-binding adapter protein 14-3-3τ [reviewed in 52 and 62]. In separate studies, it has been demonstrated that activation of PKC isoforms can trigger substantial phosphorylation of both the canine [63] and human A_{2A}ARs [GR Milne and TM Palmer, unpublished observations] to a much greater extent than agonists such as NECA. Consistent with these observations, consensus motifs for multiple AGC family kinases are present within the C-terminal domain, and include not only several PKC isoforms but also AMP-activated protein kinase (AMPK), Aurora A, Aurora B and PKA, although stimulation of A_{2A}AR-expressing rat C6 glioma cells with cyclic AMP-elevating drug forskolin failed to enhance A_{2A}AR phosphorylation under conditions in which the response to PKC was maximal [63].

While early studies demonstrated that removal of much of the C-terminal domain has minimal effects on downstream signalling to adenylyl cyclase [56, 64], it has been apparent for some time that this part of the receptor confers unique functionality. One aspect that has been well studied is the role of the C-terminal domain in mediating the observed "restricted collision coupling" of the receptor to G_s [51, 65, 66]. This is manifested as a resistance of the receptor to uncouple from G-proteins in agonist radioligand binding experiments conducted in the presence of non-hydrolysable GTP analogues [51, 66, 67] and has also been recently assessed using single particle tracking to measure accumulation of agonist-bound quantum dot-labelled A_{2A}ARs into localised membrane compartments where they display reduced mobility [68]. Interestingly, restricted collision coupling appears to require the presence of cholesterol suggesting the importance of A_{2A}AR localisation in lipid raft micro-domains [51]. Importantly, SAP102 overexpression blocked A_{2A}AR access to a compartment with restricted mobility while a C-terminal domain mutated human A_{2A}AR whose interaction with SAP102 was abolished was resistant to this effect [68]. Although not tested to date, it is likely that other A_{2A}AR C-terminal domain interactors will similarly influence receptor mobility and diffusion in specific membrane compartments. Moreover, given that receptor activation of adenylyl cyclase and ERK1,2 appear to occur in distinct membrane microdomains [51], such interactions have the potential to bias the spectrum of signalling pathways activated by agonist-occupied receptors.

Given that the A_{2A}AR is likely to be extensively phosphorylated in this region following activation of PKC, and potentially other AGC family kinases, phosphorylation is an obvious mechanism by which functionality might be dictated via modulation of interaction with specific partners. As such, an extension of the 7TMR "barcode hypothesis" [69, 70] represents an attractive, yet largely untested, model by which the A_{2A}AR could regulate downstream signalling pathway activation and their regulation. In such a model (Figure 2), phosphorylation at specific acceptor sites would dictate both the spectrum and duration of interaction of C-terminal domain-binding proteins. In support of such a model, we have preliminary evidence that the interactions of two known A_{2A}AR interactors (TRAX and 14-3-3τ) are differentially modulated by prior A_{2A}AR phosphorylation by PKC [GR Milne and TM Palmer, unpublished observations]. In addition, some interactors may co-internalise with the receptor following further stimulation with agonist, while others may be recruited to the plasma membrane then dissociate from the receptor upon its internalisation, similar to the well described interactions of arrestins with other 7TMRs [3].

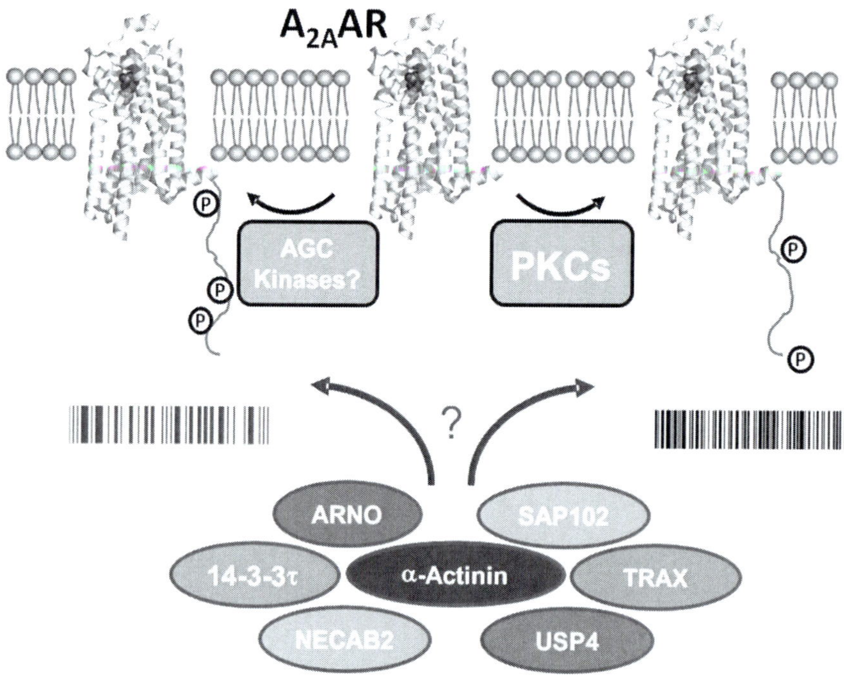

Figure 2. Modified barcode hypothesis model for regulated interaction of the $A_{2A}AR$ C-terminal domain with interacting proteins (the agonist-bound human $A_{2A}AR$ structure is from PDB 3QAK). The Ser/Thr-rich $A_{2A}AR$ C-terminal domain contains consensus sites for phosphorylation by multiple AGC family kinases, including multiple PKC isoforms [63]. Given the array of proteins known to interact with this domain of the receptor, it is likely that distinct C-terminal domain phosphorylation patterns, or "barcodes," will dictate recruitment or dissociation of specific adaptor proteins, and thereby influence the magnitude and kinetics of specific signalling pathway activation from the agonist-occupied receptors as well as their subcellular localisation.

A_{2B} Adenosine Receptor

Regulation of $A_{2B}AR$ Desensitisation and Trafficking

Like the $A_{2A}AR$, the $A_{2B}AR$ has been shown in both in vitro [53, 71] and ex vivo studies [72] to undergo rapid and substantial functional desensitisation. In the majority of these cases, desensitisation is homologous in nature, which is consistent with elegant GRK over-expression and antisense studies implicating GRK2 and/or GRK3 in the desensitisation process [53]. However the situation is likely more complex and variable between different cell types, as cyclic AMP signalling from endogenous $A_{2B}ARs$ expressed in human airway smooth muscle cells has been reported to be unaffected by disruption of GRK2/3 function [73]. Moreover, while a role for GRKs in controlling rat $A_{2B}AR$ desensitisation in HEK293 cells and the importance of a single Ser residue at position 329 have both been established [53, 74], detection of agonist-stimulated receptor phosphorylation has proved challenging [74]. Therefore, while a role for GRKs and arrestins has been proposed for $A_{2B}AR$ desensitisation and internalisation respectively (see below), the relationships between receptor phosphorylation and interactions with arrestins (and other intracellular pathways that regulate receptor trafficking) still require further study.

Several studies have demonstrated that agonist treatment mediates A_{2B}AR internalisation into early endosomal compartments via a dynamin- and clathrin-dependent mechanism that also requires arrestins [75]. Direct comparison of the abilities of arrestin-2 and arrestin-3 in transfected cells demonstrated that each isoform interacted with the A_{2B}AR uniquely and with distinct kinetics that had significant consequences for receptor re-cycling following agonist removal [74-76]. Specifically arrestin-3 promoted a significantly faster rate of recycling compared to arrestin-2, which correlated with a transient interaction of arrrestin-3 with A_{2B}ARs versus a more sustained receptor interaction with arrrestin-2, which was manifested as a co-internalisation of A_{2B}AR/arrestin-2-GFP complexes into endosomal compartments [75]. As a result it was hypothesised that a more sustained receptor/arrestin-2 interaction might inhibit receptor trafficking back to the cell surface. This could potentially occur via steric hindrance of receptor dephosphorylation by protein phosphatases or, given the difficulty in readily observing agonist-stimulated A_{2B}AR phosphorylation, an alternative mechanism. One such alternative is suggested by the presence of a class II PDZ (post-synaptic density [PSD]95 protein/Discs large/zona occludens [ZO]-1 protein) domain interaction motif at the C-terminus of the A_{2B}AR (human A_{2B}AR: Gly^{329}ValGlyLeu, rat A_{2B}AR: Ser^{329}LeuSerLeu). Importantly the A_{2B}AR has been shown to interact with PDZ domain-containing protein NHERF1 (Na^+/H^+ exchanger regulatory factor 1) via an interaction that requires the receptor's PDZ interaction motif [77]. This region is a critical determinant of A_{2B}AR internalisation and recycling, such that its removal significantly reduces the rates of both processes [76]. Thus, it is possible that sustained interactions of arrestin-2 with the A_{2B}AR might also impact PDZ domain-mediated interactions required for rapid receptor recycling back to the cell surface following agonist removal. In support of this hypothesis, interplay between arrestin- and NHERF1-mediated interactions at the C-terminal domain of the $P2Y_{12}$ purinergic receptor has been recently described [78].

CONCLUSION

A detailed molecular understanding of mechanisms responsible for the post-translational modification and intracellular trafficking of AR proteins has the potential to influence the spectrum of and duration with which specific signalling pathways are turned on by activated receptors, as well as their intracellular localisation. AR adaptive mechanisms are already known to play key roles in controlling physiologically important responses. These include the capacity to allow CD4+ and CD8+ T cells to mount an effective response to antigen by escaping from the immunosuppressive effects of A_{2A}AR activation by adenosine [54], as well as therapeutically beneficial effects of chronic A_3AR agonist treatments in mediating neuroprotection from ischaemic injury that may arise due to "functional antagonism" resulting in receptor down regulation [34, 48, 79]. However improvements in detection sensitivities achievable by mass spectrometry coupled with the advent of CRISPR/Cas9 gene editing technologies mean that key unanswered questions could now be readily addressed.

For example, using genome editing tools it is now possible to modify endogenous ARs with functional tags to study receptor regulation in cultured cell lines [80]. These would represent invaluable systems on which to apply quantitative proteomics approaches such as SILAC (stable incorporation of labelled amino acids in cell culture) to examine, for example, the impact of endogenous A_{2A}AR phosphorylation by PKCs and other AGC family kinases on the spectrum of interacting proteins capable of binding the A_{2A}AR C-terminal domain. Such approaches would also permit direct identification of the key phosphorylation sites responsible for dictating recruitment of specific adaptors. The ability to apply genome editing technology to almost any cell line is a particular advantage, as we have found that while the human A_{2A}AR is readily phosphorylated by PKC in vascular endothelial cells, phosphorylation is undetectable in transfected HEK293 cells despite similar expression levels [GR Milne and TM Palmer, unpublished observations]. In conjunction with evidence of cell type-specific regulation of the A_1AR [20, 23, 33, 36] it is clear that the choice of host cell type to examine regulatory events is critical. Finally, having identified structural determinants within AR subtypes that are responsible for controlling specific regulatory events, it will be possible to create mouse lines expressing appropriately tagged receptors containing disruptive mutations from the endogenous promoter to interrogate the consequences of dysregulation on receptor function in vivo [81]. Combined with the ongoing development of ever more selective pharmacological tools, studies such as these have the potential to define the significance of AR regulation in unprecedented detail.

Acknowledgments

Research in TMP's laboratory is supported by the British Heart Foundation (PG12/1/29276 and PG/14/32/30812) and a NHS Greater Glasgow and Clyde Research Endowment award.

References

[1] Linden J. Molecular approach to adenosine receptors: receptor-mediated mechanisms of tissue protection. *Annu Rev Pharmacol Toxicol.* 2001;41:775-87.

[2] DeWire SM, Ahn S, Lefkowitz RJ, Shenoy SK. Beta-arrestins and cell signaling. *Annu Rev Physiol.* 2007;69:483-510.

[3] Shukla AK, Xiao K, Lefkowitz RJ. Emerging paradigms of β-arrestin-dependent seven transmembrane receptor signaling. *Trends Biochem Sci.* 2011;36:457-69.

[4] Gilman AG. Nobel Lecture. G proteins and regulation of adenylyl cyclase. *Biosci Rep.* 1995;15:65-97.

[5] Neer EJ. Heterotrimeric G proteins: organizers of transmembrane signals. *Cell.* 1995;80:249-57.

[6] Krupnick JG, Benovic JL. The role of receptor kinases and arrestins in G protein-coupled receptor regulation. *Annu Rev Pharmacol Toxicol.* 1998;38:289-319.

[7] Böhm SK, Grady EF, Bunnett NW. Regulatory mechanisms that modulate signalling by G-protein-coupled receptors. *Biochem J.* 1997;322:1-18.

[8] Milligan G. Agonist regulation of cellular G protein levels and distribution: mechanisms and functional implications. *Trends Pharmacol Sci.* 1993;14:413-8.

[9] Pitcher J, Lohse MJ, Codina J, Caron MG, Lefkowitz RJ. Desensitization of the isolated beta$_2$-adrenergic receptor by beta-adrenergic receptor kinase, cAMP-dependent protein kinase, and protein kinase C occurs via distinct molecular mechanisms. *Biochemistry.* 1992;31:3193-7.

[10] Litosch I. Novel mechanisms for feedback regulation of phospholipase C-beta activity. *IUBMB Life.* 2002;54:253-60.

[11] Shenoy SK, Modi AS, Shukla AK, Xiao K, Berthouze M, Ahn S, Wilkinson KD, Miller WE, Lefkowitz RJ. Beta-arrestin-dependent signaling and trafficking of 7-transmembrane receptors is reciprocally regulated by the deubiquitinase USP33 and the E3 ligase Mdm2. *Proc Natl Acad Sci U S A.* 2009;106:6650-5.

[12] Berthouze M, Venkataramanan V, Li Y, Shenoy SK. The deubiquitinases USP33 and USP20 coordinate beta$_2$ adrenergic receptor recycling and resensitization. *EMBO J.* 2009;28:1684-96.

[13] Shenoy SK, Barak LS, Xiao K, Ahn S, Berthouze M, Shukla AK, Luttrell LM, Lefkowitz RJ. Ubiquitination of beta-arrestin links seven-transmembrane receptor endocytosis and ERK activation. *J Biol Chem.* 2007;282:29549-62.

[14] Hoffman BB, Chang H, Dall'Aglio E, Reaven GM. Desensitization of adenosine receptor-mediated inhibition of lipolysis. The mechanism involves the development of enhanced cyclic adenosine monophosphate accumulation in tolerant adipocytes. *J Clin Invest.* 1986;78:185-90.

[15] Parsons WJ, Stiles GL. Heterologous desensitization of the inhibitory A$_1$ adenosine receptor-adenylate cyclase system in rat adipocytes. Regulation of both Ns and Ni. *J Biol Chem.* 1987;262:841-7.

[16] Longabaugh JP, Didsbury J, Spiegel A, Stiles GL. Modification of the rat adipocyte A$_1$ adenosine receptor-adenylate cyclase system during chronic exposure to an A$_1$ adenosine receptor agonist: alterations in the quantity of G$_s$ alpha and G$_i$ alpha are not associated with changes in their mRNAs. *Mol Pharmacol.* 1989;36:681-8.

[17] Green A, Johnson JL, Milligan G. Down-regulation of G$_i$ sub-types by prolonged incubation of adipocytes with an A$_1$ adenosine receptor agonist. *J Biol Chem.* 1990;265:5206-10.

[18] Green A, Milligan G, Dobias SB. G$_i$ down-regulation as a mechanism for heterologous desensitization in adipocytes. *J Biol Chem.* 1992;267:3223-9.

[19] Ramkumar V, Olah ME, Jacobson KA, Stiles GL. Distinct pathways of desensitization of A$_1$- and A$_2$-adenosine receptors in DDT$_1$ MF-2 cells. *Mol Pharmacol.* 1991;40:639-47.

[20] Ciruela F, Saura C, Canela EI, Mallol J, Lluís C, Franco R. Ligand-induced phosphorylation, clustering, and desensitization of A$_1$ adenosine receptors. *Mol Pharmacol.* 1997;52:788-97.

[21] Ramkumar V, Kwatra M, Benovic JL, Stiles GL. Functional consequences of A$_1$ adenosine-receptor phosphorylation by the beta-adrenergic receptor kinase. *Biochim Biophys Acta.* 1993;1179:89-97.

[22] Ferguson G, Watterson KR, Palmer TM. Subtype-specific regulation of receptor internalization and recycling by the carboxyl-terminal domains of the human A$_1$ and rat A$_3$ adenosine receptors: consequences for agonist-stimulated translocation of arrestin3. *Biochemistry.* 2002;41:14748-61.

[23] Baines AE, Corrêa SA, Irving AJ, Frenguelli BG. Differential trafficking of adenosine receptors in hippocampal neurons monitored using GFP- and super-ecliptic pHluorin-tagged receptors. *Neuropharmacology*. 2011;61:1-11.

[24] Jajoo S, Mukherjea D, Kumar S, Sheth S, Kaur T, Rybak LP, Ramkumar V. Role of beta-arrestin1/ERK MAP kinase pathway in regulating adenosine A_1 receptor desensitization and recovery. *Am J Physiol Cell Physiol*. 2010;298:C56-65.

[25] Zhou QY, Li C, Olah ME, Johnson RA, Stiles GL, Civelli O. Molecular cloning and characterization of an adenosine receptor: the A_3 adenosine receptor. *Proc Natl Acad Sci U S A*. 1992;89:7432-6.

[26] Haskó G, Linden J, Cronstein B, Pacher P. Adenosine receptors: therapeutic aspects for inflammatory and immune diseases. *Nat Rev Drug Discov*. 2008;7:759-70.

[27] Gessi S, Merighi S, Varani K, Leung E, Mac Lennan S, Borea PA. The A_3 adenosine receptor: an enigmatic player in cell biology. *Pharmacol Ther*. 2008;117:123-40.

[28] Ali H, Cunha-Melo JR, Saul WF, Beaven MA. Activation of phospholipase C via adenosine receptors provides synergistic signals for secretion in antigen-stimulated RBL-2H3 cells. Evidence for a novel adenosine receptor. *J Biol Chem*. 1990;265:745-53.

[29] Ramkumar V, Stiles GL, Beaven MA, Ali H. The A_3 adenosine receptor is the unique adenosine receptor which facilitates release of allergic mediators in mast cells. *J Biol Chem*. 1993;268:16887-90.

[30] Apgar JR. Polymerization of actin in RBL-2H3 cells can be triggered through either the IgE receptor or the adenosine receptor but different signaling pathways are used. *Mol Biol Cell*. 1994;5:313-22.

[31] Palmer TM, Gettys TW, Stiles GL. Differential interaction with and regulation of multiple G-proteins by the rat A_3 adenosine receptor. *J Biol Chem*. 1995;270:16895-902.

[32] Palmer TM, Benovic JL, Stiles GL. Agonist-dependent phosphorylation and desensitization of the rat A_3 adenosine receptor. Evidence for a G-protein-coupled receptor kinase-mediated mechanism. *J Biol Chem*. 1995;270:29607-13.

[33] Palmer TM, Benovic JL, Stiles GL. Molecular basis for subtype-specific desensitization of inhibitory adenosine receptors. Analysis of a chimeric A_1-A_3 adenosine receptor. *J Biol Chem*. 1996;271:15272-8.

[34] Trincavelli ML, Tuscano D, Marroni M, Falleni A, Gremigni V, Ceruti S, Abbracchio MP, Jacobson KA, Cattabeni F, Martini C. A_3 adenosine receptors in human astrocytoma cells: agonist-mediated desensitization, internalization, and down-regulation. *Mol Pharmacol*. 2002;62:1373-84.

[35] Olah ME, Jacobson KA, Stiles GL. Role of the second extracellular loop of adenosine receptors in agonist and antagonist binding. Analysis of chimeric A_1/A_3 adenosine receptors. *J Biol Chem*. 1994;269:24692-8.

[36] Gao Z, Robeva AS, Linden J. Purification of A_1 adenosine receptor-G-protein complexes: effects of receptor down-regulation and phosphorylation on coupling. *Biochem J*. 1999;338:729-36.

[37] Gao Z, Ni Y, Szabo G, Linden J. Palmitoylation of the recombinant human A_1 adenosine receptor: enhanced proteolysis of palmitoylation deficient mutant receptors. *Biochem J*. 1999;342:387-95.

[38] Onorato JJ, Palczewski K, Regan JW, Caron MG, Lefkowitz RJ, Benovic JL. Role of acidic amino acids in peptide substrates of the beta-adrenergic receptor kinase and rhodopsin kinase. *Biochemistry.* 1991;30:5118-25.

[39] Palmer TM, Stiles GL. Identification of threonine residues controlling the agonist-dependent phosphorylation and desensitization of the rat A_3 adenosine receptor. *Mol Pharmacol.* 2000;57:539-45

[40] Prossnitz ER, Kim CM, Benovic JL, Ye RD. Phosphorylation of the N-formyl peptide receptor carboxyl terminus by the G protein-coupled receptor kinase, GRK2. *J Biol Chem.* 1995;270:1130-7.

[41] Eason MG, Moreira SP, Liggett SB. Four consecutive serines in the third intracellular loop are the sites for beta-adrenergic receptor kinase-mediated phosphorylation and desensitization of the alpha 2A-adrenergic receptor. *J Biol Chem.* 1995;270:4681-8.

[42] Ferguson G, Watterson KR, Palmer TM. Subtype-specific kinetics of inhibitory adenosine receptor internalization are determined by sensitivity to phosphorylation by G protein-coupled receptor kinases. *Mol Pharmacol.* 2000;57:546-52.

[43] Trincavelli ML, Tuscano D, Cecchetti P, Falleni A, Benzi L, Klotz KN, Gremigni V, Cattabeni F, Lucacchini A, Martini C. Agonist-induced internalization and recycling of the human A_3 adenosine receptors: role in receptor desensitization and resensitization. *J Neurochem.* 2000;75:1493-501.

[44] Gao ZG, Jacobson KA. Translocation of arrestin induced by human A_3 adenosine receptor ligands in an engineered cell line: comparison with G protein-dependent pathways. *Pharmacol Res.* 2008;57:303-11.

[45] Santini F, Penn RB, Gagnon AW, Benovic JL, Keen JH. Selective recruitment of arrestin-3 to clathrin coated pits upon stimulation of G protein-coupled receptors. *J Cell Sci.* 2000;113:2463-70.

[46] Palmer TM, Harris CA, Coote J, Stiles GL. Induction of multiple effects on adenylyl cyclase regulation by chronic activation of the human A_3 adenosine receptor. *Mol Pharmacol.* 1997;52:632-40.

[47] Watts VJ, Neve KA. Sensitization of adenylate cyclase by Galpha i/o-coupled receptors. *Pharmacol Ther.* 2005;106:405-21.

[48] Von Lubitz DK, Carter MF, Deutsch SI, Lin RC, Mastropaolo J, Meshulam Y, Jacobson KA. The effects of adenosine A_3 receptor stimulation on seizures in mice. *Eur J Pharmacol.* 1995;275:23-9.

[49] Safhi MM, Rutherford C, Ledent C, Sands WA, Palmer TM. Priming of signal transducer and activator of transcription proteins for cytokine-triggered polyubiquitylation and degradation by the A_{2A} adenosine receptor. *Mol Pharmacol.* 2010;77:968-78.

[50] Gsandtner I, Charalambous C, Stefan E, Ogris E, Freissmuth M, Zezula J. Heterotrimeric G protein-independent signaling of a G protein-coupled receptor. Direct binding of ARNO/cytohesin-2 to the carboxyl terminus of the A_{2A} adenosine receptor is necessary for sustained activation of the ERK/MAP kinase pathway. *J Biol Chem.* 2005;280:31898-905.

[51] Charalambous C, Gsandtner I, Keuerleber S, Milan-Lobo L, Kudlacek O, Freissmuth M, Zezula J. Restricted collision coupling of the A_{2A} receptor revisited: evidence for physical separation of two signaling cascades. *J Biol Chem.* 2008;283:9276-88.

[52] Zezula J, Freissmuth M. The A_{2A}-adenosine receptor: a GPCR with unique features? *Br J Pharmacol.* 2008;153 Suppl 1:S184-90.

[53] Mundell SJ, Benovic JL, Kelly E. A dominant negative mutant of the G protein-coupled receptor kinase 2 selectively attenuates adenosine A_2 receptor desensitization. *Mol Pharmacol.* 1997;51:991-8.

[54] Yang A, Mucsi AD, Desrosiers MD, Chen JF, Schnermann JB, Blackburn MR, Shi Y. Adenosine mediated desensitization of cAMP signaling enhances T-cell responses. *Eur J Immunol.* 2010;40:449-59.

[55] Palmer TM, Gettys TW, Jacobson KA, Stiles GL. Desensitization of the canine A_{2a} adenosine receptor: delineation of multiple processes. *Mol Pharmacol.* 1994;45:1082-94.

[56] Palmer TM, Stiles GL. Identification of an A_{2a} adenosine receptor domain specifically responsible for mediating short-term desensitization. *Biochemistry.* 1997;36:832-8.

[57] Brand F, Klutz AM, Jacobson KA, Fredholm BB, Schulte G. Adenosine A_{2A} receptor dynamics studied with the novel fluorescent agonist Alexa488-APEC. *Eur J Pharmacol.* 2008;590:36-42.

[58] Mundell SJ, Luty JS, Willets J, Benovic JL, Kelly E. Enhanced expression of G protein-coupled receptor kinase 2 selectively increases the sensitivity of A_{2A} adenosine receptors to agonist-induced desensitization. *Br J Pharmacol.* 1998;125:347-56.

[59] Willets JM, Parent JL, Benovic JL, Kelly E. Selective reduction in A_2 adenosine receptor desensitization following antisense-induced suppression of G protein-coupled receptor kinase 2 expression. *J Neurochem.* 1999;73:1781-9.

[60] Burgueño J, Blake DJ, Benson MA, Tinsley CL, Esapa CT, Canela EI, Penela P, Mallol J, Mayor F Jr, Lluis C, Franco R, Ciruela F. The adenosine A_{2A} receptor interacts with the actin-binding protein alpha-actinin. *J Biol Chem.* 2003;278:37545-52.

[61] Mundell SJ, Kelly E. The effect of inhibitors of receptor internalization on the desensitization and resensitization of three G_s-coupled receptor responses. *Br J Pharmacol.* 1998;125:1594-600.

[62] Keuerleber S, Gsandtner I, Freissmuth M. From cradle to twilight: the carboxyl terminus directs the fate of the A_{2A}-adenosine receptor. *Biochim Biophys Acta.* 2011;1808:1350-7.

[63] Palmer TM, Stiles GL. Stimulation of A_{2A} adenosine receptor phosphorylation by protein kinase C activation: evidence for regulation by multiple protein kinase C isoforms. *Biochemistry.* 1999;38:14833-42.

[64] Piersen CE, True CD, Wells JN. A carboxyl-terminally truncated mutant and nonglycosylated A_{2a} adenosine receptors retain ligand binding. *Mol Pharmacol.* 1994;45:861-70.

[65] Nanoff C, Jacobson KA, Stiles GL. The A_2 adenosine receptor: guanine nucleotide modulation of agonist binding is enhanced by proteolysis. *Mol Pharmacol.* 1991;39:130-5.

[66] Keuerleber S, Thurner P, Gruber CW, Zezula J, Freissmuth M. Reengineering the collision coupling and diffusion mode of the A_{2A}-adenosine receptor: palmitoylation in helix 8 relieves confinement. *J Biol Chem.* 2012;287:42104-18.

[67] Nanoff C, Stiles GL. Solubilization and characterization of the A_2 adenosine receptor. *J Recept Res. 1993*;13:961-73.

[68] Thurner P, Gsandtner I, Kudlacek O, Choquet D, Nanoff C, Freissmuth M, Zezula J. A two-state model for the diffusion of the A_{2A} adenosine receptor in hippocampal neurons: agonist-induced switch to slow mobility is modified by synapse-associated protein 102 (SAP102). *J Biol Chem.* 2014;289:9263-74.

[69] Butcher AJ, Prihandoko R, Kong KC, McWilliams P, Edwards JM, Bottrill A, Mistry S, Tobin AB. Differential G-protein-coupled receptor phosphorylation provides evidence for a signaling bar code. *J Biol Chem.* 2011;286:11506-18.

[70] Nobles KN, Xiao K, Ahn S, Shukla AK, Lam CM, Rajagopal S, Strachan RT, Huang TY, Bressler EA, Hara MR, Shenoy SK, Gygi SP, Lefkowitz RJ. Distinct phosphorylation sites on the β_2-adrenergic receptor establish a barcode that encodes differential functions of β-arrestin. *Sci Signal.* 2011;4:ra51.

[71] Breschi MC, Blandizzi C, Fogli S, Martinelli C, Adinolfi B, Calderone V, Camici M, Martinotti E, Nieri P. In vivo adenosine A_{2B} receptor desensitization in guinea-pig airway smooth muscle: implications for asthma. *Eur J Pharmacol.* 2007;575:149-57.

[72] Haynes J Jr, Obiako B, Babal P, Stevens T. 5-(N-ethylcarboxamido)adenosine desensitizes the A_{2b}-adenosine receptor in lung circulation. *Am J Physiol.* 1999;276:H1877-83.

[73] Kong KC, Gandhi U, Martin TJ, Anz CB, Yan H, Misior AM, Pascual RM, Deshpande DA, Penn RB. Endogenous G_s-coupled receptors in smooth muscle exhibit differential susceptibility to GRK2/3-mediated desensitization. *Biochemistry.* 2008;47:9279-88.

[74] Matharu AL, Mundell SJ, Benovic JL, Kelly E. Rapid agonist-induced desensitization and internalization of the A_{2B} adenosine receptor is mediated by a serine residue close to the COOH terminus. *J Biol Chem.* 2001;276:30199-207.

[75] Mundell SJ, Matharu AL, Kelly E, Benovic JL. Arrestin isoforms dictate differential kinetics of A_{2B} adenosine receptor trafficking. *Biochemistry.* 2000;39:12828-36.

[76] Mundell SJ, Matharu AL, Nisar S, Palmer TM, Benovic JL, Kelly E. Deletion of the distal COOH-terminus of the A_{2B} adenosine receptor switches internalization to an arrestin- and clathrin-independent pathway and inhibits recycling. *Br J Pharmacol.* 2010;159:518-33.

[77] Sitaraman SV, Wang L, Wong M, Bruewer M, Hobert M, Yun CH, Merlin D, Madara JL. The adenosine 2b receptor is recruited to the plasma membrane and associates with E3KARP and Ezrin upon agonist stimulation. *J Biol Chem.* 2002;277:33188-95.

[78] Nisar SP, Cunningham M, Saxena K, Pope RJ, Kelly E, Mundell SJ. Arrestin scaffolds NHERF1 to the $P2Y_{12}$ receptor to regulate receptor internalization. *J Biol Chem.* 2012;287:24505-15.

[79] Hu H, Lu W, Zhang M, Zhang X, Argall AJ, Patel S, Lee GE, Kim YC, Jacobson KA, Laties AM, Mitchell CH. Stimulation of the P2X7 receptor kills rat retinal ganglion cells in vivo. *Exp Eye Res.* 2010;91:425-32.

[80] Park A, Won ST, Pentecost M, Bartkowski W, Lee B. CRISPR/Cas9 allows efficient and complete knock-in of a destabilization domain-tagged essential protein in a human cell line, allowing rapid knockdown of protein function. *PLoS One.* 2014;9:e95101.

[81] H, Wang H, Shivalila CS, Cheng AW, Shi L, Jaenisch R. One-step generation of mice carrying reporter and conditional alleles by CRISPR/Cas-mediated genome engineering. *Cell.* 2013;154:1370-9.

In: Adenosine Signaling Mechanisms
Editors: V. Ramkumar and R. Paes de Carvalho
ISBN: 978-1-63483-186-4

Chapter 4

Adenosine Receptors in the Cardiovascular System

Mark E. Olah[*]
Department of Pharmaceutical and Biomedical Sciences
Raabe College of Pharmacy, Ohio Northern University, Ada, OH, US

Abstract

For over 75 years, it has been known that adenosine produces effects in the heart and vasculature. The term "retaliatory metabolite" reflects the characterization of the ability of adenosine to correct imbalances in energy supply and demand. In the cardiovascular system and elsewhere, it is now recognized that adenosine serves a more generalized protective role during stress. Adenosine produces a variety of responses through activation of four adenosine receptor (AR) subtypes, the A_1AR, $A_{2A}AR$, $A_{2B}AR$ and A_3AR, which are G protein-coupled receptors. All AR subtypes are expressed on cardiac and vascular cells and AR activation on other cell types may produce cardiovascular responses through paracrine effects. This chapter reviews advances in several areas that encompass the protective role of adenosine in the cardiovascular system. Much focus over the past ~25 years has been on AR-mediated protection of the heart from ischemia-reperfusion injury. This has substantial therapeutic potential particularly in myocardial infarction that is addressed by thrombolysis. AR activation may mediate the protective effects of ischemic preconditioning and postconditioning and AR agonists reduce infarct size and preserve cardiac function during ischemia-reperfusion. There have been significant advances in the delineation of the roles of AR subtypes in protection during ischemia-reperfusion and understanding of the mechanisms that underlie these beneficial effects. AR regulation also produces improvements in cardiac function in animal models of heart failure. The concept of adenosine serving as a compensatory molecule during stress can be extended to the vasculature. Tissues respond to hypoxia with neovascularization to increase oxygen delivery. Angiogenesis is a mechanism to develop new vasculature and promotion or inhibition of angiogenesis has substantial clinical potential. AR activation, primarily of the $A_{2A}AR$ and $A_{2B}AR$, may promote angiogenesis either directly following AR stimulation on endothelial cells or indirectly via

[*] E-mail: m-olah@onu.edu

upregulation of vascular endothelial growth factor release from multiple cell types. Endothelial progenitor cells (EPC) also have a role in neovascularization and ARs regulate events required for efficient homing of EPC to the endothelium. Animal models suggest therapeutic potential of AR regulation in the retinopathy of prematurity, wound healing and chronic kidney disease. Currently, therapeutic application of AR regulation is limited and the widespread distribution of ARs is a challenge with the possibility of undesired effects in the patient. However, continued advances in understanding mechanisms of pathology and the precise roles of AR in experimental models of multiple cardiovascular diseases will hopefully allow AR regulation to achieve full therapeutic potential.

INTRODUCTION

Cardiovascular effects of adenosine were described over 75 years ago [1] with a focus on cardiac conduction and coronary blood flow. In 1963, Berne's findings resulted in the proposal that decreased myocardial oxygen tension, occurring for example by increased myocardial work and oxygen utilization, results in generation of adenosine from adenine nucleotides [2]. Adenosine then exits the cell to the interstitial fluid to produce coronary vasodilation with increased perfusion of the myocardium and correction of the oxygen deficit [2]. Newby [3] proposed the term "retaliatory metabolite" to describe this feedback role of adenosine in general terms with reference to the metabolic status of a cell being a key determinant for the formation and release of adenosine with subsequent adenosine-induced functional effects directed at improving the energy supply and demand relationship. To perform this retaliatory role in the cardiovascular system, it is now recognized that adenosine acting on all four of the currently identified adenosine receptor (AR) subtypes, A_1AR, $A_{2A}AR$, $A_{2B}AR$ and A_3AR, regulates a wide range of functions. All four AR subtypes are expressed on various cell types in the heart [4, 5] and blood vessels [6-9]. Additionally, ARs expressed on other cell types including inflammatory cells and tumor cells may indirectly regulate cardiovascular responses by modulating release of substances that affect cardiovascular function. Activation of ARs regulates cardiovascular responses including cardiac inotropy, chronotropy and conduction and blood vessel reactivity with these topics detailed in several excellent reviews [6, 10-12]. Available and approved adenosine-based cardiovascular therapeutics currently include adenosine which acting through A_1AR stimulation is indicated for conversion of paroxysmal supraventricular tachycardia to sinus rhythm [13] and regadenoson, an $A_{2A}AR$ agonist, that is used to produce coronary vasodilation in cardiac stress imaging [14]. Dipyridamole which has vasodilatory and antiplatelet activities acts at least in part through the elevation of extracellular adenosine levels and subsequent AR activation [15]. While recognizing that advances continue to be made in the understanding of these areas, this chapter will focus on other recent areas of interest in cardiovascular effects of adenosine. Specifically, focus will be on adenosine-mediated protective effects on cardiac function during acute ischemic events and in the progression of heart failure. Additionally, studies detailing the role of adenosine in the promotion of neovascularization will be described. These areas, adenosine-mediated cardioprotection and neovascularization, represent an extension of the concept of adenosine as a retaliatory metabolite. During periods of stress, such as acute or prolonged ischemia, extracellular levels of adenosine increase and compensatory responses including protection of cardiac myocytes or development of new vasculature occur as a result of AR activation.

ADENOSINE RECEPTORS AND CARDIOPROTECTION

A vast body of literature has accumulated regarding the cardioprotective effects of AR activation whether by elevation of endogenous adenosine levels or administration of AR agonists. Specifically, interest is directed at the ability of AR activation to limit cardiac myocyte death and maintain cardiac contractile function following ischemic insult and subsequent reperfusion. The potential therapeutic application of such protection is immense with particular relevance in regard to myocardial infarction that is addressed by reperfusion via percutaneous coronary intervention or drug-induced thrombolysis. Much of the investigation into AR-mediated cardioprotection was in part stimulated by the description of the ischemic preconditioning phenomenon by Murry and coworkers [16]. This group demonstrated in canine hearts that repeated brief periods of cardiac ischemia reduced the infarct size that developed after a subsequent period of prolonged ischemia and reperfusion. Subsequently, it was found that short episodes of vascular occlusion and reperfusion, even briefer than those employed for ischemic preconditioning, at the initiation of the terminal reperfusion period were also cardioprotective [17]. This phenomenon has been termed ischemic postconditioning. From a clinical application perspective, the potential for cardioprotection provided by postconditioning may be more therapeutically relevant than ischemic preconditioning as it provides the opportunity for intervention prior to therapeutic reperfusion. With the realization of the robust cardioprotective effects of ischemic preconditioning and postconditioning, AR research in large part has focused on two overlapping areas. Specifically, investigations have been directed at identifying the role of endogenous adenosine and AR activation in ischemic preconditioning and postconditioning and to determine if cardioprotective effects can be reproduced by administration of AR agonists prior to ischemia, or perhaps more clinically relevant, prior to reperfusion.

This exploration has resulted in an immense amount of literature and several excellent and comprehensive reviews [18-22] have provided an in-depth analysis of the field particularly in regard to the precise roles of the four AR subtypes in mediating cardioprotection during the distinct stages of ischemia and reperfusion. With several studies, there have been significant conflicting results. Inconsistencies in findings may be related to species differences in response to and recovery from ischemia-reperfusion and lack of selectivity of AR ligands at employed concentrations with again possible species-dependent differences. Additionally, varying experimental models and procedures, e.g., isolated perfused heart versus open-chest animals, manner of inducing ischemia, duration of preconditioning and postconditioning protocols and duration of ligand exposure, may also contribute to inconsistency in results. Several of these methodological considerations in addition to variations in mouse strain and possible compensatory responses may contribute to varying results obtained with genetic deletion of individual AR subtypes. Nonetheless, many studies have supported the role of ARs in protection from cardiac ischemia-reperfusion injury. As described below, adenosine, as well as substances such as opioids [23, 24] and bradykinin [25], is thought to be a key endogenous mediator of ischemic preconditioning with

AR activation presumably induced by the increase in adenosine levels that occurs with ischemia [26-28].

The following is a brief overview of the contribution of individual AR subtypes in mediating cardioprotection with much of the latter focus placed on more recent findings regarding underlying signaling mechanisms. More detailed analysis of earlier studies that provide the basis for this current overview is provided elsewhere [19-21]. Generally, data support the concept that activation of the G_i-coupled A_1AR on cardiac myocytes provides a direct protective effect from ischemia-reperfusion injury and that the A_1AR can elicit protection when activated prior to ischemia [19, 20, 29]. Examination of mice overexpressing the A_1AR indicated that adenosine may be elevated maximally during ischemia and that the expression level of the cardiac A_1AR may be a limiting factor in mediating protection during ischemia-reperfusion [30]. As reviewed [19, 20], A_1AR agonists at reperfusion have been reported in various studies to have beneficial effects or offer no cardioprotection. Recently, a novel allosteric enhancer of the A_1AR as well an A_1AR partial agonist when administered during early reperfusion was reported to improve recovery of cardiac function in isolated murine hearts [31]. Allosteric enhancers offer the potential of enhancing responses to ischemia-induced local accumulation of adenosine in cardiac tissue while limiting possibly undesired effects elsewhere [32]. Though several conflicting reports exist in the literature, there is evidence indicating that the beneficial effects of ischemic preconditioning occur at least in part due to endogenous adenosine activating the A_1AR. Pharmacologic studies examining the effects of A_1AR antagonists on the cardioprotection elicited by ischemic preconditioning have produced varying results that again may reflect species-dependent differences in ligand selectivity or variations in the contribution of ARs relative to other mediators of ischemic preconditioning [19, 20]. Lankford and coworkers [33] demonstrated that genetic deletion of the A_1AR in mice abolished the cardioprotective effects of ischemic preconditioning. Furthermore, in A_1AR-overexpressing mice, infarct size was reduced relative to wild-type animals and ischemic preconditioning elicited additional protection in the transgenic line [33]. A later study employing A_1AR knockout mice did not demonstrate a similar complete dependence of the beneficial effects of ischemic preconditioning on A_1AR expression [34]. In regard to adenosine as a possible mediator of ischemic postconditioning, Xi and coworkers [35] observed that the infarct sparing effect of ischemic postconditioning was lost in A_1AR knockout mice. Interestingly, the employed postconditioning protocol did not preserve cardiac function in wild-type mice [35]. Thus, contribution of the A_1AR in postconditioning was restricted to limitation of infarct size.

The A_3AR, a G_i-coupled receptor, is expressed on cardiac myocytes at apparently low levels and this AR subtype may mediate cardioprotection when activated prior to ischemia [19, 20, 36]. A_3AR activation at the time of reperfusion may also be protective [20, 37]. An A_3AR allosteric enhancer reduced infarct size in a canine model of ischemia-reperfusion injury and the allosteric enhancer was effective both when administered prior to ischemia or prior to reperfusion [38]. Ge and coworkers [39] provided evidence that administration of the A_3AR agonist, Cl-IB-MECA, during reperfusion promoted cardioprotection at least partially via an inhibitory effect on inflammatory cells. A_3AR agonist-induced reduction of infarct size was not observed in A_3AR knockout mice or in mice in which the A_3AR was not expressed on bone marrow derived cells [39]. Examination of leukocyte infiltration into cardiac tissue and *in vitro* analysis of neutrophil migration indicated that activation of the A_3AR inhibited neutrophil chemotaxis and this may in part underlie the cardioprotective response [39]. These findings are consistent with other studies demonstrating inhibitory effects of neutrophil A_3AR stimulation [40]. While early studies provided indirect evidence for the role of the A_3AR in mediating effects of ischemic preconditioning [41, 42], there has been limited direct pharmacologic examination of this relationship. Cardioprotection provided by ischemic preconditioning was not reduced in A_3AR knockout mice relative to wild-type animals [43]. Interestingly, in non-preconditioned mice, deletion of the A_3AR reduced infarct size with decreased neutrophil accumulation in the infarcted region in A_3AR knockout mice as compared to wild-type animals [43]. These results indicate a possible deleterious role for the A_3AR in ischemia-reperfusion injury.

While there is limited support that activation of the $A_{2A}AR$ on cardiac myocytes directly protects the ischemic myocardium [44], a vast amount of evidence indicates that the primary cardioprotective effect of $A_{2A}AR$ stimulation results from inhibition of inflammatory cell function and is thus most prominent during reperfusion as opposed to the initiating ischemia [20, 21]. During reperfusion, cardiac injury results from multiple and possibly interrelated events including accumulation of reactive oxygen species, calcium overload, activation of resident inflammatory cells, infiltration by circulating inflammatory cells, and the release of cytokines from inflammatory and non-inflammatory cells [22, 45]. Ensuing cellular responses may underlie cardiac damage. The $A_{2A}AR$ inhibits function of inflammatory cells as well as pro-inflammatory responses in non-immune cells [46, 47]. With reference to cardiac ischemia-reperfusion injury, these beneficial responses to $A_{2A}AR$ activation include reductions in neutrophil infiltration [48, 49] and vascular P-selectin expression [49]. *In vitro* studies demonstrated inhibitory effects of $A_{2A}AR$ activation on neutrophil functions such as endothelial adherence and superoxide production [48]. Yang and coworkers [50] elegantly demonstrated that a key response to $A_{2A}AR$ activation is the inhibition of homing of specifically $CD4^+$ T lymphocytes to ischemic myocardium during the reperfusion period. $A_{2A}AR$-mediated inhibition of resident mast cell degranulation may also have a role in cardioprotection as demonstrated in the isolated mouse heart [51]. Additionally, an $A_{2A}AR$ agonist, ATL146e, administered hours after reperfusion demonstrated beneficial effects on cardiac function as well as anti-inflammatory effects including decreases in TNF-α and iNOS expression and inhibition of NF-κB activity in infarcted and noninfarcted myocardium [52]. Based on the cardioprotective responses of $A_{2A}AR$ activation at reperfusion, it is reasonable to predict that this receptor may have a role in the beneficial effects induced by ischemic postconditioning. This has been supported by both pharmacologic studies [37, 53, 54] and an examination of $A_{2A}AR$ knockout mice in which ischemic postconditioning did not improve recovery of cardiac function or decrease markers of cardiac damage [54].

Studies on the role of the $A_{2B}AR$ in cardioprotection lagged behind analysis of other AR subtypes chiefly due to the lack of selective agonist and antagonist ligands. Implication of the $A_{2B}AR$ typically required elimination of definitive roles for other AR subtypes through use of a nonselective AR agonist in conjunction with antagonists selective for other subtypes. However, selective ligands that are now available in addition to examination of $A_{2B}AR$ knockout mice have allowed for roles of the $A_{2B}AR$ in cardioprotection to be identified. An $A_{2B}AR$ agonist administered prior to coronary occlusion was cardioprotective in rats [55] and mice [34]. Additionally, it has rather consistently been shown that $A_{2B}AR$ activation is beneficial at the onset or during reperfusion [53, 56-58]. Interestingly, as discussed below, certain studies suggest that co-activation of the $A_{2A}AR$ may be involved in addition to $A_{2B}AR$ stimulation [53, 57]. Conflicting results have been obtained regarding the role of the $A_{2B}AR$

in mediating ischemic preconditioning. An $A_{2B}AR$ antagonist, MRS1754, administered just prior to reperfusion blocked the protective effects of ischemic preconditioning in isolated rabbit hearts [59]. In mice, Eckle and coworkers [34] reported that an ischemic preconditioning protocol elicited ~14-fold induction of cardiac $A_{2B}AR$ mRNA while expression of the A_1AR and A_3AR was unaffected and $A_{2A}AR$ mRNA decreased. In a head-to-head comparison of four lines of mice each with genetic deletion of an individual AR subtype, all receptors appeared to have at least partial involvement in the reduction in infarct size produced by ischemic preconditioning [34]. Most striking, cardioprotection induced by ischemic preconditioning was completely abolished only in mice lacking the $A_{2B}AR$ indicating a critical role for this receptor [34]. This finding contrasts with a study by Maas and coworkers [55] that employed an $A_{2B}AR$ knockout mouse line distinct from that examined by Eckle and coworkers. In both isolated heart and *in vivo* models, deletion of the $A_{2B}AR$ had no effect on the beneficial effects of ischemic preconditioning on cardiac function or infarct size following ischemia and reperfusion [55]. Lack of involvement of the $A_{2B}AR$ in ischemic preconditioning in rats was also suggested by pharmacologic studies [55]. In reference to ischemic postconditioning, Philipp and coworkers [56] demonstrated in rabbit hearts that the beneficial effects of postconditioning were blocked by an $A_{2B}AR$ antagonist but not by inhibition of the A_1AR or $A_{2A}AR$. A role for the $A_{2B}AR$ expressed on inflammatory cells in promoting cardioprotection from ischemia-reperfusion injury has also been proposed [60]. Bone marrow was transplanted either from wild-type animals into $A_{2B}AR$ knockout mice or from $A_{2B}AR$ deficient mice into wild-type animals in order to identify contributions from bone marrow derived cells versus myocytes, fibroblasts or other cardiac cells in protection from ischemia-reperfusion injury. Results indicated that $A_{2B}AR$ activation on inflammatory cells had a substantial role in cardioprotection and studies in mice in which polymorphonuclear leukocytes (PMNs) were depleted implicated specifically this cell type [60]. $A_{2B}AR$ activation did not affect PMN infiltration into the myocardium following ischemia-reperfusion but rather decreased TNF-α accumulation in cardiac tissue presumably by regulating release of this cytokine from migrated PMNs [60].

In regard to the identification of the precise roles of individual AR subtypes in cardioprotection, an additional level of complexity has been recognized with the possibility that AR subtypes may interact with each other to elicit responses. Studies demonstrating the sensitivity of A_1AR agonist-induced cardioprotection to selective $A_{2A}AR$ and $A_{2B}AR$ antagonists suggest that A_1AR responses may require concomitant activation of these additional AR subtypes [61, 62]. It is doubtful that these results were simply the result of nonselectivity of the employed ligands as A_1AR-mediated cardioprotective responses were not observed in $A_{2A}AR$ or $A_{2B}AR$ knockout mice [63]. Similarly, cardioprotection elicited by AR agonists at the onset of reperfusion may involve, and perhaps require, concomitant activation of the $A_{2A}AR$ and $A_{2B}AR$. Xi and coworkers [57] using a pharmacologic approach in isolated rat hearts with the nonselective AR agonist NECA and AR subtype selective agonists and antagonists observed that co-activation of the $A_{2A}AR$ and $A_{2B}AR$ at reperfusion was required for infarct-sparing effects. Similar findings were reported by Methner and coworkers [53] in an examination of postconditioning and AR agonist administration at reperfusion in open-chest mice. In mice with genetic deletion of CD73 in which adenosine levels do not increase with ischemia, it was found that protection afforded by either ischemic postconditioning, the $A_{2A}AR$ selective agonist CGS-21680 or the $A_{2B}AR$ selective agonist BAY60-6583 was lost [53]. However, simultaneous administration of the $A_{2A}AR$ and $A_{2B}AR$

agonists or the nonselective agonist NECA at reperfusion produced substantial reduction in infarct size [53].

Further complexity may exist as studies with A_1AR ligands and agonists and antagonists of δ opioid receptors have demonstrated an apparent interdependence among these receptor families in eliciting protection from cardiac ischemic injury in rats [64]. Similarly, an A_1AR antagonist inhibited protective effects of δ and κ opioid receptor activation in isolated cardiac myocytes exposed to ischemia-reperfusion [65]. In HEK293 cells overexpressing opioid receptors, it was shown that the A_1AR physically associates with these receptors [65]. In rat hearts, protective effects of adenosine at reperfusion were inhibited by a selective M_2 muscarinic receptor antagonist [66]. In other systems, evidence suggesting dimerization of an individual AR subtype with another AR subtype [67-69] or with other G protein-coupled receptors [70-73] has been obtained. Cardioprotection in isolated rat hearts elicited by β-adrenergic receptor agonists may be mediated by the A_3AR, $A_{2A}AR$ and $A_{2B}AR$ acting at distinct phases through the course of ischemia and reperfusion and may be at least in part attributed to the generation of adenosine following β-adrenergic receptor stimulation [74].

The signaling mechanisms and final effectors that underlie cardioprotection elicited by AR activation as well as ischemic preconditioning and postconditioning have been the focus of much investigation. From the perspective of G protein-coupled receptor signaling in cardiac myocytes, it is notable that both G_i-coupled receptors, the A_1AR and A_3AR, and G_s-coupled receptors, the $A_{2A}AR$ and $A_{2B}AR$, are cardioprotective. However, as discussed previously, not all AR-mediated cardioprotective responses occur due to direct effects on the cardiac myocyte. For example, the majority of the beneficial responses induced by $A_{2A}AR$ activation likely result from inhibitory effects of this AR subtype on inflammatory cells. Additionally, there is evidence to suggest that the $A_{2B}AR$ can couple to G proteins other than G_s [75, 76]. Nonetheless, in cardiac myocytes, it is likely that signaling pathways engaged by stimulation of various AR subtypes regardless of the proximal steps of activation converge on common key effectors. With the realization that protective effects of ischemic preconditioning and activation of ARs as well as other G protein-coupled receptors are primarily manifest at early reperfusion [77-79], similarities in the signaling mechanisms engaged by these protective triggers to signaling that is employed by ischemic postconditioning were identified. Specifically, it appears that a critical component of the signaling cascade is the activation of the reperfusion injury salvage kinase (RISK) pathway that includes extracellular signal-regulated kinase 1/2 (ERK1/2), phosphatidylinositol 3-kinase (PI3-K), Akt and glycogen synthase kinase-3β (GSK-3β) [78, 79]. Regulation of these proteins by cardioprotective stimuli is thought to ultimately result in inhibition of the mitochondrial permeability transition pore (MPTP) [79, 80]. With regard to ischemia-reperfusion injury, several deleterious and probably interrelated events including cardiac myocyte oxidative stress and calcium overload converge at the MPTP and promote opening of the channel at reperfusion [45, 81]. MPTP opening results in uncoupling of oxidative-phosphorylation, swelling and destruction of mitochondria and cell death [45, 81, 82]. Although signaling is not clearly defined and experimental inconsistencies exist, it is generally postulated that activation of RISK proteins such as ERK1/2 and PI3-K/Akt may phosphorylate and thus inhibit GSK-3β [80, 83]. GSK-3β inhibition results in inhibition of MPTP opening and subsequent cardioprotection. Using chemical inhibitors and direct activators of signaling proteins, Park and coworkers [84] provided evidence that this pathway is engaged by the A_3AR to reduce infarct size from ischemia-reperfusion in rat hearts.

Similarly, $A_{2A}AR$- and $A_{2B}AR$-mediated cardioprotection may involve GSK-3β inhibition and MPTP modulation [57]. In mice with deletion of PI3-Kγ, protection from ischemia-reperfusion injury provided by adenosine administration was lost and this occurred possibly due to impaired downstream signaling to GSK-3β [85]. Additional proteins implicated in cardioprotection, perhaps working in concert with RISK proteins, include Protein Kinase C (PKC), Protein Kinase G and mitochondrial K_{ATP} channels [18, 86]. The role of distinct PKC isoforms in cardioprotection induced by A_1AR activation [86] and ischemic preconditioning [83] has been reviewed and a rather novel relationship between the $A_{2B}AR$ and PKC is described below. As reviewed [22], mitochondrial K_{ATP} channels may be linked to PKC activation in cardioprotection and use of pharmacologic tools have indicated the involvement of these channels in AR-mediated cardioprotection [87-89]. However, Wan and coworkers [90] examining Kir6.2 knockout mice provided evidence that activation of the sarcolemmal K_{ATP} channel is required for A_3AR-mediated cardioprotection from ischemia-reperfusion injury.

As noted, ischemic preconditioning, postconditioning and activation of AR subtypes display several similarities in the regulation of the above-described signaling proteins and effectors to elicit cardioprotection during ischemia-reperfusion. This portion of the chapter will focus on several of the more recent findings regarding the mechanisms underlying AR-mediated cardioprotection. In the HL-1 murine cardiac cell line, Yitzhaki and coworkers [91] provided evidence that autophagy may have a role in A_1AR-mediated cardioprotection. In the context of ischemia-reperfusion injury, autophagy may be protective in that it permits removal of damaged or dysfunctional mitochondria that contribute to cardiac myocyte death [92]. In HL-1 cells, activation of the A_1AR induced autophagy and A_1AR-mediated protection of these cells during ischemia-reperfusion was blocked by inhibition of autophagy [91]. In mouse heart, A_1AR agonist administration induced autophagy, however, the involvement of this response in protection elicited by A_1AR activation in intact hearts was not examined [91]. Receptor tyrosine kinases have also been implicated in A_1AR-mediated cardioprotection. Williams-Pritchard and coworkers [93] demonstrated in isolated murine hearts that beneficial effects of both an A_1AR agonist and ischemic preconditioning were blocked by the epidermal growth factor receptor (EGFR) inhibitor AG1478 and by inhibition of matrix metalloproteinase activity that is necessary for generation of soluble EGF from a membrane bound form. In perfused hearts, an A_1AR agonist activated the EGFR as well as ERK1/2 and Akt in an AG1478-sensitive fashion thus supporting the concept of a requirement for EGFR transactivation in A_1AR-mediated protection [93].

In recent years, the volume of publications detailing mechanisms of $A_{2B}AR$-mediated cardioprotection has increased. As noted previously, this is in large part due to the more recent development of relatively selective $A_{2B}AR$ agonists and antagonists as well as the availability of $A_{2B}AR$ knockout mice. In an examination of signaling events that may be involved in $A_{2B}AR$-mediated cardioprotection, Eckle and coworkers [94] performed a microarray analysis of ischemic preconditioned hearts isolated from wild-type and $A_{2B}AR$ knock-out mice. This analysis revealed a marked induction of the circadian rhythm protein, Per2, with increased expression documented at the protein level with upregulation of Per2 occurring due to both transcriptional and post-translational mechanisms [94]. Increased transcription appeared to occur secondary to activation of cAMP response element binding protein (CREB) while post-translational mechanisms included an $A_{2B}AR$-mediated decrease in Per2 ubiquitination and subsequent degradation [94]. In Per2 knockout mice, cardioprotection induced by ischemic preconditioning or A_{2B}AR activation was markedly reduced relative to wild-type mice [94]. Additional data suggested that the increase in Per2 expression was manifest in the cardiac myocyte as a shift toward anaerobic glycolysis that is more favorable and cardioprotective during periods of ischemia [94].

From a receptor signal transduction viewpoint, the A_{2B}AR may have relatively novel interactions with PKC in promoting cardioprotection during the early reperfusion period. Protective effects of phorbol ester, a direct activator of PKC, at early reperfusion were blocked by A_{2B}AR antagonists [95, 96]. Analysis of the activation of protective signaling kinases, ERK1/2 and Akt, suggested that both ischemic preconditioning and PMA allowed subthreshold concentrations of an A_{2B}AR agonist to activate these kinases [96]. In this scheme, the proteins interact in a reversed role relative to the traditional signaling paradigm of receptor to kinase in that PKC activation targets the A_{2B}AR and increases the sensitivity of the receptor to concentrations of adenosine that may develop locally with ischemia and thus promote protective effects at reperfusion [96]. It is not known if this putative sensitization of the A_{2B}AR by PKC is mediated directly, for example by phosphorylation of the receptor by PKC, or secondary to modulation of other signaling proteins downstream of the A_{2B}AR that are cardioprotective. From results obtained in isolated rat heart, Terashima and coworkers [97] hypothesized that PKC activation may facilitate A_{2B}AR-mediated phosphorylation of GSK-3β to ultimately promote cardioprotection during ischemia-reperfusion. Cardioprotection elicited by activation of Protein Kinase G may also utilize this signaling pathway perhaps by activating PKC and sensitizing the A_{2B}AR [58]. Additionally, A_{2B}AR activation has been reported to decrease reactive oxygen species (ROS) production in rabbit cardiac myocytes [75]. In that ROS formation during reperfusion may promote MPTP opening, such an A_{2B}AR-mediated response may contribute to cardioprotection. The A_{2B}AR-mediated decrease in ROS formation was sensitive to inhibitors of ERK1/2 and Akt, and interestingly, was blocked by pretreatment with pertussis toxin suggesting a possible coupling of the A_{2B}AR to G_i [75]. It has been reported that in rat heart the A_{2B}AR is not expressed on the sarcolemmal surface but rather is associated with mitochondria as assessed by immunofluorescence and cell fractionation [98]. It is unknown how mitochondrial expression of the A_{2B}AR may contribute to signaling or functional responses.

Finally, with reference to the translatability of AR-mediated cardioprotective effects, various experimental models employing human cardiac tissue or cells have demonstrated a role for AR activation in protection from ischemia [99-103]. Polymorphisms in the 3'-untranslated region of the A_1AR and coding region of the A_3AR were associated with variations in infarct size in individuals with ischemic cardiomyopathy [104]. Importantly, clinical trials examining protective effects of adenosine administration subsequent to myocardial infarction [105-107] and in conjunction with various cardiac surgical procedures [108-110] have been performed. AMISTAD-I [106] and AMISTAD-II [107] were conducted primarily to assess the safety and efficacy of adenosine as adjunctive treatment in myocardial infarction patients undergoing thrombolysis and the results of these studies have been reviewed [19, 111, 112]. Briefly, in AMISTAD-I, an intravenous infusion of adenosine reduced overall myocardial infarct size by 33% relative to placebo, however, this protection occurred only in the subset of patients with anterior myocardial infarction in which infarct size was reduced by 67% [106]. A subsequent larger study, AMISTAD-II, was designed to assess efficacy of intravenous adenosine administration in specifically acute anterior myocardial infarction patients undergoing reperfusion [107]. Though adenosine at the higher dose of 70 µg/kg/min again produced a significant reduction in infarct size, there was no difference between this treatment group and placebo in clinical endpoints including death from any cause, development of new heart failure or rehospitalization for heart failure [107]. In a *post hoc* analysis of the AMISTAD-II results, it was found that patients who were reperfused with adjunct adenosine at less than 3.2 hr after symptom onset demonstrated significant improvement in clinical parameters compared to the placebo group and this benefit was not observed in patients receiving later treatment [113]. In the ADMIRE study [114], AMP 579, a compound with agonist activity at the A_1AR, $A_{2A}AR$ and $A_{2B}AR$ [115, 116], was administered to myocardial infarction patients undergoing angioplasty. Patients receiving AMP 579 did not have reduced infarct size nor demonstrate any difference in primary clinical outcomes relative to the placebo group [114]. More recent clinical studies examining intracoronary administration of adenosine following myocardial infarction have had mixed results [117, 118].

Adenosine Receptors and Cardiac Hypertrophy and Failure

The development of heart failure as a response to ischemia or hemodynamic overload is a complex process involving multiple cell types, functional effects and underlying signaling processes [119-121]. The progression to heart failure includes cardiac remodeling in which alterations in cardiac structure eventually become detrimental with a decline in function [119, 122]. Key cell types involved in remodeling include cardiac myocytes that may undergo hypertrophy and cell death [123, 124] and cardiac fibroblasts that are primarily responsible for excess deposition of extracellular matrix with ensuing fibrosis and a decrease in compliance [125, 126]. In experimental models using isolated cells, ARs have been shown to regulate these pathological processes. Activation of different AR subtypes has anti-hypertrophic effects in cardiac myocytes [127-129]. Stimulation of the $A_{2B}AR$ on cardiac fibroblasts inhibits cell proliferation [130, 131] and collagen synthesis [132]. The ability of A_2AR agonists to reduce collagen synthesis may be mediated by the exchange protein activated by cAMP (EPAC) [133]. Activation of the $A_{2A}AR$ in neutrophils inhibited release of matrix metalloproteinase-9 that may be involved in ventricular remodeling [134].

Involvement of AR activity in whole animal models of cardiac failure has been examined with roles for multiple AR subtypes identified. Liao and coworkers [135] demonstrated the beneficial effects of A_1AR agonist treatment in mice subjected to transverse aortic constriction (TAC) as both 2-chloroadenosine and N^6-cyclopentyladenosine slowed cardiac hypertrophy and improved cardiac function. In a canine model of heart failure, treatment with an A_1AR partial agonist improved left ventricular function relative to control animals and this positive effect was observed relatively soon after the initiation of therapy [136]. Additionally, the A_1AR partial agonist had beneficial effects on cardiac remodeling including decreases in cardiac myocyte hypertrophy and interstitial fibrosis [136]. However, constitutive cardiac-specific overexpression of the A_1AR in mice promoted dilated cardiomyopathy and increased mortality relative to nontransgenic mice [137]. Furthermore, when A_1AR overexpression was induced at 3 weeks followed by aortic banding, these mice developed a more severe cardiac hypertrophy and decline in cardiac function than wild-type animals with pressure overload [137]. Conversely, in A_1AR knockout mice, mortality rates induced by TAC were increased but without cardiac hypertrophy or detrimental changes in cardiac function [138]. In this study, A_3AR gene deletion was protective against TAC-induced cardiac remodeling and dysfunction [138]. Black and coworkers [139] observed a gene-dosage effect of the A_3AR on cardiac function as relatively low levels of A_3AR expression produced no cardiac dysfunction, but approximately 5-fold higher levels of overexpression resulted in cardiac dilatation, systolic dysfunction and other signs of cardiac failure. Hamad and coworkers [140] reported that inducible, cardiac-specific overexpression of the $A_{2A}AR$ in mice resulted in improved cardiac function after TAC relative to wild-type animals. Detrimental changes in cardiac dimension, fibrosis and cardiac expression of pro-hypertrophic genes and a small number of pro-inflammatory genes were attenuated with $A_{2A}AR$ overexpression [140]. A relatively high level of $A_{2A}AR$ overexpression in mice in which the A_1AR was also overexpressed improved cardiac function and increased survival relative to mice in which solely the A_1AR was overexpressed [141]. Interestingly, adriamycin-induced cardiac toxicity including conduction disturbances was more severe in mice with cardiac overexpression of the $A_{2A}AR$ than in nontransgenic animals, and an increased incidence of adriamycin-induced sudden death occurred with $A_{2A}AR$ overexpression [142]. Conversely, in mice in which induction of cardiac $A_{2A}AR$ overexpression was delayed until completion of adriamycin treatment, improved survival and cardiac function were observed when compared to nontransgenic mice receiving adriamycin [142]. A role for the $A_{2B}AR$ in regulation of the cardiac remodeling that occurs subsequent to myocardial infarction was suggested in a pharmacologic study in rats [143]. Administration of a nonselective AR agonist or dipyridamole one week after coronary occlusion resulted in improved cardiac performance and reduced fibrosis of noninfarcted ventricular tissue when evaluated after three weeks of treatment [143]. Similarly, AR activation blunted infarction-induced increases in levels of collagen type I, transforming growth factor-β1 and matrix metalloproteinase-2 [143]. Beneficial hemodynamic responses and anti-fibrotic effects induced by AR activation were blocked by the $A_{2B}AR$ selective antagonist MRS1754 but not by antagonists of other AR subtypes [143].

Activation of AR subtypes on distinct cardiac cell types has very recently been implicated in the beneficial effects of extracellular cAMP in cardiac hypertrophy and remodeling [144]. In a mouse model of isoproterenol and phenylephrine infusion-induced cardiac pressure overload, cAMP administration significantly inhibited development of cardiac hypertrophy and fibrosis [144]. Co-administration of selective AR antagonists indicated that the antihypertrophic effect of cAMP was mediated primarily by the A_1AR on cardiac myocytes while inhibition of fibrosis occurred in response to activation of the $A_{2A}AR$ and $A_{2B}AR$ on cardiac fibroblasts [144]. In conjunction with results obtained with transporter and enzyme inhibitors and genetically modified mice, the findings support a model in which cardiac myocyte cAMP, perhaps generated upon β-adrenergic receptor stimulation, is exported from the cell via ABCC transport proteins with the extracellular cAMP subsequently metabolized to adenosine [144]. In response, AR activation may occur and produce cardioprotective effects in distinct cell types and thus serve as a braking mechanism in the development of heart failure.

In the context of human heart failure, it is notable that plasma adenosine levels are elevated in heart failure patients [145, 146]. Rolofylline, an A_1AR antagonist, demonstrated disappointing results in terms of both efficacy and safety in a clinical trial of acute heart failure patients [147]. Rolofylline was examined based on its predicted beneficial effects on kidney function including enhanced diuresis and renal protective effects [148].

Adenosine Receptors and Angiogenesis

Neovascularization is generally quiescent in the healthy adult organism, however, it may be actively engaged during stressors such as ischemia, inflammation and following tissue injury. A major component of neovascularization is angiogenesis which is the development of new blood vessels from pre-existing vasculature. Therapeutic promotion of neovascularization, including angiogenesis, has significant potential in cardiac ischemia [149-151] and peripheral vascular disease [149, 152]. Conversely, in pathologies such as cancer [153, 154], certain retinopathies [155, 156] and rheumatoid arthritis [157] that rely on angiogenesis for progression, anti-angiogenic strategies are beneficial. Therapeutics that disrupt required vascularization are currently available for the treatment of certain solid tumors and ocular disorders. Angiogenesis is a well-orchestrated process that proceeds in a series of steps with the endothelium being the primary site of initial activity. Endothelial cells undergo the critical early events of proliferation and migration with subsequent tube formation [158, 159]. Hypoxia is the most physiologically relevant stimulus for angiogenesis [160]. Thus, based primarily on the elevation of adenosine levels that occurs with hypoxia, much focus has been placed on the role of AR activation in angiogenesis.

Over the past approximate 30 years, various *in vivo* models have implicated adenosine in the promotion of blood vessel growth [161-167]. Many of these findings have been described in excellent reviews [21, 168, 169]. At the isolated endothelial cell level, AR activation results in cell proliferation [170-178], migration [170, 176, 177, 179] and tube formation [176, 179, 180] which are key events in angiogenesis. In general, most studies have defined roles for the $A_{2A}AR$ and $A_{2B}AR$ expressed on endothelial cells in promoting angiogenesis or the necessary underlying signaling processes. However, a recent study reported that an A_3AR agonist inhibited endothelial cell proliferation and migration as well as pro-angiogenic signaling pathways [181]. As described below, all four AR subtypes expressed on a variety of cell types may have a paracrine role in regulating angiogenesis.

In regard to the ability of AR activation to promote angiogenesis or angiogenesis-related responses in endothelial cells, effects of receptor stimulation may be direct or indirect. Specifically, direct effects occur subsequent to activation of ARs on endothelial cells and engagement of intracellular signaling pathways that ultimately produce responses such as cell proliferation and migration. Conversely, indirect angiogenic responses result from the ability of AR activation on endothelial or other cell types to induce or inhibit the release of pro- or anti-angiogenic mediators, respectively. In this regard, most focus has been directed at the ability of AR activation to increase expression and secretion of vascular endothelial growth factor (VEGF) which is the most extensively studied pro-angiogenic signaling molecule. Via activation of receptor tyrosine kinases, VEGF induces endothelial cell proliferation, migration, tube formation and other responses [182-184]. VEGF Receptor-2 is the primary endothelial cell receptor for VEGF that is responsible for initiating pro-angiogenic signaling [182, 183].

In isolated human umbilical vein endothelial cells (HUVEC), the ability of ARs to activate ERK1/2, a key event in cell proliferation, appears to occur independently of VEGF signaling. ERK1/2 activation elicited by stimulation of the $A_{2A}AR$ [185] and $A_{2B}AR$ [186] was insensitive to PKC inhibitors that substantially reduced the response to VEGF [186, 187]. Additionally, stimulation of ERK1/2 mediated by the $A_{2A}AR$ [185] and $A_{2B}AR$ [186] did not require transactivation of the EGFR. Mechanisms underlying direct mitogenic effects of AR activation in endothelial cells have not been completely defined. In HUVEC [186] and human retinal endothelial cells [176], $A_{2B}AR$-mediated activation of ERK1/2 was not sensitive to inhibition of PKA and in HUVEC the response may be mediated by the cAMP effector EPAC [186]. In HUVEC, EPAC activation was demonstrated to be involved in pro-angiogenic signaling and cell proliferation induced by forskolin treatment [188] that like activation of the $A_{2B}AR$ results in an elevation of intracellular cAMP. Conversely, it has been reported that cAMP inhibits HUVEC proliferation [173].

AR agonists may increase expression and secretion of VEGF and many studies have identified the $A_{2B}AR$ as the responsible subtype. $A_{2B}AR$ activation induces VEGF upregulation in endothelial cells [175, 189], tumor cells [190], mast cells [191], glomerular podocytes [192], cardiac stromal cells [193] and skeletal muscle [194]. Ryzhov and coworkers [195] found that the $A_{2B}AR$-mediated increase in VEGF expression in endothelial cells occurred independently of PKA activation but required phospholipase C stimulation and activation of signaling that ultimately resulted in a JunB-mediated increase in VEGF gene transcription. Increases in VEGF expression occur upon activation of the A_1AR in monocytes [196], the $A_{2A}AR$ in macrophages [197-199] and thyroid cells [200] and the A_3AR in HT29 colon tumor cells [201] and glioblastoma cells [202]. More recently, it was shown that $A_{2B}AR$ and A_3AR activation increased VEGF expression in a model of atherosclerotic foam cells [203]. The hypoxia-induced increase in VEGF expression in certain cells may occur at least in part due to adenosine accumulation and subsequent AR activation [204-206]. Some studies have demonstrated that angiogenic responses produced by AR activation can be inhibited by anti-VEGF antibodies thus implicating a necessary role for VEGF activation of its receptor [175, 196, 207]. Caffeine inhibited angiogenesis in a mouse ischemic hind limb model with this response presumed to occur at least partially due to blockade of AR-induced upregulation of VEGF [208]. AR stimulation is not always associated with VEGF upregulation as activation of the $A_{2A}AR$ in PC12 cells [209, 210] and the $A_{2B}AR$ in neutrophils [211] promoted a decrease in VEGF expression with the response in PC12 cells involving deactivation of PKCζ [212].

Expression of receptors for VEGF may also be subject to AR regulation. In retinal endothelial cells, the downregulation of VEGF Receptor-2 that occurred in response to short-term hypoxia was in part mediated by an A_2AR [213]. Activation of the $A_{2A}AR$ on macrophages decreased production of a soluble form of VEGF Receptor-1 (sFlt-1) [214]. sFlt-1 is an anti-angiogenic protein that acts via binding of VEGF and prevention of VEGF Receptor-2 activation [215]. Thus, this downregulation of sFlt-1 by the $A_{2A}AR$ may promote angiogenesis. However, AR activation in rat placental villous explants was reported to increase sFlt-1 expression [216]. Other than VEGF, angiogenic factors such as basic fibroblast growth factor [217], angiopoietins [218, 219], interleukin-8 [203, 208, 217] and CXCL1 [220] are regulated by AR activation. Desai and coworkers [180] provided evidence that activation of the $A_{2A}AR$ on human microvascular endothelial cells promotes angiogenic responses by decreasing the expression and secretion of the anti-angiogenic protein thrombospondin-1. In HUVEC, $A_{2B}AR$ activation also decreased thrombospondin-1 expression [221].

Importantly, the functional consequences of AR-mediated stimulation of angiogenesis, whether dependent or independent of VEGF, have been demonstrated. Promotion of angiogenesis by AR activation may favor disease progression or have therapeutic potential. One context of pathologic AR-promoted angiogenesis is the retinopathy of prematurity (ROP). ROP is a complication of preterm birth that in its most severe form may result in retinal detachment and blindness [65]. ROP is a complex pathology that appears to involve initial impairment of retinal vascular development followed by a perhaps compensatory abnormal vasoproliferation [222]. The abnormalities in vascularization may be related to several factors including genetics, the immaturity in retinal vascular development associated with preterm birth, treatment with supplemental oxygen and associated changes in relative oxygen tension [222]. In a mouse pup model of ROP, Grant & coworkers demonstrated that chemical antagonists of the $A_{2B}AR$ [165] as well as hammerhead ribozymes that decrease $A_{2B}AR$ expression [164] inhibit oxygen-induced retinal neovascularization. Liu and coworkers [223] employed $A_{2A}AR$ knockout mice in the examination of ROP. Genetic deletion of the $A_{2A}AR$ did not affect normal developmental vascularization of the retina [223]. However, in response to hyperoxia followed by normoxia, significantly decreased parameters of pathologic angiogenesis were observed in $A_{2A}AR$ knockout mice relative to wild-type animals [223]. Additionally, in response to the hyperoxia-normoxia treatment protocol, the increase in retinal expression of VEGF mRNA that occurred in wild-type mice was not observed in $A_{2A}AR$ knockout mice [223]. The basis for the discrepancy of the identification of the $A_{2B}AR$ versus the $A_{2A}AR$ as the receptor subtype involved in the pathologic angiogenesis associated with ROP in mice is not clear. Perhaps both AR subtypes have a role in ROP and this was not detected. Additionally, differences may lie in the experimental protocols as pharmacologic and genetic approaches were taken in these studies. In a canine model of ROP, a role for the $A_{2A}AR$ was suggested based on regulation of receptor expression by hyperoxia followed by normoxia [224, 225]. In a large clinical trial to assess the efficacy of caffeine therapy in infants with the apnea of prematurity, it was noted that the incidence of severe ROP was lower in caffeine-treated infants relative to the placebo group [226]. It may be proposed that the effects of caffeine occurred due to blockade of retinal ARs, however, the authors of this study speculated that the reduced incidence of ROP may reflect the decreased exposure to positive airway pressure and supplemental oxygen in caffeine-treated infants [226].

Recently, regulation of VEGF expression by the $A_{2B}AR$ has been proposed to have a role in the progression of diabetic nephropathy. Overproduction of VEGF is associated with the glomerulopathy that develops with diabetes and VEGF may promote dysfunction via pathologic angiogenesis as well as other mechanisms [227]. Induction of diabetes via streptozocin administration increased renal expression of the $A_{2B}AR$ in both rats [228] and mice [229]. However, additional findings in regard to the beneficial effects of $A_{2B}AR$ regulation in these models varied considerably. Treatment of diabetic rats with the $A_{2B}AR$ antagonist MRS1754 lessened mesangial cell activation and proteinuria [228] and these findings are consistent with those demonstrating improvement of renal function in diabetic animals receiving MRS1754 [230]. Conversely, in diabetic mice, parameters of diabetic nephropathy were improved by administration of the $A_{2B}AR$ agonist BAY 60-6583 [229]. Additionally, mice with endothelial cell specific deletion of the $A_{2B}AR$ demonstrated a more severe nephropathy than wild-type animals [229]. Differences were also noted in these studies upon examination of the regulation of VEGF expression by the A_{2B}AR. In isolated rat glomeruli, pharmacologic activation of the A_{2B}AR promoted VEGF production [228]. However, in A_{2B}AR deficient mice, the increase in VEGF induced by diabetes was more marked than that observed in wild-type mice [229]. Though perhaps independent of any involvement of VEGF, upregulation of the glomerular A_{2B}AR has also recently been implicated in contributing to the hypertension associated with chronic kidney disease (CKD). In an *in vivo* murine model of CKD, Zhang and coworkers [231] observed an upregulation of both renal CD73 and the A_{2B}AR. Data supported a paradigm of increased glomerular adenosine formation driving an upregulation of A_{2B}AR expression that in turn promoted enhanced production and release of endothelin-1 with resulting hypertension [231]. Interestingly, both the adenosine-mediated upregulation of A_{2B}AR expression and the A_{2B}AR-induced increased expression of endothelin-1 required HIF-1α activation [231]. It should be noted that A_{2B}AR signaling may contribute to CKD through other mechanisms including increased renal fibrosis [232].

Promotion of angiogenesis via AR activation has clinical potential in certain conditions including wound healing. Cronstein and coworkers [163, 233, 234] have shown that topical application of A_{2A}AR agonists promotes wound healing in multiple animal models and this response was lost in A_{2A}AR knockout mice [235]. Additionally, A_{2A}AR-mediated healing of excisional wounds in mice correlated with markedly enhanced vascularization of granulation tissue [235]. This A_{2A}AR agonist-induced neovascularization may occur due to both angiogenesis and vasculogenesis with the latter suggested by an increase in endothelial progenitor cell (EPC) recruitment in the wound granulation tissue [166]. Macedo and coworkers [236] observed that A_{2A}AR-mediated healing and accompanying increased vascularization of dermal wounds was impaired in mice with genetic deletion of the signaling molecule MyD88 and this perhaps may involve loss of VEGF upregulation. Analysis of tissue repair induced by polydeoxyribonucleotide (PDRN) administration has indicated involvement of A_{2A}AR-mediated upregulation of VEGF. PDRN is a therapeutic mixture of deoxyribonucleotide polymers that may have A_{2A}AR agonist activity [237, 238]. In various animal models, PDRN promoted wound healing and this was accompanied by increases in VEGF expression and microvascular density with beneficial responses inhibited by co-administration of an A_{2A}AR antagonist [239, 240]. A recent clinical trial reported that PDRN promoted healing of chronic foot ulcers in diabetic patients [241]. Additionally, in a rat hindlimb ischemia model, PDRN increased VEGF expression and promoted revascularization, and these responses were blocked by an A_{2A}AR antagonist [242].

In addition to promotion of angiogenesis via administration of pro-angiogenic molecules, transplantation of bone marrow-derived EPC has been a more recently explored approach directed toward therapeutic neovascularization [243, 244]. Successful application of this strategy requires efficient homing and retention of EPC to the targeted vasculature. Roles for AR activation in promoting adhesion of EPC to vascular endothelium have been identified. Ryzhov and coworkers [245] examined the function of AR subtypes in an *in vitro* model of EPC adhesion with results suggesting pro-adhesive roles for the A_1AR expressed on EPC and the A_{2B}AR expressed on endothelial cells. Retention of infused EPC in the coronary vasculature of isolated murine hearts was enhanced by adenosine administration presumably through activation of the individual AR subtypes on the respective cells [245]. Blockade of the interaction between P-selectin and P-selectin glycoprotein reduced AR-mediated adhesion of EPC to the endothelium, and activation of the $A_{2B}AR$ on endothelial cells induced cell surface expression of P-selectin [245]. These findings suggest a possible underlying mechanism for the pro-adhesive response. Rolland-Turner and coworkers [246] observed that AR activation enhanced migration of human EPC toward SDF-1α as well as conditioned medium from cultured human cardiac fibroblasts. In rats subjected to myocardial infarction induced by coronary ligation, long-term treatment with AR agonists promoted recruitment of EPC and revascularization in the infarct border zone [246]. Mechanistic studies indicated a role for the $A_{2B}AR$ on EPC with receptor activation producing an increase in cell surface expression of CXCR4, a pro-migratory receptor [246]. As noted above, in an *in vivo* murine model of wound healing, activation of the $A_{2A}AR$ promoted recruitment of bone marrow-derived EPCs to the neovasculature of the dermal wound [166].

In addition to the well-described ability of hypoxia to promote elevation of adenosine levels, low oxygen conditions regulate AR expression in vascular endothelium and this may in turn modulate AR-mediated responses. Hypoxia increased expression of the $A_{2B}AR$ but no other AR subtypes in the human microvascular endothelial cell line HMEC-1 and this response was mediated by activation of HIF-1α [247, 248]. HUVEC responded to 3 hour of hypoxia with an increase in $A_{2B}AR$ mRNA and a substantial decrease in $A_{2A}AR$ mRNA and this selective modulation had effects on cellular responses such as regulation of VEGF expression [189]. A similar hypoxia-induced switching in AR expression may have pathologic relevance in preeclamptic pregnancy. Escudero and coworkers found that relative to HUVEC obtained from individuals with normal pregnancy, $A_{2B}AR$ expression was increased with preeclampsia [178] while $A_{2A}AR$ expression was decreased in patients with early-onset but not late-onset preeclampsia [249]. The underlying mechanism for this differential regulation is not known though the change in relative expression may have impact on angiogenesis and other functional consequences in pregnancy as recently reviewed [250]. Conversely, in human microvascular lung endothelial cells (HMVLEC), hypoxia increased the expression of the $A_{2A}AR$ but had no effect on $A_{2B}AR$ levels with upregulation of the $A_{2A}AR$ occurring via HIF-2α activation [251]. To explore the impact of upregulation of ARs on angiogenic processes, effects of forced overexpression of the receptors have been examined. Overexpression of the $A_{2B}AR$ in HMEC-1 promoted tube formation [248] and $A_{2A}AR$ overexpression in HMVLEC increased proliferation and migration [251, 252]. There are limited *in vivo* findings available to assess the physiologic relevance of a hypoxia-induced upregulation of endothelial cell ARs. Interestingly, it should be noted that activation of the $A_{2B}AR$ inhibits proliferation [253-256] and promotes apoptosis [257] of vascular smooth muscle cells.

CONCLUSION

This chapter has sought to summarize recent findings regarding the compensatory effects of AR activation in the cardiovascular system. Particular emphasis has been placed on the ability of AR activation to preserve myocardial function during ischemia-reperfusion and heart failure as well as to regulate neovascularization, a process that may be beneficial in ischemic disorders or detrimental in disease progression. Unfortunately, space limitation prevented review of important recent advances in the understanding of the roles of ARs in atherosclerosis [258, 259], protection from vascular injury such as restenosis [260-262], regulation of vascular barrier function [263, 264] and sickle cell disease [265, 266] as well as in cardiovascular development [267, 268]. Many of these areas encompass regulation of ARs on cells that comprise the cardiovascular system as well other cell types such as those of the immune system. In general, the current therapeutic translation of AR regulation has been quite limited and confined to short-term indications such as the conversion of cardiac arrhythmia and diagnostic coronary vasodilation. It is the widespread distribution of the four AR subtypes, not only throughout the cardiovascular system but other organs systems as well, that has perhaps been the most significant barrier to achieving broader clinical use of AR ligands. The prevalence of AR expression and thus the possible development of adverse effects is a substantial challenge. The increasing development of potent and subtype selective AR ligands has resulted in the availability of potential drugs. Strategies to overcome the hurdles in the approval of adenosine-based therapies for multiple disease states have recently been reviewed [269]. Others have identified possible limitations in clinical trials that may have obscured the ability to observe clear beneficial effects of AR agonists or other strategies of cardioprotection in ischemia-reperfusion injury [19, 22, 270]. With further advances in identifying the mechanisms underlying cardiovascular pathologies and the utility of AR regulation in addressing these diseases, clinical use of AR ligands will hopefully expand. The indications for use may be short-term, for example as an adjunct during reperfusion interventions following myocardial infarction, or for chronic conditions such as heart failure. Bringing AR-based therapeutics to the clinic will remain challenging as studies in isolated cells, animal models and healthy humans do not replicate the dynamics encountered in affected patients perhaps receiving concomitantly administered drugs and displaying other complexities. However, the successful therapeutic targeting of G protein-coupled receptors in a myriad of diseases provides encouragement for the continued exploration of AR regulation in humans.

REFERENCES

[1] Drury, A. N.; Szent-Gyorgyi, A. The physiological activity of adenine compounds with especial reference to their action upon the mammalian heart. *J. Physiol.* 1929, 68, 213-37.

[2] Berne, R. M. Cardiac nucleotides in hypoxia: possible role in regulation of coronary blood flow. *Am. J. Physiol.* 1963, 204, 317-22.

[3] Newby, A. C.; Worku, Y.; Holmquist, C. A. Adenosine formation. Evidence for a direct biochemical link with energy metabolism. *Adv. Myocardiol.* 1985, 6, 273-84.

[4] Morrison, R. R.; Teng, B.; Oldenburg, P. J.; Katwa, L. C.; Schnermann, J. B.; Mustafa, S. J. Effects of targeted deletion of A1 adenosine receptors on postischemic cardiac function and expression of adenosine receptor subtypes. *Am. J. Physiol. Heart Circ. Physiol.* 2006, 291, H1875-82.

[5] Ashton, K. J.; Nilsson, U.; Willems, L.; Holmgren, K.; Headrick, J. P. Effects of aging and ischemia on adenosine receptor transcription in mouse myocardium. *Biochem. Biophys. Res. Commun.* 2003, 312, 367-72.

[6] Mustafa, S. J.; Morrison, R. R.; Teng, B.; Pelleg, A. Adenosine receptors and the heart: role in regulation of coronary blood flow and cardiac electrophysiology. *Handb. Exp. Pharmacol.* 2009, 193, 161-88.

[7] Olanrewaju, H. A.; Qin, W.; Feoktistov, I.; Scemama, J. L.; Mustafa, S. J. Adenosine A(2A) and A(2B) receptors in cultured human and porcine coronary artery endothelial cells. *Am. J. Physiol. Heart Circ. Physiol.* 2000, 279, H650-6.

[8] Zatta, A. J.; Matherne, G. P.; Headrick, J. P. Adenosine receptor-mediated coronary vascular protection in post-ischemic mouse heart. *Life Sci.* 2006, 78, 2426-37.

[9] Sato, A.; Terata, K.; Miura, H.; Toyama, K.; Loberiza, F. R., Jr.; Hatoum, O. A.; Saito, T.; Sakuma, I.; Gutterman, D. D. Mechanism of vasodilation to adenosine in coronary arterioles from patients with heart disease. *Am. J. Physiol. Heart Circ. Physiol.* 2005, 288, H1633-40.

[10] Olsson, R. A.; Pearson, J. D. Cardiovascular purinoceptors. *Physiol. Rev.* 1990, 70, 761-845.

[11] Shryock, J. C.; Belardinelli, L. Adenosine and adenosine receptors in the cardiovascular system: biochemistry, physiology, and pharmacology. *Am. J. Cardiol.* 1997, 79, 2-10.

[12] Headrick, J. P.; Peart, J. N.; Reichelt, M. E.; Haseler, L. J. Adenosine and its receptors in the heart: regulation, retaliation and adaptation. *Biochim. Biophys. Acta.* 2011, 1808, 1413-28.

[13] Wilbur, S. L.; Marchlinski, F. E. Adenosine as an antiarrhythmic agent. *Am. J. Cardiol.* 1997, 79, 30-7.

[14] Palani, G.; Ananthasubramaniam, K. Regadenoson: review of its established role in myocardial perfusion imaging and emerging applications. *Cardiol. Rev.* 2013, 21, 42-8.

[15] Harker, L. A.; Kadatz, R. A. Mechanism of action of dipyridamole. *Thromb. Res. Suppl.* 1983, 4, 39-46.

[16] Murry, C. E.; Jennings, R. B.; Reimer, K. A. Preconditioning with ischemia: a delay of lethal cell injury in ischemic myocardium. *Circulation.* 1986, 74, 1124-36.

[17] Zhao, Z. Q.; Corvera, J. S.; Halkos, M. E.; Kerendi, F.; Wang, N. P.; Guyton, R. A.; Vinten-Johansen, J. Inhibition of myocardial injury by ischemic postconditioning during reperfusion: comparison with ischemic preconditioning. *Am. J. Physiol. Heart Circ. Physiol.* 2003, 285, H579-88.

[18] Cohen, M. V.; Downey, J. M. Adenosine: trigger and mediator of cardioprotection. *Basic Res. Cardiol.* 2008, 103, 203-15.

[19] Headrick, J. P.; Lasley, R. D. Adenosine receptors and reperfusion injury of the heart. *Handb. Exp. Pharmacol.* 2009, 193, 189-214.

[20] McIntosh, V. J.; Lasley, R. D. Adenosine receptor- mediated cardioprotection: are all 4 subtypes required or redundant? *J. Cardiovasc. Pharmacol. Ther.* 2012, 17, 21-33.

[21] Headrick, J. P.; Ashton, K. J.; Rose'Meyer, R. B.; Peart, J. N. Cardiovascular adenosine receptors: expression, actions and interactions. *Pharmacol. Ther.* 2013, 140, 92-111.

[22] Cohen, M. V.; Downey, J. M. Signaling pathways and mechanisms of protection in pre- and postconditioning: historical perspective and lessons for the future. *Br. J. Pharmacol.* 2014, 1-19.

[23] Gross, G. J. Role of opioids in acute and delayed preconditioning. *J. Mol. Cell. Cardiol.* 2003, 35, 709-18.

[24] Peart, J. N.; Gross, E. R.; Gross, G. J. Opioid-induced preconditioning: recent advances and future perspectives. *Vascul. Pharmacol.* 2005, 42, 211-8.

[25] Baxter, G. F.; Ebrahim, Z. Role of bradykinin in preconditioning and protection of the ischaemic myocardium. *Br. J. Pharmacol.* 2002, 135, 843-54.

[26] Lasley, R. D.; Konyn, P. J.; Hegge, J. O.; Mentzer, R. M., Jr. Effects of ischemic and adenosine preconditioning on interstitial fluid adenosine and myocardial infarct size. *Am. J. Physiol.* 1995, 269, H1460-6.

[27] Schulz, R.; Rose, J.; Post, H.; Heusch, G. Involvement of endogenous adenosine in ischaemic preconditioning in swine. *Pflugers Arch.* 1995, 430, 273-82.

[28] Headrick, J. P. Ischemic preconditioning: bioenergetic and metabolic changes and the role of endogenous adenosine. *J. Mol. Cell. Cardiol.* 1996, 28, 1227-40.

[29] Cohen, M. V.; Downey, J. M. Adenosine: trigger and mediator of cardioprotection. *Basic Res. Cardiol.* 2008, 103, 203-215.

[30] Matherne, G. P.; Linden, J.; Byford, A. M.; Gauthier, N. S.; Headrick, J. P. Transgenic A1 adenosine receptor overexpression increases myocardial resistance to ischemia. *Proc. Natl. Acad. Sci. U.S.A.* 1997, 94, 6541-6.

[31] Butcher, A.; Scammells, P. J.; White, P. J.; Devine, S. M.; Rose'meyer, R.B. An allosteric modulator of the adenosine A1 receptor improves cardiac function following ischaemia in murine isolated hearts. *Pharmaceuticals (Basel)* 2013, 6, 546-556.

[32] Gao, Z. G.; Kim, S. K.; Ijzerman, A. P.; Jacobson, K. A. Allosteric modulation of the adenosine family of receptors. *Mini Rev. Med. Chem.* 2005, 5, 545-53.

[33] Lankford, A. R.; Yang, J. N.; Rose'Meyer, R.; French, B. A.; Matherne, G. P.; Fredholm, B. B.; Yang, Z. Effect of modulating cardiac A1 adenosine receptor expression on protection with ischemic preconditioning. *Am. J. Physiol. Heart Circ. Physiol.* 2006, 290, H1469-73.

[34] Eckle, T.; Krahn, T.; Grenz, A.; Kohler, D.; Mittelbronn, M.; Ledent, C.; Jacobson, M. A.; Osswald, H.; Thompson, L. F.; Unertl, K.; Eltzschig, H. K. Cardioprotection by ecto-5'-nucleotidase (CD73) and A2B adenosine receptors. *Circulation* 2007, 115, 1581-90.

[35] Xi, L.; Das, A.; Zhao, Z. Q.; Merino, V. F.; Bader, M.; Kukreja, R. C. Loss of myocardial ischemic postconditioning in adenosine A1 and bradykinin B2 receptors gene knockout mice. *Circulation* 2008, 118, S32-7.

[36] Headrick, J. P.; Peart, J. A3 adenosine receptor-mediated protection of the ischemic heart. *Vascul. pharmacol.* 2005, 42, 271-9.

[37] Kin, H.; Zatta, A. J.; Lofye, M. T.; Amerson, B. S.; Halkos, M. E.; Kerendi, F.; Zhao, Z. Q.; Guyton, R. A.; Headrick, J. P.; Vinten-Johansen, J. Postconditioning reduces infarct size via adenosine receptor activation by endogenous adenosine. *Cardiovasc. Res.* 2005, 67, 124-33.

[38] Du, L.; Gao, Z. G.; Nithipatikom, K.; Ijzerman, A. P.; Veldhoven, J. P.; Jacobson, K. A.; Gross, G. J.; Auchampach, J. A. Protection from myocardial ischemia/reperfusion injury by a positive allosteric modulator of the A(3) adenosine receptor. *J. Pharmacol. Exp. Ther.* 2012, 340, 210-7.

[39] Ge, Z. D.; van der Hoeven, D.; Maas, J. E.; Wan, T. C.; Auchampach, J. A. A(3) adenosine receptor activation during reperfusion reduces infarct size through actions on bone marrow-derived cells. *J. Mol. Cell. Cardiol.* 2010, 49, 280-6.

[40] van der Hoeven, D.; Wan, T. C.; Auchampach, J. A. Activation of the A(3) adenosine receptor suppresses superoxide production and chemotaxis of mouse bone marrow neutrophils. *Mol. Pharmacol.* 2008, 74, 685-96.

[41] Armstrong, S.; Ganote, C. E. Adenosine receptor specificity in preconditioning of isolated rabbit cardiomyocytes: evidence of A3 receptor involvement. *Cardiovasc. Res.* 1994, 28, 1049-56.

[42] Liu, G. S.; Richards, S. C.; Olsson, R. A.; Mullane, K.; Walsh, R. S.; Downey, J. M. Evidence that the adenosine A3 receptor may mediate the protection afforded by preconditioning in the isolated rabbit heart. *Cardiovasc. Res.* 1994, 28, 1057-61.

[43] Guo, Y.; Bolli, R.; Bao, W.; Wu, W. J.; Black, R. G., Jr.; Murphree, S. S.; Salvatore, C. A.; Jacobson, M. A.; Auchampach, J. A. Targeted deletion of the A3 adenosine receptor confers resistance to myocardial ischemic injury and does not prevent early preconditioning. *J. Mol. Cell. Cardiol.* 2001, 33, 825-30.

[44] Lasley, R. D.; Jahania, M. S.; Mentzer, R. M., Jr. Beneficial effects of adenosine A(2a) agonist CGS-21680 in infarcted and stunned porcine myocardium. *Am. J. Physiol. Heart Circ. Physiol.* 2001, 280, H1660-6.

[45] Hausenloy, D. J.; Yellon, D. M. Myocardial ischemia-reperfusion injury: a neglected therapeutic target. *J. Clin. Invest.* 2013, 123, 92-100.

[46] Linden, J. Regulation of leukocyte function by adenosine receptors. *Adv. Pharmacol.* 2011, 61, 95-114.

[47] Milne, G. R.; Palmer, T. M. Anti-inflammatory and immunosuppressive effects of the A2A adenosine receptor. *ScientificWorldJournal.* 2011, 11, 320-39.

[48] Jordan, J. E.; Zhao, Z. Q.; Sato, H.; Taft, S.; Vinten-Johansen, J. Adenosine A2 receptor activation attenuates reperfusion injury by inhibiting neutrophil accumulation, superoxide generation and coronary endothelial adherence. *J. Pharmacol. Exp. Ther.* 1997, 280, 301-9.

[49] Glover, D. K.; Riou, L. M.; Ruiz, M.; Sullivan, G. W.; Linden, J.; Rieger, J. M.; Macdonald, T. L.; Watson, D. D.; Beller, G. A. Reduction of infarct size and postischemic inflammation from ATL-146e, a highly selective adenosine A2A receptor agonist, in reperfused canine myocardium. *Am. J. Physiol. Heart Circ. Physiol.* 2005, 288, H1851-8.

[50] Yang, Z.; Day, Y. J.; Toufektsian, M. C.; Xu, Y.; Ramos, S. I.; Marshall, M. A.; French, B. A.; Linden, J. Myocardial infarct-sparing effect of adenosine A2A receptor activation is due to its action on CD4+ T lymphocytes. *Circulation.* 2006, 114, 2056-64.

[51] Rork, T. H.; Wallace, K. L.; Kennedy, D. P.; Marshall, M. A.; Lankford, A. R.; Linden, J. Adenosine A2A receptor activation reduces infarct size in the isolated, perfused mouse heart by inhibiting resident cardiac mast cell degranulation. *Am. J. Physiol. Heart Circ. Physiol.* 2008, 295, H1825-H1833.

[52] Toufektsian, M. C.; Yang, Z.; Prasad, K. M.; Overbergh, L.; Ramos, S. I.; Mathieu, C.; Linden, J.; French, B. A. Stimulation of A2A-adenosine receptors after myocardial infarction suppresses inflammatory activation and attenuates contractile dysfunction in the remote left ventricle. *Am. J. Physiol. Heart Circ. Physiol.* 2006, 290, H1410-8.

[53] Methner, C.; Schmidt, K.; Cohen, M. V.; Downey, J. M.; Krieg, T. Both A2a and A2b adenosine receptors at reperfusion are necessary to reduce infarct size in mouse hearts. *Am. J. Physiol. Heart Circ. Physiol.* 2010, 299, H1262-4.

[54] Morrison, R. R.; Tan, X. L.; Ledent, C.; Mustafa, S. J.; Hofmann, P. A. Targeted deletion of A2A adenosine receptors attenuates the protective effects of myocardial postconditioning. *Am. J. Physiol. Heart Circ. Physiol.* 2007, 293, H2523-9.

[55] Maas, J. E.; Wan, T. C.; Figler, R. A.; Gross, G. J.; Auchampach, J. A. Evidence that the acute phase of ischemic preconditioning does not require signaling by the A 2B adenosine receptor. *J. Mol. Cell. Cardiol.* 2010, 49, 886-93.

[56] Philipp, S.; Yan, X. M.; Cui, L.; Davis, A. M.; Downey, J. M.; Cohen, M. V. Postconditioning protects rabbit hearts through a protein kinase C-adenosine A2b receptor cascade. *Cardiovasc. Res.* 2006, 70, 308-314.

[57] Xi, J.; McIntosh, R.; Shen, X.; Lee, S.; Chanoit, G.; Criswell, H.; Zvara, D. A.; Xu, Z. Adenosine A2A and A2B receptors work in concert to induce a strong protection against reperfusion injury in rat hearts. *J Mol. Cell. Cardiol.* 2009, 47, 684-90.

[58] Kuno, A.; Solenkova, N. V.; Solodushko, V.; Dost, T.; Liu, Y.; Yang, X. M.; Cohen, M. V.; Downey, J. M. Infarct limitation by a protein kinase G activator at reperfusion in rabbit hearts is dependent on sensitizing the heart to A2b agonists by protein kinase C. *Am. J. Physiol. Heart Circ. Physiol.* 2008, 295, H1288-h1295.

[59] Solenkova, N. V.; Solodushko, V.; Cohen, M. V.; Downey, J. M. Endogenous adenosine protects preconditioned heart during early minutes of reperfusion by activating Akt. *Am. J. Physiol. Heart Circ. Physiol.* 2006, 290, H441-9.

[60] Koeppen, M.; Harter, P. N.; Bonney, S.; Bonney, M.; Reithel, S.; Zachskorn, C. Mittelbronn, M.; Eckle, T. Adora2b signaling on bone marrow derived cells dampens myocardial ischemia-reperfusion injury. *Anesthesiol.* 2012, 116, 1245-57.

[61] Urmaliya, V. B.; Pouton, C. W.; Ledent, C.; Short, J. L.; White, P. J. Cooperative cardioprotection through adenosine A1 and A2A receptor agonism in ischemia-reperfused isolated mouse heart. *J. Cardiovas. Pharmacol.* 2010, 56, 379-88.

[62] Lasley, R. D.; Kristo, G.; Keith, B. J.; Mentzer, R. M., Jr. The A2a/A2b receptor antagonist ZM-241385 blocks the cardioprotective effect of adenosine agonist pretreatment in in vivo rat myocardium. *Am. J. Physiol. Heart Circ. Physiol.* 2007, 292, H426-31.

[63] Zhan, E.; McIntosh, V. J.; Lasley, R. D. Adenosine A2A and A2B receptors are both required for adenosine A1 receptor-mediated cardioprotection. *Am. J. Physiol. Heart Circ. Physiol.* 2011, 301, H1183-H1189.

[64] Peart, J. N.; Gross, G. J., Adenosine and opioid receptor-mediated cardioprotection in the rat: evidence for cross-talk between receptors. *Am. J. Physiol. Heart Circ. Physiol.* 2003, 285, H81-9.

[65] Surendra, H.; Diaz, R. J.; Harvey, K.; Tropak, M.; Callahan, J.; Hinek, A.; Hossain, T.; Redington, A.; Wilson, G. J. Interaction of delta and kappa opioid receptors with adenosine A1 receptors mediates cardioprotection by remote ischemic preconditioning. *J. Mol. Cell. Cardiol.* 2013, 60, 142-50.

[66] Sun, L.; Li, D. L.; Zhao, M.; He, X.; Yu, X. J.; Miao, Y.; Wang, H.; Ren, J.; Zang, W. J. The role of muscarinic receptors in the beneficial effects of adenosine against myocardial reperfusion injury in rats. *PloS One.* 2011, 6, e25618.

[67] Ciruela, F.; Casado, V.; Rodrigues, R. J.; Lujan, R.; Burgueno, J.; Canals, M.; Borycz, J.; Rebola, N.; Goldberg, S. R.; Mallol, J.; Cortes, A.; Canela, E. I.; Lopez-Gimenez, J. F.; Milligan, G.; Lluis, C.; Cunha, R. A.; Ferre, S.; Franco, R. Presynaptic control of striatal glutamatergic neurotransmission by adenosine A1-A2A receptor heteromers. *J Neurosci.* 2006, 26, 2080-7.

[68] Ciruela, F.; Ferre, S.; Casado, V.; Cortes, A.; Cunha, R. A.; Lluis, C.; Franco, R. Heterodimeric adenosine receptors: a device to regulate neurotransmitter release. *Cell. Mol. Life Sci.* 2006, 63, 2427-31.

[69] Moriyama, K.; Sitkovsky, M. V. Adenosine A2A receptor is involved in cell surface expression of A2B receptor. *J. Biol. Chem.* 2010, 285, 39271-88.

[70] Canals, M.; Marcellino, D.; Fanelli, F.; Ciruela, F.; de Benedetti, P.; Goldberg, S. R.; Neve, K.; Fuxe, K.; Agnati, L. F.; Woods, A. S.; Ferre, S.; Lluis, C.; Bouvier, M.; Franco, R. Adenosine A2A-dopamine D2 receptor-receptor heteromerization: qualitative and quantitative assessment by fluorescence and bioluminescence energy transfer. *J. Biol. Chem.* 2003, 278, 46741-9.

[71] Franco, R.; Ferre, S.; Agnati, L.; Torvinen, M.; Gines, S.; Hillion, J.; Casado, V.; Lledo, P.; Zoli, M.; Lluis, C.; Fuxe, K. Evidence for adenosine/dopamine receptor interactions: indications for heteromerization. *Neuropsychopharmacology.* 2000, 23, S50-9.

[72] Chandrasekera, P. C.; Wan, T. C.; Gizewski, E. T.; Auchampach, J. A.; Lasley, R. D. Adenosine A1 receptors heterodimerize with beta1- and beta2-adrenergic receptors creating novel receptor complexes with altered G protein coupling and signaling. *Cell Signal.* 2013, 25, 736-42.

[73] Vidi, P. A.; Chemel, B. R.; Hu, C. D.; Watts, V. J. Ligand-dependent oligomerization of dopamine D(2) and adenosine A(2A) receptors in living neuronal cells. *Mol. Pharmacol.* 2008, 74, 544-51.

[74] Salie, R.; Moolman, J. A.; Lochner, A. The mechanism of beta-adrenergic preconditioning: roles for adenosine and ROS during triggering and mediation. *Basic Res. Cardiol.* 2012, 107, 281.

[75] Yang, X.; Xin, W.; Yang, X. M.; Kuno, A.; Rich, T. C.; Cohen, M. V.; Downey, J. M. A2B adenosine receptors inhibit superoxide production from mitochondrial complex I in rabbit cardiomyocytes via a mechanism sensitive to Pertussis toxin. *Br. J. Pharmacol.* 2011, 163, 995-1006.

[76] Ryzhov, S.; Zaynagetdinov, R.; Goldstein, A. E.; Matafonov, A.; Biaggioni, I.; Feoktistov, I. Differential role of the carboxy-terminus of the A(2B) adenosine receptor in stimulation of adenylate cyclase, phospholipase Cbeta, and interleukin-8. *Purinergic Signal.* 2009, 5, 289-98.

[77] Hausenloy, D. J.; Yellon, D. M. New directions for protecting the heart against ischaemia-reperfusion injury: targeting the Reperfusion Injury Salvage Kinase (RISK)-pathway. *Cardiovasc. Res.* 2004, 61, 448-60.

[78] Hausenloy, D. J.; Tsang, A.; Mocanu, M. M.; Yellon, D. M. Ischemic preconditioning protects by activating prosurvival kinases at reperfusion. *Am. J. Physiol. Heart Circ. Physiol.* 2005, 288, H971-6.

[79] Hausenloy, D. J.; Yellon, D. M. Reperfusion injury salvage kinase signaling: taking a RISK for cardioprotection. *Heart Fail Rev.* 2007, 12, 217-234.

[80] Ong, S. B.; Dongworth, R.; Cabrera-Fuentes, H. A.; Hausenloy, D. The role of the MPTP in conditioning the heart - translatability and mechanism. *Br. J. Pharmacol.* 2014. [Epub ahead of print]

[81] Halestrap, A. P. A pore way to die: the role of mitochondria in reperfusion injury and cardioprotection. *Biochem. Soc. Trans.* 2010, 38, 841-60.

[82] Javadov, S.; Karmazyn, M.; Escobales, N. Mitochondrial permeability transition pore opening as a promising therapeutic target in cardiac diseases. *J. Pharmacol. Exp. Ther.* 2009, 330, 670-8.

[83] Yang, X.; Cohen, M. V.; Downey, J. M. Mechanism of cardioprotection by early ischemic preconditioning. *Cardiovasc. Drugs Ther.* 2010, 24, 225-34.

[84] Park, S. S.; Zhao, H.; Jang, Y.; Mueller, R. A.; Xu, Z. N6-(3-iodobenzyl)-adenosine-5'-N-methylcarboxamide confers cardioprotection at reperfusion by inhibiting mitochondrial permeability transition pore opening via glycogen synthase kinase 3 beta. *J. Pharmacol. Exp. Ther.* 2006, 318, 124-31.

[85] Ban, K.; Cooper, A. J.; Samuel, S.; Bhatti, A.; Patel, M.; Izumo, S.; Penninger, J. M.; Backx, P. H.; Oudit, G. Y.; Tsushima, R. G. Phosphatidylinositol 3-kinase gamma is a critical mediator of myocardial ischemic and adenosine-mediated preconditioning. *Circ. Res.* 2008, 103, 643-53.

[86] Peart, J. N.; Headrick, J. P. Adenosinergic cardioprotection: multiple receptors, multiple pathways. *Pharmacol. Ther.* 2007, 114, 208-21.

[87] Baxter, G. F.; Yellon, D. M. ATP-sensitive K+ channels mediate the delayed cardioprotective effect of adenosine A1 receptor activation. *J. Mol. Cell. Cardiol.* 1999, 31, 981-9.

[88] Tracey, W. R.; Magee, W.; Masamune, H.; Oleynek, J. J.; Hill, R. J. Selective activation of adenosine A3 receptors with N6-(3-chlorobenzyl)-5'-N-methylcarboxamidoadenosine (CB-MECA) provides cardioprotection via KATP channel activation. *Cardiovasc. Res.* 1998, 40, 138-45.

[89] Auchampach, J. A.; Ge, Z. D.; Wan, T. C.; Moore, J.; Gross, G. J. A3 adenosine receptor agonist IB-MECA reduces myocardial ischemia-reperfusion injury in dogs. *Am. J. Physiol. Heart Circ. Physiol.* 2003, 285, H607-13.

[90] Wan, T. C.; Ge, Z. D.; Tampo, A.; Mio, Y.; Bienengraeber, M. W.; Tracey, W. R.; Gross, G. J.; Kwok, W. M.; Auchampach, J. A. The A3 adenosine receptor agonist CP-532,903 [N6-(2,5-dichlorobenzyl)-3'-aminoadenosine-5'-N-methylcarboxamide] protects against myocardial ischemia/reperfusion injury via the sarcolemmal ATP-sensitive potassium channel. *J. Pharmacol. Exp. Ther.* 2008, 324, 234-43.

[91] Yitzhaki, S.; Huang, C.; Liu, W.; Lee, Y.; Gustafsson, A. B.; Mentzer, R. M., Jr.; Gottlieb, R. A. Autophagy is required for preconditioning by the adenosine A1 receptor-selective agonist CCPA. *Basic Res. Cardiol.* 2009, 104, 157-67.

[92] Gustafsson, A. B.; Gottlieb, R. A. Autophagy in ischemic heart disease. *Circ. Res.* 2009, 104, 150-8.

[93] Williams-Pritchard, G.; Knight, M.; Hoe, L. S.; Headrick, J. P.; Peart, J. N. Essential role of EGFR in cardioprotection and signaling responses to A1 adenosine receptors and ischemic preconditioning. *Am. J. Physiol. Heart. Circ. Physiol.* 2011, 300, H2161-8.

[94] Eckle, T.; Hartmann, K.; Bonney, S.; Reithel, S.; Mittelbronn, M.; Walker, L. A.; Lowes, B. D.; Han, J.; Borchers, C. H.; Buttrick, P. M.; Kominsky, D. J.; Colgan, S. P.; Eltzschig, H. K. Adora2b-elicited Per2 stabilization promotes a HIF-dependent metabolic switch crucial for myocardial adaptation to ischemia. *Nat Med.* 2012, 18, 774-82.

[95] Philipp, S.; Yang, X. M.; Cui, L.; Davis, A. M.; Downey, J. M.; Cohen, M. V. Postconditioning protects rabbit hearts through a protein kinase C-adenosine A2b receptor cascade. *Cardiovasc. Res.* 2006, 70, 308-14.

[96] Kuno, A.; Critz, S. D.; Cui, L.; Solodushko, V.; Yang, X. M.; Krahn, T.; Albrecht, B.; Philipp, S.; Cohen, M. V.; Downey, J. M. Protein kinase C protects preconditioned rabbit hearts by increasing sensitivity of adenosine A2b-dependent signaling during early reperfusion. *J. Mol. Cell. Cardiol.* 2007, 43, 262-71.

[97] Terashima, Y.; Sato, T.; Yano, T.; Maas, O.; Itoh, T.; Miki, T.; Tanno, M.; Kuno, A.; Shimamoto, K.; Miura, T. Roles of phospho-GSK-3beta in myocardial protection afforded by activation of the mitochondrial K ATP channel. *J. Mol. Cell. Cardiol.* 2010, 49, 762-70.

[98] Grube, K.; Rüdebusch, J.; Xu, Z.; Böckenholt, T.; Methner, C.; Müller, T.; Cuello, F.; Zimmermann, K.; Yang, X.; Felix, S. B.; Cohen, M. V.; Downey, J. M.; Krieg, T. Evidence for an intracellular localization of the adenosine A2B receptor in rat cardiomyocytes. *Basic Res. Cardiol.* 2011, 106, 385-396.

[99] Ikonomidis, J. S.; Shirai, T.; Weisel, R. D.; Derylo, B.; Rao, V.; Whiteside, C. I.; Mickle, D. A.; Li, R. K. Preconditioning cultured human pediatric myocytes requires adenosine and protein kinase C. *Am. J. Physiol.* 1997, 272, H1220-30.

[100] Cohen, G.; Shirai, T.; Weisel, R. D.; Rao, V.; Merante, F.; Tumiati, L. C.; Mohabeer, M. K.; Borger, M. A.; Li, R. K.; Mickle, D. A. Optimal myocardial preconditioning in a human model of ischemia and reperfusion. *Circulation.* 1998, 98, II184-94; discussion II194-6.

[101] Riksen, N. P.; Zhou, Z.; Oyen, W. J.; Jaspers, R.; Ramakers, B. P.; Brouwer, R. M.; Boerman, O. C.; Steinmetz, N.; Smits, P.; Rongen, G. A. Caffeine prevents protection in two human models of ischemic preconditioning. *J. Am. Coll. Cardiol.* 2006, 48, 700-7.

[102] Tomai, F.; Crea, F.; Gaspardone, A.; Versaci, F.; De Paulis, R.; Polisca, P.; Chiariello, L.; Gioffre, P. A. Effects of A1 adenosine receptor blockade by bamiphylline on ischaemic preconditioning during coronary angioplasty. *Eur. Heart J.* 1996, 17, 846-53.

[103] Carr, C. S.; Hill, R. J.; Masamune, H.; Kennedy, S. P.; Knight, D. R.; Tracey, W. R.; Yellon, D. M. Evidence for a role for both the adenosine A1 and A3 receptors in protection of isolated human atrial muscle against simulated ischaemia. *Cardiovasc. Res.* 1997, 36, 52-9.

[104] Tang, Z.; Diamond, M. A.; Chen, J. M.; Holly, T. A.; Bonow, R. O.; Dasgupta, A.; Hyslop, T.; Purzycki, A.; Wagner, J.; McNamara, D. M.; Kukulski, T.; Wos, S.; Velazquez, E. J.; Ardlie, K.; Feldman, A. M. Polymorphisms in adenosine receptor genes are associated with infarct size in patients with ischemic cardiomyopathy. *Clin. Pharmacol. Ther.* 2007, 82, 435-40.

[105] Fokkema, M. L.; Vlaar, P. J.; Vogelzang, M.; Gu, Y. L.; Kampinga, M. A.; de Smet, B. J.; Jessurun, G. A.; Anthonio, R. L.; van den Heuvel, A. F.; Tan, E. S.; Zijlstra, F. Effect of high-dose intracoronary adenosine administration during primary percutaneous coronary intervention in acute myocardial infarction: a randomized controlled trial. *Circ. Cardiovasc. Interv.* 2009, 2, 323-9.

[106] Mahaffey, K. W.; Puma, J. A.; Barbagelata, N. A.; DiCarli, M. F.; Leesar, M. A.; Browne, K. F.; Eisenberg, P. R.; Bolli, R.; Casas, A. C.; Molina-Viamonte, V.; Orlandi, C.; Blevins, R.; Gibbons, R. J.; Califf, R. M.; Granger, C. B. Adenosine as an adjunct to

thrombolytic therapy for acute myocardial infarction: results of a multicenter, randomized, placebo-controlled trial: the Acute Myocardial Infarction STudy of ADenosine (AMISTAD) trial. *J. Am. Coll. Cardiol.* 1999, 34, 1711-20.

[107] Ross, A. M.; Gibbons, R. J.; Stone, G. W.; Kloner, R. A.; Alexander, R. W. A randomized, double-blinded, placebo-controlled multicenter trial of adenosine as an adjunct to reperfusion in the treatment of acute myocardial infarction (AMISTAD-II). *J. Am. Coll. Cardiol.* 2005, 45, 1775-80.

[108] Jin, Z.; Duan, W.; Chen, M.; Yu, S.; Zhang, H.; Feng, G.; Xiong, L.; Yi, D. The myocardial protective effects of adenosine pretreatment in children undergoing cardiac surgery: a randomized controlled clinical trial. *Eur. J. Cardiothorac. Surg.* 2011, 39, e90-6.

[109] Wei, M.; Kuukasjarvi, P.; Laurikka, J.; Honkonen, E. L.; Kaukinen, S.; Laine, S.; Tarkka, M. Cardioprotective effect of adenosine pretreatment in coronary artery bypass grafting. *Chest.* 2001, 120, 860-5.

[110] Liu, R.; Xing, J.; Miao, N.; Li, W.; Liu, W.; Lai, Y. Q.; Luo, Y.; Ji, B. The myocardial protective effect of adenosine as an adjunct to intermittent blood cardioplegia during open heart surgery. *Eur. J. Cardiothorac. Surg.* 2009, 36, 1018-23.

[111] Gerczuk, P. Z.; Kloner, R. A. An update on cardioprotection: a review of the latest adjunctive therapies to limit myocardial infarction size in clinical trials. *J. Am. Coll. Cardiol.* 2012, 59, 969-78.

[112] Dominguez-Rodriguez, A.; Abreu-Gonzalez, P.; Reiter, R. J. Cardioprotection and pharmacological therapies in acute myocardial infarction: Challenges in the current era. *World J. Cardiol.* 2014, 6, 100-6.

[113] Kloner, R. A.; Forman, M. B.; Gibbons, R. J.; Ross, A. M.; Alexander, R. W.; Stone, G. W. Impact of time to therapy and reperfusion modality on the efficacy of adenosine in acute myocardial infarction: the AMISTAD-2 trial. *Eur. Heart J.* 2006, 27, 2400-5.

[114] Kopecky, S. L.; Aviles, R. J.; Bell, M. R.; Lobl, J. K.; Tipping, D.; Frommell, G.; Ramsey, K.; Holland, A. E.; Midei, M.; Jain, A.; Kellett, M.; Gibbons, R. J. A randomized, double-blinded, placebo-controlled, dose-ranging study measuring the effect of an adenosine agonist on infarct size reduction in patients undergoing primary percutaneous transluminal coronary angioplasty: the ADMIRE (AmP579 Delivery for Myocardial Infarction REduction) study. *Am. Heart J.* 2003, 146, 146-52.

[115] Liu, Y.; Yang, X.; Yang, X. M.; Walker, S.; Forster, K.; Cohen, M. V.; Krieg, T.; Downey, J. M. AMP579 is revealed to be a potent A2b-adenosine receptor agonist in human 293 cells and rabbit hearts. *Basic Res. Cardiol.* 2010, 105, 129-37.

[116] Smits, G. J.; McVey, M.; Cox, B. F.; Perrone, M. H.; Clark, K. L. Cardioprotective effects of the novel adenosine A1/A2 receptor agonist AMP 579 in a porcine model of myocardial infarction. *J. Pharmacol. Exp. Ther.* 1998, 286, 611-8.

[117] Niccoli, G.; Rigattieri, S.; De Vita, M. R.; Valgimigli, M.; Corvo, P.; Fabbiocchi, F.; Romagnoli, E.; De Caterina, A. R.; La Torre, G.; Lo Schiavo, P.; Tarantino, F.; Ferrari, R.; Tomai, F.; Olivares, P.; Cosentino, N.; D'Amario, D.; Leone, A. M.; Porto, I.; Burzotta, F.; Trani, C.; Crea, F. Open-label, randomized, placebo-controlled evaluation of intracoronary adenosine or nitroprusside after thrombus aspiration during primary percutaneous coronary intervention for the prevention of microvascular obstruction in acute myocardial infarction: the REOPEN-AMI study (Intracoronary Nitroprusside

Versus Adenosine in Acute Myocardial Infarction). *JACC. Cardiovasc. Interv.* 2013, 6, 580-9.
[118] Desmet, W.; Bogaert, J.; Dubois, C.; Sinnaeve, P.; Adriaenssens, T.; Pappas, C.; Ganame, J.; Dymarkowski, S.; Janssens, S.; Belmans, A.; Van de Werf, F. High-dose intracoronary adenosine for myocardial salvage in patients with acute ST-segment elevation myocardial infarction. *Eur. Heart J.* 2011, 32, 867-77.
[119] Burchfield, J. S.; Xie, M.; Hill, J. A. Pathological ventricular remodeling: mechanisms: part 1 of 2. *Circulation.* 2013, 128, 388-400.
[120] Braunwald, E. Heart failure. *JACC. Heart Fail.* 2013, 1, 1-20.
[121] Swynghedauw, B. Molecular mechanisms of myocardial remodeling. *Physiol. Rev.* 1999, 79, 215-62.
[122] Shah, A. M.; Mann, D. L. In search of new therapeutic targets and strategies for heart failure: recent advances in basic science. *Lancet.* 2011, 378, 704-12.
[123] Frey, N.; Katus, H. A.; Olson, E. N.; Hill, J. A. Hypertrophy of the heart: a new therapeutic target? *Circulation.* 2004, 109, 1580-9.
[124] Anversa, P.; Leri, A.; Kajstura, J.; Nadal-Ginard, B. Myocyte growth and cardiac repair. *J. Mol. Cell. Cardiol.* 2002, 34, 91-105.
[125] Porter, K. E.; Turner, N. A. Cardiac fibroblasts: at the heart of myocardial remodeling. *Pharmacol. Ther.* 2009, 123, 255-78.
[126] Creemers, E. E.; Pinto, Y. M. Molecular mechanisms that control interstitial fibrosis in the pressure-overloaded heart. *Cardiovasc. Res.* 2011, 89, 265-72.
[127] Gan, X. T.; Rajapurohitam, V.; Haist, J. V.; Chidiac, P.; Cook, M. A.; Karmazyn, M. Inhibition of phenylephrine-induced cardiomyocyte hypertrophy by activation of multiple adenosine receptor subtypes. *J. Pharmacol. Exp. Ther.* 2005, 312, 27-34.
[128] Pang, T.; Gan, X. T.; Freeman, D. J.; Cook, M. A.; Karmazyn, M. Compensatory upregulation of the adenosine system following phenylephrine-induced hypertrophy in cultured rat ventricular myocytes. *Am. J. Physiol. Heart Circ. Physiol.* 2010, 298, H545-53.
[129] Liao, Y.; Lin, L.; Lu, D.; Fu, Y.; Bin, J.; Xu, D.; Kitakaze, M. Activation of adenosine A1 receptor attenuates tumor necrosis factor-alpha induced hypertrophy of cardiomyocytes. *Biomed. Pharmacother.* 2011, 65, 491-5.
[130] Dubey, R. K.; Gillespie, D. G.; Mi, Z.; Jackson, E. K. Exogenous and endogenous adenosine inhibits fetal calf serum-induced growth of rat cardiac fibroblasts: role of A2B receptors. *Circulation.* 1997, 96, 2656-66.
[131] Dubey, R. K.; Gillespie, D. G.; Zacharia, L. C.; Mi, Z.; Jackson, E. K. A(2b) receptors mediate the antimitogenic effects of adenosine in cardiac fibroblasts. *Hypertension.* 2001, 37, 716-21.
[132] Dubey, R. K.; Gillespie, D. G.; Jackson, E. K. Adenosine inhibits collagen and protein synthesis in cardiac fibroblasts: role of A2B receptors. *Hypertension.* 1998, 31, 943-8.
[133] Villarreal, F.; Epperson, S. A.; Ramirez-Sanchez, I.; Yamazaki, K. G.; Brunton, L. L. Regulation of cardiac fibroblast collagen synthesis by adenosine: roles for Epac and PI3K. *Am. J. Physiol. Cell Physiol.* 2009, 296, C1178-84.
[134] Ernens, I.; Rouy, D.; Velot, E.; Devaux, Y.; Wagner, D. R. Adenosine inhibits matrix metalloproteinase-9 secretion by neutrophils: implication of A2a receptor and cAMP/PKA/Ca2+ pathway. *Circ. Res.* 2006, 99, 590-7.

[135] Liao, Y.; Takashima, S.; Asano, Y.; Asakura, M.; Ogai, A.; Shintani, Y.; Minamino, T.; Asanuma, H.; Sanada, S.; Kim, J.; Ogita, H.; Tomoike, H.; Hori, M.; Kitakaze, M. Activation of adenosine A1 receptor attenuates cardiac hypertrophy and prevents heart failure in murine left ventricular pressure-overload model. *Circ. Res.* 2003, 93, 759-66.

[136] Sabbah, H. N.; Gupta, R. C.; Kohli, S.; Wang, M.; Rastogi, S.; Zhang, K.; Zimmermann, K.; Diedrichs, N.; Albrecht-Kupper, B. E. Chronic therapy with a partial adenosine A1-receptor agonist improves left ventricular function and remodeling in dogs with advanced heart failure. *Circ. Heart Fail.* 2013, 6, 563-71.

[137] Funakoshi, H.; Chan, T. O.; Good, J. C.; Libonati, J. R.; Piuhola, J.; Chen, X.; MacDonnell, S. M.; Lee, L. L.; Herrmann, D. E.; Zhang, J.; Martini, J.; Palmer, T. M.; Sanbe, A.; Robbins, J.; Houser, S. R.; Koch, W. J.; Feldman, A. M. Regulated overexpression of the A1-adenosine receptor in mice results in adverse but reversible changes in cardiac morphology and function. *Circulation.* 2006, 114, 2240-50.

[138] Lu, Z.; Fassett, J.; Xu, X.; Hu, X.; Zhu, G.; French, J.; Zhang, P.; Schnermann, J.; Bache, R. J.; Chen, Y. Adenosine A3 receptor deficiency exerts unanticipated protective effects on the pressure-overloaded left ventricle. *Circulation.* 2008, 118, 1713-21.

[139] Black, R. G., Jr.; Guo, Y.; Ge, Z. D.; Murphree, S. S.; Prabhu, S. D.; Jones, W. K.; Bolli, R.; Auchampach, J. A. Gene dosage-dependent effects of cardiac-specific overexpression of the A3 adenosine receptor. *Circ. Res.* 2002, 91, 165-72.

[140] Hamad, E. A.; Zhu, W.; Chan, T. O.; Myers, V.; Gao, E.; Li, X.; Zhang, J.; Song, J.; Zhang, X. Q.; Cheung, J. Y.; Koch, W.; Feldman, A. M. Cardioprotection of controlled and cardiac-specific over-expression of A(2A)-adenosine receptor in the pressure overload. *PloS One.* 2012, 7, e39919.

[141] Chan, T. O.; Funakoshi, H.; Song, J.; Zhang, X. Q.; Wang, J.; Chung, P. H.; DeGeorge, B. R., Jr.; Li, X.; Zhang, J.; Herrmann, D. E.; Diamond, M.; Hamad, E.; Houser, S. R.; Koch, W. J.; Cheung, J. Y.; Feldman, A. M. Cardiac-restricted overexpression of the A(2A)-adenosine receptor in FVB mice transiently increases contractile performance and rescues the heart failure phenotype in mice overexpressing the A(1)-adenosine receptor. *Clin. Transl. Sci.* 2008, 1, 126-33.

[142] Hamad, E. A.; Li, X.; Song, J.; Zhang, X. Q.; Myers, V.; Funakoshi, H.; Zhang, J.; Wang, J.; Li, J.; Swope, D.; Madonick, A.; Farber, J.; Radice, G. L.; Cheung, J. Y.; Chan, T. O.; Feldman, A. M. Effects of cardiac-restricted overexpression of the A(2A) adenosine receptor on adriamycin-induced cardiotoxicity. *Am. J. Physiol. Heart. Circ. Physiol.* 2010, 298, H1738-47.

[143] Wakeno, M.; Minamino, T.; Seguchi, O.; Okazaki, H.; Tsukamoto, O.; Okada, K.; Hirata, A.; Fujita, M.; Asanuma, H.; Kim, J.; Komamura, K.; Takashima, S.; Mochizuki, N.; Kitakaze, M. Long-term stimulation of adenosine A2b receptors begun after myocardial infarction prevents cardiac remodeling in rats. *Circulation.* 2006, 114, 1923-32.

[144] Sassi, Y.; Ahles, A.; Truong, D. J.; Baqi, Y.; Lee, S. Y.; Husse, B.; Hulot, J. S.; Foinquinos, A.; Thum, T.; Muller, C. E.; Dendorfer, A.; Laggerbauer, B.; Engelhardt, S. Cardiac myocyte-secreted cAMP exerts paracrine action via adenosine receptor activation. *J. Clin. Invest.* 2014, 124, 5385-97.

[145] Funaya, H.; Kitakaze, M.; Node, K.; Minamino, T.; Komamura, K.; Hori, M. Plasma adenosine levels increase in patients with chronic heart failure. *Circulation.* 1997, 95, 1363-5.
[146] Franceschi, F.; Deharo, J. C.; Giorgi, R.; By, Y.; Monserrat, C.; Condo, J.; Ibrahim, Z.; Saadjian, A.; Guieu, R. Peripheral plasma adenosine release in patients with chronic heart failure. *Heart.* 2009, 95, 651-5.
[147] Massie, B. M.; O'Connor, C. M.; Metra, M.; Ponikowski, P.; Teerlink, J. R.; Cotter, G.; Weatherley, B. D.; Cleland, J. G.; Givertz, M. M.; Voors, A.; DeLucca, P.; Mansoor, G. A.; Salerno, C. M.; Bloomfield, D. M.; Dittrich, H. C. Rolofylline, an adenosine A1-receptor antagonist, in acute heart failure. *N. Engl. J Med.* 2010, 363, 1419-28.
[148] Givertz, M. M.; Massie, B. M.; Fields, T. K.; Pearson, L. L.; Dittrich, H. C. The effects of KW-3902, an adenosine A1-receptor antagonist,on diuresis and renal function in patients with acute decompensated heart failure and renal impairment or diuretic resistance. *J. Am. Coll. Cardiol.* 2007, 50, 1551-60.
[149] Gupta, R.; Tongers, J.; Losordo, D. W. Human studies of angiogenic gene therapy. *Circ. Res.* 2009, 105, 724-36.
[150] Silvestre, J. S.; Smadja, D. M.; Levy, B. I. Postischemic revascularization: from cellular and molecular mechanisms to clinical applications. *Physiol. Rev.* 2013, 93, 1743-802.
[151] Simons, M. Angiogenesis: where do we stand now? *Circulation.* 2005, 111, 1556-66.
[152] Raval, Z.; Losordo, D. W. Cell therapy of peripheral arterial disease: from experimental findings to clinical trials. *Circ. Res.* 2013, 112, 1288-302.
[153] Gasparini, G.; Longo, R.; Toi, M.; Ferrara, N. Angiogenic inhibitors: a new therapeutic strategy in oncology. *Nat. Clin. Pract. Oncol.* 2005, 2, 562-77.
[154] Vasudev, N. S.; Reynolds, A. R. Anti-angiogenic therapy for cancer: current progress, unresolved questions and future directions. *Angiogenesis.* 2014, 17, 471-94.
[155] Haller, J. A. Current anti-vascular endothelial growth factor dosing regimens: benefits and burden. *Ophthalmol.* 2013, 120, S3-7.
[156] Miller, J. W.; Le Couter, J.; Strauss, E. C.; Ferrara, N. Vascular endothelial growth factor a in intraocular vascular disease. *Ophthalmol.* 2013, 120, 106-14.
[157] Maracle, C. X.; Tas, S. W. Inhibitors of angiogenesis: Ready for prime time? *Best practice & research. Clin. Rheumatol.* 2014, 28, 637-649.
[158] Conway, E. M.; Collen, D.; Carmeliet, P. Molecular mechanisms of blood vessel growth. *Cardiovasc. Res.* 2001, 49, 507-21.
[159] Lamalice, L.; Le Boeuf, F.; Huot, J. Endothelial cell migration during angiogenesis. *Circ. Res.* 2007, 100, 782-94.
[160] Rey, S.; Semenza, G. L. Hypoxia-inducible factor-1-dependent mechanisms of vascularization and vascular remodelling. *Cardiovasc. Res.* 2010, 86, 236-42.
[161] Dusseau, J. W.; Hutchins, P. M.; Malbasa, D. S. Stimulation of angiogenesis by adenosine on the chick chorioallantoic membrane. *Circ. Res.* 1986, 59, 163-70.
[162] Adair, T. H.; Montani, J. P.; Strick, D. M.; Guyton, A. C. Vascular development in chick embryos: a possible role for adenosine. *Am. J. Physiol.* 1989, 256, H240-6.
[163] Montesinos, M. C.; Gadangi, P.; Longaker, M.; Sung, J.; Levine, J.; Nilsen, D.; Reibman, J.; Li, M.; Jiang, C. K.; Hirschhorn, R.; Recht, P. A.; Ostad, E.; Levin, R. I.; Cronstein, B. N. Wound healing is accelerated by agonists of adenosine A2 (G alpha s-linked) receptors. *J. Exp. Med.* 1997, 186, 1615-20.

[164] Afzal, A.; Shaw, L. C.; Caballero, S.; Spoerri, P. E.; Lewin, A. S.; Zeng, D.; Belardinelli, L.; Grant, M. B. Reduction in preretinal neovascularization by ribozymes that cleave the A2B adenosine receptor mRNA. *Circ. Res.* 2003, 93, 500-506.

[165] Mino, R. P.; Spoerri, P. E.; Caballero, S.; Player, D.; Belardinelli, L.; Biaggioni, I.; Grant, M. B. Adenosine receptor antagonists and retinal neovascularization in vivo. *Invest. Ophthalmol. Vis. Sci.* 2001, 42, 3320-3324.

[166] Montesinos, M. C.; Shaw, J. P.; Yee, H.; Shamamian, P.; Cronstein, B. N. Adenosine A(2A) receptor activation promotes wound neovascularization by stimulating angiogenesis and vasculogenesis. *Am. J. Pathol.* 2004, 164, 1887-92.

[167] Rizvi, Y. Q.; Mehta, C. S.; Oyekan, A. Interactions of PPAR-alpha and adenosine receptors in hypoxia-induced angiogenesis. *Vasc. Pharmacol.* 2013, 59, 144-51.

[168] Adair, T. H. Growth regulation of the vascular system: an emerging role for adenosine. *Am. J. Physiol. Regul. Integr. Comp. Physiol.* 2005, 289, R283-296.

[169] Feoktistov, I.; Biaggioni, I.; Cronstein, B. N. Adenosine receptors in wound healing, fibrosis and angiogenesis. *Handb. Exp. Pharmacol.* 2009, 193, 383-397.

[170] Meininger, C. J.; Schelling, M. E.; Granger, H. J. Adenosine and hypoxia stimulate proliferation and migration of endothelial cells. *Am. J. Physiol.* 1988, 255, H554-62.

[171] Meininger, C. J.; Granger, H. J. Mechanisms leading to adenosine-stimulated proliferation of microvascular endothelial cells. *Am. J. Physiol.* 1990, 258, H198-206.

[172] Ethier, M. F.; Chander, V.; Dobson, J. G., Jr. Adenosine stimulates proliferation of human endothelial cells in culture. *Am. J. Physiol.* 1993, 265, H131-8.

[173] Sexl, V.; Mancusi, G.; Baumgartner-Parzer, S.; Schutz, W.; Freissmuth, M. Stimulation of human umbilical vein endothelial cell proliferation by A2-adenosine and beta 2-adrenoceptors. *Br. J. Pharmacol.* 1995, 114, 1577-1586.

[174] Ethier, M. F.; Dobson, J. G., Jr. Adenosine stimulation of DNA synthesis in human endothelial cells. *Am. J. Physiol.* 1997, 272, H1470-9.

[175] Grant, M. B.; Tarnuzzer, R. W.; Caballero, S.; Ozeck, M. J.; Davis, M. I.; Spoerri, P. E.; Feoktistov, I.; Biaggioni, I.; Shryock, J. C.; Belardinelli, L. Adenosine receptor activation induces vascular endothelial growth factor in human retinal endothelial cells. *Circ. Res.* 1999, 85, 699-706.

[176] Grant, M. B.; Davis, M. I.; Caballero, S.; Feoktistov, I.; Biaggioni, I.; Belardinelli, L., Proliferation, migration, and ERK activation in human retinal endothelial cells through A(2B) adenosine receptor stimulation. *Invest. Ophthalmol. Vis. Sci.* 2001, 42, 2068-2073.

[177] Dubey, R. K.; Gillespie, D. G.; Jackson, E. K. A(2B) adenosine receptors stimulate growth of porcine and rat arterial endothelial cells. *Hypertension.* 2002, 39, 530-535.

[178] Acurio, J.; Troncoso, F.; Bertoglia, P.; Salomon, C.; Aguayo, C.; Sobrevia, L.; Escudero, C. Potential role of A2B adenosine receptors on proliferation/migration of fetal endothelium derived from preeclamptic pregnancies. *Biomed. Res. Int.* 2014, 2014, 274507.

[179] Lutty, G. A.; Mathews, M. K.; Merges, C.; McLeod, D. S. Adenosine stimulates canine retinal microvascular endothelial cell migration and tube formation. *Curr. Eye Res.* 1998, 17, 594-607.

[180] Desai, A.; Victor-Vega, C.; Gadangi, S.; Montesinos, M. C.; Chu, C. C.; Cronstein, B. N. Adenosine A2A receptor stimulation increases angiogenesis by down-regulating

production of the antiangiogenic matrix protein thrombospondin 1. *Mol. Pharmacol.* 2005, 67, 1406-1413.

[181] Kim, G. D.; Oh, J.; Jeong, L. S.; Lee, S. K. Thio-Cl-IB-MECA, a novel A(3) adenosine receptor agonist, suppresses angiogenesis by regulating PI3K/AKT/mTOR and ERK signaling in endothelial cells. *Biochem. Biophys. Res. Commun.* 2013, 437, 79-86.

[182] Veikkola, T.; Karkkainen, M.; Claesson-Welsh, L.; Alitalo, K. Regulation of angiogenesis via vascular endothelial growth factor receptors. *Cancer Res.* 2000, 60, 203-12.

[183] Kliche, S.; Waltenberger, J. VEGF receptor signaling and endothelial function. *IUBMB Life.* 2001, 52, 61-6.

[184] Ferrara, N. Role of vascular endothelial growth factor in physiologic and pathologic angiogenesis: therapeutic implications. *Semin.Ooncol.* 2002, 29, 10-4.

[185] Sexl, V.; Mancusi, G.; Holler, C.; Gloria-Maercker, E.; Schutz, W.; Freissmuth, M. Stimulation of the mitogen-activated protein kinase via the A2A-adenosine receptor in primary human endothelial cells. *J. Biol. Chem.* 1997, 272, 5792-5799.

[186] Fang, Y.; Olah, M. E. Cyclic AMP-dependent, protein kinase A-independent activation of extracellular signal-regulated kinase 1/2 following adenosine receptor stimulation in human umbilical vein endothelial cells: role of exchange protein activated by cAMP 1 (Epac1). *J. Pharmacol. Exp. Ther.* 2007, 322, 1189-200.

[187] Yashima, R.; Abe, M.; Tanaka, K.; Ueno, H.; Shitara, K.; Takenoshita, S.; Sato, Y. Heterogeneity of the signal transduction pathways for VEGF-induced MAPKs activation in human vascular endothelial cells. *J. Cell. Physiol.* 2001, 188, 201-10.

[188] Namkoong, S.; Kim, C.-K.; Cho, Y.-L.; Kim, J.-H.; Lee, H.; Ha, K.-S.; Choe, J.; Kim, P.-H.; Won, M.-H.; Kwon, Y.-G.; Shim, E. B.; Kim, Y.-M. Forskolin increases angiogenesis through the coordinated cross-talk of PKA-dependent VEGF expression and Epac-mediated PI3K/Akt/eNOS signaling. *Cell. Signal.* 2009, 21, 906-915.

[189] Feoktistov, I.; Ryzhov, S.; Zhong, H.; Goldstein, A. E.; Matafonov, A.; Zeng, D.; Biaggioni, I. Hypoxia modulates adenosine receptors in human endothelial and smooth muscle cells toward an A2B angiogenic phenotype. *Hypertension.* 2004, 44, 649-654.

[190] Ryzhov, S.; Novitskiy, S. V.; Zaynagetdinov, R.; Goldstein, A. E.; Carbone, D. P.; Biaggioni, I.; Dikov, M. M.; Feoktistov, I. Host A(2B) adenosine receptors promote carcinoma growth. *Neoplasia.* 2008, 10, 987-995.

[191] Feoktistov, I.; Ryzhov, S.; Goldstein, A. E.; Biaggioni, I. Mast cell-mediated stimulation of angiogenesis: cooperative interaction between A2B and A3 adenosine receptors. *Circ. Res.* 2003, 92, 485-92.

[192] Valladares, D.; Quezada, C.; Montecinos, P.; Concha, II; Yanez, A. J.; Sobrevia, L.; San Martin, R. Adenosine A(2B) receptor mediates an increase on VEGF-A production in rat kidney glomeruli. *Biochem. Biophys. Res. Commun.* 2008, 366, 180-5.

[193] Ryzhov, S.; Goldstein, A. E.; Novitskiy, S. V.; Blackburn, M. R.; Biaggioni, I.; Feoktistov, I. Role of A2B adenosine receptors in regulation of paracrine functions of stem cell antigen 1-positive cardiac stromal cells. *J. Pharmacol. Exp. Ther.* 2012, 341, 764-74.

[194] Hoier, B.; Olsen, K.; Nyberg, M.; Bangsbo, J.; Hellsten, Y. Contraction-induced secretion of VEGF from skeletal muscle cells is mediated by adenosine. *Am. J. Physiol. Heart Circ. Physiol.* 2010, 299, H857-62.

[195] Ryzhov, S.; Biktasova, A.; Goldstein, A. E.; Zhang, Q.; Biaggioni, I.; Dikov, M. M.; Feoktistov, I. Role of JunB in adenosine A2B receptor-mediated vascular endothelial growth factor production. *Mol. Pharmacol.* 2014, 85, 62-73.

[196] Clark, A. N.; Youkey, R.; Liu, X.; Jia, L.; Blatt, R.; Day, Y. J.; Sullivan, G. W.; Linden, J.; Tucker, A. L. A1 adenosine receptor activation promotes angiogenesis and release of VEGF from monocytes. *Circ. Res.* 2007, 101, 1130-8.

[197] Leibovich, S. J.; Chen, J. F.; Pinhal-Enfield, G.; Belem, P. C.; Elson, G.; Rosania, A.; Ramanathan, M.; Montesinos, C.; Jacobson, M.; Schwarzschild, M. A.; Fink, J. S.; Cronstein, B. Synergistic up-regulation of vascular endothelial growth factor expression in murine macrophages by adenosine A(2A) receptor agonists and endotoxin. *Am. J. Pathol.* 2002, 160, 2231-44.

[198] Ernens, I.; Leonard, F.; Vausort, M.; Rolland-Turner, M.; Devaux, Y.; Wagner, D. R., Adenosine up-regulates vascular endothelial growth factor in human macrophages. *Biochem. Biophys. Res. Commun.* 2010, 392, 351-6.

[199] Pinhal-Enfield, G.; Ramanathan, M.; Hasko, G.; Vogel, S. N.; Salzman, A. L.; Boons, G. J.; Leibovich, S. J. An angiogenic switch in macrophages involving synergy between Toll-like receptors 2, 4, 7, and 9 and adenosine A(2A) receptors. *Am. J. Pathol.* 2003, 163, 711-21.

[200] Zhang, L.; Xu, J.; Sun, N.; Cai, H.; Ren, M.; Zhang, J.; Yu, C.; Wang, Z.; Gao, L.; Zhao, J. The presence of adenosine A2a receptor in thyrocytes and its involvement in Graves' IgG-induced VEGF expression. *Endocrinol.* 2013, 154, 4927-38.

[201] Merighi, S.; Benini, A.; Mirandola, P.; Gessi, S.; Varani, K.; Simioni, C.; Leung, E.; Maclennan, S.; Baraldi, P. G.; Borea, P. A. Caffeine inhibits adenosine-induced accumulation of hypoxia-inducible factor-1alpha, vascular endothelial growth factor, and interleukin-8 expression in hypoxic human colon cancer cells. *Mol. Pharmacol.* 2007, 72, 395-406.

[202] Merighi, S.; Benini, A.; Mirandola, P.; Gessi, S.; Varani, K.; Leung, E.; Maclennan, S.; Borea, P. A. Adenosine modulates vascular endothelial growth factor expression via hypoxia-inducible factor-1 in human glioblastoma cells. *Biochem. Pharmacol.* 2006, 72, 19-31.

[203] Gessi, S.; Fogli, E.; Sacchetto, V.; Merighi, S.; Varani, K.; Preti, D.; Leung, E.; Maclennan, S.; Borea, P. A. Adenosine modulates HIF-1{alpha}, VEGF, IL-8, and foam cell formation in a human model of hypoxic foam cells. *Arterioscler. Thromb. Vasc. Biol.* 2010, 30, 90-7.

[204] Fischer, S.; Knoll, R.; Renz, D.; Karliczek, G. F.; Schaper, W. Role of adenosine in the hypoxic induction of vascular endothelial growth factor in porcine brain derived microvascular endothelial cells. *Endothelium.* 1997, 5, 155-65.

[205] Takagi, H.; King, G. L.; Robinson, G. S.; Ferrara, N.; Aiello, L. P. Adenosine mediates hypoxic induction of vascular endothelial growth factor in retinal pericytes and endothelial cells. *Invest. Ophthalmol. Vis. Sci.* 1996, 37, 2165-76.

[206] Hashimoto, E.; Kage, K.; Ogita, T.; Nakaoka, T.; Matsuoka, R.; Kira, Y. Adenosine as an endogenous mediator of hypoxia for induction of vascular endothelial growth factor mRNA in U-937 cells. *Biochem. Biophys. Res. Commun.* 1994, 204, 318-24.

[207] Burns, P. A.; Wilson, D. J. Angiogenesis mediated by metabolites is dependent on vascular endothelial growth factor (VEGF). *Angiogenesis.* 2003, 6, 73-7.

[208] Ryzhov, S.; McCaleb, J. L.; Goldstein, A. E.; Biaggioni, I.; Feoktistov, I. Role of adenosine receptors in the regulation of angiogenic factors and neovascularization in hypoxia. *J. Pharmacol. Exp. Ther.* 2007, 320, 565-572.

[209] Olah, M. E.; Roudabush, F. L. Down-regulation of vascular endothelial growth factor expression after A(2A) adenosine receptor activation in PC12 pheochromocytoma cells. *J. Pharmacol. Exp. Ther.* 2000, 293, 779-87.

[210] Kobayashi, S.; Millhorn, D. E. Stimulation of expression for the adenosine A2A receptor gene by hypoxia in PC12 cells. A potential role in cell protection. *J. Biol. Chem.* 1999, 274, 20358-65.

[211] Wakai, A.; Wang, J. H.; Winter, D. C.; Street, J. T.; O'Sullivan, R. G.; Redmond, H. P. Adenosine inhibits neutrophil vascular endothelial growth factor release and transendothelial migration via A2B receptor activation. *Shock.* 2001, 15, 297-301.

[212] Gardner, A. M.; Olah, M. E. Distinct protein kinase C isoforms mediate regulation of vascular endothelial growth factor expression by A2A adenosine receptor activation and phorbol esters in pheochromocytoma PC12 cells. *J. Biol. Chem.* 2003, 278, 15421-8.

[213] Takagi, H.; King, G. L.; Ferrara, N.; Aiello, L. P. Hypoxia regulates vascular endothelial growth factor receptor KDR/Flk gene expression through adenosine A2 receptors in retinal capillary endothelial cells. *Invest. Ophthalmol. Vis. Sci.* 1996, 37, 1311-21.

[214] Leonard, F.; Devaux, Y.; Vausort, M.; Ernens, I.; Rolland-Turner, M.; Wagner, D. R. Adenosine modifies the balance between membrane and soluble forms of Flt-1. *J. Leukoc. Biol.* 2011, 90, 199-204.

[215] Wu, F. T.; Stefanini, M. O.; Mac Gabhann, F.; Kontos, C. D.; Annex, B. H.; Popel, A. S. A systems biology perspective on sVEGFR1: its biological function, pathogenic role and therapeutic use. *J. Cell. Mol. Med.* 2010, 14, 528-52.

[216] George, E. M.; Cockrell, K.; Adair, T. H.; Granger, J. P. Regulation of sFlt-1 and VEGF secretion by adenosine under hypoxic conditions in rat placental villous explants. *Am. J. Physiol. Regul. Integr. Comp. Physiol.* 2010, 299, R1629-33.

[217] Feoktistov, I.; Goldstein, A. E.; Ryzhov, S.; Zeng, D.; Belardinelli, L.; Voyno-Yasenetskaya, T.; Biaggioni, I. Differential expression of adenosine receptors in human endothelial cells: role of A2B receptors in angiogenic factor regulation. *Circ. Res.* 2002, 90, 531-538.

[218] Feoktistov, I.; Ryzhov, S.; Goldstein, A. E.; Biaggioni, I. Mast cell-mediated stimulation of angiogenesis: cooperative interaction between A2B and A3 adenosine receptors. *Circ. Res.* 2003, 92, 485-492.

[219] Hormozi, M.; Talebi, S.; Zarnani, A. H.; Jeddi-Tehrani, M.; Gohari, L. H.; Soltanghoraei, H.; Jafarabadi, M.; Akhondi, M. M. 5'-(N-ethylcarboxamido) adenosine improves angiogenesis in transplanted human ovarian tissue. *Fertil. Steril.* 2011, 95, 2560-3.e1-5.

[220] Mohsenin, A.; Burdick, M. D.; Molina, J. G.; Keane, M. P.; Blackburn, M. R. Enhanced CXCL1 production and angiogenesis in adenosine-mediated lung disease. *FASEB J.* 2007, 21, 1026-36.

[221] Li, H.; Jin, S. Y.; Son, H. J.; Seo, J. H.; Jeong, G. B. Caffeine-induced endothelial cell death and the inhibition of angiogenesis. *Anat. Cell Biol.* 2013, 46, 57-67.

[222] Sapieha, P.; Joyal, J. S.; Rivera, J. C.; Kermorvant-Duchemin, E.; Sennlaub, F.; Hardy, P.; Lachapelle, P.; Chemtob, S. Retinopathy of prematurity: understanding ischemic retinal vasculopathies at an extreme of life. *J. Clin. Invest.* 2010, 120, 3022-32.

[223] Liu, X. L.; Zhou, R.; Pan, Q. Q.; Jia, X. L.; Gao, W. N.; Wu, J.; Lin, J.; Chen, J. F. Genetic inactivation of the adenosine A2A receptor attenuates pathologic but not developmental angiogenesis in the mouse retina. *Invest. Ophthalmol. Vis. Sci.* 2010, 51, 6625-32.

[224] Taomoto, M.; McLeod, D. S.; Merges, C.; Lutty, G. A. Localization of adenosine A2a receptor in retinal development and oxygen-induced retinopathy. *Invest. Ophthalmol. Vis. Sci.* 2000, 41, 230-43.

[225] Lutty, G. A.; McLeod, D. S., Retinal vascular development and oxygen-induced retinopathy: a role for adenosine. *Prog. Retin. Eye Res.* 2003, 22, 95-111.

[226] Schmidt, B.; Roberts, R. S.; Davis, P.; Doyle, L. W.; Barrington, K. J.; Ohlsson, A.; Solimano, A.; Tin, W. Long-term effects of caffeine therapy for apnea of prematurity. *N Engl. J. Med.* 2007, 357, 1893-902.

[227] Tufro, A.; Veron, D. VEGF and podocytes in diabetic nephropathy. *Semin. Nephrol.* 2012, 32, 385-93.

[228] Cardenas, A.; Toledo, C.; Oyarzun, C.; Sepulveda, A.; Quezada, C.; Guillen-Gomez, E.; Diaz-Encarnacion, M. M.; Pastor-Anglada, M.; San Martin, R. Adenosine A(2B) receptor-mediated VEGF induction promotes diabetic glomerulopathy. *Lab. Invest.* 2013, 93, 135-44.

[229] Tak, E.; Ridyard, D.; Kim, J. H.; Zimmerman, M.; Werner, T.; Wang, X. X.; Shabeka, U.; Seo, S. W.; Christians, U.; Klawitter, J.; Moldovan, R.; Garcia, G.; Levi, M.; Haase, V.; Ravid, K.; Eltzschig, H. K.; Grenz, A. CD73-dependent generation of adenosine and endothelial Adora2b signaling attenuate diabetic nephropathy. *J. Am. Soc. Nephrol.* 2014, 25, 547-63.

[230] Patel, L.; Thaker, A. The effects of adenosine A2B receptor inhibition on VEGF and nitric oxide axis-mediated renal function in diabetic nephropathy. *Ren. Fail.* 2014, 36, 916-24.

[231] Zhang, W.; Zhang, Y.; Wang, W.; Dai, Y.; Ning, C.; Luo, R.; Sun, K.; Glover, L.; Grenz, A.; Sun, H.; Tao, L.; Zhang, W.; Colgan, S. P.; Blackburn, M. R.; Eltzschig, H. K.; Kellems, R. E.; Xia, Y. Elevated ecto-5'-nucleotidase-mediated increased renal adenosine signaling via A2B adenosine receptor contributes to chronic hypertension. *Circ. Res.* 2013, 112, 1466-78.

[232] Dai, Y.; Zhang, W.; Wen, J.; Zhang, Y.; Kellems, R. E.; Xia, Y. A2B adenosine receptor-mediated induction of IL-6 promotes CKD. *J. Am. Soc. Nephrol.* 2011, 22, 890-901.

[233] Victor-Vega, C.; Desai, A.; Montesinos, M. C.; Cronstein, B. N. Adenosine A2A receptor agonists promote more rapid wound healing than recombinant human platelet-derived growth factor (Becaplermin gel). *Inflammation.* 2002, 26, 19-24.

[234] Valls, M. D.; Cronstein, B. N.; Montesinos, M. C. Adenosine receptor agonists for promotion of dermal wound healing. *Biochem. Pharmacol.* 2009, 77, 1117-24.

[235] Montesinos, M. C.; Desai, A.; Chen, J. F.; Yee, H.; Schwarzschild, M. A.; Fink, J. S.; Cronstein, B. N. Adenosine promotes wound healing and mediates angiogenesis in response to tissue injury via occupancy of A(2A) receptors. *Am. J. Pathol.* 2002, 160, 2009-18.

[236] Macedo, L.; Pinhal-Enfield, G.; Alshits, V.; Elson, G.; Cronstein, B. N.; Leibovich, S. J. Wound healing is impaired in MyD88-deficient mice: a role for MyD88 in the regulation of wound healing by adenosine A2A receptors. *Am. J. Pathol.* 2007, 171, 1774-88.

[237] Altavilla, D.; Bitto, A.; Polito, F.; Marini, H.; Minutoli, L.; Di Stefano, V.; Irrera, N.; Cattarini, G.; Squadrito, F. Polydeoxyribonucleotide (PDRN): a safe approach to induce therapeutic angiogenesis in peripheral artery occlusive disease and in diabetic foot ulcers. *Cardiovasc. Hematol. Agents Med. Chem.* 2009, 7, 313-21.

[238] Thellung, S.; Florio, T.; Maragliano, A.; Cattarini, G.; Schettini, G. Polydeoxyribonucleotides enhance the proliferation of human skin fibroblasts: involvement of A2 purinergic receptor subtypes. *Life Sci.* 1999, 64, 1661-74.

[239] Polito, F.; Bitto, A.; Galeano, M.; Irrera, N.; Marini, H.; Calo, M.; Squadrito, F.; Altavilla, D. Polydeoxyribonucleotide restores blood flow in an experimental model of ischemic skin flaps. *J. Vasc. Surg.* 2012, 55, 479-88.

[240] Galeano, M.; Bitto, A.; Altavilla, D.; Minutoli, L.; Polito, F.; Calo, M.; Lo Cascio, P.; Stagno d'Alcontres, F.; Squadrito, F. Polydeoxyribonucleotide stimulates angiogenesis and wound healing in the genetically diabetic mouse. *Wound Repair Regen.* 2008, 16, 208-17.

[241] Squadrito, F.; Bitto, A.; Altavilla, D.; Arcoraci, V.; De Caridi, G.; De Feo, M. E.; Corrao, S.; Pallio, G.; Sterrantino, C.; Minutoli, L.; Saitta, A.; Vaccaro, M.; Cucinotta, D. The effect of PDRN, an adenosine receptor A2A agonist, on the healing of chronic diabetic foot ulcers: results of a clinical trial. *J. Clin. Endocrinol. Metab.* 2014, 99, E746-53.

[242] Bitto, A.; Polito, F.; Altavilla, D.; Minutoli, L.; Migliorato, A.; Squadrito, F. Polydeoxyribonucleotide (PDRN) restores blood flow in an experimental model of peripheral artery occlusive disease. *J. Vasc. Surg.* 2008, 48, 1292-300.

[243] Asahara, T. Cell therapy and gene therapy using endothelial progenitor cells for vascular regeneration. *Handb. Exp. Pharmacol.* 2007, 180, 181-94.

[244] Jujo, K.; Ii, M.; Losordo, D. W. Endothelial progenitor cells in neovascularization of infarcted myocardium. *J. Mol. Cell. Cardiol.* 2008, 45, 530-44.

[245] Ryzhov, S.; Solenkova, N. V.; Goldstein, A. E.; Lamparter, M.; Fleenor, T.; Young, P. P.; Greelish, J. P.; Byrne, J. G.; Vaughan, D. E.; Biaggioni, I.; Hatzopoulos, A. K.; Feoktistov, I. Adenosine receptor-mediated adhesion of endothelial progenitors to cardiac microvascular endothelial cells. *Circ. Res.* 2008, 102, 356-363.

[246] Rolland-Turner, M.; Goretti, E.; Bousquenaud, M.; Leonard, F.; Nicolas, C.; Zhang, L.; Maskali, F.; Marie, P. Y.; Devaux, Y.; Wagner, D. Adenosine stimulates the migration of human endothelial progenitor cells. Role of CXCR4 and microRNA-150. *PloS One.* 2013, 8, e54135.

[247] Eltzschig, H. K.; Ibla, J. C.; Furuta, G. T.; Leonard, M. O.; Jacobson, K. A.; Enjyoji, K.; Robson, S. C.; Colgan, S. P. Coordinated adenine nucleotide phosphohydrolysis and nucleoside signaling in posthypoxic endothelium: role of ectonucleotidases and adenosine A2B receptors. *J. Exp. Med.* 2003, 198, 783-96.

[248] Kong, T.; Westerman, K. A.; Faigle, M.; Eltzschig, H. K.; Colgan, S. P. HIF-dependent induction of adenosine A2B receptor in hypoxia. *FASEB J.* 2006, 20, 2242-50.

[249] Escudero, C.; Bertoglia, P.; Hernadez, M.; Celis, C.; Gonzalez, M.; Aguayo, C.; Acurio, J. Impaired A2A adenosine receptor/nitric oxide/VEGF signaling pathway in fetal

endothelium during late- and early-onset preeclampsia. *Purinergic Signal.* 2013, 9, 215-26.

[250] Escudero, C.; Roberts, J. M.; Myatt, L.; Feoktistov, I. Impaired adenosine-mediated angiogenesis in preeclampsia: potential implications for fetal programming. *Front. Pharmacol.* 2014, 5, 134.

[251] Ahmad, A.; Ahmad, S.; Glover, L.; Miller, S. M.; Shannon, J. M.; Guo, X.; Franklin, W. A.; Bridges, J. P.; Schaack, J. B.; Colgan, S. P.; White, C. W. Adenosine A2A receptor is a unique angiogenic target of HIF-2alpha in pulmonary endothelial cells. *Proc. Natl. Acad. Sci. U. S. A.* 2009, 106, 10684-9.

[252] Ahmad, A.; Schaack, J. B.; White, C. W.; Ahmad, S. Adenosine A2A receptor-dependent proliferation of pulmonary endothelial cells is mediated through calcium mobilization, PI3-kinase and ERK1/2 pathways. *Biochem. Biophys. Res. Commun.* 2013, 434, 566-71.

[253] Jackson, E. K.; Ren, J.; Gillespie, D. G. 2',3'-cAMP, 3'-AMP, and 2'-AMP inhibit human aortic and coronary vascular smooth muscle cell proliferation via A2B receptors. *Am. J. Physiol. Heart Circ. Physiol.* 2011, 301, H391-401.

[254] Mayer, P.; Hinze, A. V.; Harst, A.; von Kugelgen, I. A(2)B receptors mediate the induction of early genes and inhibition of arterial smooth muscle cell proliferation via Epac. *Cardiovasc. Res.* 2011, 90, 148-56.

[255] Kimura, T. E.; Duggirala, A.; Hindmarch, C. C.; Hewer, R. C.; Cui, M. Z.; Newby, A. C.; Bond, M. Inhibition of Egr1 expression underlies the anti-mitogenic effects of cAMP in vascular smooth muscle cells. *J. Mol. Cell. Cardiol.* 2014, 72, 9-19.

[256] Dubey, R. K.; Gillespie, D. G.; Mi, Z.; Jackson, E. K. Adenosine inhibits growth of human aortic smooth muscle cells via A2B receptors. *Hypertension.* 1998, 31, 516-21.

[257] Peyot, M. L.; Gadeau, A. P.; Dandre, F.; Belloc, I.; Dupuch, F.; Desgranges, C. Extracellular adenosine induces apoptosis of human arterial smooth muscle cells via A(2b)-purinoceptor. *Circ. Res.* 2000, 86, 76-85.

[258] Koupenova, M.; Johnston-Cox, H.; Ravid, K. Regulation of atherosclerosis and associated risk factors by adenosine and adenosine receptors. *Curr. Atheroscler. Rep.* 2012, 14, 460-8.

[259] Reiss, A. B.; Cronstein, B. N. Regulation of foam cells by adenosine. *Arterioscler. Thromb. Vasc. Biol.* 2012, 32, 879-86.

[260] Johnston-Cox, H. A.; Koupenova, M.; Ravid, K. A2 adenosine receptors and vascular pathologies. *Arterioscler. Thromb. Vasc. Biol.*2012, 32, 870-8.

[261] Yang, D.; Koupenova, M.; McCrann, D. J.; Kopeikina, K. J.; Kagan, H. M.; Schreiber, B. M.; Ravid, K. The A2b adenosine receptor protects against vascular injury. *Proc. Natl. Acad. Sci. U. S. A.* 2008, 105, 792-6.

[262] Bot, I.; de Vries, H.; Korporaal, S. J.; Foks, A. C.; Bot, M.; van Veldhoven, J.; Ter Borg, M. N.; van Santbrink, P. J.; van Berkel, T. J.; Kuiper, J.; Ijzerman, A. P. Adenosine A(2)B receptor agonism inhibits neointimal lesion development after arterial injury in apolipoprotein E-deficient mice. *Arterioscler. Thromb. Vasc. Biol.* 2012, 32, 2197-205.

[263] Eckle, T.; Faigle, M.; Grenz, A.; Laucher, S.; Thompson, L. F.; Eltzschig, H. K. A2B adenosine receptor dampens hypoxia-induced vascular leak. *Blood.* 2008, 111, 2024-35.

[264] Umapathy, N. S.; Fan, Z.; Zemskov, E. A.; Alieva, I. B.; Black, S. M.; Verin, A. D., Molecular mechanisms involved in adenosine-induced endothelial cell barrier enhancement. *Vasc. Pharmacol.* 2010, 52, 199-206.

[265] Zhang, Y.; Xia, Y., Adenosine signaling in normal and sickle erythrocytes and beyond. *Microbes Infect.* 2012, 14, 863-73.

[266] Field, J. J.; Nathan, D. G.; Linden, J. The role of adenosine signaling in sickle cell therapeutics. *Hematol. Oncol. Clin. North Am.* 2014, 28, 287-99.

[267] Buscariollo, D. L.; Fang, X.; Greenwood, V.; Xue, H.; Rivkees, S. A.; Wendler, C. C. Embryonic caffeine exposure acts via A1 adenosine receptors to alter adult cardiac function and DNA methylation in mice. *PloS One.* 2014, 9, e87547.

[268] Rivkees, S. A.; Wendler, C. C. Regulation of cardiovascular development by adenosine and adenosine-mediated embryo protection. *Arterioscler. Thromb. Vasc. Biol.* 2012, 32, 851-5.

[269] Chen, J. F.; Eltzschig, H. K.; Fredholm, B. B. Adenosine receptors as drug targets--what are the challenges? *Nat. Rev. Drug Discov.* 2013, 12, 265-86.

[270] Kloner, R. A.; Schwartz Longacre, L. State of the science of cardioprotection: Challenges and opportunities--proceedings of the 2010 NHLBI Workshop on Cardioprotection. *J. Cardiovasc. Pharmacol. Ther.* 2011, 16, 223-32.

In: Adenosine Signaling Mechanisms
Editors: V. Ramkumar and R. Paes de Carvalho

ISBN: 978-1-63483-186-4

Chapter 5

ADENOSINE AND CENTRAL NERVOUS SYSTEM DEVELOPMENT

Roberto Paes-de-Carvalho[1*], Mariana R. Pereira[1], Camila C. Portugal[2], Renato Socodato[2] and Nádia A. de Oliveira[1]

[1] Program of Neurosciences and Department of Neurobiology, Fluminense Federal University, Niterói, Brazil

[2] Instituto de Investigação e Inovação em Saúde and Instituto de Biologia Molecular e Celular (IBMC), Universidade do Porto, Porto, Portugal

ABSTRACT

Adenosine regulates several functions during Central Nervous System (CNS) development such as neuronal survival, neurite growth and synapse formation. Here we reviewed different aspects of adenosine metabolism, transport, receptor regulation and cell signaling mechanisms during CNS development. Neuroprotection by adenosine observed in ischemia and different neurodegenerative pathologies is also highlighted. Development of adenosine receptors and signaling mechanisms during retinal development is greatly emphasized, especially those related to neuronal survival and protection against glutamate excitotoxicity operating during avian retina development in vivo and in culture.

INTRODUCTION

Different molecules regulate several aspects of cell survival, differentiation and drive cell communication during development. In the central nervous system (CNS), such molecules include neurotransmitters, neuromodulators and trophic factors. Indeed, neurotransmitters as dopamine and serotonin are known to regulate neurite extension and growth cone motility in

[*] All authors equally contributed to this chapter; Correspondence E-mail: robpaes@vm.uff.br.

developing neurons [1, 2], and trophic factors such as BDNF (brain-derived neurotrophic fator) have tremendous impact on neuronal development and function [3]. In this chapter we will focus on the evidence that adenosine also plays important functions during CNS development.

SYNTHESIS OF ADENOSINE DURING DEVELOPMENT

The main pathway of adenosine synthesis in the CNS is from the enzymatic breakdown of AMP into adenosine by 5'-nucleotidase. Different works have shown the presence of this enzyme since early CNS development. For example, Schoen et al., [4] have documented the immunocytochemical staining for this enzyme in the cat visual cortex since early development in a distribution related to that of ocular dominance columns. Interestingly, ecto 5'-nucleotidase showed a transient appearance during specific stages of postnatal synaptogenesis in the rat cerebellum [5]. These authors found that in adult cerebellum all synapses were devoid of enzymatic activity, and concluded that 5'-nucleotidase is present in excitatory cerebellar synapses during part of their generation period. They also suggested that the transient nature of this phenomenon indicates that 5'-nucleotidase may serve as a marker for a specific state of synaptic maturation and that a role of this enzyme in purinergic neuromodulation and cellular contact formation could be significant.

In a study using a cytochemical method to measure the distribution of 5'-nucleotidase activity with electron microscopy at different postnatal stages of cerebellum development, Fenoglio et al., [6] suggested that, during cerebellar histogenesis, 5'-nucleotidase may be involved in the mechanisms of cell migration and proliferation but that, in adulthood, localization of enzyme activity on neuronal elements indicated a role for adenosine in neuromodulation processes.

The distribution of 5'-nucleotidase was also investigated in the developing mouse retina [7]. During embryonic development, immunoreactivity could be initially associated with proliferating cells, and in the adult stage to Müller glial cells, cellular processes in the outer nuclear layer and, to a minor extent, to the inner nuclear layer. Indeed, the presence of 5'-nucleotidase appears to be predominantly found in glial cells as compared to other ecto-nucleotidases, which can be found in neurons or glial cells as well [8].

In a more recent study, Stanojevic et al., [9] showed the ontogenesis of 5'-nucleotidase in rat brain synaptic plasma membranes. These authors investigated the enzyme activity, kinetic properties concerning AMP hydrolysis and the expression of enzyme protein in synaptic plasma membrane preparations from the rat brain at different ages. They found an increase of enzyme activity during maturation but, interestingly, a decrease of enzyme expression up to the adult stage. Additionally, they found a co-localization of 5'-nucleotidase and the pre-synaptic marker syntaxin, providing the first evidence for the presence of this enzyme in the pre-synaptic terminal. These studies indicate that a conversion of extracellular ATP to AMP and adenosine may be operational since early stages of development as in the mature tissue.

Adenosine Catabolism during Development: Involvement of Adenosine Deaminase and Adenosine Kinase

The regulation of adenosine concentration is also expected to be an important factor contributing to the adenosine actions during CNS development. This regulation is performed by the enzymes adenosine kinase and adenosine deaminase, as well as by nucleoside transporters (see next section).

The ontogenesis of adenosine deaminase (ADA) activity in the rat brain was first studied by Geiger and Nagy [10]. They found a striking variation of enzyme activity in different brain regions and at different developmental stages. Except for hypothalamus, where ADA activity increased nearly twofold in rats between the ages of 1 and 50 days, significant decreases of as much as fivefold were found in whole brain, superior colliculus, cortex, hippocampus, cerebellum, olfactory bulbs and olfactory nucleus. On the other hand, enzyme activity in pons and subcortex remained relatively constant throughout the developmental period. Although they have not found any correspondence to particular processes such as cell proliferation, cell death, synaptogenesis or myelination, or to known developmental profiles of transmitters, their receptors, or their metabolic enzymes, they suggested that the complex changes in ADA activity during ontogenesis indicate an important role of ADA at early stages of development.

Senba et al., [11] studied the localization of ADA immunohistochemically in brains of embryonic, early postnatal and young adult rats. Immunolabeling was first detected on embryonic day 15 in the hypoglossal motor nucleus, and on day 18 in cingulate, retrosplenial and visual cortex, as well as in the posterior basal hypothalamus and in the facial motor nucleus. However, on postnatal days 10 and 15, immunostained neurons appeared in several subcortical structures whereas the number of these decreased in the anterior olfactory nucleus and some related cortical areas. The intensity of immunostaining continued to increase in 25 day-old animals, essentially producing the adult pattern in most areas. They suggested that the restricted pattern of ADA-immunostaining observed in adult rat brain is generated over a long period of development, various stages of which are characterized predominantly by the expression of ADA in greater abundance. Studying the development of ADA immunoreactivity in the olfactory system Semba et al., [12] also suggested that the functions of ADA in the olfactory system might be related to the precocious development of a "purinergic neurotransmission" within this system.

One interesting new class of ADA-related molecules was recently found and constitute a novel family of growth factors with sequence similarity to classical cellular ADA, and then called "ADA-related growth factors" (ADGF) or CECR1 in vertebrates. Although genes for ADGF/CECR1 have been identified in both invertebrates as well as vertebrates, their in vivo functions in vertebrates remain unknown. In a recent work, Lijima et al., [13] isolated cDNA clones from Xenopus laevis and the recombinant Xenopus CECR1s exhibited ADA and growth factor activity, and the ADA activity was found to be indispensable for growth factor activity. The Xenopus CECR1s are expressed in the somites, pronephros, eyes, cement gland, neural tube, and neural floor plate of the embryos. Knockdown of these genes caused a reduction in the body size and abnormalities of the body axis in the Xenopus embryos, accompanied by selective changes in the expression of developmental marker genes. Noteworthy, injection of adenosine, adenosine receptor agonists or ADA inhibitor into embryos resulted in developmental defects similar to those caused by the knock-down, showing the involvement of CECR1s in the regulation of extracellular adenosine concentrations and activation of adenosine receptors in vertebrate embryogenesis.

In addition, adenosine and its subsequent conversion to inosine was shown to be involved in stimulating axon outgrowth, expression of growth associated protein (GAP-43), and reverses inhibition of axon outgrowth due to 6-thioguanine in goldfish and rat retinal ganglion cells [14]. Inosine also stimulated extensive sprouting of pyramidal cells into the denervated spinal cord white matter [15], resulting in the formation of new corticospinal tract synapses in the denervated spinal cord [16]. These data suggest that inosine is involved in neuronal reorganization and restoration of essential circuitry after CNS injury.

Adenosine kinase is probably the most important enzyme involved in adenosine metabolism, but very little is known on its distribution and regulation in the CNS. Most of the adenosine taken up by CNS cells can be found as nucleotides in several areas, including the retina [17]. Moreover, inhibition of adenosine kinase activity is known to substantially decrease adenosine uptake and increase adenosine release stimulated by NMDA [18], indicating that this enzyme activity is one important driving force for equilibrative nucleoside transport [17]. However, there are very few studies on the development of adenosine kinase in the CNS. Studer et al., [19] investigated adenosine kinase expression and cellular localization during postnatal development of the mouse brain. They found that the enzyme is present at early postnatal stages in neurons from the cerebral cortex and hippocampus. Interestingly, the enzyme gradually disappeared from neurons and appeared in newly developed nestin- and glial fibrillary acidic protein (GFAP)-positive astrocytes in the hippocampus, coinciding with the onset of myelination. The transition from neuronal to astrocytic adenosine kinase expression was complete after postnatal day 14, except in a subset of neurons that retained the enzyme until adulthood in regions such as striatum. Accordingly, recordings of excitatory field potentials in hippocampus revealed a reduced adenosinergic inhibition at P14 compared with adult, suggesting distinct roles for adenosine in the developing and adult brain.

In line with the importance of adenosine kinase in the developing CNS, Shen et al., [20] found that the increase of adenosine by pharmacologic inhibition of adenosine kinase exerted antipsychotic-like activity in mice and overexpression of this enzyme in transgenic mice was associated with attentional impairments linked to schizophrenia. Interestingly, Aronica et al., [21] found that the expression of astrocytic adenosine kinase was increased in the hippocampus and temporal cortex of patients with temporal lobe epilepsy and suggested that dysregulation of this enzyme in astrocytes is a common pathologic hallmark of this type of epilepsy.

Adenosine Inactivation by Re-Uptake: The Development of Nucleoside Transporters

The nucleoside transporters are key elements for the regulation of adenosine extracellular levels and the availability of this nucleoside for activation of its receptors and related signaling pathways. The activity of these transporters can be especially important during development as differential activation of adenosine receptors at specific developmental stages is potentially important in events such as neuronal precursor survival, proliferation, and differentiation (see below). Two classes of nucleoside transporters were described, the equilibrative and the concentrative transporters. Equilibrative transporters are bidirectional and sodium-independent and appear to be the most important in the CNS. Although more subtypes were already cloned, it seems that the most prominent are the ENT 1 and 2, which are related to es and ei, which are respectively sensitive or insensitive to the drug nitrobenzylthioinosine (NBI or NBMPR). Adenosine uptake sites as labeled with [^{3}H]nitrobenzylthioinosine ([^{3}H] NBI) were first examined by Geiger [22] using radioligand binding and membrane preparations from whole brain and 4 brain regions of rats between the postnatal ages from one day up to adulthood. In whole brain, cerebral cortex and cerebellum binding was higher in 6-day-old than in 50-day-old rats, but the reverse was found in hypothalamus and superior colliculus. Accordingly, [^{3}H] adenosine accumulation was higher in brain cells from one-day-old than from adult rat brains. In the rat forebrain, by the other hand, no variation in [^{3}H] NBI binding was observed as a function of developmental stage, as opposed to what happens with A1 receptors (A1Rs) as measured with [^{3}H] CHA binding [23]. In an elegant study, Deckert et al., [24] showed differences in the binding of [^{3}H] NBI and [^{3}H] dipyridamol during guinea pig brain development and suggested that part of the [^{3}H] dipyridamol binding sites, i.e., the NBI-insensitive ones, develops later than [^{3}H] NBI binding.

The nucleoside transporter blocker dipyridamol was also found to have profound influence on cerebellar development [25]. In this tissue, dipyridamol reduced the number of calbindin D-28k-positive Purkinje cells as well as decreased the thickness of the granule cell layer. In the caudate/putamen, dipyridamol also decreased tyrosine hydroxylase expression. These observations suggest that elevations of adenosine levels in brain observed when blocking nucleoside transporters leads to neurochemical and cellular changes in different brain regions during development.

Adenosine Receptor Expression during CNS Development

Adenosine is an important neuromodulator in the CNS regulating several processes like neurotransmitter release, synaptic plasticity and cellular survival [26]. The four types of adenosine metabotropic receptors A1, A2a, A2b and A3 present a differential distribution in mature CNS areas. A1Rs are widely distributed with high levels of expression in cortex, cerebellum and hippocampus while A2aRs are highly expressed in striatum, nucleus accumbens and olfactory tubercles [27]. A2b and A3 receptors present low levels in CNS [28]. Adenosine receptors are expressed during CNS development and modulate different processes such as proliferation, differentiation, survival, neuritogenesis and axonal growth [29].

In a rat brain, A1R mRNA is not detected until embryonic day 14 (E14). At this age, A1R mRNA is localized in thalamus, hypothalamus, hippocampus, midbrain, cerebellum, pontine area and spinal cord. A1R expression increases at E17 and at E20 presents a similar distribution pattern as observed in the adult [30, 31]. A1Rs are also expressed in the developing guinea pig brain and detected in forebrain at E36 [32]. Similar results were observed by Aden and collaborators [33].

A1Rs are also present in the visual cortex. This receptor was detected at high levels in lamina VI in 2 month-old cats and may not be necessary to plasticity and ocular dominance because cats dark-reared since birth presented no alteration in the distribution of A1 receptors when compared to control [34]. A2aR mRNA is widely distributed in brain and mainly detected in rat striatum at E14 [33, 35]. However, the binding with an A2aR agonist could only be detected at E21 in caudate/putamen [33].

Modulation of Adenosine Receptor Expression during CNS Development

Exogenous factors can modify several processes during development including the expression of adenosine receptors. Caffeine and theophyline, adenosine receptor antagonists present in many popular beverages as coffee and tea, modify the adenosine receptor expression. The administration of these methylxanthines to pregnant rats from gestation day 2 (GD2) until the end of pregnancy produced a decrease of [^{3}H] DPCPX binding to A1R in brain plasma membranes of both mother and fetus. In addition, treatment with caffeine or theophyline increased A1R mRNA only in fetal brain. No alteration of A2aR mRNA was detected in mother and fetus brain [36]. This A1R down-regulation induced by caffeine and theophyline produced an inhibition of receptor function with a decrease of $\alpha G_{i1,2}$ protein levels and a decrease of inhibition of adenylyl cyclase activity induced by CHA, an A1R agonist [37]. However, another study demonstrated that caffeine administration to pregnant rats produced no alterations in A1R mRNA and [^{3}H] DPCPX binding in cerebral cortex, cerebellar cortex, hippocampus and thalamus at E14, E18 and E21 fetuses. Similar results were observed for A2aR with no alterations in mRNA and [^{3}H] CGS21680 or [^{3}H] SCH58261 binding in caudate/putamen at E14, E18 and E21 [33]. Moreover, the administration of R-PIA, an A1R agonist, to pregnant rats produced a decrease of A1Rs detection and adenylyl cyclase inhibition in fetus brain [38].

Besides modifying adenosine receptor expression, caffeine can interfere in neural development. The administration of caffeine to pregnant mice induced impairment in the migration of GABAergic neurons into hippocampus. The neuronal excitability was also increased and the pups demonstrated more vulnerability to seizures when exposed to a convulsive agent [39].

Cocaine is another psychostimulant that when administrated during pregnancy promotes alterations in neurological functions. The treatment of pregnant mice with cocaine from E8 to E14 induced an increase of cAMP levels in cortex at E15 that was reduced by an A2aR antagonist. In addition, adenosine uptake was also reduced by cocaine in striatum. These results suggest that cocaine modulates the adenosine transporters and A2aR signaling [40].

Adenosine Receptor Signaling Regulating CNS Development

Adenosine receptors regulate different processes during neural development such as proliferation of progenitors, myelinization and neurite outgrowth [41-44]. These observations demonstrate that adenosine is an important factor for CNS development.

Modulation of Proliferation and Differentiation of Progenitor Cells by Adenosine Receptors

Adenosine receptors affect the proliferation of neural stem cells (NSCs) which are multipotent cells able to differentiate in neuron or glial cells and then play an important role in neural development (Figure 1). NSCs isolated from mice subventricular zone (SVZs) express mainly A1 and A2bRs. In these cells, adenosine increased cell proliferation in a concentration and time-dependent manner. This effect was mimicked by CHA, an A1R agonist, blocked by DPCPX, an A1R antagonist, and was dependent on MEK and Akt activation. These results suggest that A1Rs are important to NSCs proliferation, neural development and neuroregeneration [41].

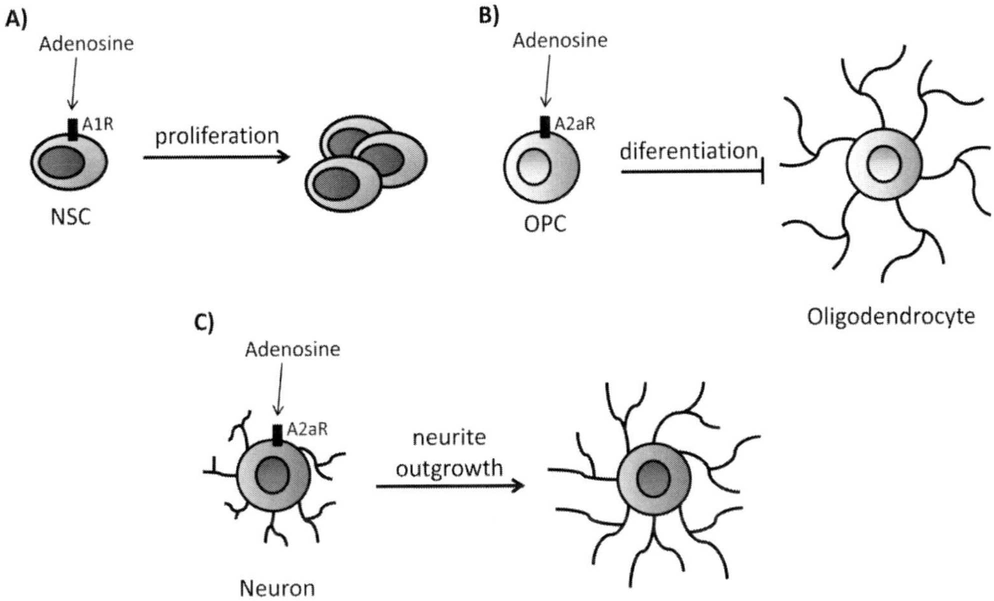

Figure 1. *Adenosine regulates different processes during neural development.* A) The activation of A1 receptors promotes proliferation of NSCs (neural stem cells). B) A2a receptor activation inhibits the differentiation of OPCs (oligodendrocyte progenitor cells) in oligodendrocytes. C) A2a receptor activation induces neurite outgrowth.

Adenosine also regulates the proliferation of glial progenitors and their differentiation in oligodendrocytes and Schwann cells (Figure 1). Oligodendrocyte progenitor cells (OPCs) express all adenosine receptors and the stimulation of A2aRs did not affect the proliferation but inhibited the differentiation of these cells [45]. In addition, the differentiation of Schwann cells is regulated by adenosine. Premyelinating Schwann cells in monoculture express A2aRs coupled to Gs whose activation increases intracellular cAMP levels and promotes ERK activation in these cells. In addition, A2aR activation inhibited PDGF-induced Schwann cells proliferation but had no inhibitory effect on its differentiation [46].

Myelinization is a process dependent on neuronal electric activity and release of different signaling molecules as adenosine. The treatment of OPCs with adenosine or NECA, non-selective adenosine receptor agonists, inhibited proliferation, induced differentiation and altered the morphology of these cells with an increase in cell surface area, perimeter length, outer cell radius and branching (Figure 1). Moreover, treatment of dorsal root ganglion (DRG)/OPC co-cultures with adenosine promoted an increase in myelinanting markers and myelination of DRGs [42].

Modulation of Neurite Outgrowth by Adenosine Receptors

Neurite outgrowth is another developmental process regulated by adenosine (Figure 1). Treatment of retinal ganglion cells in culture with adenosine induced an increase of neurite length and this effect was not blocked by A1R or A2aR antagonists neither mimicked by cAMP. In addition, inhibition of adenosine deaminase, an enzyme that metabolizes adenosine to inosine, decreased neurite length and cell survival. Treatment with inosine was able to induce neurite outgrowth in these cells [47]. Moreover, several studies demonstrated that adenosine receptors are involved in the modulation of neurite outgrowth. A2aR activation induced an increase of neurite length in PC12 cells and this effect was not dependent on PKA [48]. Another study using PC12 cells demonstrated that neurite outgrowth induced by NGF was reduced by ERK inhibition and this effect was rescued by A2aR and PKA/CREB activation. However, activation of this receptor did not increase the p-ERK levels suggesting that this protein is not involved in neurite outgrowth induced by A2aR activation [43]. Neurite outgrowth in PC12 cells can also be induced by the nucleoside N^6-methyldeoxyadenosine (MDA). This effect was dependent on A2aR, cAMP and ERK activation but independent of PKA [49]. On the other hand, A1R activation was demonstrated to be necessary to collapse the outgrowth of retinal ganglion cells axons demonstrating that these receptors are important to formation of retinotectal maps [50].

The neurite outgrowth is also regulated by netrins, proteins that are important for axon elongation. Netrins modulate growth cone by attraction through its interaction with DCC (deleted in colorectal cancer) or repulsion via its interaction with UNC [51]. Several studies demonstrated that A2bRs regulate this process mediated by netrins. Corset and collaborators [52] demonstrated that, in presence of netrin-1, the A2bR interacts with DCC in HEK293 cells. In addition, netrin-1 or the adenosine agonist NECA induced an increase of cAMP intracellular levels and the A2bR antagonist inhibited this effect induced by netrin-1. These results suggest that the effect of netrin-1 on neurite outgrowth is dependent on A2bR activation. However, another study using rat dorsal spinal cord explants demonstrated that neurite outgrowth induced by netrin-1 and DCC was not dependent on A2bRs [53].

McKenna and collaborators demonstrated that activation of A2bRs induced a down-regulation of UNC5A and this effect was dependent on PKC activation. Thus, A2bR activation induces a conversion of growth cone behavior from repulsion to attraction mediated by netrin-1 [54]. A2bRs are also related to retinal axon migration to the optic tectum [55]. In Xenopus, during retinal axon migration to tectum, the initial attraction induced by netrin-1 is converted to repulsion. This effect is dependent on the down-regulation of DCC and A2bRs and decrease of cAMP intracellular levels [56].

SIGNALING PATHWAYS REGULATED BY ADENOSINE DURING DEVELOPMENT

Adenosine A1 Receptor

A1R is the most abundant adenosine receptor in the brain [57, 58]. Classically, it couples to members of the pertussis toxin-sensitive family of G proteins, namely, G_{i1}; G_{i2}; G_{i3} and G_0 [59, 60]. The interaction between the A_1R and the G protein induces the split of the protein in two different components, the Gα subunit and the βγ complex, which could regulate several effectors. The most known effector of $G\alpha_i$ subunit is the inhibition of adenylyl cyclase. This enzyme is responsible for the production of the second messenger cAMP. The reduction in cAMP levels induced by A_1R activation leads to less activation of cAMP-dependent protein kinase (PKA), and sometimes inhibition of synaptic transmission [61] and release of neurotransmitters and neuropeptides [62, 63].

To better illustrate the importance of this signaling pathway during development, Jeong and colleagues [62] demonstrated in immature rat hippocampal CA1 neurons that activation of presynaptic A1Rs by adenosine modulates the probability of spontaneous GABA release via inhibition of the cAMP-PKA signaling pathway. This A_1R-mediated modulation of GABAergic transmission may play an important role in the regulation of excitability of immature hippocampal CA1 neurons. It is also important to mention that this effect is observed only during development and adenosine had no effect on the frequency of GABAergic synaptic currents in postnatal neurons [62].

Other downstream targets of A1R-induced $G\alpha_i$ activation include K^+ and Ca^{++} channels. The interaction between Gαi and Ca^{++} channel inhibits this channel, reducing Ca^{++} influx. During development it has been described in cultured rat dorsal root ganglion neurons that 2-chloroadenosine reduces the Ca^{++} inward current (I_{Ca}) activation by a direct effect on A1Rs [64]. Regarding K^+ channels, it has been described the interaction between $G\alpha_i$ and K^+ channel, mainly with inwardly rectifying K^+ channels (GIRKs). The activation of these channels leads to a K^+ outward current, explaining the slow hyperpolarization effects induced by adenosine and 2-chloroadenosine observed in cultured embryonic striatal neurons, although the authors did not elucidate which adenosine receptor type was involved in the phenomenon [65].

As previously described in this section, the βγ complex is also capable of stimulating downstream signaling pathways and the best example is the activation of phospholipase C (PLC). It has been described, that the exposure of cultured embryonic neurons to R-PIA, an A1R agonist, induces a short-lived enhancement in PLC activity [66]. This happens not only in embryonic neurons, but also in embryonic astrocytes in which activation of A1Rs is capable of inducing PLC activation [67].

An important aspect in A1R functioning is the interaction between this receptor and other proteins. It has been described that adenosine deaminase (ADA) is able to interact within the A1R extracellular loops in the brain [68]. Although this interaction was not described during CNS development, it has been shown that human ADA, apart from reducing the adenosine concentration and thus preventing adenosine A_1R desensitization, binds to A1R behaving as an allosteric effector that markedly enhances agonist affinity and increases receptor functionality [69]. The physiological role of such interaction could be to bring A1Rs more sensitive to adenosine, an effect that could potentially influence the role of these receptors in brain development.

Adenosine A2a Receptor

The A2aR is expressed at high amounts in few brain regions, namely the striatum, olfactory tubercle and nucleus accumbens [70]. Classically, this receptor type couples to G_s/G_{olf} proteins [71] and stimulation of $G\alpha_s$ protein activates adenylyl cyclase, leading to cAMP accumulation and PKA activation. This signaling pathway is not only involved in neurite outgrowth, as described in previous sections, but also in neurotransmitter release. It has been observed that A2aR stimulation induces PKA activation and stimulates the release of acetylcholine in the rat motor endplate [72]. Moreover, in SH-SY5Y cells differentiated into a sympathetic neuronal cell type phenotype, the activation of the cAMP/PKA pathway by A2aR has been reported to control the release of endogenous noradrenaline [73].

A2aR-induced PKA activation has also been demonstrated to downregulate PKC function, facilitating GABA uptake through the GAT-1 transporter in hippocampal synaptosomal preparations [74]. Furthermore, in the corticostriatal pathway, A2aR activates the cAMP/PKA signaling to control D2 dopamine receptor-induced calcium influx through NMDA receptors [75]. In the hippocampus, AMPA mediated synaptic transmission in the CA1 region is controlled by an A2aR-induced PKA activation with the concomitant phosphorylation of the GluN1 AMPA subunit at Ser845 [76]. A2aR-mediated PKA activation also controls retinal cells functions by modulating connexin phosphorylation and photoreceptor coupling [77].

The A2aR also presents a peculiar characteristic in comparison with other adenosine receptors since its intracellular carboxyl-terminal tail is unusually longer than that of other adenosine receptors, presenting 122 amino acids in contrast to the 34 residues of A1R, 39 of A2bR or 34 of A3R. This long C-terminal tail can interact with many described partners, as for example NECAB2, ARNO, TRAX and many others (for review see [78]). This protein-protein interaction can regulate neurite outgrowth, an important developmental phenomenon. It has already been demonstrated that the ability of A2aR to regulate NGF-induced neurite outgrowth is TRAX-dependent [79].

Adenosine A2b Receptor

Although the A2bR is widely distributed but with a relative low expression in the brain [58], very little is known about its function in CNS development. It has been described that this receptor can couple to many different types of G proteins, namely G_s, G_q [80] and G_{14} in circumvallate taste buds [81]. As occurs in the A2aR stimulation, the classical signaling pathway activated by A2bR is stimulation of $G_{\alpha s}$ protein and adenylyl cyclase, leading to cAMP accumulation and PKA activation. Regarding development, it has already been described that adenosine signaling mediated by cAMP promotes the differentiation of primary neural crest cells into neurons. As these cells express both A2aR and A2bR it was not clear which one was the responsible for primary neural crest cells differentiation [82].

The A2bR is also capable of regulating Ca^{++} channels [83]. This regulation can be mediated by $G_{\alpha s}$ interaction or, indirectly, by PKA activation [83]. In acutely isolated pyramidal neurons from the CA3 region of guinea pig hippocampus, it has been described that A2bR stimulation potentiates a P-type Ca^{++} current [84]. Moreover, it was also described that A2bR stimulation induces long-term potentiation in the CA1 region of rat hippocampus [85].

It was observed that extracellular ADA is also capable of interacting with the A2bR [86]. As described for A1R, this interaction does not have a clear involvement in CNS development but, as described by Herrera and colleagues [86], this interaction can increase the affinity of A2bR and potentiates the receptor-mediated cAMP accumulation. This phenomenon can potentially regulate some cAMP-dependent processes during development. Therefore, cell surface ADA, apart from degradation of extracellular adenosine, can also regulate adenosine actions via A2bR activation.

Another interesting potential regulation of A2bR during CNS development can be mediated by the transcription factor NF-κB. It was previously described that NF-κB is involved in the generation and morphological differentiation of neurons in developmental processes such as neurogenesis, axon initiation, axon guidance and others (for review see [87]). Recently, Sun and colleagues described a direct binding of the p105 NF-kB family member to the C-terminus of the A2bR, independent of receptor activation. This interaction inhibits NF-kB activation by physically interacting with it, blocking its polyubiquitylation and activation [88].

Adenosine A3 Receptor

The A3R has a low, but ubiquitous, expression in the brain [58, 89]. It has been described that stimulation of A3Rs can induce the coupling with G_{i2}, G_{i3} and G_q proteins, leading to adenylyl cyclase inhibition and stimulation of phospholipase C, thus reducing cAMP levels and increasing IP3, intracellular Ca^{++} and DAG. Moreover, the βγ complex can stimulate PI3K/AKT and ERK signaling pathways. PLC/PKC or PI3K/Ras/MEK pathways can mediate A3R-ERK stimulation [89].

Regarding the modulation of Ca^{++} currents, it has been described in acutely dissociated guinea pig hippocampal CA3 pyramidal neurons that administration of A3R agonists leads to a potentiation of a Ca^{++} inward current (I_{Ca}). This potentiation appears to be PKA-dependent and PKC-independent since this phenomenon was blocked by PKA but not by PKC inhibitors [90]. Moreover, it has been shown that A3R stimulation in brain slices from young Wistar rats exerts a modulatory effect on excitatory amino acid projection. The authors observed that A3R agonist diminished NMDA-induced I_{Ca} in spiny neurons [91].

Donoso and colleagues described an interesting model of small and large synaptic vesicles release in sympathetic nerve endings. It was demonstrated that a simultaneous activation of A2aR and A3R induces additive reductions in noradrenaline release and only the combination of the two respective antagonists fully blocked the adenosine-induced inhibition. Moreover, they also described that pre-synaptic A2aR and A3R activation modulates only sympathetic co-transmission by exclusively inhibiting noradrenaline release, without affecting the release of large vesicles containing neuropeptide Y, suggesting separate mechanisms for large and small vesicular release modulation [92]. This modulation of large and small vesicular release was not described during development. However, a differential release of small neurotransmitters and neuropeptides regulated by A3Rs could modulate the maturation and functioning of synaptic circuits during CNS development.

Signaling with Adenosine Receptors in Developing Neurons

Adenosine A1 Receptor

A striking feature of the immature CNS is widespread developmental neuronal cell death, which is also important for shaping the neuronal parenchyma establishing accurate and integrated neuronal circuits in different brain areas. Adenosine, by activating A1Rs, has been demonstrated to prevent the cell death of developing cortical neurons via the transactivation of EGF receptors [93]. Such regulatory mechanism has been demonstrated to be AKT and Src-dependent, downstream of A1R activation. Moreover, A1R associates with EGFR at the membrane of developing cortical neurons, promoting EGFR activation in the absence of EGF, which is sufficient to relay the neurotrophic action of A1R activation in developing cortical neurons challenged by oxygen/glucose deprivation [93].

Differently from developing cortical neurons, activation of A1R in mature hippocampal neurons seems not to engage in the classical signaling pathway regulated by the activation of neurotrophin receptors [94]. Interestingly, inhibition of A1R in the mature hippocampal circuit by *in vivo* administration of an A1R antagonist leads to increased production of BDNF and canonical BDNF signaling [94], which is suggestive that the crosstalk between A1Rs with growth factors and their receptors might be strikingly different among different brain regions and developmental stages.

Cytokines have also been shown to participate in the crosstalk with A1Rs in developing neurons. That is the case for the cytokines IL-6 and oncostatin M (a member of the IL-6 family), which display neuroprotective actions via A1Rs [95]. Both cytokines exert their function via activation of the canonical gp130 co-receptor in developing cortical neurons, which upregulates A1Rs in neuronal membranes dramatically reducing excitotoxic damage mediated by extracellular glutamate [95]. On the other hand, it is well known that IL-6 production and release are dramatically increased under neuropathological conditions in the mature brain. These data set the stage for a mechanism relating the A1R-mediated signaling in the developing brain under excitotoxic stress, with a beneficial impact for the survival of developing cortical neurons, which might be significantly different in mature neurons chronically exposed to IL-6.

Information processing in the developing CNS is also dependent on precise formation of neuronal connections and involves the signaling of axon guidance molecules that can both repel advancing growth cones from inappropriate areas and attract growth cones to appropriate target zones. In the developing primary visual projection in the optic tract, retinal ganglion cell axons from the nasal retina project to the posterior optic tectum/colliculus and axons from the temporal retina project to the anterior tectum/colliculus so that the retinal temporal-nasal axis connects to the anterior-posterior axis of the tectum/superior colliculus, giving rise to a topographic map responsible for integrating the visual stimuli in the mature

CNS. The Ephrins are classical growth-related molecules serving as cues for the axons of retinal ganglion cells for path finding the right target in the superior colliculus.

Evidence has now emerged showing that Engrailed, an homeodomain transcription factor that directly control the synthesis of Ephrins and induces the release of ATP from the growth cone of developing retinal ganglion cells [96]. The hydrolysis of this extracellular ATP in the superior colliculus promotes A1Rs-dependent growth cone collapse. Such Engrailed/Ephrin /A1R crosstalk has been proposed as a non-cell autonomous mechanism for refining the retino-tectal map [96], which has a huge impact for the precise development and functioning of the primary visual pathway. Within the retina, A1Rs in immature retinal ganglion cells mediate the neuroprotective role of IL-6 treatment [97], which may be further modulated by a signaling crosstalk involving a BDNF-induced IL-6 release and A2aR activation [98]. Therefore, one might conclude that the crosstalk between A1Rs and growth factors is not exclusive for the developing brain but is also important for shaping the development of the neuronal parenchyma and the connectivity of neurons in CNS regions outside the brain.

Adenosine A2a Receptor

Neurotrophins play prominent roles in nervous system development by influencing neuronal survival, differentiation, and proliferation. These molecules also display acute regulatory effects on neurotransmitter release and synaptic strength in mature neuronal networks. Besides promoting axonal and dendritic branching, neurotrophins serve as chemoattractants for extending growth cones *in vitro*, indicating that such molecules are also guiding cues for sprouting axons during nervous system differentiation. Neurotrophins actions are mediated by neurotrophins interaction with two separate receptor classes, the Trk family (tyrosine kinase receptors) and the p75 neurotrophin receptor.

Until the pioneering work from Lee and Chao [99] there was an absolute lack of experimental evidence relating the function of neurotrophins receptors with the activation of the A2aR pathway in neurons. Lee and Chao described a fancy mechanism by which the activation of A2aR would exert trophic actions in developing hippocampal neurons via transactivation of the high-affinity BDNF receptor (TrkB) in the absence of the ligand. The authors showed that A2aR activation would engage in downstream Src modulation and transactivation of TrkB receptor. A2aR-mediated TrkB transactivation has important implications for the survival of developing hippocampal neurons [99] by subsequently activating AKT and not MAPKs (the canonical pathway stimulated by BDNF-TrkB signaling).

The A2aR-induced Trk transactivation mechanism in developing neurons has been shown to persist associated with signaling vesicles near the Golgi apparatus in the absence of neurotrophins [100] and this seems to be somewhat different from the canonical BDNF engagement of TrkB receptors in mature neurons integrating synaptic circuits. In line with this, the presynaptic activation of A2aRs might facilitate the BDNF-mediated synaptic potentiation in the mature CA1 hippocampal circuit [101]. This A2aR-mediated regulation of BDNF actions is also prominent in the aged hippocampus [102] and spinal cervical tract [103]. Furthermore, as could be expected, activation of A2aRs has also been described to promote BDNF synthesis [104].

In PC12 cells, a somewhat artificial paradigm for studying the differentiation of neuronal cells, activation of A2aRs has been demonstrated to cooperate with NGF to promote neurite outgrowth via PKA and CREB [43]. This neurotrophic action of adenosine in modulating NGF responses also occurs in sympathetic neurons of the rat superior cervical ganglion [105]. Moreover, direct interaction between A2a and FGF receptors has been demonstrated to induce neurite extension in PC12 cells and spine morphogenesis in developing hippocampal neurons [106]. Altogether, these findings suggest that alternative approaches for stimulating trophic functions in the developing nervous system might be achieved by the linkage between A2aR and neurotrophin receptors in different immature neuronal populations.

Adenosine A2b Receptor

During CNS development, neurite extension is important for accurate formation of synaptic contacts and, consequently functional maturation of neuronal networks. Different set of experiments indicated that axons are often subjected to opposing gradients, an attractive and a repellant source both of which must be integrated in the growth cones. Four classes of molecules serving as instructive guidance cues for the growth cones of a sprouting axon exist. Such cues might attract or repel axons depending on the class of membrane receptors present in migrating growth cones. These cues include the Netrins/UNC-6, Slits, Semaphorins and Ephrins. Netrin is a laminin-like extracellular cue expressed in a wide variety of cells, serving as guidepost in the developing nervous system. Direct evidence for a role of A2bR in relaying trophic actions in the developing CNS came from experiments in immature spinal dorsal neurons showing that the effect of netrin-1 in promoting neurite outgrowth in cultures from these cells was mediated in fact by netrin-1 binding to and activating A2bRs, leading to the accumulation of cAMP [52]. Therefore, netrin-1/A2bR-dependent axon outgrowth in immature neurons indicate that adenosine receptors may have a function in the developing nervous system, which might be far away from their well known modulation of synaptic transmission in the mature CNS.

Neuroprotection by Adenosine during Development and Disease

Hypoglycemia

Acute or chronic hypoglycemic events can lead to neurological impairment in both newborn infants and adults [107], with loss of cognitive function [108], elevated risk of epilepsy [109] and injury. Children with diabetes mellitus can develop learning deficiency [110]. It was believed that mature brains were more sensitive to hypoglycemia than those under development, but clinical studies have indicated that developing brains also present sensitivity to low glucose levels [107].

During hypoglycemia, adenosine release is exacerbated [111, 112], and is known that in adulthood the activation of A1Rs may be neuroprotective [113]. However, some evidences show that in developing brain the activation of A1Rs may promote injury.

When neonatal murine at postnatal day 7 (P7) and P21 were submitted to acute hypoglycemia, more injury was observed in younger mice [114]. Otherwise, when hippocampal slices were treated with CPA or DPCPX (A1R agonist and antagonist, respectively) the amount of cell death due to hypoglycemia was changed. CPA enhanced TUNEL labeling (cells undergoing apoptosis) induced by low glucose levels, while DPCPX blocked the hypoglycemia effect. The same results were observed in hippocampal slices from $A1R^{-/-}$ mice [114]. In cultured cortical neurons from rat embryos (E18), adenosine is released from neurons by hypoglycemic insult and the A1R antagonist DPCPX prevented neuronal damage by rescuing the intracellular Ca^{2+} previously depleted by the hypoglycemia [115]. There is additional evidence suggesting that, during hypoglycemia in developing CNS, adenosine is released and by activating A1Rs can deregulate the intracellular Ca^{++} concentrations and induce neuronal apoptosis [116].

Hypoxia-Ischemia

As in hypoglycemia, during hypoxia-ischemic (HI) events occurs a rise in extracellular adenosine levels and this is considered as a protective event. Generally, adenosine achieves neuroprotection by decreasing excitatory amino acids release and Ca^{++} influx, hyperpolarizing post-synaptic membrane, as well as inducing vasodilatation, among other ways.

A prior treatment with theophylline (a non-selective adenosine receptor antagonist) reduced brain damage caused by 60-90 minutes HI in rat pups at P7 [117]. While DPCPX had no such effect, blocking A2aR with SCH 58261 showed a moderate protection when administered during reperfusion [117]. Drugs that increase extracellular adenosine levels, such as deoxycoformycin and propentofylline (adenosine deaminase inhibitor and adenosine uptake blocker, respectively) can provide protection from HI-induced injury in perinatal rats [118]. In primary cortical cultures from E16 rat embryos submitted to ischemic insult through oxygen-glucose deprivation (OGD), the chronic pre-treatment with Cl-IB-MECA, a selective A3R agonist, was capable to increase the cellular viability, protecting those neurons from the ischemic injury [119].

Periventricular white matter injury (PWMI) is the main form of brain injury in survivors of premature birth [120]. PWMI can manifest as cerebral paralysis, cognitive and learning disabilities [121], and presents reduced cerebral myelination and depletion of premyelinating oligodendrocytes [122, 123]. One of the significant causes of PWMI is HI.

Mice at postnatal days 3 to 12 submitted to HI conditions presented hypomyelination due to abnormal maturation of oligodendrocytes, with reduced expression of myelin basic protein (MBP) and development of ventriculomegaly [124]. Interestingly, animals that received caffeine, which is a non-selective adenosine receptor antagonist, through breast-feeding did not show enlargement of ventricles and reduced expression of MBP [124]. In premature infants, caffeine is widely used clinically to stimulate respiration [125] and the activation of A1R usually protects against damage insults on adult brain [126]. Some data have shown that either adenosine via A1R activation or the hypoxia-induced premature and accelerated maturation of oligodendrocytes, event that is associated with hypomyelination in PWMI [127]. Accordingly, treatment of neonatal rats from P3 to P14 with the A1R agonist CPA resulted in reductions in white matter volume, ventriculomegaly and neuronal loss. Treated animals also presented lower levels of MBP [128]. Moreover, ventriculomegaly was observed in neonate mice deficient in the A_1R gene or lacking the enzyme ADA [129].

Sympathetic neurons from 11-day-old chick embryos submitted to chronic adenosine presented nearly to 80% of death through apoptotic mechanisms and reduced neurite growth [130]. These toxic effects were lost when adenosine was given to mature neurons [131]. Adenosine seems to inhibit both RNA and protein synthesis, and its toxic effect can be due to its phosphorylation to form ATP [132]. Additionally, mesencephalic cultures from E15.5 embryo rats spontaneously lose dopaminergic neurons as the culture progresses. Adenosine was able to prevent this loss possibly through activation of ARs [133].

ADENOSINE AND RETINAL DEVELOPMENT

As described in a previous chapter in this book, the retina is an excellent model for neurochemical studies of the CNS and many neurotransmitter systems found in other parts of the CNS are also present in the retina. In particular, the retina from *Gallus gallus* is especially suitable as a model for developmental studies because of its well-known histogenesis and the tissue availability during a major part of the embryonic period [134, 135, 136]. The development of the chick embryo is complete in 21 days when occurs the hatching phenomenon (Figure 2). The retina is easily accessed and dissected free of contamination with other neural and non-neural tissues since 6 days of incubation (E6), or even earlier, up to the post-hatching period. Another interesting characteristic of the chick retina is the possibility to make cultures where many of the properties of the intact tissue are preserved as, for example, the development of neurotransmitter synthetic machineries and receptors [137].

DEVELOPMENT OF ADENOSINE RECEPTORS IN THE RETINA

Adenosine Regulates cAMP Production in the Developing Retina

Adenosine promotes cAMP accumulation in the chick embryo retina. This effect is first observed in 14-day-old embryo retinas (E14), increases up to E17 and decreases in the post-hatching period [138] (Figure 2).

Endogenous adenosine is released during incubation of retinas as detected by the difference between cAMP levels observed with the distinct phosphodiesterase inhibitors IBMX (an adenosine receptor antagonist) and RO20-1724 and the decreased levels observed incubating retinas with RO-201724 plus adenosine deaminase [138]. Adenosine-elicited cAMP accumulation was also observed in cultured retinal cells and the responsiveness to adenosine is diminished upon a chronic incubation of cultures with adenosine [138, 139]. Adenosine also inhibits the accumulation of cAMP induced by dopamine in the chick embryo retina and interestingly this effect is observed since E10 (Figure 2), a developmental stage in which no direct accumulation of cAMP by adenosine was observed [140]. Pharmacological experiments showed that the accumulation of cAMP is mediated by A2 receptors (A2a and A2b were not distinguished between them at that time) while the inhibition of dopamine-stimulated cAMP accumulation is mediated by A1Rs. It could be concluded that A1R and A2Rs had a differential ontogenesis in the chick embryo retina. One interesting finding was that the A1R-mediated inhibition was observed in the embryonic period but not in post-hatched retinas and, on the other hand, was not observed on apomorphine-stimulated cAMP accumulation which was similar at all stages of retinal development (Figure 2). These findings raised the possibility of existence of two different types of dopamine D1 receptors coupled to stimulation of adenylyl cyclase in the chick embryo retina: One stimulated by apomorphine in a similar degree at the whole period of development including the post-hatching period and not regulated by the A1R. A second one named D1E could exist only in the embryonic period, is not expressed in the post-hatching period and is regulated by A1R. The presence of D1 receptors in the chick embryo retina was previously shown to regulate neurite extension and growth cone motility in embryonic chick retina cells [2], suggesting that D1E receptors and A1Rs can regulate important functions during retinal development.

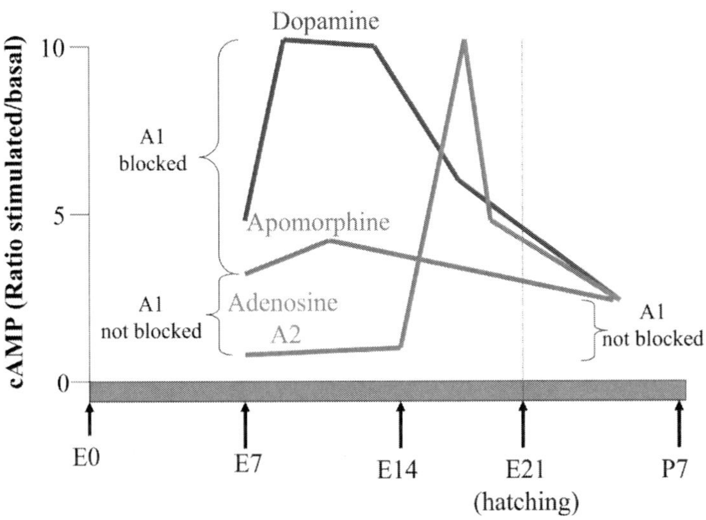

Figure 2. Summary of changes of cyclic AMP levels stimulated by activation of dopamine and adenosine receptors during development of the chick retina. Dopamine receptors were stimulated by dopamine or the partial agonist apomorphine. Note the difference in the stimulated/basal ratio when retinas were exposed to dopamine or apomorphine in the embryonic period while the ratio is similar at the post-hatching period. The activation of A1 adenosine receptors blocks the stimulation by dopamine, but not apomorphine, in the embryonic retina. At the post-hatching period, the effects of both dopamine or apomorphine are not regulated by A1 receptors. This figure was obtained from the reference [158] with permission.

Binding Studies for A1R in the Developing Retina

One question regarding the development of neurotransmitter receptors and their coupling to different effectors and signaling pathways during development is whether there is a correlation between receptor and signaling pathway ontogeny. In order to study this interaction related to A1R in the developing chick retina, binding measurements were performed using the selective A1R CHA [141]. The binding was developmentally regulated, increasing from E10 to E17 and decreasing thereafter up to the post-hatching period.

Noteworthy, binding was still detected in post-hatched retinas as opposed to the inhibition of dopamine-dependent cAMP accumulation that is no longer detected at this period. Interestingly, forskolin-induced cAMP increase could be inhibited by A1R stimulation in post-hatched retinas [141], suggesting that the receptors are still present and coupled to adenylyl cyclase at this period but unable to inhibit D1 dopamine receptor-mediated cAMP accumulation. Pereira et al., [142] have studied the development of A1Rs in chick embryo retina cells in culture and found that 1) the expression of A1Rs is dependent on cell aggregation since it was very low in monolayer dispersed cultures and much higher in aggregate cultures which show an histotypic organization of cells in a similar way as the intact tissue in vivo; 2) the expression of A1Rs in dispersed cultures could be increased by permeable cAMP analogs; and 3) activation of A2aRs could also promote an increase in A1R expression. These results showed for the first time that the expression of one class of adenosine receptors is under regulation of another class of adenosine receptors, strongly suggesting the idea that a concatenated expression of different classes of receptors can have important impact on CNS development. In a more recent report, Brito et al., [143] found that the expression of A1R is under the control of A2aR activation in the developing in vivo retina. Furthermore, they found that endogenous adenosine levels regulated by nucleoside transporters are important partners in this process. A1Rs were localized in the developing chick retina by receptor autoradiography and found to be present at both outer and inner plexiform layers since E12 with increasing labeling at the same regions up to the post-hatched period [144], suggesting that these receptors are closely related to synapses in this tissue since early stages of development.

NUCLEOSIDE TRANSPORT IN THE DEVELOPING RETINA

Detection of Endogenous Adenosine and Localization of Nucleoside Transporters in the Developing Retina

Endogenous adenosine was localized in the retina by immunocytochemistry [145]. In the developing chick retina, adenosine was detected as soon as E12 in cells localized in the inner nuclear and ganglion cell layers as well as in photoreceptors [144]. Nucleoside transporters were also studied in the developing retina [144]. Binding of [^{3}H] NBI was studied in E15 retinas and showed a homogenous class of sites with high affinity for this compound. Binding was also displaced by dipyridamol with higher concentrations but not with the A1R agonist CHA. Interestingly, autoradiographic analysis of these binding sites revealed the presence of transporter sites since E8, a stage in which no A1Rs were present, with increased levels at E12, E18 and post-hatched retinas over the outer and inner plexiform layers in a localization similar to that of A1Rs.

Adenosine Uptake and Release in Purified Retinal Cultures

Adenosine labeling was also found in a proportion of retinal neurons and all photoreceptors cultured in the absence of glial cells [146]. Interestingly, adenosine uptake was also detected autoradiographically in a proportion of neurons and all photoreceptors in these purified retinal neuronal cultures [146]. Noteworthy, imunocytochemical and autoradiographic double labeling showed a majority of multipolar neurons and all photoreceptors with both labels and a small proportion not labeled for any marker. The uptake by cultured developing neurons was intense and could be blocked by the inhibitors NBI or dipyridamol. Moreover, in cultures incubated with [^{3}H] adenosine, the taken up radioactivity could be released by depolarization with high KCl concentration in a calcium-dependent manner. However, inosine was found to be the major component released under these conditions. A work using mixed cultures of developing chick retinal cells containing neurons and glial cells showed that the release could also be stimulated by activation of glutamate receptors, inhibited by NBI as well as by calcium/calmodulin-dependent kinase blockers [147]. Adenosine uptake was also found in purified retinal glial cultures and both purified and glial cultures showed the majority of taken up radioactivity as nucleotides [17]. Additionally, the uptake was blocked by iodotubercidin, an adenosine kinase inhibitor, demonstrating that this enzyme is a major driving force for adenosine uptake in retinal cells. Moreover, adenosine uptake was inhibited by MAP kinase inhibitors, indicating that the transporter is regulated, directly or indirectly, by this signaling pathway.

Regulation of Neuronal Survival by Adenosine in the Retina

The Refeeding Paradigm

Purified cultures of chick retinal neurons undergo an extensive cell death when refed with fresh medium [148,149]. This cell death can be blocked by a pre-incubation with adenosine, permeable cyclic AMP analogs or A2aR agonists [149]. This effect is observed only with a long-term pre-incubation and the maximal survival is obtained with 24 hours of pre-incubation. Interestingly, cell death is not observed when cultures were reefed with a conditioned medium from a sister culture, suggesting that cultured cells release a survival factor that accumulates in the extracellular medium. Noteworthy, cells in culture, including the photoreceptors, undergo a similar death when cultures were incubated with adenosine deaminase even without refeeding with fresh medium, indicating that adenosine is an important factor regulating neuronal survival in these cultures.

Blocking Glutamate Excitotoxicity

High concentrations of the neurotransmitter glutamate are toxic to neurons and promote neurodegeneration through a phenomenon called glutamate excitotoxicity, which is involved in several neuropathologies in the CNS [reviewed in 150, 151, 152]. The addition of glutamate to purified cultures of chick retinal neurons promotes an extensive cell death that is prevented by a long-term preincubation with adenosine and A2aR agonists, but not A1 agonists [153]. The effect of glutamate is blocked by NMDA or AMPA/kainate antagonists and observed in concentrations as low as 50 μM. As in the case of cell death induced by refeeding, the incubation with adenosine needs a previous pre-incubation time of 24 hours and is mimicked by permeable cAMP analogs, suggesting a mechanism involving long-term events such as protein synthesis.

Shift of Signaling Pathways during Development

Adler and Hatlee [154] showed that the proportion of photoreceptors in neuronal cultures obtained from E6 or E8 chick embryos is approximately 80% and 30% respectively. In order to explain this difference, they proposed the "default pathway" hypothesis arguing that neuroblasts are committed to differentiate in photoreceptors except when they receive signals indicating to differentiate in a different retinal neuronal type. Then, this hypothesis predicts that developmental signals released during the E6-E8 period indicate that each neuronal sub-population is shifted from the photoreceptor default pathway [154].

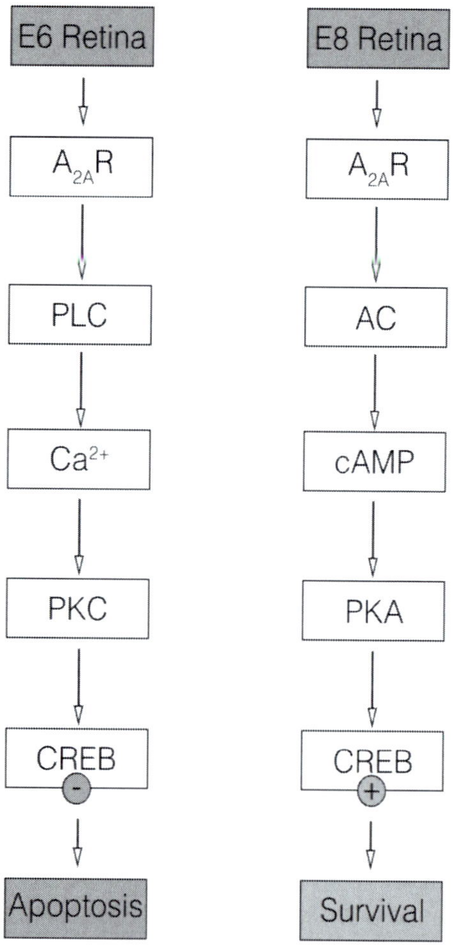

Figure 3. Summary of the signaling pathways triggered by A2a adenosine receptor activation and leading to apoptosis or neuronal survival in cultures of developing retinal cells obtained from E6 or E8 chick embryos.

Noteworthy, when the refeeding paradigm as well as incubation with adenosine deaminase was applied to E6 cultures, no cell death was observed as reported to E8 cultures (Paes-de-Carvalho, R. and Socodato, R., unpublished data). Interestingly, addition of adenosine to E6 cultures 2 hours after seeding the cells (C0), but not after 1 day in culture (C1), promoted an extensive cell death [155]. Addition of adenosine to E8 cultures never produced cell death, even at C0 or C1 and, on the other hand, promotes cell survival when cultures were reefed with fresh medium, as described above. One important finding was that activation of A2a receptors promotes CREB phosphorylation in E8 cultures and CREB dephosphorylation in E6 cultures, and two distinct signaling pathways mediated these different effects. In E6 cultures, CREB phosphorylation involves the phospholipase C/PKC pathway whereas in E8 cultures stimulation of CREB phosphorylation depends on the cAMP/PKA pathway. However, both pathways were triggered by activation of A2aRs, indicating that a shift in signaling pathways stimulated by A2aR acvtivation is occurring in the developmental period between E6 and E8 (Figure 3). These results were also observed in the in vivo developing retina and highlight the possibility that adenosine is regulating the survival of different cell progenitors during retinal development [155].

REFERENCES

[1] Haydon, P.G.; McCobb, D.P.; Kater, S.B. Serotonin selectively inhibits growth cone motility and synaptogenesis of specific identified neurons. *Science* 1984, 226(4674), 561-564.

[2] Lankford, K.L.; de Mello, F.G.; Klein, W.L. D1-type dopamine receptors inhibit growth cone motility in cultured retina neurons: evidence that neurotransmitters act as morphogenic growth regulators in the developing central nervous system. *Proc. Natl. Acad. Sci. USA.* 1988, 85(8), 2839-2843.

[3] Greenberg, M.E.; Xu, B.; Lu, B.; Hempstead, B.L. New insights in the biology of BDNF synthesis and release: implications in CNS function. *J. Neurosci.* 2009, 29(41), 12764-12767.

[4] Schoen S.W.; Leutenecker B.; Kreutzberg G.W.; Singer W. Ocular dominance plasticity and developmental changes of 5'-nucleotidasedistributions in the kitten visual cortex. *J. Comp. Neurol.* 1990, 296(3), 379-392.

[5] Schoen S.W.; Graeber M.B.; Tóth L.; Kreutzberg G.W. Synaptic 5'-nucleotidase is transient and indicative of climbing fiber plasticity during the postnatal development of rat cerebellum. *Brain Res. Dev. Brain Res.* 1991, 61(1), 125-138.

[6] Fenoglio C.; Scherini E.; Vaccarone R.; Bernocchi G. A re-evaluation of the ultrastructural localization of 5'-nucleotidase activity in the developing rat cerebellum, with a cerium-based method. *J. Neurosci. Methods* 1995, 59(2), 253-263.

[7] Braun N.; Brendel P.; Zimmermann H. Distribution of 5'-nucleotidase in the developing mouse retina. *Brain Res. Dev. Brain Res.* 1995, 88(1), 79-86.

[8] Zimmermann H. Biochemistry, localization and functional roles of ecto-nucleotidases in the nervous system. *Prog. Neurobiol.* 1996, 49(6), 589-618.

[9] Stanojević I.; Bjelobaba I.; Nedeljković N.; Drakulić D.; Petrović S.; Stojiljković M.; Horvat A. Ontogenetic profile of ecto-5'-nucleotidase in rat brain synaptic plasma membranes. *Int. J. Dev. Neurosci.* 2011, 29(4), 397-403.

[10] Geiger J.D.; Nagy J.I.; Ontogenesis of adenosine deaminase activity in rat brain. *J. Neurochem.* 1987, 48(1), 147-153.

[11] Senba E.; Daddona P.E.; Nagy J.I.; Development of adenosine deaminase-immunoreactive neurons in the rat brain. *Brain Res.* 1987, 428(1), 59-71.

[12] Senba E.; Daddona P.E.; Nagy J.I.; Adenosine deaminase-containing neurons in the olfactory system of the rat during development. *Brain Res. Bull.* 1987, 18(5), 635-648.

[13] Iijima R.; Kunieda T.; Yamaguchi S.; Kamigaki H.; Fujii-Taira I.; Sekimizu K.; Kubo T.; Natori S.; Homma K.J. The extracellular adenosine deaminase growth factor, ADGF/CECR1, plays a role in Xenopus embryogenesis via the adenosine/P1 receptor. *J. Biol. Chem.* 2008, 283(4), 2255-2264.

[14] Benowitz L.I.; Jing Y.; Tabibiazar R.; Jo S.A.; Petrausch B.; Stuermer C.A.O.; Rosenberg P.A.; Irwin N. Axon outgrowth is regulated by an intracellular purine-sensitive mechanism in retinal ganglion cells. *J. Biol. Chem.* 1998, 273(45), 29626–29634.

[15] Benowitz L.I.; Goldberg D.E.; Madsen J.R.; Soni D.; Irwin N. Inosine stimulates extensive axon collateral growth in the rat corticospinal tract after injury. *Proc. Natl. Acad. Sci. USA* 1999, 96(23), 13486–13490.

[16] Soni D.; Aoki C.; Yang M.; Black P.M.; Irwin N.; Benowitz L.I. Inosine induces the growth of new corticospinal tract synapses in the injured rat spinal cord. *Soc. Neurosci. Abstr.* 2000, 26, 863.

[17] Santos-Rodrigues A.; Ferreira J.M.; Paes-de-Carvalho R. Differential adenosine uptake in mixed neuronal/glial or purified glial cultures of avian retinal cells: Modulation by adenosine metabolism and the ERK cascade. *Biochem. Biophys. Res. Comm.* 2011, 414(1), 175-180.

[18] Hebb M.O.; White T.D. Co-administration of adenosine kinase and deaminase inhibitors produces supra-additive potentiation of N-methyl-D-aspartate-evoked adenosine formation in cortex. *Eur. J. Pharmacol.* 1998, 344 (2-3), 121-125.

[19] Studer F.E.; Fedele D.E.; Marowsky A.; Schwerdel C.; Wernli K.; Vogt K.; Fritschy J.M.; Boison D. Shift of adenosine kinase expression from neurons to astrocytes during postnatal development suggests dual functionality of the enzyme. *Neuroscience* 2006, 142(1), 125-137.

[20] Shen H.Y.; Singer P.; Lytle N.; Wei C.J.; Lan J.Q.; Williams-Karnesky R.L.; Chen J.F.; Yee B.K.; Boison D. Adenosine augmentation ameliorates psychotic and cognitive endophenotypes of schizophrenia. *J. Clin. Invest.* 2012, 122(7), 2567-2577.

[21] Aronica E.; Zurolo E.; Iyer A.; de Groot M.; Anink J.; Carbonell C.; van Vliet E.A.; Baayen J.C.; Boison D.; Gorter J.A. Upregulation of adenosine kinase in astrocytes in experimental and human temporal lobe epilepsy. *Epilepsia* 2011, 52(9), 1645-1655.

[22] Geiger J.D. Adenosine uptake and [3H] nitrobenzylthioinosine binding in developing rat brain. *Brain Res.* 1987, 436(2), 265-272.

[23] Morgan P.F.; Montgomery P.; Marangos P.J. Ontogenetic profile of the adenosine uptake sites in rat forebrain. *J. Neurochem.* 1987, 49(3), 852-855.

[24] Deckert J.; Morgan P.F.; Daval J.L.; Nakajima T.; Marangos P.J. Ontogeny of adenosine uptake sites in guinea pig brain: differential profile of [3H]

nitrobenzylthioinosine and [3H] dipyridamole binding sites. *Brain Res.* 1988, 470(2), 313-316.

[25] Hefner K.E.; Butler R.; Ravindra A.; Ahmad Z.; Molina D.; Turner C.P. Dipyridamole promotes changes in calbindin-D28k and tyrosine hydroxylase expression in neonatal rats. *Neonatology.* 2007, 91(4), 222-232.

[26] Cunha R.A. Adenosine as a neuromodulator and as a homeostatic regulator in the nervous system: different roles, different sources and different receptors. *Neurochem. Int.* 2001, 38(2), 107-125.

[27] Moreau J.L.; Huber G. Central adenosine A(2A) receptors: an overview. *Brain Res. Brain Res. Rev.* 1999, 31(1), 65-82.

[28] Dixon A.K.; Gubitz A.K.; Sirinathsinghji D.J.; Richardson P.J.; Freeman T.C. Tissue distribution of adenosine receptor mRNAs in the rat. *Br. J. Pharmacol.* 1996, 118(6), 1461-1468.

[29] Rivkees S.A.; Zhao Z.; Porter G.; Turner C. Influences of adenosine on the fetus and newborn. *Mol. Genet. Metab.* 2001, 74(1-2), 160-171.

[30] Rivkees S.A. The ontogeny of cardiac and neural A1 adenosine receptor expression in rats. *Brain Res. Dev. Brain Res.* 1995, 89(2), 202-213.

[31] Weaver D.R. A1-adenosine receptor gene expression in fetal rat brain. *Brain Res. Dev. Brain Res.* 1996, 94(2), 205-223.

[32] Morgan P.F.; Deckert J.; Nakajima T.; Daval J.L.; Marangos P.J. Late ontogenetic development of adenosine A1 receptor coupling to associated G-proteins in guinea pig cerebellum but not forebrain. *Mol. Cell Biochem.* 1990, 92(2), 169-176.

[33] Aden U.; Herlenius E.; Tang L.Q.; Fredholm B.B. Maternal caffeine intake has minor effects on adenosine receptor ontogeny in the rat brain. *Pediatr. Res.* 2000, 48(2), 177-183.

[34] Aoki C. Development of the A1 adenosine receptors in the visual cortex of cats, dark-reared and normally reared. *Brain Res.* 1985, 354(1), 125-133.

[35] Weaver D.R. A2a adenosine receptor gene expression in developing rat brain. *Brain Res. Mol. Brain Res.* 1993, 20(4), 313-327.

[36] Leon D.; Albasanz J.L.; Ruiz M.A.; Fernandez M.; Martin M. Adenosine A1 receptor down-regulation in mothers and fetal brain after caffeine and theophylline treatments to pregnant rats. *J. Neurochem.* 2002, 82(3), 625-634.

[37] Leon D.; Albasanz J.L.; Ruiz M.A.; Martin M. Chronic caffeine or theophylline intake during pregnancy inhibits A1 receptor function in the rat brain. *Neuroscience* 2005, 131(2), 481-489.

[38] Leon D.A.; Castillo C.A.; Albasanz J.L.; Martin M. Reduced expression and desensitization of adenosine A1 receptor/adenylyl cyclase pathway after chronic (-)N6-phenylisopropyladenosine intake during pregnancy. *Neuroscience* 2009, 163(2), 524-532.

[39] Silva C.G.; Metin C.; Fazeli W.; Machado N.J.; Darmopil S.; Launay P.S.; Ghestem A.; Nesa M.P.; Bassot E.; Szabó E.; Baqi Y.; Müller C.E.; Tomé A.R.; Ivanov A.; Isbrandt D.; Zilberter Y.; Cunha R.A.; Esclapez M.; Bernard C. Adenosine receptor antagonists including caffeine alter fetal brain development in mice. *Sci. Transl. Med.* 2013, 5(197), 197ra104.

[40] Kubrusly R.C.; Bhide P.G. Cocaine exposure modulates dopamine and adenosine signaling in the fetal brain. *Neuropharmacology* 2010, 58(2), 436-443.

[41] Migita H.; Kominami K.; Higashida M.; Maruyama R.; Tuchida N.; McDonald F.; Shimada F.; Sakurada K. Activation of adenosine A1 receptor-induced neural stem cell proliferation via MEK/ERK and Akt signaling pathways. *J Neurosci Res.* 2008, 86(13), 2820-2828.

[42] Stevens B.; Porta S.; Haak L.L.; Gallo V.; Fields R.D. Adenosine: a neuron-glial transmitter promoting myelination in the CNS in response to action potentials. *Neuron.* 2002, 36(5), 855-868.

[43] Cheng H.C.; Shih H.M.; Chern Y. Essential role of cAMP-response element-binding protein activation by A2A adenosine receptors in rescuing the nerve growth factor-induced neurite outgrowth impaired by blockage of the MAPK cascade. *J. Biol. Chem.* 2002, 277(37), 33930-33942.

[44] Shaban M.; Smith R.A.; Stone T.W. Adenosine receptor-mediated inhibition of neurite outgrowth from cultured sensory neurons is via an A1 receptor and is reduced by nerve growth factor. *Brain Res. Dev. Brain Res.* 1998, 105(2), 167-173.

[45] Coppi E.; Cellai L.; Maraula G.; Pugliese A.M.; Pedata F. Adenosine A(2)A receptors inhibit delayed rectifier potassium currents and cell differentiation in primary purified oligodendrocyte cultures. *Neuropharmacology* 2013, 73, 301-310.

[46] Stevens B.; Ishibashi T.; Chen J.F.; Fields R.D. Adenosine: an activity-dependent axonal signal regulating MAP kinase and proliferation in developing Schwann cells. *Neuron Glia Biol.* 2004, 1(1), 23-34.

[47] Benowitz L.I.; Jing Y.; Tabibiazar R.; Jo S.A.; Petrausch B.; Stuermer C.A.; Rosemberg P.A.; Irwin N. Axon outgrowth is regulated by an intracellular purine-sensitive mechanism in retinal ganglion cells. *J. Biol. Chem.* 1998, 273(43), 29626-29634.

[48] Sun C.N.; Chuang H.C.; Wang J.Y.; Chen S.Y.; Cheng Y.Y.; Lee C.F.; Chern Y. The A2A adenosine receptor rescues neuritogenesis impaired by p53 blockage via KIF2A, a kinesin family member. *Dev. Neurobiol.* 2010, 70(8), 604-621.

[49] Charles M.P.; Adamski D.; Kholler B.; Pelletier L.; Berger F.; Wion D. Induction of neurite outgrowth in PC12 cells by the bacterial nucleoside N6-methyldeoxyadenosine is mediated through adenosine A2a receptors and via cAMP and MAPK signaling pathways. *Biochem. Biophys. Res. Comm.* 2003, 304(4), 795-800.

[50] Stettler O.; Joshi R.L.; Wizenmann A.; Reingruber J.; Holcman D.; Bouillot C.; Castagner F.; Prochiantz A.; Moya K.L. Engrailed homeoprotein recruits the adenosine A1 receptor to potentiate ephrin A5 function in retinal growth cones. *Development* 2012, 139(1), 215-224.

[51] Moore S.W.; Tessier-Lavigne M.; Kennedy T.E. Netrins and their receptors. Adv. *Exp. Med. Biol.* 2007, 621, 17-31.

[52] Corset V.; Nguyen-Ba-Charvet K.T.; Forcet C.; Moyse E.; Chedotal A.; Mehlen P. Netrin-1-mediated axon outgrowth and cAMP production requires interaction with adenosine A2b receptor. *Nature* 2000, 407(6805), 747-750.

[53] Stein E.; Zou Y.; Poo M.; Tessier-Lavigne M. Binding of DCC by netrin-1 to mediate axon guidance independent of adenosine A2B receptor activation. *Science* 2001, 291(5510), 1976-1982.

[54] McKenna W.L.; Wong-Staal C.; Kim G.C.; Macias H.; Hinck L.; Bartoe J.L. Netrin-1-independent adenosine A2b receptor activation regulates the response of axons to

netrin-1 by controlling cell surface levels of UNC5A receptors. *J. Neurochem.* 2008, 104(4), 1081-1090.

[55] Dingwell K.S.; Holt C.E.; Harris W.A. The multiple decisions made by growth cones of RGCs as they navigate from the retina to the tectum in Xenopus embryos. *J. Neurobiol.* 2000, 44(2), 246-259.

[56] Shewan D.; Dwivedy A.; Anderson R.; Holt C.E. Age-related changes underlie switch in netrin-1 responsiveness as growth cones advance along visual pathway. *Nat. Neurosci.* 2002, 5(10), 955-962.

[57] Reppert S.M.; Weaver D.R.; Stehle J.H.; Rivkees S.A. Molecular Cloning and Characterization of a Rat A1-Adenosine Receptor that is Widely Expressed in *Brain and Spinal Cord. Mol. Endocrinol.* 1991, 5(8), 1037-1048.

[58] Dixon A.K.; Gubitz A.K.; Sirinathsinghji D.J.; Richardson P.J.; Freeman T.C. Tissue distribution of adenosine receptor mRNAs in the rat. *Br. J. Pharmacol.* 1996, 118(6), 1461-1468.

[59] Munshi R.; Pang I.H.; Sternweis P.; Linden J. A1 adenosine receptors of bovine brain couple to guanine nucleotide-binding proteins Gi1, Gi2, and Go. *J. Biol. Chem.* 1991, 266(33), 22285-22289.

[60] Freissmuth M.; Schütz W.; Linder M. Interactions of the bovine brain A1-adenosine receptor with recombinant G protein alpha-subunits. Selectivity for rGi alpha-3. *J. Biol. Chem.* 1991, 266(27), 17778-17783.

[61] Gundlfinger A.; Bischofberger J.; Johenning F.; Torvinen M.; Schmitz D.; Breustedt J. Adenosine modulates transmission at the hippocampal mossy fibre synapse via direct inhibition of presynaptic calcium channels. *J. Physiol.* 2007, 582(1), 263-277.

[62] Jeong H.J.; Jang I.S.; Nabekura J.; Akaike N. Adenosine A1 receptor-mediated presynaptic inhibition of GABAergic transmission in immature rat hippocampal CA1 neurons. *J. Neurophysiol.* 2003, 89(3), 1214-1222.

[63] Carruthers A.M.; Sellers L.A.; Jenkins D.W.; Jarvie E.M.; Feniuk W.; Humphrey P.P. Adenosine A1 receptor-mediated inhibition of protein kinase A-induced calcitonin gene-related peptide release from rat trigeminal neurons. *Mol. Pharmacol.* 2001, 59(6), 1533-1541.

[64] Dolphin A.; Forda S.; Scott R. Calcium-dependent currents in cultured rat dorsal root ganglion neurones are inhibited by an adenosine analogue. *J. Physiol.* 1986, 373(1), 47-61.

[65] Trussell L.O.; Jackson M.B. Adenosine-activated potassium conductance in cultured striatal neurons. *Proc. Natl. Acad. Sci. USA* 1985, 82(14), 4857-4861.

[66] Rogel A.; Bromberg Y.; Sperling O.; Zoref-Shani E. Phospholipase C is involved in the adenosine-activated signal transduction pathway conferring protection against iodoacetic acid-induced injury in primary rat neuronal cultures. *Neurosci. Lett.* 2005, 373(3), 218-221.

[67] Biber K.; Klotz K.N.; Berger M.; Gebicke-Härter P.J.; Van-Calker D. Adenosine A1 receptor-mediated activation of phospholipase C in cultured astrocytes depends on the level of receptor expression. *J. Neurosci.* 1997, 17(13), 4956-4964.

[68] Saura C.; Ciruela F.; Casadó V.; Canela E.I.; Mallol J.; Lluis C.; Franco R. Adenosine deaminase interacts with A1 adenosine receptors in pig brain cortical membranes. *J. Neurochem.* 1996, 66(4), 1675-1682.

[69] Gracia E.; Cortés A.; Meana J.J.; García-Sevilla J.; Herhsfield M.S.; Canela E.I.; Mallol J.; Lluís C.; Franco R.; Casadó V. Human adenosine deaminase as an allosteric modulator of human A1 adenosine receptor: abolishment of negative cooperativity for [3H](R)-pia binding to the caudate nucleus. *J. Neurochem.* 2008, 107(1), 161-170.

[70] Fredholm B.B. Purinoceptors in the Nervous System. *Pharmacol. Toxicol.* 1995, 76(4), 228-239.

[71] Marala R.B.; Mustafa S.J. Direct evidence for the coupling of A2-adenosine receptor to stimulatory guanine nucleotide-binding-protein in bovine brain striatum. *J. Pharmacol. Exp. Therapeut.* 1993, 266(1), 294-300.

[72] Oliveira L.; Correia-de-Sá P. Protein kinase A and Ca(v)1 (L-Type) channels are common targets to facilitatory adenosine A2A and muscarinic M1 receptors on rat motoneurons. *Neurosignals* 2005, 14(5), 262-272.

[73] Ibrisimovic E.; Drobny H.; Yang Q.; Hofer T.; Boehm S.; Nanoff C.; Schicker K. Constitutive activity of the A2A adenosine receptor and compartmentalised cyclic AMP signalling fine-tune noradrenaline release. *Purinergic Signal.* 2012, 8(4), 677-692.

[74] Cristovão-Ferreira S.; Vaz S.H.; Ribeiro J.A.; Sebastião A.M. Adenosine A2A receptors enhance GABA transport into nerve terminals by restraining PKC inhibition of GAT-1. *J. Neurochem.* 2009, 109(2), 336-347.

[75] Higley M.J.; Sabatini B.L. Competitive regulation of synaptic Ca2+ influx by D2 dopamine and A2A adenosine receptors. *Nat. Neurosci.* 2010, 13(8), 958-966.

[76] Dias R.B.; Ribeiro J.A.; Sebastião A.M. Enhancement of AMPA currents and GluR1 membrane expression through PKA-coupled adenosine A(2A) receptors. *Hippocampus* 2012, 22(2), 276-291.

[77] Li H.; Zhang Z.; Blackburn M.R.; Wang S.W.; Ribelayga C.P.; O'Brien J. Adenosine and dopamine receptors coregulate photoreceptor coupling via gap junction phosphorylation in mouse retina. *J Neurosci.* 2013, 33(7), 3135-3150.

[78] Ciruela F.; Albergaria C.; Soriano A.; Cuffí L.; Carbonell L.; Sánchez S.; Gandía J.; Fernández-Dueñas V. Adenosine receptors interacting proteins (ARIPs): behind the biology of adenosine signaling. *Biochim. Biophys. Acta* 2010, 1798(1), 9-20.

[79] Sun C.N.; Cheng H.C.; Chou J.I.; Lee S.Y.; Lin Y.W.; Lai H.L.; Chen H.M.; Chern Y. Rescue of p53 blockage by the A2A adenosine receptor via a novel interacting protein, translin-associated protein X. *Mol. Pharmacol.* 2006, 70(2), 454-466.

[80] Aherne C.M.; Kewley E.M.; Eltzschig H.K. The resurgence of A2B adenosine receptor signaling. *Biochim. Biophys. Acta* 2011, 1808(5), 1329-1339.

[81] Kataoka S.; Baquero A.; Yang D.; Shultz N.; Vandenbeuch A.; Ravid K.; Kinnamon S.C.; Finger T.E. A2BR adenosine receptor modulates sweet taste in circumvallate taste buds. *PloS one.* 2012, 7(1), e30032.

[82] Bilodeau M.L.; Ji M.; Paris M.; Andrisani O.M. Adenosine signaling promotes neuronal, catecholaminergic differentiation of primary neural crest cells and CNS-derived CAD cells. *Mol. Cel. Neurosci.* 2005, 29(3), 394-404.

[83] Feoktistov I.; Biaggioni I. Adenosine A2B receptors. *Pharmacol. Rev.* 1997, 49(4), 381-402.

[84] Mogul D.J.; Adams M.E.; Fox A.P. Differential activation of adenosine receptors decreases N type but potentiates P type Ca2+ current in hippocampal CA3 neurons. *Neuron* 1993, 10(2), 327-334.

[85] Kessey K.; Trommer B.L.; Overstreet L.S.; Ji T.; Mogul D.J. A role for adenosine A2 receptors in the induction of long-term potentiation in the CA1 region of rat hippocampus. *Brain Res.* 1997, 756(1), 184-190.

[86] Herrera C.; Casadó V.; Ciruela F.; Schofield P.; Mallol J.; Lluis C.; Franco R. Adenosine A2B receptors behave as an alternative anchoring protein for cell surface adenosine deaminase in lymphocytes and cultured cells. *Mol. Pharmacol.* 2001, 59(1), 127-134.

[87] Gutierrez H.; Davies A.M. Regulation of neural process growth, elaboration and structural plasticity by NF-kappaB. *Trends Neurosci.* 2011, 34(6), 316-325.

[88] Sun Y.; Duan Y.; Eisenstein A.S.; Hu W.; Quintana A.; Lam W.K.; Wang Y.; Wu Z.; Ravid K.; Huang P. A novel mechanism of control of NF-kappaB activation and inflammation involving A2B adenosine receptors. *J. Cell Sci.* 2012, 125(19), 4507-4517.

[89] Zhou Q.Y.; Li C.; Olah M.E.; Johnson R.A.; Stiles G.L.; Civelli O. Molecular cloning and characterization of an adenosine receptor: the A3 adenosine receptor. *Proc. Natl. Acad. Sci. USA 1*992, 89(16), 7432-7436.

[90] Fleming K.; Mogul D. Adenosine A3 Receptors Potentiate Hippocampal Calcium Current by a PKA-dependent/PKC-independent Pathway. *Neuropharmacology* 1997, 36(3), 353-362.

[91] Nörenberg W.; Wirkner K.; Illes P. Effect of adenosine and some of its structural analogues on the conductance of NMDA receptor channels in a subset of rat neostriatal neurones. *Br. J. Pharmacol.* 1997, 122(1), 71-80.

[92] Donoso M.V.; Aedo F.; Huidobro-Toro J.P. The role of adenosine A2A and A3 receptors on the differential modulation of norepinephrine and neuropeptide Y release from peripheral sympathetic nerve terminals. *J. Neurochem.* 2006, 96(6), 1680-1695.

[93] Xie K.Q.; Zhang L.M.; Cao Y.; Zhu J.; Feng L.Y. Adenosine A(1) receptor-mediated transactivation of the EGF receptor produces a neuroprotective effect on cortical neurons in vitro. *Acta Pharmacol. Sin.* 2009, 30(7), 889-898.

[94] Katoh-Semba R.; Kaneko R.; Kitajima S.; Tsuzuki M.; Ichisaka S.; Hata Y.; Yamada H.; Miyazaki N.; Takahashi Y.; Kato K. Activation of p38 mitogen-activated protein kinase is required for in vivo brain-derived neurotrophic factor production in the rat hippocampus. *Neuroscience* 2009, 163(1), 352-361.

[95] Moidunny S.; Dias R.B.; Wesseling E.; Sekino Y.; Boddeke H.W.; Sebastião A.M.; Biber K. Interleukin-6-type cytokines in neuroprotection and neuromodulation: oncostatin M, but not leukemia inhibitory factor, requires neuronal adenosine A1 receptor function. *J. Neurochem.* 2010, 114(6), 1667-1677.

[96] Stettler O.; Joshi R.L.; Wizenmann A.; Reingruber J.; Holcman D.; Bouillot C.; Castagner F.; Prochiantz A.; Moya K.L. Engrailed homeoprotein recruits the adenosine A1 receptor to potentiate ephrin A5 function in retinal growth cones. *Development* 2012, 139(1), 215-224.

[97] Perigolo-Vicente R.; Ritt K.; Pereira M.R.; Torres P.M.; Paes-de-Carvalho R.; Giestal-de-Araujo E. IL-6 treatment increases the survival of retinal ganglion cells in vitro: the role of adenosine A1 receptor. *Biochem. Biophys. Res. Commun.* 2013, 430(2), 512-518.

[98] Perigolo-Vicente R.; Ritt K.; Gonçalves-de-Albuquerque C.F.; Castro-Faria-Neto H.C.; Paes-de-Carvalho R.; Giestal-de-Araujo E. IL-6, A1 and A2aR: a crosstalk that

modulates BDNF and induces neuroprotection. *Biochem. Biophys. Res. Commun.* 2014, 449(4), 477-482.

[99] Lee F.S.; Chao M.V. Activation of Trk neurotrophin receptors in the absence of neurotrophins. *Proc. Natl. Acad. Sci. USA* 2001, 98(6), 3555-3560.

[100] Rajagopal R.; Chen Z.Y.; Lee F.S.; Chao M.V. Transactivation of Trk neurotrophin receptors by G-protein-coupled receptor ligands occurs on intracellular membranes. *J Neurosci.* 2004, 24(30), 6650-6658.

[101] Diogenes M.J.; Fernandes C.C.; Sebastião A.M.; Ribeiro J.A. Activation of adenosine A2A receptor facilitates brain-derived neurotrophic factor modulation of synaptic transmission in hippocampal slices. *J. Neurosci.* 2004, 24(12), 2905-2913.

[102] Diogenes M.J.; Assaife-Lopes N.; Pinto-Duarte A.; Ribeiro J.A.; Sebastião A.M. Influence of age on BDNF modulation of hippocampal synaptic transmission: interplay with adenosine A2A receptors. *Hippocampus* 2007, 17(7), 577-585.

[103] Golder F.J.; Ranganathan L.; Satriotomo I.; Hoffman M.; Lovett-Barr M.R.; Watters J.J.; Baker-Herman T.L.; Mitchell G.S. Spinal adenosine A2a receptor activation elicits long-lasting phrenic motor facilitation. *J. Neurosci.* 2008, 28(9), 2033-2042.

[104] Jeon S.J.; Rhee S.Y.; Ryu J.H.; Cheong J.H.; Kwon K.; Yang S.I.; Park S.H.; Lee J.; Kim H.Y.; Han S.H.; Ko K.H.; Shin C.Y. Activation of adenosine A2A receptor up-regulates BDNF expression in rat primary cortical neurons. *Neurochem. Res.* 2011, 36(12), 2259-2269.

[105] Ramirez S.H.; Fan S.; Maguire C.A.; Perry S.; Hardiek K.; Ramkumar V.; Gelbard H.A.; Dewhurst S.; Maggirwar S.B. Activation of adenosine A2A receptor protects sympathetic neurons against nerve growth factor withdrawal. *J. Neurosci. Res.* 2004, 77(2), 258-269.

[106] Flajolet M.; Wang Z.; Futter M.; Shen W.; Nuangchamnong N.; Bendor J.; Wallach I.; Nairn A.C.; Surmeier D.J.; Greengard P. FGF acts as a co-transmitter through adenosine A(2A) receptor to regulate synaptic plasticity. *Nat. Neurosci.* 2008, 11(12), 1402-1409.

[107] Vannucci R.C.; Vannucci S.J. Hypoglycemic brain injury. *Semin. Neonatol.* 2001, 6(2), 147-155.

[108] Ryan C.; Vega A.; Drash A. Cognitive deficits in adolescents who developed diabetes early in life. *Pediatrics* 1985, 75(5), 921-927.

[109] Kaufman F.R. Diabetes in children and adolescents. Areas of controversy. *Med. Clin. North Am.* 1998, 82(4), 721-738.

[110] Kaufman F.R.; Epport K.; Engilman R.; Halvorson M. Neurocognitive functioning in children diagnosed with diabetes before age 10 years. *J. Diabetes Complications* 1999, 13(1), 31-38.

[111] Dunwiddie T.V.; Masino S.A. The role and regulation of adenosine in the central nervous system. *Ann. Rev. Neurosci.* 2001, 24, 31-55.

[112] Chapman A.G.; Westerberg E.; Siesjo B.K. The metabolism of purine and pyrimidine nucleotides in rat cortex during insulin-induced hypoglycemia and recovery. *J. Neurochem.* 1981, 36(1), 179-189.

[113] de Mendonça A.; Sebastião A.M.; Ribeiro J.A. Adenosine: does it have a neuroprotective role after all? *Brain Res. Brain Res. Rev.* 2000, 33(2-3), 258-274.

[114] Kim M.; Yu Z.X.; Fredholm B.B.; Rivkees S.A. Susceptibility of the developing brain to acute hypoglycemia involving A1 adenosine receptor activation. *Am. J. Physiol. Endocrinol. Metab.* 2005, 289(4), E562-E569.

[115] Turner C.P.; Blackburn M.R.; Rivkees S.A.; A1 adenosine receptors mediate hypoglycemia-induced neuronal injury. *J. Mol. Endocrinol.* 2004, 32(1), 129-144.

[116] Turner C.P.; Pulciani D.; Rivkees S.A. Reduction in intracellular calcium levels induces injury in developing neurons. *Exp. Neurol.* 2002, 178(1), 21-32.

[117] Bona E.; Aden U.; Gilland E.; Fredholm B.B.; Hagberg H. Neonatal cerebral hypoxia-ischemia: the effect of adenosine receptor antagonists. *Neuropharmacology* 1997, 36(9), 1327-1338.

[118] Gidday J.M.; Fitzgibbons J.C.; Shah A.R.; Kraujalis M.J.; Park T.S. Reduction in cerebral ischemic injury in the newborn rat by potentiation of endogenous adenosine. *Pediatric Res.* 1995, 38(3), 306-311.

[119] Chen G.J.; Harvey B.K.; Shen H.; Chou J.; Victor A.; Wang Y. Activation of adenosine A3 receptors reduces ischemic brain injury in rodents. *J. Neurosci. Res.* 2006, 84(8), 1848-1855.

[120] Hack M.; Taylor H.G.; Drotar D.; Schluchter M.; Cartar L.; Andreias L.; Wilson-Costello D.; Klein N. Chronic conditions, functional limitations, and special health care needs of school-aged children born with extremely low-birth-weight in the 1990s. *J.A.M.A.* 2005, 294(3), 318-325.

[121] Litt J.; Taylor H.G.; Klein N.; Hack M. Learning disabilities in children with very low-birth-weight: prevalence, neuropsychological correlates, and educational interventions. *J. Learning Disabilities* 2005, 38(2), 130-141.

[122] Back S.A.; Rivkees S.A. Emerging concepts in periventricular white matter injury. *Sem. Perinatol.* 2004, 28(6), 405-414.

[123] Back S.A.; Luo N.L.; Mallinson R.A.; O'Malley J.P.; Wallen L.D.; Frei B.; Morrow J.D.; Petito C.K.; Roberts C.T.Jr.; Murdoch G.H.; Montine T.J. Selective vulnerability of preterm white matter to oxidative damage defined by F2-isoprostanes. *Ann. Neurol.* 2005, 58(1), 108-120.

[124] Back S.A.; Craig A.; Luo N.L.; Ren J.; Akundi R.S.; Ribeiro I.; Rivkees S.A.; Protective effects of caffeine on chronic hypoxia-induced perinatal white matter injury. *Ann. Neurol.* 2006, 60(6), 696-705.

[125] Millar D.; Schmidt B. Controversies surrounding xanthine therapy. *Semin. Neonatol.* 2004, 9(3), 239-244.

[126] Fredholm B.B. Astra Award Lecture. Adenosine, adenosine receptors and the actions of caffeine. *Pharmacol. Toxicol.* 1995, 76(2), 93-101.

[127] Rivkees S.A.; Wendler C.C. Adverse and protective influences of adenosine on the newborn and embryo: implications for preterm white matter injury and embryo protection. *Pediatr. Res.* 2011, 69(4), 271-278.

[128] Turner C.P.; Yan H.; Schwartz M.; Othman T.; Rivkees S.A. A1 adenosine receptor activation induces ventriculomegaly and white matter loss. *Neuroreport.* 2002, 13(9), 1199-1204.

[129] Turner C.P.; Seli M.; Ment L.; Stewart W.; Yan H.; Johansson B.; Fredholm B.B.; Blackburn M.; Rivkees S.A. A1 adenosine receptors mediate hypoxia-induced ventriculomegaly. *Proc. Natl. Acad. Sci. USA* 2003, 100(20), 11718-11722.

[130] Wakade T.D.; Palmer K.C.; McCauley R.; Przywara D.A.; Wakade A.R. Adenosine-induced apoptosis in chick embryonic sympathetic neurons: a new physiological role for adenosine. *J. Physiol.* 1995, 488(Pt1), 123-138.

[131] Wakade A.R.; Wakade T.D.; Kulkarni J.S. Adenosine protects chick embryonic sympathetic neurons from apoptotic death by 2'-deoxyadenosine -- Importance of ATP in apoptosis. *Neurosci. Lett.* 1998, 252(3), 163-166.

[132] Kulkarni J.S.; Przywara D.A.; Wakade T.D.; Wakade A.R. Adenosine induces apoptosis by inhibiting mRNA and protein synthesis in chick embryonic sympathetic neurons. *Neurosci. Lett.* 1998, 248(3), 187-190.

[133] Michel P.P.; Marien M.; Ruberg M.; Colpaert F.; Agid Y. Adenosine prevents the death of mesencephalic dopaminergic neurons by a mechanism that involves astrocytes. *J. Neurochem.* 1999, 72(5); 2074-2082.

[134] Fujita S.; Horii M. Analysis of cytogenesis in chick retina by tritiated thymidine autoradiography. *Arch. Histol. Jpn.* 1963, 23, 359-366.

[135] Kahn AJ. An autoradiographic analysis of the time of appearance of neurons in the developing chick neural retina. *Dev. Biol.* 1974, 38(1), 30-40.

[136] Hughes W.F.; LaVelle A. On the synaptogenic sequence in the chick retina. *Anat. Rec.* 1974, 179(3), 297-301.

[137] Kim S.U. Neuronal types in long-term culture of avian retina. *Experientia* 1971, 27(11), 1319-1320.

[138] Paes de Carvalho R.; de Mello F.G. Adenosine-elicited accumulation of adenosine 3'-5' - cyclic-monophosphate in the chick embryo retina. *J. Neurochem.* 1982, 38(2), 493-500.

[139] de Mello M.C.; Ventura A.L.; Paes de Carvalho R.; Klein W.L; de Mello F.G. Regulation of dopamine and adenosine-dependent adenylate cyclase systems of chicken embryo retina cells in culture. *Proc. Natl. Acad. Sci. USA* 1982, 79(18), 5708-5712.

[140] Paes de Carvalho R.; de Mello F.G. Expression of A1 adenosine receptors modulating dopamine-dependent cyclic AMP accumulation in the chick embryo retina. *J. Neurochem.* 1985, 44(3), 845-851.

[141] Paes de Carvalho R. Development of A1 adenosine receptors in the chick embryo retina. *J. Neurosci. Res.* 1990, 25(2), 236-242.

[142] Pereira M.R.; Hang V.R.; Vardiero E.; de Mello F.G.; Paes-de-Carvalho R. Modulation of A1 adenosine receptor expression by cell aggregation and long-term activation of A2a receptors in cultures of avian retinal cells: Involvement of the cyclic AMP/PKA pathway. *J. Neurochem.* 2010, 113(3), 661-673.

[143] Brito R.; Pereira M.R.; Paes-de-Carvalho R.; Calaza K.C. Development of A1 adenosine receptors in the avian retina: In vivo modulation by A2a receptors and endogenous adenosine. *J. Neurochem.* 2012, 123(2), 239-249.

[144] Paes de Carvalho R.; Braas K.M.; Adler R.; Snyder S.H. Developmental regulation of adenosine A1 receptors, uptake sites and endogenous adenosine in the chick retina. *Dev. Brain Res.* 1992, 70(1), 87-95.

[145] Braas K.M.; Zarbin M.A; Snyder S.H. Endogenous adenosine and adenosine receptors localized to ganglion cells of the retina. *Proc. Natl. Acad. Sci. USA* 1987, 84(11), 3906-3910.

[146] Paes de Carvalho R.; Braas K.M.; Snyder S.H.; Adler R. Analysis of adenosine immunoreactivity, uptake and release in purified cultures of developing chick embryo retinal neurons and photoreceptors. *J. Neurochem.* 1990, 55(5), 1603-1611.

[147] Paes de Carvalho R.; Dias B.V.; Martins R.A.; Pereira M.R.; Portugal C.C.; Lanfredi C. Activation of glutamate receptors promotes a calcium-dependent and transporter-mediated release of purines in cultured avian retinal cells: Possible involvement of calcium/calmodulin-dependent protein kinase II. *Neurochem. Int.* 2005, 46(6), 441-451.

[148] Mejía-García T.A.; Paes-de-Carvalho R. Nitric oxide regulates cell survival in purified cultures of avian retinal neurons: Involvement of multiple transduction pathways. *J. Neurochem*. 2007, 100(2); 382-394.

[149] Paes-de-Carvalho R.; Maia G.A.; Ferreira J.M. Adenosine regulates the survival of avian retinal neurons and photoreceptors in culture. *Neurochem. Res*. 2003, 28(10), 1583-1590.

[150] Plitman E.; Nakajima S.; de la Fuente-Sandoval C.; Gerretsen P.; Chakravarty M.M.; Kobylianskii J.; Chung J.K.; Caravaggio F.; Iwata Y.; Remington G.; Graff-Guerrero A. Glutamate-mediated excitotoxicity in schizophrenia: A review. *Eur. Neuropsychopharmacol.* 2014, 24(10), 1591-1605.

[151] Mehta A.; Prabhakar M.; Kumar P.; Deshmukh R.; Sharma P.L. Excitotoxicity: bridge to various triggers in neurodegenerative disorders. *Eur. J. Pharmacol.* 2013, 698(1-3), 6-18.

[152] Choi D.W. Excitotoxic cell death. *J. Neurobiol*. 1992, 23(9); 1261-1276.

[153] Ferreira J.M.; Paes-de-Carvalho R. Long-term activation of adenosine A2a receptors blocks glutamate excitotoxicity in cultures of avian retinal neurons. *Brain Res.* 900(2), 169-176.

[154] Adler R.; Hatlee M. Plasticity and differentiation of embryonic retinal cells after terminal mitosis. *Science* 1989, 243(4889), 391-393.

[155] Socodato R.; Brito R.; Calaza K.C.; Paes-de-Carvalho R. Developmental regulation of neuronal survival by adenosine in the in vitro and in vivo avian retina depends on a shift of signaling pathways leading to CREB phosphorylation or dephosphorylation. *J. Neurochem*. 2011, 116(2), 227-239.

[156] Paes-de-Carvalho R. Adenosine as a signaling molecule in the retina: Biochemical and developmental aspects. *An. Acad. Bras. Cienc*. 2002, 74(3), 437-451.

In: Adenosine Signaling Mechanisms
Editors: V. Ramkumar and R. Paes de Carvalho
ISBN: 978-1-63483-186-4

Chapter 6

ADENOSINE IN THE RETINA: ROLES AND PERSPECTIVES

Rafael Brito, Danniel Pereira-Figueiredo, Elisa Guimarães-Souza, Alexandre dos Santos-Rodrigues and Karin C. Calaza*

Neuroscience Program, Neurobiology Department,
Fluminense Federal University, Brazil

ABSTRACT

The vertebrate retina is a very well conserved tissue responsible for transducing light stimulus and pre-processing some aspects of visual information such as contrast, shape, movement and so on. The retina shares the same embryological origin of the brain and because of that is a central nervous system tissue. According to this, the retinal structure, functional organization and chemical mediators are similar to the brain. Adenosine is a nucleoside that plays its biological actions through four metabotropic receptors classified as A_1, A_{2A}, A_{2B} and A_3. A_1 and A_3 receptors may be coupled to $G_{i/o}$ protein, while A_{2A} and A_{2B} receptors may be coupled to $G_{s/olf}$ protein. Therefore, these receptors regulate the levels of intracellular cAMP in different ways. Unlike most chemical mediators present in the central nervous system, adenosine is released into the extracellular environment primarily through its transporters, which are classified into equilibrative and concentrative. The former transports adenosine in a bidirectional manner, without energy expenditure and in favor of its concentration gradient, and the latter transports adenosine to the intracellular environment against the concentration gradient in an energy-dependent manner. All the components of the adenosinergic system have been demonstrated in the retinal tissue of several vertebrate species. The presence of this complex system in the developing and mature retina suggests an important role for adenosine in retinal functionality. In fact, a number of evidences have reported the involvement of adenosine in the retinal events related to ionic conductance, circadian cycle, neuroprotection and release of chemical mediators. Furthermore, the adenosinergic

* Corresponding author:Dr. Karin C. Calaza, Neurobiology Department, Biology Institute, Fluminense Federal University, 24020-140, Niterói. Rio de Janeiro, Brasil. Email: karincalaza@gmail.com

system seems to be important in pathological conditions such as glaucoma and diabetic retinopathy. Thus, adenosine adds one more step to the understanding of how the retina works.

INTRODUCTION

One of the very well-studied models of the great neuroscientist and Nobel Prized Santiago Ramon-y-Cajal was the retina. This transparent and 200 μm thick tissue is the only light-sensitive part of the whole visual system. Apparently, the retina seemed to be a very simple tissue containing only five major classes of nerve cells but it turned out to be as complex as the brain. Another interesting point is the fact that the retina constitutes a whole neural circuitry. In this tissue, the external stimulus (light) is transduced, inducing a receptor potential in the light-sensitive receptor cells (photoreceptors) that send an afferent signal to retinal interneurons which, in turn, transmit to the efferent cells (ganglion cells) responsible to transfer the pre-processed information from retina to visual brain nuclei.

The vertebrate retina is a highly conserved tissue and consists of three layers of cell bodies, mainly neuronal cells (nuclear layers) separated by two layers with virtually no cell bodies, but enriched in synapses (synaptic layers). The photomicrograph showed in Figure 1A is a transverse section of a rat retina oriented in a way that the inside of the ocular globe, filled in by aqueous humor, is at the bottom. Following this arrangement, light from the visual scene would come from underneath. This is the convention used in visual science; clinicians usually present retinal images in an opposite way with photoreceptor appearing at the bottom [1].

The retina is also widely used as a model for studying the central nervous system since it has the same embryological origin, the neural ectoderm, and similar organization. During embryogenesis, the eye develops from the diencephalon (forebrain), which evaginates to form the optic vesicle. This last structure continues to be connected to the central nervous system from the optic stalk that later develops in the optic nerve. Finally, the retinal pigmented epithelium and the retina derive from the outer and inner part of the optic cup, respectively. A proper vision relies on the correct development of the retina which, in turn, depends on the accurate progress of neurogenesis, migration, natural cell death, synaptogenesis and differentiation process.

As mentioned above, the mature retina is organized in nuclear layers, called outer nuclear layer (ONL), inner nuclear layer (INL) and ganglion cell layer (GCL), interspersed by two synaptic layers, named outer plexiform layer (OPL) and inner plexiform layer (IPL). This laminar organization is represented in Figure 1B. The cell bodies of photoreceptors, which contain light-sensitive molecules (visual pigments), are localized in the ONL. There are two types of photoreceptors: rods and cones. It is amazing how sensitive the visual system is with respect to light intensity range (Figure 2). This is so because rods and cones have very different ranges of light intensity sensitivity due to specializations to low (rods) or high (cones) light levels (Figure 2). Thus, rods are highly sensitive to light stimulus and are very well adapted to dim light conditions (scotopic), being specialized to nocturnal vision with no discrimination of colors. On the other hand, cones have higher spatial resolution in expense of light sensitivity, and because of that operate in photopic conditions (higher levels of light like daylight).

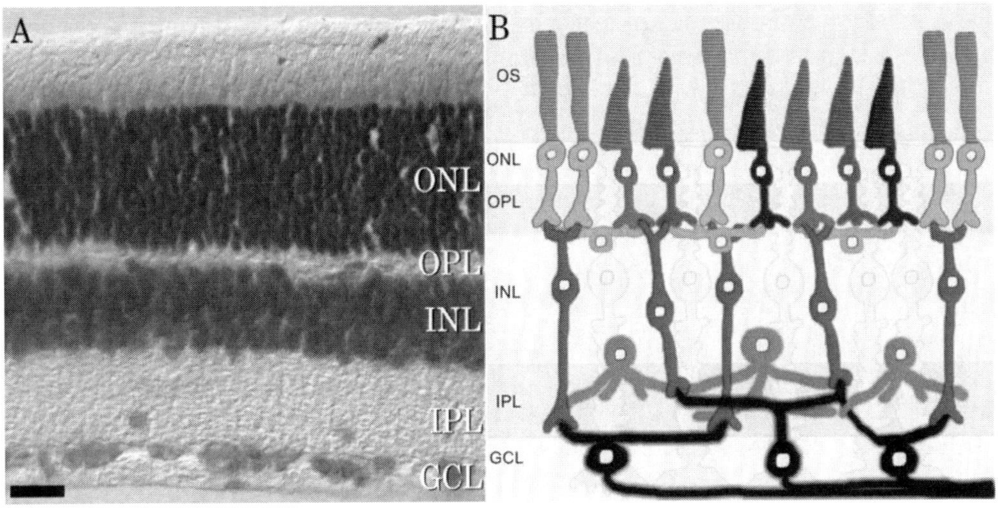

Figure 1. *Morphological organization of the retina.* A, Photomicrograph of radial section of rat retina counterstained with violet cresyl showing the nuclear layers intercalated with synaptic layers. Scale bar= 20 μm. B, Schematic drawing of the retinal layers: outer segments (OS), outer nuclear layer (ONL), outer plexiform layer (OPL), inner nuclear layer (INL), inner plexiform layer (IPL), ganglion cell layer (GCL). Retinal cell types are also represented as follows: rods in grey; cones in green, red and blue; horizontal cells in orange; bipolar cells in cyan; amacrine cells in magenta; ganglion cells in navy blue and Müller glial cells in yellow.

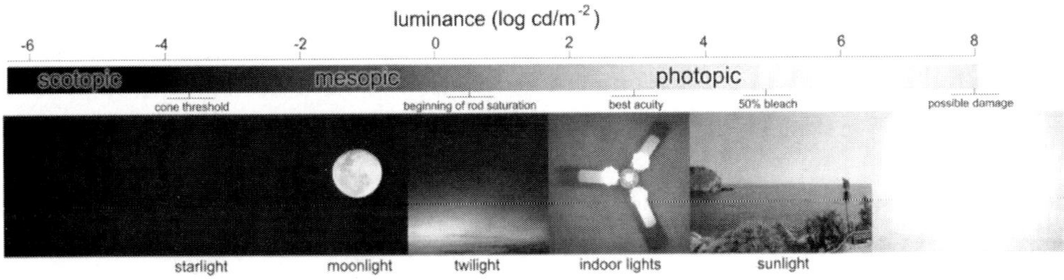

Figure 2. *The range of luminance over which the visual system works.* At scotopic condition, only rods respond to the low amount of photons. Cones begin to contribute to visual perception at the level of starlight and are the only photoreceptors that function under high levels of light (photopic condition).

So the visual acuity is given by the cone system while light sensitivity is given by the rod system. Another very important property of the cone system is the ability to extract color information. Primates have three different types of cones (named S, M, and L), and each one has a specific visual pigment that is better stimulated by a distinct portion of the visual spectrum of light. In this visual range, Short cones (S cones) are more sensitive to short-wavelength photons (or higher-frequencies); Medium cones (M cones) to medium-wavelength (or medium-frequencies), and Long cones (L cones) to longer wavelength (or low-frequencies) photons. Animals that have three types of cones are called trichromats while animals expressing two types of cones are named dichromats. The latter are still able to discriminate colors even though in a different way of trichromats, discriminating fewer colors. The majority of non-primate mammals have S cones and a type that could be referred as M/L cone, and based on this characteristic they are considered dichromats. Animals that have only one type of cone, named monochromats, are colorblind since color perception will depend on a comparison between activities from two or more cone types.

The light stimulus is transduced by photoreceptors in a way that, in the presence of light, they become hyperpolarized. The higher the intensity of light, the greater the hyperpolarization that photoreceptors experience. On the other hand, in the dark, photoreceptors are depolarized and release its neurotransmitter, glutamate. Photoreceptors synapse onto bipolar and horizontal cells in the OPL. Two types of cone bipolar cells are present and respond differently to glutamate released by photoreceptors. In dark conditions, when there is more glutamate, ON-bipolar cells are inhibited by stimulation of metabotropic glutamate receptors, whereas OFF-bipolar cells are excited by activation of ionotropic glutamate receptors. In light conditions, ON-bipolar cells are stimulated and release glutamate, the neurotransmitter of both types of bipolar cells. Finally, each type of bipolar cells forms excitatory synapses, using glutamate as neurotransmitter as well, onto a specific type of ganglion cells (e.g., ON-bipolar with ON-ganglion cells). Therefore, the retinal circuitry has two parallel visual channels: 1) the OFF channel, that responds to stimuli that are darker than the background; 2) the ON channel, that detects stimuli that are brighter than the background. So, vertebrate vision is clearly based on the perception of contrast between images and their backgrounds, like reading these black letters against the white background using the OFF channels that start in the retina. When it comes to the dim light conditions in which rods work, the most important thing is to be able to catch the presence of few photons. Indeed, rods are able to respond to only one photon! In agreement with that, the rod system has only one rod-bipolar cell that is always ON-type. These ON rod-bipolar cells transmit information to ON-ganglion cells indirectly through amacrine cells (AII) and ON-cone bipolar cells. Some other alternative routes to rod signals have been described, like gap junctions [2] between rods and cones and synapses with OFF-bipolar cells [3, 4].

The INL is mainly composed by the cell bodies of horizontal cells, bipolar cells, amacrine cells, and the Müller cell, the major retinal glial cell. Both horizontal and amacrine cells participate in the so called lateral/horizontal pathway, that generates lateral inhibition in the retina. Horizontal cells provide feedback inhibition to the photoreceptor synaptic terminal via GABA release and/or ephatic modulation [5, 6], enhancing the response to the edges of visual stimuli and reducing the response to areas of uniform brightness [7]. Amacrine cells synapse onto ganglion cells, bipolar cells and other amacrine cells in the IPL and are involved with several functions in the retina such as center-surround interactions and direction-selective light responses of some ganglion cells. Finally, ganglion cells send the pre-processed visual information (contrast, size, movement, wavelength and so on) to different brain nuclei through the optic nerve.

Most neurotransmitters and/or neuromodulators described in the central nervous system are found in the retina, specifically in amacrine cells. Most amacrine cells are GABAergic, but several other molecules are found in this cell population such as acetylcholine, dopamine, serotonin, VIP, substance P, nitric oxide and so on. Besides these mediators, adenosine is also found in the retina.

Adenosine in the Retina

Adenosine is a constitutive nucleoside present within each cell being a part of their metabolism. An increasing number of works concerned with adenosine functions in the brain have postulated that this nucleoside has important roles in different mechanisms. First, it can act as a neuromodulator either presynaptically, postsynaptically or even nonsynaptically, regulating the action of neurotransmitters [8]. Second, this nucleoside participates in the processes of synaptic plasticity that involve facilitation or inhibition of LTP [9]. Finally, this molecule is known to display neuroprotective effects by activating its receptors [10].

As mentioned before, adenosine can also be found in the retina to exert the same roles cited above. One of the first papers focusing in adenosine in the retinal tissue was the work of Ehinger and Perez (1984) [11]. Back then, they demonstrated that [^{3}H]-adenosine was able to accumulate inside retinal cells of rabbit and goldfish. In the mammal model, glial cells were the best accumulators, but ganglion cells, amacrine cells and bipolar cells displayed some radioactivity as well. In the case of the goldfish, outer segments of photoreceptors had the strongest radioactivity labeling, but their cell bodies displayed some staining as well. Moderate to weak labeling was observed in horizontal cells and cell bodies in the outer plexiform layer [11]. They did not actually show by immunostaining that adenosine was present endogenously in normal conditions, but it was a kick off leading to that.

In 1987, Braas and co-workers, using immunohistochemical techniques, observed the presence of endogenous adenosine in human, monkey, rat and guinea pig retinas. In guinea pigs and rats, the nucleoside was present mainly in ganglion cells and their nerve fibers, but some diffuse staining was also detected in the inner plexiform layer and inner nuclear layer. This expression pattern was similar in both humans and monkeys, but some photoreceptors were also highlighted in these species [12].

Two years later, Blazynski and co-workers demonstrated the presence of endogenous adenosine in retinas of ground squirrel, rabbits and mouse. The staining was observed in ganglion cell layer and cells in the inner nuclear layer for all three species [13]. In another paper, these same authors demonstrated by autoradiography that the cat retina was able to accumulate adenosine as well, mainly in the ganglion cell layer [14].

In 1990, Paes-de-Carvalho and colleagues showed in an *in vitro* model of purified cultures of avian retinal neurons that, in the absence of glial cells, practically all photoreceptors were labeled, but a less extent of other neurons were stained [15]. Soon after, another paper from Paes-de-Carvalho (1992) [16] showed the presence of adenosine throughout the development of avian retina.

At embryonic day eight (E8), no labeling could be seen. At this stage, retinogenesis and migration of cells in this tissue are not completed, and more mature retinal organization could be an important step leading to adenosine accumulation. At E12, ganglion cells, amacrine cells and photoreceptors were labeled, and this pattern only became stronger throughout embryo development.

At a post-hatched stage (PH5), staining became more intense in the ganglion cell layer in comparison to others. But there were populations of cells unlabeled at this layer, and it was unclear whether this heterogeneity would reflect differences between ganglion and displaced amacrine cells or between subpopulations of ganglion cells. Intense immunoreactivity was also observed at the inner plexiform layer, a less intense labeling was detected in amacrine cells, and a lesser yet in horizontal cells. Staining in photoreceptors could be evidenced as well, mainly in its inner segments [16].

Together these studies demonstrated that adenosine has a wide expression and distribution in the retina that is probably related to the important roles of this nucleoside in this tissue.

ADENOSINE RECEPTORS IN THE RETINA

When released in the extracellular space, adenosine will perform its physiological roles through four metabotropic receptors that have been cloned and classified into A_1, A_{2A}, A_{2B} and A_3 [17]. Classically, A_1 and A_3 receptors are coupled to $G_{i/o}$ protein, while A_{2A} and A_{2B} receptors are coupled to $G_{s/olf}$ protein. Thus, the activation of A_{2A} and A_{2B} adenosine receptors increases the production of cyclic adenosine monophosphate (cAMP) leading to activation of protein kinase A (PKA) and consequent activation of the transcription factor CREB (cAMP response element-binding protein). On the other hand, the activation of A_1 and A_3 receptors inhibits such signaling cascade [17]. In the CNS, A_1 receptors are widely distributed on neurons in the cortex, hippocampus and cerebellum [17]. On the other hand, the A_{2A} receptors have a more restrictive localization in the striatum and olfactory bulb [17]. Low levels of A_{2B} receptors are expressed on neuronal and glial cells [18, 19] as well as A_3 receptors, which are sparsely detectable in the hippocampus, cortex, cerebellum and striatum [17].

A series of studies in the literature points to the presence of adenosine receptors in the developing retina, indicating a fundamental importance of this nucleoside in the formation of the visual system [20, 21, 22]. As mentioned above, classically adenosine regulates the cAMP pathway and Paes-de-Carvalho and de Mello were the first to observe this phenomenon in chick embryo retinas. Retinas of animals with 17 embryonic days (E17) were able to accumulate cAMP when challenged with adenosine or its non-hydrolysable analog, 2-chloroadenosine [20]. However, the accumulation of cAMP varied according to the stage studied. Retinas from E8 and E13 were unable to produce cAMP when treated with adenosine. Only from E14 on, adenosine could lead to an increase in the accumulation of cAMP, achieving a peak between E17/E18. In post-hatched animals, adenosine can also induce an accumulation of cAMP in the retina, although less efficiently when compared to animals from E15 to E18 [20]. In retinas from E12, it was also shown that dopamine, through the activation of D_1 receptors, has the ability to induce the production of cAMP. Interestingly, this accumulation is inhibited by increasing concentrations of adenosine or by incubation with 2-chloroadenosine [21]. Besides the chick retina, other vertebrate retinas also accumulate cAMP when challenged with adenosine or its analogs. Retinas of rabbits stimulated with adenosine, dopamine and norepinephrine exhibited an accumulation of cAMP. However, only the effect of adenosine was blocked by IBMX, an antagonist of adenosine receptors when used in higher concentrations. Furthermore, the individual effect of each neuromodulator was enhanced by the other when they were added together, indicating that in the rabbit retina the control of adenylyl cyclase activity by adenosinergic, dopaminergic and noradrenergic systems is independent [23].

The expression of A_1 adenosine receptors has also been well characterized throughout the retinal development. Autoradiography studies using [^{3}H]-CHA (N6-cyclohexyladenosine), a selective A_1 receptor agonist, has demonstrated the expression of these receptors in homogenates of E10, but at low concentrations [22]. However, at later stages of development, A_1 receptor levels increase, reaching a peak of expression at E16. Peculiarly, A_1 receptor levels drop dramatically in retinas of post-hatched animals, presenting an expression similar to that found in E12 retinas [22]. Later, autoradiographic studies unveiled the cellular localization of these receptors in the chicken retina. Autoradiography labeling for L-[^{3}H]-PIA (R-N6-phenylisopropyladenosine, selective A_1 receptor agonist) was first seen in the inner plexiform layer at E12, gradually increasing up to E18. However, in the retina of post-hatched animals, the expression was lower when compared to initial stages of development [16]. In a similar manner to that found with [^{3}H]-CHA, autoradiography revealed no labeling at E8. In addition, very little specific binding was found in the inner and outer nuclear layer and in the ganglion cell layer [16]. Besides the chick retina, other vertebrate retinas have been described to possess adenosine A_1 receptors. Autoradiography studies using L-[^{3}H]-PIA and ^{125}I-HPIA, both A_1 receptor agonists, showed the expression of adenosine A_1 receptors in the optic fiber layer, ganglion cells, inner plexiform layer, as well as in the inner portion of the inner nuclear layer of rat retinas [12]. In monkey retinas, binding sites for L-[^{3}H]-PIA were diffusely distributed from the optic fiber layer to outer plexiform layer. However, in the human retina, the autoradiographic sites were abundant in the inner retina, particularly in the optic fiber layer [12]. Although the autoradiographic grain densities varied in retinas from rats, monkeys and humans, the ganglion cell layer and the optic fiber layer showed higher densities of binding sites in all three species. Homogenates of rabbit retina showed an increase in adenylyl cyclase activity when challenged with forskolin, a direct activator of this enzyme. Peculiarly, pre-incubation of the homogenate with CHA or R-PIA, both A_1 receptor agonists, led to inhibition of adenylyl cyclase activity induced by forskolin, indicating that rabbit retinas possess A_1 receptors [24]. In fact, later on, it was shown that both mice and rabbit retinas express A_1 receptors in all retinal layers, especially in the inner nuclear and outer plexiform layer [25]. Recently, Li and colleagues took advantage of various molecular tools and showed heterologous expression of A_1 receptor expression in retinas of transgenic mice [26].

In the chick retina, recent evidence demonstrated that the intracellular pathway coupled to A_{2A} receptor activation, and therefore its function, is age-dependent. E6 embryos treated for three days with DPMA, A_{2A} agonist, had an increase of cell death in the inner nuclear layer and ganglion cell layer. However, eight–day-old embryos subjected to the same experimental paradigm did not show an increase in cell death compared to control [27]. This dual effect of A_{2A} receptor activation depends directly on the level of phosphorylation of the transcription factor CREB. In E6, DPMA induced lower levels of CREB phosphorylation, whereas higher levels were detected in E8 [27]. It is possible that during normal retinal development the intracellular signaling pathway coupled to A_{2A} receptor activation changes. This could lead to different effects on the levels of CREB phosphorylation, which is directly related to the survival of retinal neurons. Corroborating the data cited above, other results point to the expression of A_{2A} receptors along the chicken retina (personal data) and Brito and collaborators showed the expression of A_{2A} receptors in E16. Animals treated for two days with CGS 21680 (A_{2A} receptor agonist) and ZM 241385 (A_{2A} receptor antagonist) showed a decrease and increase, respectively, of the A_{2A} receptor compared to control [28]. On the other hand, treatment with A_{2A} receptor agonist (CGS 21680) was able to increase the expression of A_1 receptor while selective antagonists (ZM 241385 and SCH 58261) reduced its expression. These data indicate that endogenous adenosine, released in this microenvironment, is capable of regulating the expression of A_1 receptor by activation of A_{2A} receptor. This hypothesis is ensured by the fact that blockade of adenosine transporters also leads to a decrease in the expression of A_1 receptors [28]. The signaling pathway behind the regulation of A_{2A} receptor by A_1 seems to be the classical formation of cAMP and activation of PKA [29].

Besides the chicken retina, A_{2A} receptors have been characterized in salamander [30], rat [31-33] mouse [25, 34, 35], guinea pig [36], dog [37] retinas. In the rat retina, the mRNA for A_{2A} receptor was found in the ganglion cell layer and inner and outer nuclear layer as well as in the choriocapillaris [31]. Other studies revealed the expression of the protein in the inner plexiform layer and ganglion cell layer as well as in non-neuronal tissue, such as the retinal pigmented epithelium and choroid [33]. Moreover, immunoreactivity to A_{2A} receptors was observed in starburst amacrine cells (SACs), in retinal sections and in dissociated cell culture [33]. Binding assays for [^{3}H]-NECA in retinas of mice revealed the presence of A_{2A} receptors in the retinal pigmented epithelium and in the inner and outer segment of the photoreceptors [25]. On the other hand, the location of the A_{2A} receptor in guinea pig retina was restricted to the cytoplasm and the extracellular matrix of the sclera and retina [36].

Although a pioneer study using in situ hybridization found no expression of A_3 receptor mRNA in rat retina [31], there is now evidence of its expression in retinal ganglion cells of rats [38] and guinea pigs [36]. Therefore, subsequent experiments need to be developed to confirm the presence of A_3 receptors in the retina. As well as the A_3 receptor, there is little information about the A_{2B} receptor. It is known that the A_{2B} receptor is expressed in retina and sclera of guinea pigs [36] and its mRNA has been identified in the ciliary body of the rat retina, but in no other structure [31].

The presence of adenosine receptors in the retina of different species suggests a possible importance of this nucleoside in its development, as well as in retinopathies, and both subjects will be addressed in the following topics.

NUCLEOSIDE TRANSPORTERS

Adenosine levels in extracellular and intracellular compartments are regulated by a group of membrane transport proteins known as nucleoside transporters. There are two families of nucleoside transporters: 1) solute carrier family 29 (SLC29), named as equilibrative nucleoside transporters (ENT) family; 2) solute carrier family 28 (SLC28), known as concentrative nucleoside transporters (CNT) family.

Equilibrative Nucleoside Transporters (ENTs) transport nucleosides bidirectionally according to their concentrations in extracellular and intracellular milieu, while CNTs promote influx of nucleosides against the concentration gradient using the energy from sodium concentration gradient across cellular membranes [39]. At present, there are four isoforms of ENTs described, which are referred to as ENT1, ENT2, ENT3 and ENT4. ENT1 and ENT2 are the most studied isoforms of nucleoside transporters family.

ENT1 is a widely expressed transporter, mainly found at the plasma membrane, being considered the main regulator of homeostatic maintenance of adenosine levels and also the major transporter in the CNS [40, 41, 42]. In the literature, there are several reports that prove that ENT1 inhibition can potentiate neuroprotective and cardioprotective effects induced by adenosine [42]. These transporters were cloned from several species, including rats, mice, humans and canines [43].

Concentrative Nucleoside Transporters (CNTs) promote the influx of nucleosides against concentration gradient using the energy from sodium concentration gradient across cell membranes [39]. These transporters are sub-divided in three isoforms, known as CNT1, CNT2 and CNT3.

Nucleoside Transporters in the Retina

Several works in the literature report the presence of nucleoside transporters in retinas from different species [44, 12, 45]. Retinas from goldfish [46], rabbits [47], rats [44] and chick [15, 48], among others, are known to present an adenosine uptake system. There is also a work showing the presence of ENT1 through binding assays in a cultured human retinal cell line with pharmacological features very similar to human retinal tissue [49].

It is important to point out that most studies about nucleoside transporters in the retina were made by looking for ENT1 binding sites, mainly because there is a great tool to perform these studies: [^{3}H]-NBMPR (also called NBTI), a small molecule inhibitor of equilibrative nucleoside transporter that binds to ENT1 with high affinity. So far, there are no reports showing the presence of CNTs in significant amount in the retina and because of this, we decided to focus this part of the chapter on description of ENTs in the retina, especially ENT1.

In the chicken retina, Paes-de-Carvalho and colleagues (1992) [16] found a large labeling of ENT1 sites throughout the E8 retinal neuroblastic layer, in accordance with the lack of cellular organization in specific layers at this stage. The analysis of retinas at older stages of development as E12, E15, E18 and post-hatched animals also showed ENT1 sites labeling. However, older stages presented a lower labeling compared to E8, which was restricted to the outer and inner plexiform layers [16]. In cultured chick retinal neurons, it was described that ENT1 is the main transporter expressed in these cells [15].

Repérant and colleagues (1982) [50] described axonal transport of [^{3}H]-adenosine in the visual system of the lamprey following intraocular injection of this nucleoside. They showed that this transport was bidirectional and transcellular and although they did not describe the profile of nucleoside transporters in this retina, this work suggested that adenosine was being taken up by nucleoside transporters in axonal endings located in the retina.

In the rabbit retina, most of [^{3}H]-adenosine taken up by retina is converted to adenine nucleotides [47], but when they induced a release of purines with a high potassium stimulus, they observed that the most part of released radioactivity was identified as hypoxanthine, xanthine and inosine. They also observed that this increase in purines release was prevented by a nucleoside transport inhibitor, dipyridamole. Later, this same group demonstrated that this release of purines induced by depolarization is dependent of extracellular calcium [51]. Similar results were obtained in chicken retinal cultured cells, such as purified cultures of retinal neurons [15], and mixed cultures of neurons and glial cells [52].

Thereza and colleagues (1989) [53] showed that different neurotransmitters such as dopamine, acetylcholine, GABA and NMDA were able to induce an increase of purines release from the rabbit retina, all of them in an extracellular calcium-dependent manner. In cultured chick retinal neurons, there is also evidence showing that dopamine is able to promote an increase of purines release [54]. In this same work, Paes de Carvalho described that a long term treatment of mixed cultures (neurons and glial cells) from chick retina with adenosine plus EHNA, an adenosine deaminase inhibitor, induced a significant decrease in ENT1 levels. In mixed cultures from chick retina, like previously demonstrated in rabbit retina [53], it was described that the release of purines from cultured retinal cells is stimulated by activation of NMDA or AMPA/kainate glutamate receptors [52]. They showed that this release was also dependent of extracellular calcium, ENTs activity and calcium/calmodulin-dependent protein kinase II (CAMK II) activity.

Other work from this same group [55] described that glutamate excitotoxicity in purified cultures of retinal neurons is completely blocked in the presence of NBTI, an adenosine uptake blocker. A similar protective effect induced by NBTI was observed in the death induced by refeeding the cultures with fresh medium [56]. Other authors showed that IL-6 can increase the survival of rat retinal ganglion cells [57], but this neuroprotective effect was blocked by NBTI, suggesting that the effect of IL-6 could be dependent on the release of adenosine.

Another mediator that is described to regulate ENT1 transporter activity in the rat retina is cannabidiol, a nonpsychotropic cannabinoid [58]. In this study, they described that this cannabinoid compound has anti-inflammatory effects partially through inhibition of adenosine uptake via ENT1. Such inhibition induced an increase of extracellular adenosine, which led to a higher activation of A_{2A} adenosine receptors in retinal microglial cells and this caused an inhibition of TNF-α production in response to LPS. Other compound that has anti-inflammatory effects mediated in some degree by purinergic system is the glucocorticoid triamcinolone [59]. This group demonstrated that triamcinolone induces inhibition of the swelling of retinal glial cells of diabetic rats and this effect was completely blocked in the presence of a selective inhibitor of nucleoside transporters, NBTI.

In terms of inner blood–retinal barrier (inner BRB), there is a work with rat retinal capillary endothelial cell line (TR-iBRB2), which is used as an in vitro model of the inner BRB, where they showed that although this cell line expresses ENT1, ENT2, CNT2 and CNT3 mRNAs, most of adenosine uptake is mediated by ENT2 activity [60]. They also showed that *in vivo* blood-to-retina influx transport of adenosine is likely mediated by ENT2. More recently, this same group published some results showing that ENT2 is also the main mediator of adenosine uptake in rat Müller cells from the retinal interstitial fluid [61]. They suggest Müller cells are important to convert this adenosine into hypoxanthine, which would be eliminated across the inner BRB. This could be important to decrease levels of retinal oxidative stress, since reactive oxygen species (ROS) can be produced via hypoxanthine metabolism to xanthine.

There are few studies available about the intracellular mechanisms regulating activity and/or expression of nucleoside transporters in the retina. There is a report showing involvement of CAMK II, since its inhibition decreased both the release stimulated by glutamate and the uptake of adenosine by cultured chick retinal cells [52]. However, it is not clear how this regulation happens in the system, neither if it is an effect that modulates specifically ENT1, ENT 2 or both transporters. More recently, similar results demonstrated a role of ERKs in the regulation of uptake of adenosine in the same cells [62], but further studies are required to better understand molecular details about this regulation.

ADENOSINE FUNCTION IN THE RETINA

As discussed above, it is clear that adenosine might have a great importance for the retinal physiology, because it is widely expressed in this tissue and even more concentrated in both photoreceptor and ganglion cell.

Adenosine expression appears to be highly modulated by the circadian clock and by light stimulus. It is known that darkness stimulates ATP release from photoreceptors, and that ectonucleotidases transform it into adenosine, increasing its concentration in the extracellular space [63]. In addition to that, during the night, there is an increase in energy consumption [64] that might lead to higher AMP availability inside the cells, which could be used to produce adenosine. The intracellular accumulation of adenosine could prevent adenosine uptake from the extracellular pool, keeping higher concentration outside the cell during the night. On the other hand, in the light, photoreceptors release less ATP, and extracellular adenosine levels are diminished [63].

Still regarding photoreceptors, adenosine is also able to inhibit glutamate exocytosis from both cone and rod terminals by inhibiting L-type voltage-gated calcium channels, likely through activation of A2-like receptors [65, 66, 67, 68]. Since there is more glutamate release in the dark, the concomitant increase in the presence of adenosine in this condition and its ability to modulate glutamate exocytosis may represent a neuroprotective feature of the nucleoside in the first retinal synapse.

Adenosine is also involved in photoreceptor coupling through GAP junctions. Activation of A_{2A} receptors leads to cyclic AMP increases, leading to enhanced PKA activity and phosphorylation of Cx36 at S293, thus increasing photoreceptor coupling [35]. Dephosphorylation of Cx36 at S293 causes decreased coupling of photoreceptors, and may be achieved by either blocking A_{2A} receptor activation or stimulating dopamine D4 receptors. Considering that dopamine is released in the presence of light [69], while adenosine is released in the absence of light, it can be inferred that these neurotransmitters have opposite roles in photoreceptor coupling and retinal adaptation to ambient illumination.

Literature lacks studies about the function of adenosine in horizontal cells. Concerning localization, one study describes [^{3}H]-adenosine uptake by rod horizontal cells in the goldfish retina, which might make adenosine a good marker for those cells in this species [46]. Another histologic study shows co-localization of adenosine and GABA in a few horizontal cells in chicken and rabbit retina [48].

Bipolar cells are also affected by adenosine. It has been shown that rat retinal bipolar cell cultures treated with the adenosine analogue 2-chloroadenosine show diminished NMDA receptor responses, putatively through the activation of A_1 receptors [70].

Adenosine is also able to modulate neurotransmitter release in the retina. In amacrine cell retinal cultures, it has been shown that acetylcholine tonic release is inhibited by endogenous levels of adenosine through activation of A_1 receptors and consequent inhibition of N-type calcium channels. In addition, depolarization of amacrine cells with KCl causes adenosine accumulation on the extracellular space due to its release, and the conversion of nucleotides, such as ATP into adenosine [71, 72].

In the mammalian retina, amacrine cells participate in a phenomenon known as retinal wave, specifically during stage II. It happens in the early postnatal days (P0-9) in rats, and is important for the correct development of retinal projections to the brain [73]. During the retinal waves, acetylcholine and GABA are released from amacrine cells and reach the ganglion cell layer, generating a wave-patterned oscillation of Ca^{2+} ions and depolarization across the retinal layers [74, 75]. Adenosine appears to be essential for the correct establishment of retinal waves frequency through the activation of the A_{2A} receptor [33].

Ganglion cell activity is also modulated by adenosine. Stimulation of A_1 receptors in ganglion cells promotes the activation of two distinct K^+ currents: one that is mediated by GIRK channels, and other that depends on IP_3-mediated Ca^{2+} release from internal stores, and leads to the prolonged opening of SK channels. This increased K^+ conductance promotes the hyperpolarization of ganglion cells [76]. Adenosine also regulates calcium-related signaling in ganglion cells. A_1 receptors are capable of directly inhibiting voltage-gated Ca^{2+} channels through its Gβγ subunit [77], and A_3 receptors decrease Ca^{2+} influx following NMDA receptor activation in cultured rat retinal ganglion cells, probably through the same mechanism [78].

In addition to neurons, glial cells have an important role in retinal physiology, modulating neuron spiking, synaptic transmission and neurotransmitter transport [79]. Müller glia cells express A_1, A_{2A} and A_{2B} adenosine receptors [80]. Retinal stimulation with flickering light, ATP or adenosine increases the calcium transient in Muller glia [81].

Intense synaptic transmission causes an increase in ion concentrations and water influx into neurons, which is compensated by K^+ release into the extracellular space [82]. Such increase in extracellular K^+ could lead to increasing depolarization, which is avoided by the Müller glial cells ability to take up and store K^+. This capacity to store K^+ is linked to the glial cell ability to prevent its own osmotic swelling, a phenomenon promoted by A_1 receptor activation that leads to the opening of K^+ and Cl^- channels in the cell, allowing ion outflow and avoiding swelling [59].

As mentioned above, adenosine has important physiological functions, but its true hallmark is neuroprotection. Adenosine can be released by transporters or can be produced from ATP in the extracellular compartment, but its concentration is highly increased by the occurrence of cell rupture, injury or death. The activation of adenosine receptors has been shown to prevent or decrease cell injury and death in several experimental models in different organs and tissues, including the retina. The latter is an interesting model for studying the neuroprotective effects of adenosine because it has a high oxygen demand, and low oxygen tension appears to be a condition in which adenosine release is increased [83].

In chick embryo purified retinal neurons cultures, cell death can be induced by medium exchange or an eight hours incubation with 1mM glutamate. Pretreating the cultures with adenosine deaminase inhibitors, A_{2A} receptor agonists, adenosine uptake inhibitors or a cyclic AMP analog prevented cell death induced by both medium exchange and glutamate-induced excitotoxicity [84, 55, 56]. Therefore, adenosine appears to be a very important endogenous neuroprotective agent in the retina.

Adenosine appears to be a key molecule in the phenomenon known as ischemic preconditioning, characterized by previous tissue exposure to short periods of ischemia that produce a tolerance to subsequent more important ischemia/reperfusion insults that would cause severe injury otherwise [85]. Ischemic preconditioning has been shown to protect the retina against increased intraocular pressure-induced ischemia, effect that can be decreased by treatment with adenosine A_1 or A_{2A} receptors antagonists, ATP-sensitive K^+ blockers and protein synthesis inhibitors [86, 87]. Apparently, brief ischemia leads to adenosine release and its receptors activation change protein expression patterns through PKC and K^+ ATP-sensitive channels, increasing protective proteins whereas decreasing killer proteins [85].

Studies with retinal ganglion cell cultures have shown an interesting crosstalk among adenosine receptors, IL-6 cytokine and BDNF neurotrophic factor signaling pathways. Cultured retinal ganglion cells suffer a 50% cell death rate after 48h in culture. This cell death is prevented by treatment with either A_1 or A_{2A} receptors agonists. However, in the presence of either IL-6 or BDNF neutralizing antibodies, the action of adenosine receptors is completely blocked. The same studies have shown that adenosine induces the release of IL-6 and BDNF, which are acting synergistically at their receptors to prevent cell death [57, 88].

The adenosine A_3 receptor is able to provide retinal ganglion cells protection against Ca^{2+}-induced cell death. It has been shown that the activation of P2X7 ATP receptor in cultured retinal ganglion cells leads to Ca^{2+}-induced cell death, and pre-exposure to an A_3 receptor agonist leads to decreased intracellular Ca^{2+} signal and consequently less cell death [89].

Therefore, it is clear that studies concerning adenosine function and neuroprotective features are essential for the advance of therapies and pharmacological interventions to help prevent ischemia-induced cell death.

Adenosine and Retinal Disease

Adenosine is an important nucleoside that orchestrates different mechanisms of cell signaling throughout the body. It is well known that this molecule is increased extracellularly in response to tissue injury, especially following hypoxia or ischemia. This increase occurs through enzyme- mediated ATP breakdown once it is released to extracellular space in response to injury. Adenosine acts on membrane adenosine receptors, which are subjected to altered regulation following injury. The roles of adenosine during these stressful events seem to be opposed to one another considering acute and chronic disease states, mediating beneficial effects during the first, and detrimental responses during the second [90].

In retinal tissue, adenosine will also display a crucial role in mediating injury responses during ischemic events, glaucoma and different retinopathies. Retinal ischemia can be induced by central artery occlusion, diabetes, glaucoma and retinopathy of prematurity. Using rat retinas as a model, Anna K. Larsen and Neville N. Osborne were looking for potential beneficial roles of adenosine during retinal ischemic situations [91]. They could observe that upon direct treatment of the adenosine deaminase inhibitor (EHNA), or intraperitoneal injection of adenosine A_1 receptor agonist (R-PIA), ischemia-induced changes in the b-wave and calretinin immunoreactivity could be largely prevented, suggesting that activation of A_1 receptors leads to neuroprotective effects [91].

In accordance to the work previously described, Ghiardi and colleagues demonstrated, also in rat retinas, that the administration of DPCPX, the selective adenosine A_1 receptor antagonist, 5 or 30 minutes before ischemia (mild and moderate, respectively) decreased the recovery of a- and b-waves, but had no effect when given prior to 60 minutes ischemia (severe). Also, the use of CSC, adenosine A_{2A} receptor antagonist, 30 or 60 minutes before ischemia, resulted in almost complete recovery in both, histological and electrical aspects [92]. These data demonstrated that A_1 receptors promote neuroprotective effects whereas A_{2A}

receptor activation goes the opposite way. However, the mechanism that each receptor triggers their effects was not clear.

As discussed above, adenosine itself and its agonists may have therapeutic potential. However, their usage is limited by their systemic side effects, which includes: bradycardia, hypotension and sedation [93]. Also, if adenosine is available in the extracellular space for too long, enzymes like Xanthine dehydrogenase (XDH) and Xanthine oxidase (XO) start to act on products of its metabolism, like xanthines and hypoxanthines, generating $O2^-$. It is well known that superoxide radicals can be very dangerous to the tissue. Some works using rat and cat retinas demonstrated the protective effects of allopurinol and oxypurinol (both XO inhibitors) in improving the recovery of electrical responsiveness of retinal cells after ischemia [94, 95].

Diabetic retinopathy is among the most common blinding causes in working-age adults [96]. As mentioned before, chronic ischemia is a common event in the context of diabetic retinopathy and retinopathy of prematurity. Blood flow decrease and oxygen deprivation are common aspects of these pathological states, and biological systems respond creating new vascularization to compensate these problems. Adenosine seems to participate in this process of neovascularization in the retina by acting on the A_{2A} receptors, modulating the synthesis of VEGF and its receptors [97, 98, 99]. However, these new vessels are not functional and do not lead to a regenerating process. Instead, the neovasculature creates some problems to the tissue, eventually leading to blindness [100].

Besides this problematic effect, adenosine modulation of retinal blood vessels physiology is not always a bad deal. The vasodilatation capacity of this molecule has already been described at least in the retina of rabbits, marmosets and cats [101, 102], and it seems that A2 receptors activation is required [103] as well as ATP-sensitive K^+ channels opening [104]. Knowing that this molecule is able to increase retinal blood flow, therapeutic agents that raise its availability could be a nice strategy during and after oxygen and substrate deprivation events.

Other therapeutic aspects of the adenosinergic system usage in the retina have been enlightened.

Activation of A_{2A} receptor seems to play an important role in diabetic retinopathy, diminishing the release of TNF-α, abrogating the inflammatory response induced by microglial cells, and decreasing hyperglycemia-induced cell death in mice retina [105]. In humans and porcine diabetic retinas, adenosine deaminase-2 (ADA2) has higher patterns of expression and activity. This abnormal behavior was correlated with increasing TNF-α release by microglia and impaired anti-inflammatory adenosine A_{2A} receptors activity [106]. Adenosine kinase has also been described as a possible target for diabetic retinopathy treatment [107]. Interestingly, an epidemiological study showed that genetic variants of adenosine A_{2A} receptor offer significant protection against proliferative diabetic retinopathy incidence in humans with type 1 diabetes [108].

Glaucoma is another of the leading blinding causes worldwide, progressively resulting in retinal ganglion cells death and defects in visual fields [109].

Although the main risk factor is aging, elevated intraocular pressure (IOP) could also be related to some types of glaucoma. Therefore, lowering this pressure is important to prevent and reduce glaucoma progression. IOP depends directly on the aqueous humor formation and drainage, and adenosine, acting on A_{2A}, A_1 and A_3 adenosine receptors, is able to modulate this equilibrium in several species, regulating IOP, as reviewed by Zhong and coworkers

(2013) [110]. This mechanism could be explored as a therapeutic target for the treatment of glaucoma as already discussed by Fishman and colleagues (2013) [111].

Several pathological situations discussed above can generate a parallel complication, termed retinal edema. One of the main mechanisms by which retinal edema is formed, in humans and other species, is through intracellular water accumulation resulting in cell swelling [112].

Adenosine also seems to be involved in a protective mechanism for the prevention of cell swelling. The adenosine A_1 receptor, coupled to a nonconventional intracellular cell signaling pathway, has been described as a crucial receptor for prevention of glial Müller cells swelling by interacting with potassium and chloride channels, allowing ionic extrusion and cellular volume maintenance [113, 114].

Conclusion

The visual perception depends on a number of factors that culminate in the transduction of light into an electrical stimulus in the retina. Visual processing is orchestrated by the interaction between different cell types in the retina through the release of neurotransmitters such as glutamate and GABA. However, other systems of neuromodulators participate in this intricate phenomenon. The existence of adenosine and its system of transporters and receptors in almost all cell types of the retina underlines the importance of this nucleoside in the homeostatic control of this tissue. In fact, aspects such as neurotransmitter release and control of the circadian cycle critical for maintaining the periodicity of sleep and wakefulness are regulated by adenosine. Imbalances in this system are still responsible for visual conditions such as those found in diabetic retinopathy. Thus, adenosine is certainly an important molecule for retinal functionality.

References

[1] Rodieck, RW. The first steps in seeing. Sinauer Associates, Inc, 1998.

[2] DeVries, S. H; Baylor, D. A. An alternative pathway for signal flow from rod photoreceptors to ganglion cells in mammalian retina. *Proc. Natl. Acad. Sci. U. S. A.* 1995, 92, 10658–10662.

[3] Soucy, E; Wang, Y; Nirenberg, S; Nathans, J; Meister, M. A novel signaling pathway from rod photoreceptors to ganglion cells in mammalian retina. *Neuron,* 1998, 21, 481–493.

[4] Hack, I; Peichl, L; Brandstätter, J. H. An alternative pathway for rod signals in the rodent retina: rod photoreceptors, cone bipolar cells, and the localization of glutamate receptors. *Proc. Natl. Acad. Sci. U. S. A.* 1999, 96, 14130–14135.

[5] Verweij, J; Kamermans, M; Spekreijse, H. Horizontal cells feedback to cones by shifting the cone calcium-current activation range. *Vision Res.* 1996, 36(24), 3943-3953.

[6] Fahrenfort, I; Klooster, J; Sjoerdsma, T; Kamermans, M. The involvement of glutamate-gated channels in negative feedback from horizontal cells to cones. *Prog. Brain Res.* 2005, 147, 219-229.

[7] Wässle, H. Parallel processing in the mammalian retina. *Nat. Rev. Neurosci.* 2004, 5(10), 747-757

[8] Sebastião, A. M; Ribeiro, J. A. Fine-tuning neuromodulation by adenosine. *Trends Pharmacol. Sci.* 2000, 21(9), 341-6.

[9] Ribeiro, J.A. Purinergic inhibition of neurotransmitter release in the central nervous system. *Pharmacol. Toxicol.* 1995, 77(5), 299-305.

[10] Cunha, R. A. Adenosine as a neuromodulator and as a homeostatic regulator in the nervous system: different roles, different sources and different receptors. *Neurochem. Int.* 2001, 38(2), 107-125.

[11] Ehinger, B; Perez, M. T. Autoradiography of nucleoside uptake into the retina. *Neurochem. Int.* 1984, 6(3), 369-81.

[12] Braas, K. M; Zarbin, M. A; Snyder, S. H. Endogenous adenosine and adenosine receptors localized to ganglion cells of the retina. *Proc. Natl. Acad. Sci. U. S. A.* 1987, 84(11), 3906-3910.

[13] Blazynski, C; Mosinger, J. L; Cohen, A. L. Comparison of adenosine uptake and endogenous adenosine-containing cells in mammalian retina. *Vis. Neurosci.* 1989, 2(2), 109-16.

[14] Blazynski, C; Cohen, A. L; Früh, B; Niemeyer, G. Adenosine: autoradiographic localization and electrophysiologic effects in the cat retina. *Invest. Ophthalmol. Vis. Sci.* 1989, 30(12), 2533-6.

[15] Paes de Carvalho, R; Braas, K. M; Snyder, S. H; Adier, R. Analysis of adenosine immunoreactivity, uptake, and release in purified cultures of developing chick embryo retinal Neurons and Photoreceptors. *J. Neurochem.* 1990, 55(5), 1603-1611.

[16] Paes de Carvalho. R; Braas, K. M; Adler, R; Snyder, S. H. Developmental regulation of adenosine A_1 receptors, uptake sites and endogenous adenosine in the chick retina. *Dev. Brain Res.* 1992, 70(1), 87-95.

[17] Fredholm, B. B; AP, I. J; Jacobson, K. A; Linden, J; Muller, C. E. International Union of Basic and Clinical Pharmacology. LXXXI. Nomenclature and classification of adenosine receptors—An update. *Pharmacol. Rev.* 2011, 63(1), 1-34.

[18] Allaman, I; Lengacher, S; Magistretti, P. J; Pellerin, L. A2B receptor activation promotes glycogen synthesis in astrocytes through modulation of gene expression. *Am. J. Physiol. Cell Physiol.* 2003, 284(3), C696-C704.

[19] Sebastiao, A. M; Ribeiro, J. A. Adenosine A2 receptor-mediated excitatory actions on the nervous system. *Prog. Neurobiol.* 1996, 48(3), 167–189.

[20] Paes-de-Carvalho, R; de Mello, F. G. Adenosine-elicited accumulation of adenosine 3', 5'-cyclic monophosphate in the chick embryo retina. *J. Neurochem.* 1982, 38(2), 493-500.

[21] Paes-de-Carvalho, R; de Mello, F. G. Expression of A1 adenosine receptor modulating dopamine-dependent cyclic AMP accumulation in the chick embryo retina. *J. Neurochem.* 1985, 44(3), 845-851.

[22] Paes-de-Carvalho, R. Development of A1 adenosine receptors in the chick embryo retina. *J. Neurosci. Res.* 1990, 25(2), 236-242.

[23] Blazynski, C.; Kinscherf, D. A; Geary, K. M.; Ferrendelli, J. A. Adenosine-mediated regulation of cyclic AMP levels in isolated incubated retinas. *Brain Res.* 1986, 366(1-2), 224-229.
[24] Blazynski, C. Adenosine A1 receptor-mediated inhibition of adenylate cyclase in rabbit retina. *J. Neurosci.* 1987, 7(8), 2522-2528.
[25] Blazynski, C. Discrete distributions of adenosine receptors in mammalian retina. *J. Neurochem.* 1990, 54(2), 648-655.
[26] Li, N.; Salom, D.; Zhang, L.; Harris, T.; Ballesteros, J. A.; Golczak, M.; Jastrzebska, B.; Palczewski, K.; Kurahara. C.; Juan, T.; Jordan, S.; Salon, J. A. Heterologous expression of the adenosine A1 receptor in transgenic mouse retina. *Biochem.* 2007, 46(28), 8350-8359.
[27] Socodato. R.; Brito, R.; Calaza, K. C.; Paes-de-Carvalho, R. Developmental regulation of neuronal survival by adenosine in the in vitro and in vivo avian retina depends on a shift of signaling pathways leading to CREB phosphorylation or dephosphorylation. *J. Neurochem.* 2011, 116(2), 227-239.
[28] Brito, R.; Pereira, M. R.; Paes-de-Carvalho, R.; Calaza, Kda. C. Expression of A1 adenosine receptors in the developing avian retina: in vivo modulation by A(2A) receptors and endogenous adenosine. *J. Neurochem.* 2012, 123(2), 239-249.
[29] Pereira, M. R.; Hang, V. R.; Vardiero, E.; de Mello, F. G.; Paes-de-Carvalho, R. Modulation of A1 adenosine receptor expression by cell aggregation and long-term activation of A2a receptors in cultures of avian retinal cells: Involvement of the cyclic AMP/PKA pathway. *J. Neurochem.* 2010, 113(3), 661–673
[30] Alfinito, P. D.; Alli, R.; Townes-Anderson, E. Adenosine A(2a) receptor-mediated inhibition of rod opsin mRNA expression in tiger salamander. *J. Neurochem.* 2002, 83(3), 665-672.
[31] Kvanta, A.; Seregard, S.; Sejersen, S.; Kull, B.; Fredholm B. B. Localization of adenosine receptor messenger RNAs in the rat eye. *Exp. Eye Res.* 1997, 65(5), 595-602.
[32] Dreixler, J. C.; Hemmert, J. W.; Shenoy, S. K.; Shen, Y.; Lee, H. T.; Shaikh, A. R.; Rosenbaum, D. M.; Roth, S. The role of Akt/protein kinase B subtypes in retinal ischemic preconditioning. *Exp. Eye Res.* 2009, 88(3), 512-521.
[33] Huang, P. C.; Hsiao, Y. T.; Kao, S. Y.; Chen, C. F.; Chen, Y. C.; Chiang, C. W.; Lee, C. F.; Lu, J. C.; Chern, Y.; Wang, C. T. Adenosine A(2A) receptor up-regulates retinal wave frequency via starburst amacrine cells in the developing rat retina. *PLoS One.* 2014, 9(4), e95090.
[34] Blazynski, C.; Perez, M. T. Adenosine in vertebrate retina: localization, receptor characterization, and function. *Cell. Mol. Neurobiol.* 1991, 11(5), 463-484.
[35] Li, H.; Zhang, Z.; Blackburn, M. R.; Wang, S. W.; Ribelayga, C. P.; & O'Brien, J. Adenosine and dopamine receptors coregulate photoreceptor coupling via gap junction phosphorylation in mouse retina. *J. Neurosci.* 2013, 33(7), 3135-3150.
[36] Cui, D.; Trier, K.; Zeng, J.; Wu, K.; Yu, M.; Ge, J. Adenosine receptor protein changes in guinea pigs with form deprivation myopia. *Acta Ophthalmol.* 2010, 88(7), 759-765.
[37] Taomoto, M.; McLeod, D. S.; Merges, C.; Lutty, G. A. Localization of adenosine A2a receptor in retinal development and oxygen-induced retinopathy. *Invest. Ophthalmol. Vis. Sci.* 2000, 41(1), 230-243.

[38] Zhang, M.; Budak, M. T.; Lu, W.; Khurana, T. S.; Zhang, X.; Laties, A. M.; Mitchell, C. H. Identification of the A3 adenosine receptor in rat retinal ganglion cells. *Mol. Vision*, 2006, 16(12), 937-948.

[39] Podgorska, M.; Kocbuch, K.; Pawelczyk, T. Recent advances in studies on biochemical and structural properties of equilibrative and concentrative nucleoside transporters. *Acta Biochim. Pol.* 2005, 52(4), 749-758.

[40] Anderson, C. M.; Xiong, W.; Geiger, J. D.; Young, J. D.; Cass, C. E.; Baldwin, S. A.; Parkinson, F. E. Distribution of equilibrative, nitrobenzylthioinosine-sensitive nucleoside transporters (ENT1) in brain. *J. Neurochem.* 1999, 73(2), 867-873.

[41] Chu, S.; Xiong, W.; Zhang, D.; Soylu, H.; Sun, C.; Albensi, B. C.; Parkinson, F. E. Regulation of adenosine levels during cerebral ischemia. *Acta Pharmacol. Sin.* 2013, 34(1), 60-66.

[42] Bone, D. B.; Robillard, K. R.; Stolk, M.; Hammond, J. R. Differential regulation of mouse equilibrative nucleoside transporter 1 (mENT1) splice variants by protein kinase CK2. *Mol. Membr. Biol.* 2007, 24(4), 294-303.

[43] Hammond, J. R.; Stolk, M.; Archer, R. G.; McConnell, K. Pharmacological analysis and molecular cloning of the canine equilibrative nucleoside transporter 1. *Eur. J. Pharmacol.* 2004, 491(1), 9-19.

[44] Schaeffer, J. M.; Anderson, S. M. Nucleoside uptake by rat retina cells. *Life Sci.* 1981, 29(9), 939-946.

[45] Blazynski, C. The accumulation of [3H]phenylisopropyl adenosine ([3H]PIA) and [3H]adenosine into rabbit retinal neurons is inhibited by nitrobenzylthioinosine (NBI). *Neurosci. Lett.* 1991, 121(1-2), 1-4.

[46] Studholme, K. M.; Yazulla, S. 3H-adenosine uptake selectively labels rod horizontal cells in goldfish retina. *Vis. Neurosci.* 1997, 14(2), 207-212.

[47] Perez, M. T.; Ehinger, B. E.; Lindstrom, K.; Fredholm, B. B. Release of endogenous and radioactive purines from the rabbit retina. *Brain Res.* 1986, 398(1), 106-112.

[48] Perez, M. T.; Bruun, A. Colocalization of (3H)-adenosine accumulation and GABA immunoreactivity in the chicken and rabbit retinas. *Histochem.* 1987, 87(5), 413-417.

[49] Williams, E. F.; Ezeonu, I.; Dutt, K. Nucleoside transport sites in a cultured human retinal cell line established by SV-40 T antigen gene. *Curr. Eye Res.* 1994, 13(2), 109-118.

[50] Reperant, J.; Vesselkin, N. P.; Kenigfest, N. B.; Miceli, D.; Rio, J. P. Bidirectional axonal and transcellular transport of [3H]adenosine from the lamprey retina. *Brain Res.* 1982, 246(2), 305-310.

[51] Perez, M. T.; Arner, K.; Ehinger, B. Stimulation-evoked release of purines from the rabbit retina. *Neurochem. Int.* 1988, 13(3), 307-318.

[52] Paes-de-Carvalho, R.; Dias, B. V.; Martins, R. A.; Pereira, M. R.; Portugal, C. C.; Lanfredi, C. Activation of glutamate receptors promotes a calcium-dependent and transporter-mediated release of purines in cultured avian retinal cells: possible involvement of calcium/calmodulin-dependent protein kinase II. *Neurochem. Int.* 2005, 46(6), 441-451.

[53] Thereza, M.; Perez, R.; Ehinger, B. Multiple neurotransmitter systems influence the release of adenosine derivatives from the rabbit retina. *Neurochem. Int.* 1989, 15(4), 411-420.

[54] Paes-De-Carvalho, R. Adenosine as a signaling molecule in the retina: biochemical and developmental aspects. *An. Acad. Bras. Ciênc.* 2002, 74(3), 437-451.

[55] Ferreira, J. M.; Paes-de-Carvalho, R. Long-term activation of adenosine A(2a) receptors blocks glutamate excitotoxicity in cultures of avian retinal neurons. *Brain Res.* 2001, 900(2), 169-176.

[56] Paes-de-Carvalho, R.; Maia, G. A.; Ferreira, J. M. Adenosine regulates the survival of avian retinal neurons and photoreceptors in culture. *Neurochem. Res.* 2003, 28(10), 1583-1590.

[57] Perígolo-Vicente, R.; Ritt, K.; Pereira, M. R.; Torres, P. M.; Paes-de-Carvalho, R. Giestal-de-Araujo, E. IL-6 treatment increases the survival of retinal ganglion cells in vitro: the role of adenosine A1 receptor. *Biochem Biophys. Res. Commun.* 2013, 430(2), 512-518.

[58] Liou, G. I.; Auchampach, J. A.; Hillard, C. J.; Zhu, G.; Yousufzai, B.; Mian, S.; Khan, S.; Khalifa, Y. Mediation of cannabidiol anti-inflammation in the retina by equilibrative nucleoside transporter and A2A adenosine receptor. *Invest. Ophthalmol. Vis. Sci.* 2008, 49(12), 5526-5531.

[59] Wurm, A.; Iandiev, I.; Hollborn, M.; Wiedemann, P.; Reichenbach, A.; Zimmermann, H.; Bringmann, A.; Pannicke, T. Purinergic receptor activation inhibits osmotic glial cell swelling in the diabetic rat retina. *Exp. Eye Res.* 2008, 87(4), 385-393.

[60] Nagase, K.; Tomi, M.; Tachikawa, M.; Hosoya, K. Functional and molecular characterization of adenosine transport at the rat inner blood-retinal barrier. *Biochim. Biophys. Acta.* 2006, 1758(1), 13-19.

[61] Akanuma, S.; Soutome, T.; Hisada, E.; Tachikawa, M.; Kubo, Y.; Hosoya, K. Na+-independent nucleoside transporters regulate adenosine and hypoxanthine levels in Muller cells and the inner blood-retinal barrier. *Invest. Ophthalmol. Vis. Sci.* 2013, 54(2), 1469-1477.

[62] dos Santos-Rodrigues, A.; Ferreira, J. M.; Paes-de-Carvalho, R. Differential adenosine uptake in mixed neuronal/glial or purified glial cultures of avian retinal cells: modulation by adenosine metabolism and the ERK cascade. *Biochem. Biophys. Res. Commun.* 2011, 414(1), 175-180.

[63] Ribelayga, C.; Mangel, S. C. A circadian clock and light/dark adaptation differentially regulate adenosine in the mammalian retina. *J. Neurosci.* 2005, 25(1), 215–222.

[64] Dmitriev, A. V.; Mangel, S. C. Retinal pH reflects retinal energy metabolism in the day and night. *J. Neurophysiol.* 2004, 91(6), 2404–2412.

[65] Stella, S. L. Jr.; Bryson, E. J.; Thoreson, W. B. A2 adenosine receptors inhibit calcium influx through L-type calcium channels in rod photoreceptors of the salamander retina. *J. Neurophysiol.* 2002, 87(1), 351-360.

[66] Stella, S. L. Jr.; Bryson, E. J.; Cadetti, L.; Thoreson, W. B. Endogenous adenosine reduces glutamatergic output from rods through activation of A2-like adenosine receptors. *J. Neurophysiol.* 2003, 90(1), 165-174.

[67] Stella, S. L. Jr.; Hu, W. D.; Vila, A; Brecha, N. C. Adenosine inhibits voltage-dependent Ca2+ influx in cone photoreceptor terminals of the tiger salamander retina. *J. Neurosci. Res.* 2007, 85(5), 1126-37.

[68] Stella, S. L. Jr.; Hu, W. D.; Brecha, N. C. Adenosine suppresses exocytosis from cone terminals of the salamander retina. *Neuroreport.* 2009, 20(10), 923-929.

[69] Doyle, S. E.; McIvor, W. E.; Menaker, M. Circadian rhythmicity in dopamine content of mammalian retina: role of the photoreceptors. *J. Neurochem.* 2002, 83(1), 211-219.

[70] Costenla, A. R.; De Mendonça, A.; Sebastião, A.; Ribeiro, J. A. An adenosine analogue inhibits NMDA receptor-mediated responses in bipolar cells of the rat retina. *Exp. Eye Res.,* 1999, 68(3), 367-370.

[71] Santos, P. F.; Santos, M. S.; Carvaiho, A. P.; Duarte, C. B. Modulation of [3H] Acetylcholine Release from Cultured Amacrine-Like Neurons by Adenosine A 1 Receptors. *J. Neurochem.* 1998, 71(3), 1086–1094.

[72] Santos, P. F.; Caramelo, O. L.; Carvalho, P.; Duarte, C. B. Adenosine A1 receptors inhibit Ca2+ channels coupled to the release of ACh, but not of GABA, in cultured retina cells. *Brain Res.* 2000, 852(1), 10-15.

[73] Dunn, T. A.; Wang, C. T.; Colicos, M. A.; Zaccolo, M.; DiPilato, L. M.; Zhang, J.; Tsien, R. Y.; Feller, M. B. Imaging of cAMP levels and protein kinase A activity reveals that retinal waves drive oscillations in second-messenger cascades. *J. Neurosci.* 2006, 26(49), 12807-12815.

[74] Wang, C. T.; Blankenship, A. G.; Anishchenko, A.; Elstrott, J.; Fikhman, M.; Nakanishi, S.; Feller, M. B. GABA(A) receptor-mediated signaling alters the structure of spontaneous activity in the developing retina. *J. Neurosci.* 2007, 27(34), 9130-9140.

[75] Lu, J. C.; Hsiao, Y. T.; Chiang, C. W.; Wang, C. T. GABAA receptor-mediated tonic depolarization in developing neural circuits. *Mol. Neurobiol.* 2014, 49(2), 702-723.

[76] Clark, B. D.; Kurth-Nelson, Z. L.; Newman, E. A. Adenosine-evoked hyperpolarization of retinal ganglion cells is mediated by G-protein-coupled inwardly rectifying K+ and small conductance Ca2+-activated K+ channel activation. *J. Neurosci.* 2009, 29(36), 11237-11245.

[77] Sun, X.; Barnes, S.; Baldridge, W. H. Adenosine inhibits calcium channel currents via A1 receptors on salamander retinal ganglion cells in a mini-slice preparation. *J. Neurochem.* 2002, 81(3), 550-556.

[78] Zhang, M.; Hu, H.; Zhang, X.; Lu, W.; Lim, J.; Eysteinsson, T.; Jacobson, K. A.; Laties, A. M.; Mitchell, C. H. The A3 adenosine receptor attenuates the calcium rise triggered by NMDA receptors in retinal ganglion cells. *Neurochem. Int.* 2010, 56(1), 35-41.

[79] Newman, E. A. A dialogue between glia and neurons in the retina: modulation of neuronal excitability. *Neuron Glia Biol.* 2004, 1(3), 245-252.

[80] Wurm, A.; Pannicke, T.; Iandiev, I.; Francke, M.; Hollborn, M.; Wiedemann, P.; Reichenbach, A.; Osborne, N. N.; Bringmann, A. Purinergic signaling involved in Müller cell function in the mammalian retina. *Prog. Ret. Eye Res.* 2011, 30(5), 324-342.

[81] Newman, E. A. Calcium increases in retinal glial cells evoked by light-induced neuronal activity. *J. Neurosci.* 2005, 25(23), 5502-5510.

[82] Uckermann, O.; Vargová, L.; Ulbricht, E.; Klaus, C.; Weick, M.; Rillich, K.; Wiedemann, P.; Reichenbach, A.; Syková, E.; Bringmann, A. Glutamate-Evoked Alterations of Glial and Neuronal Cell Morphology in the Guinea Pig Retina. *J. Neurosci.* 2004, 24(45), 10149-10158.

[83] Lutty, G. A.; McLeod, D. S. Retinal vascular development and oxygen-induced retinopathy: a role for adenosine. *Prog. Retin. Eye Res.* 2003, 22(1), 95-111.

[84] Paes-de-Carvalho, R.; Maia, G. A. Adenosine modulates neuronal survival in purified avian retinal cultures. *Soc. Neurosci. Abstracts,* 1996, 22(2), 1485.

[85] Roth, S. Endogenous neuroprotection in the retina. *Brain Res. Bull.* 2004, 62(6), 461–466.

[86] Li, B.; Roth, S. Retinal ischemic preconditioning in the rat: requirement for adenosine and repetitive induction. *Invest. Ophthalmol Vis. Sci.* 1999, 40(6), 1200-1216.

[87] Sakamoto, K.; Kuwagata, M.; Nakahara, T.; Ishii, K. Late preconditioning in rat retina: involvement of adenosine and ATP-sensitive K+ channel. *Eur. J. Pharmacol. 2001,* 418(1-2), 89–93.

[88] Perígolo-Vicente, R.; Ritt, K.; Gonçalves-de-Albuquerque, C. F.; Castro-Faria-Neto, H. C.; Paes-de-Carvalho, R.; Giestal-de-Araujo, E. IL-6, A1 and A2aR: a crosstalk that modulates BDNF and induces neuroprotection. *Biochem. Biophys. Res. Commun.* 2014, 449(4), 477–482.

[89] Zhang, X.; Zhang, M.; Laties, A. M.; Mitchell, C. H. Balance of purines may determine life or death of retinal ganglion cells as A3 adenosine receptors prevent loss following P2X7 receptor stimulation. *J. Neurochem.* 2006, 98(2), 566–575.

[90] Karmouty-Quintana, H.; Xia, Y.; Blackburn, M. R. Adenosine signaling during acute and chronic disease states. *J. Mol. Med.* 2013, 91(2), 173–181.

[91] Larsen, A, K.; Osborne, N. N. Involvement of Adenosine in Retinal Ischemia. Studies on rat. *Invest. Ophthalmol. Vis. Sci.* 1996. 37(13), 2603-11.

[92] Ghiardi, G. J.; Gidday, J. M.; Roth, S. The purine nucleoside adenosine in retinal ischemia-reperfusion injury. *Vision Res.* 1999, 39(15), 2519–2535.

[93] Williams, M. Challenges in developing P2 purinoceptor-based therapeutics. *Ciba Found. Symp.* 1996, 198, 309–21.

[94] Peachey, N. S.; Green, D. J.; Ripps, H. Ocular ischemia and the effects of allopurinol on functional recovery in the retina of the arterially perfused cat eye. *Invest. Ophthalmol. Vis. Sci.* 1993, 34(1), 58–65.

[95] Roth, S.; Park, S. S.; Sikorski, C.; Osinski, J.; Chan, R.; Loomis, K. Concentration of adenosine and its metabolites in the rat retina/choroid during reperfusion after ischemia. *Curr. Eye Res.* 1997, 16(9), 875–885.

[96] Klein, R.; Klein, B. E.; Moss, S. E.; Davis, M. D.; DeMets, D. L. The Wisconsin epidemiologic study of diabetic retinopathy. II. Prevalence and risk of diabetic retinopathy when age at diagnosis is less than 30 years. *Arch. Ophthalmol.* 1984, 102(4), 520–6.

[97] Lutty, G. A.; Mathews, M. K.; Merges, C.; McLeod, D. S. Adenosine stimulates canine retinal microvascular endothelial cell migration and tube formation. *Curr. Eye Res.* 1998, 17(6), 594–607.

[98] Takagi, H.; King, G. L.; Ferrara, N.; Aiello, L. P. Hypoxia regulates vascular endothelial growth factor receptor KDR/Flk gene expression through adenosine A2 receptors in retinal capillary endothelial cells. *Invest. Ophthalmol. Vis. Sci.* 1996, 37(7), 1311–1321.

[99] Takagi, H.; King, G. L.; Robinson, G. S.; Ferrara, N.; Aiello, L. P. Adenosine mediates hypoxic induction of vascular endothelial growth factor in retinal pericytes and endothelial cells. *Invest. Ophthalmol. Vis. Sci.* 1996, 37(11), 2165–2176.

[100] Lim, L. S.; Mitchell, P.; Seddon, J. M.; Holz, F. G.; Wong, T. Y. Age-related macular degeneration. *Lancet.* 2012, 379(9827), 1728–1738.

[101] Braunagel, S. C.; Xaio, J. G.; Chiou, G. C. The potential role of adenosine in regulating blood flow in the eye. *J. Ocul. Pharmacol.* 1988, 4(1), 61–73.

[102] Campochiaro, P. A.; Sen, H. A. Adenosine and its agonists cause retinal vasodilation and hemorrhage. Implications for ischemic retinopathies. *Arch. Ophthalmol.* 1989, 107(3), 412–416.

[103] Gidday, J. M.; Park, T. S. Adenosine-mediated autoregulation of retinal arteriolar tone in the piglet. *Invest. Ophthalmol. Vis. Sci.* 1993, 34(9), 2713–2719.

[104] Gidday, J. M.; Maceren, R. G.; Shah, A. R.; Meier, J. A.; Zhu, Y. K_{ATP} channels mediate adenosine-induced hyperemia in retina. *Invest. Ophthalmol. Vis. Sci.* 1996, 37(13), 2624–2633.

[105] Ibrahim, A. S.; El-shishtawy, M. M.; Zhang, W.; Caldwell, R. B.; Liou, G. I. A2A Adenosine Receptor (A2AAR) as a Therapeutic Target in Diabetic Retinopathy. *Am. J. Pathol.* 2011, 178(5), 2136-45.

[106] Elsherbiny, N. M.; Naime, M.; Ahmad, S.; Elsherbini, A. M.; Mohammad, S.; Fulzele. S.; El-Remessy, A. B.; Al-Gayyar, M. M.; Eissa, L. A.; El-Shishtawy, M. M.; Han, G.; White, R.; Haroldo, T. F.; Liou, G. I. Potential roles of adenosine deaminase-2 in diabetic retinopathy. *Biochem. Biophys. Res. Commun.* 2013, 436(3), 355–361.

[107] Elsherbiny, N. M.; Ahmad, S.; Naime, M.; Elsherbini, A. M.; Fulzele, S.; Al-Gayyar, M. M.; Eissa, L. A.; El-Shishtawy, M. M.; Liou, G. I. ABT-702, an adenosine kinase inhibitor, attenuates inflammation in diabetic retinopathy. *Life Sci.* 2013, 93(2-3), 78–88.

[108] Charles, B. A.; Conley, Y. P.; Chen, G.; Miller, R. G.; Dorman, J. S.; Gorin, M. B.; Ferrell, R. E.; Sereika, S. M.; Rotimi, C. N.; Orchard, T. J. Variants of the adenosine A2A receptor gene are protective against proliferative diabetic retinopathy in patients with type 1 diabetes. *Ophthalmic. Res.* 2010, 46(1), 1–8.

[109] Foster, P. J.; Buhrmann, R.; Quigley, H. A.; Johnson, G. J. The definition and classification of glaucoma in prevalence surveys. *Br. J. Ophthalmol.* 2002, 86(2), 238–242.

[110] Zhong, Y.; Yang, Z.; Huang, W. C.; Luo, X. Adenosine, adenosine receptors and glaucoma: An updated overview. *Biochim. Biophys. Acta*, 2013, 1830(4), 2882–2890.

[111] Fishman, P.; Cohen, S.; Bar-Yehuda, S. Targeting the A_3 adenosine receptor for glaucoma treatment (Review). *Mol. Med. Rep.* 2013, 7(6), 1723-1725.

[112] Kimelberg, H. K. Current concepts of brain edema. Review of laboratory investigations. *J. Neurosurg.* 1995, 83(6), 1051-9.

[113] Uckermann, O.; Kutzera, F.; Wolf, A.; Pannicke, T.; Reichenbach, A.; Wiedemann, P.; Wolf, S.; Bringmann, A. The glucocorticoid triamcinolone acetonide inhibits osmotic swelling of retinal glial cells via stimulation of endogenous adenosine signaling. *J. Pharmacol. Exp. Ther.* 2005, 315(3), 1036–1045.

[114] Wurm, A.; Lipp, S.; Pannicke, T.; Linnertz, R.; Färber, K.; Wiedemann, P.; Reichenbach, A.; Bringmann, A. Involvement of A1 adenosine receptors in osmotic volume regulation of retinal glial cells in mice. *Mol. Vis.* 2009, 15, 1858-1867.

In: Adenosine Signaling Mechanisms
Editors: V. Ramkumar and R. Paes de Carvalho
ISBN: 978-1-63483-186-4

Chapter 7

The Adenosinergic System and Its Interactions with Cytokines

Rafael Perígolo-Vicente[1,2*]
and Elizabeth Giestal-de-Araujo[2#]
[1] Universidade Federal do Rio de Janeiro
Rio de Janeiro, Brazil
[2] Universidade Federal Fluminense
Niterói, Brazil

Abstract

In the past few years the relationship between adenosine and cytokines, especially interleukins, in both physiological and pathological conditions have been investigated.

Most of the data indicate a protective role of adenosine during the inflammatory responses by regulating the levels of pro and anti-inflammatory cytokines.

Accordingly, in this chapter we try to answer two major questions: What is known about adenosine and cytokines? Why should we study the interactions between them?

Introduction

Cytokines are pleiotropic small proteins involved in intercellular communication. They act through cell-surface receptors, triggering downstream signalling pathways and modulating the cell function and phenotype. Cytokines encompass a large group of polypeptides, including chemokines, interferons, interleukins, transforming growth factors, trophic factors, neurotrophins and tumour necrosis factors. Their classification was first based on its presumed function, cell of secretion or target of action. However, based on their redundancy

* Correspondence to: rafaelperigolo@gmail.com
# egiestal@vm.uff.br

and pleiotropism, with few exceptions, such distinctions are currently considered inaccurate [1, 2, 3].

Cytokines are often involved in several developmental processes during embryogenesis [4, 5]. They are also crucial for fighting off infections as well as other immune responses. Pro-inflammatory cytokines are released after a variety of stimuli, which includes infections and tissue damage. They play an important role in eliminating pathogens or damaged and dying cells [6, 7]. At the same time, anti-inflammatory cytokines play a key role in controlling the intensity of the immune response contributing to the homeostasis. Tumour necrosis factor-alpha (TNF-alpha) is, for instance, a key pro-inflammatory cytokine involved in the inflammatory pathologies such as Crohn's disease, ulcerative colitis and rheumatoid arthritis. Anti-TNF therapy is a gold standard for the treatment of these diseases [8].

Another group of cytokines, the interleukins, was classically defined as pro- or anti-inflammatory molecules. It has been demonstrated that the balance between pro and anti-inflammatory molecules determines the level of control of inflammation [8]. IL-6 is classically defined as an important pro-inflammatory cytokine and its expression is related to many inflammatory disorders [9]. After an infection or organ inflammation, the levels of IL-6 rapidly increase. Therefore, measuring its levels is an efficient way to evaluate inflammatory conditions, especially in sepsis [10]. In patients suffering from rheumatoid arthritis, for instance, high levels of IL-6 have been associated with chronic inflammation, and clinical improvements have been reported with the administration of anti-IL-6 to these patients [11].

The mechanisms controlling the interplay between pro and anti-inflammatory cytokines are crucial for the homeostasis and therefore needs to be sharp and precise [8]. An imbalance in these mechanisms can have deleterious effects on the tissue. On the one hand, the excess of pro-inflammatory mediators can cause organ damage, killing healthy cells [9, 11]. On the other hand, if the anti-inflammatory cytokines production persists, the suppression of the immune response against malignant cells fails, which may ensue tumour growth [12]. After myocardial infarction, in animal models, an increase in the pro-inflammatory and a decrease in the anti-inflammatory cytokine levels are correlated with the size of tissue lesion and may point to a worse clinical outcome [13].

It is now known that cytokines are more than just modulators of the inflammatory responses. In fact, they act as chemical mediators between immune, endocrine and nervous systems [14, 15, 16]. In some inflammatory conditions, when peripheral immune reaction is excessive, for example, the exacerbated response is associated with several alterations in the brain function. It is believed that the vagus nerve, a major component of the autonomic nervous system, plays an important role in the neuro-endocrine-immune axis to maintain homeostasis. After stimulation of the vagus nerve at low frequency (5Hz), the cholinergic anti-inflammatory pathways are recruited, decreasing the levels of TNF-alpha, one of the first cytokine to be discovered, in the periphery [15]. Recently, some authors showed that peripheral inflammation evokes a brain region-specific cytokine response as well as cholinergic signalling modulation, which are known to regulate peripheral cytokine levels. In mice intraperitoneal lipopolysaccharide injection resulted in high levels of IL-1beta and IL-6 mRNA in the cortex, cerebellum and striatum [17]. In the nervous system, cytokines seem to play an important role regulating the balance between cell survival and cell death. Many authors refer to them as new trophic factors acting also on neuronal growth and cell migration. In this context, much attention has been focused on the functional role of neuropoietic cytokines, such as IL-6, in the central nervous system (CNS) [18]. IL-6 modulates cellular function in the nervous system by acting as a chemical mediator [19]. It also promotes the differentiation of immature into mature oligodendrocytes and induces the differentiation of cortical precursor cells into astrocytes. IL-6 can stimulate the activation of astrocytes and induce survival and differentiation of peripheral and central neurons [20, 21]. Additionally, data from the literature show that IL-6 intravitreal injection increases the survival and regeneration of retinal ganglion cells after optic nerve crush *in vivo* [22].

ADENOSINE AND CYTOKINES

Adenosine (Ado) has been described as a modulator of the inflammatory response in different tissues. This modulation is based on tipping the balance between the production of pro-inflammatory and anti-inflammatory cytokines [23]. The levels of adenosine depend on either the amount of Ado synthesized in the extracellular space or its release from the intracellular compartment [24]. It has been shown that adenosine is able to stimulate the release of IL-6 in different tissues. During ischemia, the levels of IL-6 in the brain increase due to a release of this molecule by astrocytes [25]. It is already known that adenosine plays an important role in the airway epithelia and some evidence point to a relation between adenosine and lung inflammation. In fact, it was demonstrated that adenosine triggers the release of IL-6 and IL-1 beta. The authors claim that extracellular adenosine might be inducing tissue inflammation [26]. Ado is known to be released in response to stress and cell damage. The increase in the Ado levels can indicate not only the level of tissue injury, but also can induce a protective tissue response [27]. Under some conditions, such as ischemia and inflammation, adenosine production increases, contributing to protection of the injured tissue [28]. After spinal cord trauma, for instance, the damaged neurons release high levels of ATP, which is quickly converted into Ado. The amplification of the Ado levels indicates that adenosinergic system can decrease the death of injured central neurons [29]. During the development, Ado also plays an important role controlling the cytokine levels. It is well known that human newborns show impairment at Th1-type pro-inflammatory cytokines production. Some authors showed that newborn plasma contains high concentrations of Ado and low levels of TNF-alpha, compared to adult plasma. After an inflammatory stimulus, when Ado is, experimentally, degraded by Ado deaminase, the production of the cytokine increases in neonatal, but not in adult mononuclear cells. The authors claimed that adenosine protects the infant from the excessive inflammatory responses, but could increase the risk of infections [30]. Ado mediates its effects through the signalling evoked by the activation of A1, A2a, A2b and A3 receptors. Each of them can trigger a different signalling pathway, leading to the modulation of distinct cytokines levels [29]. Herein we grouped the evidence of the interactions between adenosine and cytokine in sections, based on the receptor to which adenosine can bind.

Adenosine A1 Receptors

Multiple sclerosis (MS) is a chronic disease characterized by an intense neuroinflammation, which causes demyelination and axonal injury [31]. Cytokines, such as

TNF alpha and IL-6 have been related to the pathogenesis of MS and some authors associate the levels of these cytokines to the activation of the purinergic system. Peripheral blood mononuclear cells (PBMC) from patients suffering with MS, exhibit lower plasmatic levels of TNF alpha and adenosine compared to the controls. Accordingly, the same profile was found in PBMC cultures from MS or control patients. In addition, PBMCs from patients with MS did not show differences in the A1 receptors (A1R). However, when these cells were treated with A1R agonist, the production of IL-6 but not TNF alpha, decreased. In healthy PBMCs, A1R activation inhibited mitogen-stimulated production of TNF alpha and failed at decreasing IL-6 production [32]. These data suggest that the role of Ado in controlling the levels of TNF alpha is impaired during MS. More recently, it was demonstrated that the knockout of A1R in animals developing experimental allergic encephalomyelitis showed an increase in axonal injury, demyelination and activation of microglia. In addition, these animals also present an enhancement in spinal cord pro-inflammatory gene expression, as well as an increase in IL-1 beta levels in macrophages [31]. Taken together, these results suggest that A1R may play an important role at controlling the cytokine production during axonal injuries.

In mice with atherosclerotic lesion, high levels of pro-inflammatory cytokines are associated to tissue damage. Recent evidence suggests that A1R activation mediates increase of IL-13 levels. In the apoliprotein E knockout animals, the levels of IL-13 were significantly increased. In contrast, in double knockouts for apoliprotein E and A1R, the levels of IL-13 were low [33]. These data suggest a strong relationship between A1R activation and IL-13 production.

Preclinical studies using both knockout and pharmacological approaches suggest that A1R may serve as a useful treatment target for sepsis. In mice with sepsis, evoked by cecal ligation and puncture, A1R blockade increased mortality. In addition, A1R knockout animals showed higher levels of pro-inflammatory cytokines in the plasma during the sepsis. These effects were related to injuries caused by intense inflammation in the liver and kidney [34].

It was already shown that IL-6 induces an increase in mRNA expression of A1R in astrocytes [35]. In this study the authors also demonstrate that IL-6 can increase phenylephrine activity leading to the accumulation of phosphatidylinositol. These results are in agreement with a previous one from the same group, which demonstrated that following Ado stimulation, astrocytes accumulated phosphatidylinositol [36]. More recently it was showed that Ado can stimulate IL-6 release in hypoxic conditions, leading to an increase in A1 receptor activity [37]. Torres and Araujo showed that IL-6 treatment protects retinal ganglion cells from death *in vitro* [38]. The authors demonstrated that IL-6 trophic effect was mediated by the release of polypeptides from the Golgi. Additionally, it has been shown that IL-6 effect depends on and modulates BDNF levels. After stimulation with IL-6 for 48h, BDNF levels decreased significantly. Similarly to IL-6, A1R activation also protected retinal ganglion cells from death and downregulates BDNF in the same time frame [39].

Another study showed that A1 and A2a receptor (A2aR) activation increases the levels of IL-6 inside neurons. The authors revealed that after treatment with either CHA (A1R agonist) the retinal cell cultures presented higher levels of IL-6 than the untreated group. What was intriguing in this study was that after CHA treatment the levels of IL-6 in the extracellular medium decreased drastically [40].

In support of these data, we found that treatment of mixed retinal cell cultures with CHA significantly increased the number of IL-6 positive cells (Figure 1). These cultures were kept for 48h under 95% of air and 5% of CO_2 and evaluated for IL-6 expression.

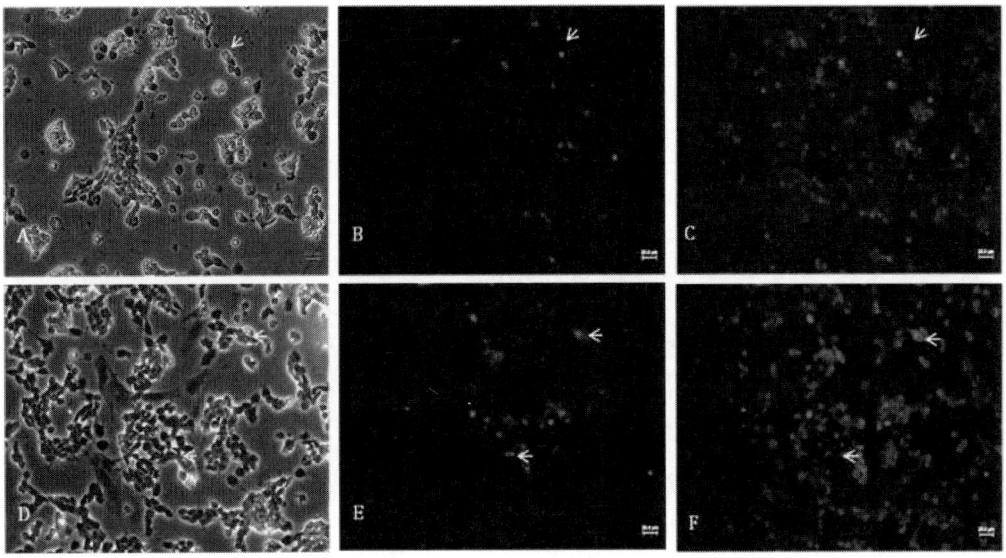

Figure 1. A1R activation increases the expression IL-6 in retinal cultures after 48h. Retinal cells from newborn rats were kept in dissociated culture for 48 hours accordingly with the protocol previously described [39,40]. In both conditions, the same density of retinal cells was plated ($1{,}25x10^6$ cells/mL). A, B and C – Control cultures. D, E and F – Cultures treated with 1nM CHA (A1AR antagonist). A and B – Phase contrast images. B and E – Immunocytochemistry for IL-6. Primary antibody anti-IL-6; secondary antibody goat anti-rabbit alexa 568 IgG. C and F – Merge with DAPI. The arrows indicate retinal cells expressing IL-6. Scale bar: 20μm.

Figure 1 also shows that although both cultures were plated at the same density ($1{,}25x10^6$ cells/mL), those that received treatment with CHA, clearly showed a higher number of cells. Taken together these data suggest that A1R activation might protect not only the ganglion cells but also other retinal cells.

Adenosine A2a Receptors

Data from the literature suggest that prolonged inflammation is the major cause of several diseases. It has been showed that A2a receptor (A2aR) plays an important role regulating pro-inflammatory cytokine production. Sub-threshold doses of an inflammatory stimulus were able to induce extensive tissue damage and high levels of cytokines in animals presenting A2aR deficiency. Interestingly, no other mechanism known to control inflammation were able to substitute for the absence of A2aR *in vivo* [41].

A2aR are expressed in immature dendritic cells. Activation of these receptors in these cells was unable to trigger a downstream signalling pathway coupled to increases in cAMP production. As the maturation of the dendritic cells takes place, the receptors became functional regulating the migratory response induced by inflammatory stimulus. Activation of

A2aR, in mature dendritic cells, induces a shift into their cytokine profile from a pro-inflammatory to an anti-inflammatory, reducing IL-12 and IL-6 production but increasing IL-10 [27].

There is a consensus opinion that A2aR plays an important role in lymphocyte responses. During inflammation, cytokines contribute to lymphocyte recruitment to the lesion site. Tissue damage activates natural killer T cells (NKT) that rapidly produce IFN-gamma. In a mouse model of sickle cell anemia, NKT cells activation is increased as well as the levels of IFN-gamma. When A2aR agonist is infused into these mice, the production of IFN-gamma production ceases, as well as lung inflammation and lung injury [42]. In A2aR knockout animals, the Ado receptor activation inhibited IL-2 release by T cells, reducing their proliferation. Moreover, A2aR activation suppressed the production of both IFN gamma and IL-4 [43, 44]. In a different study, A2aR activation in T cells reduced IL-2 levels but not IL-4. The authors showed that in Tc1 and Tc2 cells, A2aR activation decreased IL-2 and TNF alpha secretion. In addition, under the same condition, Tc2 reduced IL-10 production while IL-4 and IL-5 levels were kept unchanged [45].

In macrophages A2aR are also involved in cytokine production. Macrophage-derived cytokines are the triggers of an immune response. The presence of *Escherichia coli* normally induces macrophages to release IL-10. However, in the absence of A2aR, the IL-10 secretion was not suppressed [46]. Additionally, A2aR activation also promoted release of macrophage inflammatory proteins such as: CCL3, CCl4, CXCL2 and CCL20 [47].

Ado can also alter the endocrine function via cytokine release. Stimulation of adrenal zona glomerulosa cells with adenosine increased the levels of IL-6 but decreased the levels of TNF. The authors demonstrated that these effects were mediated by A2aR activation, since its activation leaded to IL-6 release [48]. In the central nervous system, A2aR activation leads to neuroprotection via IL-6. Treatment of retinal cells with A2aR agonist, *in vitro*, increases cell survival via an IL-6 signaling pathway. When IL-6 is neutralized, the neurotrophic effect of A2aR activation is completely lost. Additionally, CGS21680 treatment (A2aR agonist) induced a robust IL-6 release by retinal cells compared to the control [40].

It was already discussed in this chapter that Ado plays an important effect in airway epithelia modulating IL-6 release. Bronchoalveolar cells predominantly express low A2aR and high A2bR levels. Treatment with NECA (a nonspecific adenosine receptor agonist) increased IL-6 mRNA levels in horses presenting with airway inflammation. However, in healthy animals, the levels remained unchanged. When A2aR was antagonized with SCH-58261, treatment with NECA was able to increase IL-6 expression in both airway-compromised and healthy horses [49].

Adenosine A2b Receptors

The activation of A2bR, requires high levels of adenosine in the extracellular medium (EC_{50} of 24μm) [50]. As physiological concentration of Ado is lower than 1μM, the activation of A2bR would happen only under pathological conditions [51].

Experimentally, LPS stimulus produced high levels of IL-10, among other cytokines, in macrophages. It has been suggested that A2bR mediates this effect. Accordingly, when macrophages were exposed to an A2bR antagonist, adenosine failed to boost IL-10 production when stimulated by LPS [52].

It has been suggested that Ado plays a role in inflammatory airway diseases such as asthma and chronic obstructive pulmonary disease. It has been shown that the A2bR was involved in the production of IL-13, an important cytokine in asthma. In the mast cells from wild type mice, Ado, acting via A2bR increased the production of IL-13, which was abrogated in A2bR knockout cells [53]. In human mastocytoma (HMC-1) cells, Ado was also found to induce IL-8 release [54]. In bronchial smooth muscle cells, Ado increased the release of IL-6 via activation of A2bR [55]. Moreover, the activation of A2bR induces an increase in IL-19 release. In addition, the amount of the interleukin released by these cells was sufficient to activate a human monocytic cell line and increase TNF-alpha secretion. Furthermore, TNF-alpha was able to upregulate A2bR. The authors claimed that adenosine might initiate and amplify inflammatory response in lung diseases [56].

The important characteristic of the chronic inflammatory airway diseases is the subepithelial fibroblast activation, myofibroblast hyperplasia, hypoxia and an increase in the interstitial adenosine concentrations. A2bR activation, under normoxic conditions, increases the release of IL-6 by 14 fold and induces the differentiation of human lung fibroblast into myofibroblast. When IL-6 was neutralized with selective antibody, this effect was completely abolished. During hypoxia, IL-6 release is increased by 3-fold, and is associated with increased expression of A2bR. It seems that there is a synergy between hypoxia and the A2bR activation and that hypoxia amplifies the effect of Ado in releasing IL-6 and cell differentiation by upregulating the expression of A2bR [57].

During lumen intestinal inflammation, Ado is formed from neutrophil-derived 5'AMP. Stimulation of monolayers of T84 cells, an intestinal epithelial cell line, with Ado, increased IL-6 secretion via A2bR, which is polarized towards the luminal compartment. In addition, intestinal cells release IL-6 in luminal compartment following *Samonella typhimurium* infection, exposure to TNF alpha or following forskolin treatment. It was already demonstrated that Ado induces IL-6 production and release via cAMP-mediated activation of the transcript factor cyclic AMP response element binding (CREB) protein [58].

Adenosine A3 Receptors

The Ado A3 receptors (A3R) are a well-known and potent anti-inflammatory target. There is extensive evidence pointing to A3R involvement in several pathophysiological processes including, ischemic conditions of central nervous system, inflammation, rheumatoid arthritis and Crohn's disease [59]. Several studies have provided evidence to support the idea that A3R activation plays a major role in modulating cytokine production, especially during pro-inflammatory events. In patients presenting with rheumatoid arthritis, an overexpression of A3R was observed in both peripheral blood mononuclear and synovial cells. Antibodies against IL-12 or TNF alpha inhibited the overexpression of A3R and NF-kappaB. In addition, treatment with IB-MECA, a highly selective A3R agonist, reduced the inflammation by decreasing TNF release and inducing apoptosis of inflammatory cells [60, 61]. MIP-alpha is a cytokine that induces the migration of neutrophil into the inflamed tissue. In macrophage cultures, treatment with IB-MECA suppressed the production of MIP-1-alpha as well as IL-12 and IL-6. In a model of collagen-induced arthritis, IB-MECA administrated orally reduced the inflammation but decreased the production of MIP-alpha and IL-12 in the paws as well as suppressed neutrophil infiltration [62]. In agreement with these findings, another study showed that A3R activation in macrophage lineage with LPS treatment decreased nitric oxide synthase, IL-1 beta and TNF-alpha levels [63].

Intestinal cells also modulate the inflammatory condition through A3R activation. During the pathophysiology of colitis, there is a significant increase in inflammatory cell infiltration into the colon, which is associated with an increase in the levels of cytokines. In this experimental condition, activation of A3R receptors decreases the levels of pro-inflammatory cytokines, thereby impairing the progression of the tissue inflammation [64]. High levels of TNF alpha are produced during inflammation of colonic epithelial cells, inducing an increase in IL-8 and IL-1 beta production and leading to cell damage. However, if the A3Rs were activated, the levels of IL-8 and IL-1 beta mRNA decreases and the damage is reduced [59]. Additionally, renal cells in A3R knockout mice exhibit high levels of plasma TNF-alpha as well as an increase in mRNA encoding pro-inflammatory cytokines during sepsis [65].

The activation of microglial cells also plays an important role during inflammation. In microglial cells the stimulation of TNF-alpha production by LPS is reduced by A3AR activation through inhibition of the PI3-kinase/Akt pathway. The authors suggest that A3R ligands may have therapeutic potential modulating neuroinflammatory conditions [64]. LJ529, a selective A3R agonist, was found to reduce the inflammatory responses during ischemic injury to the cerebral artery. This effect of A3R activation on reduced microglia migration was mediated by the inhibition of IL-1 beta, TNF-alpha and MCP-1 production [66].

Although many studies point to an anti-inflammatory effect of A3R activation, in mast cells these receptors mediate the release of inflammatory compounds [67]. Also, IB-MECA treatment enhances LPS-induced TNF-alpha release in peritoneal macrophages, while inhibition of A3R significantly abrogated cytokine release. Moreover, the pro-inflammatory effect of A3R activation involved the activation of PI3-kinase signal transduction pathway because its inhibitor abolished IB-MECA-mediated effect [68].

CONCLUSION

The anti-inflammatory properties of Ado have been known for the past several years and based on the evidences that have been produced, an intimate relation between Ado and cytokine production has been revealed. In the central nervous system, for instance, the adenosinergic system increases neuronal survival mainly by controlling the neuroinflammation. When faced with tissue insults, Ado works as a signaling molecule modulating cytokine production via its receptors to preserve tissue function and to prevent further injury. In this sense, this chapter attempts to link a variety of information available from different physiological and pathological studies, concerning adenosine receptors activation with cytokine release and production. In order to more easily follow the discussion, please refer to the 'Summary Table" showing the effects of Ado on cytokine production.

There is a paucity of data relating A1 receptors and cytokine production. However, the Ado anti-inflammatory properties seem to be mainly obtained through the activation of A1 and A3R. Although the activation of each Ado receptor subtype may lead to a different signaling platform, there is often an agreement about cytokine modulation. The TNF alpha production decreased in response to Ado in most of the experimental paradigms discussed in this chapter. Notably, Ado decreases TNF-alpha in human newborns plasma [30], acts on A1R in plasma during sepsis [34], on A2aR in Tc1 and Tc2 cells [45], on A2b and on A3R in peripheral blood mononuclear cells [60,61], and on A3R in macrophages (RAW264.7) [63] in plasma during sepsis [65] and microglial cells [65,66]. Nonetheless, some studies have found that Ado could also increases TNF alpha in other tissues such as bronchial smooth muscle cells mediated via A2bR activation [56] and in peritoneal macrophages through A3R activation [68].

Summary Table. Adenosine receptors and its effect on cytokine production

	EFFECT	CYTOKINE	SOURCE	REFERENCE
Ado	Increases	IL-6	Astrocytes	[26]
Ado	Increases	IL-6	Astrocytes	[37]
Ado	Increases	IL-6	Airway epithelia	[26]
Ado	Decreases	IL-1β	Airway epithelia	[26]
Ado	Decreases	TNF-α	Newborn plasma	[30]
Ado	Increases	IL-2	T cells	[43]
A1	Decreases	IL-1β	Macrophages	[31]
A1	Decreases	IL-1β	Spinal cord	[31]
A1	Increases	IL-13	Vascular tissue	[33]
A1	Decreases	TNF-α	Plasma	[34]
A1	Increases	IL-6	Retinal cells	[40]
A1	Decreases	IL-6	PBMCs from MS patients	[32]
A2A	Decreases	IFN-Y	NK T cells from lungs	[42]
A2A	Decreases	IL-6	Dendritc cells	[27]
A2A	Decreases	IL-12	Dendritc cells	[27]
A2A	Increases	IL-10	Dendritc cells	[27]
A2A	Decreases	IFN-Y	T cells	[44]
A2A	Decreases	IL-4	T cells	[44]
A2A	Decreases	IL-2	Tc1 and Tc2 cells	[45]
A2A	Decreases	TNF-α	Tc1 and Tc2 cells	[45]
A2A	No effect	IL-5	Tc2	[45]
A2A	No effect	IL-4	Tc2	[45]
A2A	Decreases	IL-10	Tc2	[45]
A2A	Increases	IL-10	Macrophages	[46]
A2A	Increases	CCL3	Macrophages	[47]
A2A	Increases	CCL4	Macrophages	[47]
A2A	Increases	CXCL2	Macrophages	[47]
A2A	Increases	CCL20	Macrophages	[47]
A2A	Increases	IL-6	Adrenal zona glomerulosa	[48]
A2A	Decreases	TNF-α	Adrenal zona glomerulosa	[48]
A2A	Increases	IL-6	Retinal cells	[40]
A2A	Decreases	IL-6	Bronchoalveolar cells	[49]

Summary Table (Continued)

	EFFECT	CYTOKINE	SOURCE	REFERENCE
A2B	Increases	IL-10	Macrophages	[52]
A2B	Increases	IL-13	Mast cells	[53]
A2B	Increases	IL-8	Mastocytoma lineage	[54]
A2B	Increases	IL-6	Bronchial smooth muscle cells	[55]
A2B	Increases	IL-19	Bronchial smooth muscle cells	[56]
A2B	Increases	TNF-α	Bronchial smooth muscle cells	[56]
A2B	Increases	IL-6	Lung fibroblast	[57]
A2B	Increases	IL-6	Intestinal epithelial cell	[58]
A2B	Decreases	TNF-α	PBMCs	[60]
A3	Decreases	TNF-α	PBMCs	[61]
A3	Decreases	MIP- α	Macrophages	[62]
A3	Decreases	MIP- α	Paw	[62]
A3	Decreases	IL-12	Paw	[62]
A3	Decreases	IL-12	Macrophages	[62]
A3	Decreases	IL-6	Macrophages	[62]
A3	Decreases	TNF-α	Macrophage lineage	[63]
A3	Decreases	IL-1β	Macrophage lineage	[63]
A3	Decreases	IL-1β	Colonic epithelial cells	[59]
A3	Decreases	IL-8	Colonic epithelial cells	[59]
A3	Decreases	TNF-α	Plasma	[65]
A3	Decreases	TNF-α	Microglial cells	[65]
A3	Decreases	TNF-α	Microglial cells	[66]
A3	Decreases	IL-1β	Microglial cells	[66]
A3	Decreases	MCP-1	Microglial cells	[66]
A3	Increases	TNF-α	Macrophages	[68]

As showed in the summary table, the production of IL-1 beta is increased in airway epithelia by Ado, which could induce lung inflammation [26]. However, in all the articles we reviewed, Ado was found to downregulate IL-1 beta production through A1R activation (in macrophages and spinal cord) [31,69] and through A3R in macrophages RAW264.7 [63], colonic epithelial cells [59] and microglial cells [66].In contrast, there is no agreement about Ado and IL-6 production. Ado increases IL-6 production in astrocytes [25,37] and airway epithelia [26]. Through A1R activation the purine also decreases IL-6 release in peripheral blood monuclear cells from patients suffering from multiple sclerosis [32] and increases IL-6 levels in retinal cells culture [40]. Interestingly, A2aR also modulates the IL-6, decreasing its levels in dendritic cells [27] and in bronchoalveolar cells [70], whereas increasing its levels in adrenal zona glomerulosa [48] and retinal cells culture [40]. A2bR activation leads to an increase in IL-6 release in lung fibroblasts [57] and intestinal epithelial cell lines [58] while

A3R activation decreases IL-6 levels as seen in macrophages [62]. One possible explanation for such disagreement would be the dual inflammatory effect of this cytokine. IL-6 can act both as a pro-inflammatory and an anti-inflammatory molecule and therefore, in order to properly modulate the cytokine, the adenosinergic system would have to act depending on the cellular context [8].

Another cytokine that has been often associated to the adenosinergic system is the IL-10, which often induces anti-inflammatory and regenerative responses [71]. Most of the papers demonstrated that Ado might act by downregulating cytokine production through A2aR activation. Haskó and colleagues in 2010 showed that activating A2aR, adenosine increases IL-10 production in dendritic cells [27]. In macrophages, A2aR regulates the IL-10 production during *E. coli* infection [62]. However, in Tc2 cells A2aR activation decreased IL-10 release [45]. Although most of the data related A2aR of the cytokine production, other Ado receptors may also modulate IL-10 production. For instance, A2bR increased IL-10 production in macrophages after LPS stimulus [52]. IL-12 is also frequently associated with adenosine receptors and its production promotes T helper immunity and seems to be related with tumor growth [72]. In dendritic cells, A2aR activation leads to an increase of the cytokine production [27] whilst A3R increased it in both inflamed paw and macrophages [62].

Over the years, advances in understanding the adenosinergic system and cytokine production have been observed. However, the mechanisms responsible for this intricate array are not well known. One possible explanation could be the capacity of adenosine receptors to modulate cAMP production. Increase in intracellular cAMP has been known for long time to have immunosuppressive effects. In this sense, the anti-inflammatory effects of A2a and A2bR, for instance, are largely mediated through cAMP upregulation [73].

Ado itself in the form of adenocard and adenoscan or its derivatives has been used in the clinical practice, for the treatment of supraventricular tachycardia since 1940 and its systemic side effects are well known. There is other evidence that points toward a clinical role for adenosine as typified by the dipyridamole, which impairs uptake of adenosine and impair thrombus formation and the methotrexate, which is used for the treatment of reumathoid arthritis.

The utility of the latter compound is mediated by Ado. Recently, the US FDA approved the use of an A2aR antagonist to treat Parkinson's disease [73, 74]. As extensively discussed in this chapter, cytokines plays an important role in several pathologies. Understanding the involvement of adenosinergic system in these pathologies associated with cytokine unbalance may lead to new therapies [8, 12, 11].

However, there are some challenges regarding the use of adenosine receptors as target for new therapies. Ado receptors are wildly expressed in the body and therefore, its systemic activation leads to several side effects. Also, many of adenosinergic agonists are highly lipophilic, a property, which would facilitate diffusion of the drug to adenosine receptor targets in several tissues [73, 74]. In this sense, it is also necessary to develop new Ado ligands that could overcome those problems by activating the Ado receptors selectively without diffusing outside the tissue.

References

[1] Bauer, S.; Kerr, B.J.; Patterson, P.H. The neuropoietic cytokine family in development, plasticity, disease and injury. *Nat. Rev. Neurosci.* 2007, 8(3), 221–232.

[2] Arango Duque, G.; Descoteaux, A. Macrophage cytokines: involvement in immunity and infectious diseases. *Front. Immunol.* 2014, 5, 491.

[3] Nedoszytko, B.; Sokołowska-Wojdyło, M.: Ruckemann-Dziurdzińska, K.: Roszkiewicz, J.; Nowicki, R.J. Chemokines and cytokines network in the pathogenesis of the inflammatory skin diseases: atopic dermatitis, psoriasis and skin mastocytosis. *Postepy. Dermatol. Alergol.* 2014, 31(2), 84–91.

[4] Mirantes, C.; Passegué, E.; Pietras, E.M. Pro-inflammatory cytokines: Emerging players regulating HSC function in normal and diseased hematopoiesis. *Exp. Cell. Res.* 2014. [Epub ahead of print]

[5] Endele, M.; Etzrodt, M.; Schroeder, T. Instruction of hematopoietic lineage choice by cytokine signaling. *Exp. Cell. Res.* 2014. [Epub ahead of print]

[6] Glatman Zaretsky, A.; Engiles, JB.; Hunter, C.A. Infection-induced changes in hematopoiesis. *J. Immunol.* 2014, 192(1), 27–33.

[7] Cayrol, C.; Girard, J.P. IL-33: an alarmin cytokine with crucial roles in innate immunity, inflammation and allergy. *Curr. Opin. Immunol.* 2014, 31C:31–7.

[8] Fujiwara, N.; Kobayashi, K. Macrophages in inflammation. *Curr. Drug. Targets. Inflamm. Allergy*. 2005, 4(3), 281–6.

[9] Tanaka, T.; Narazaki, M.; Kishimoto, T. IL-6 in Inflammation, Immunity, and Disease. *Cold Spring Harb. Perspect. Biol.* 2014, 6(10), a016295.

[10] Jekarl, D.W.; Lee, S.Y.; Lee, J.; Park, Y.J.; Kim, Y.; Park, J.H.; Wee, J.H.; Choi, S.P. Procalcitonin as a diagnostic marker and IL-6 as a prognostic marker for sepsis. *Diagn. Microbiol. Infect. Dis.* 2013, 75(4), 342–7.

[11] Kang, S.; Tanaka, T.; Kishimoto, T. Therapeutic uses of anti-interleukin-6 receptor antibody. *Int. Immunol.* 2014. [Epub ahead of print]

[12] Burkholder, B.; Huang, R.Y.; Burgess, R.; Luo, S.; Jones, V.S.; Zhang, W.; Lv, Z.Q.; Gao, C.Y.; Wang, B.L.; Zhang, Y.M.; Huang, R.P.. Tumor-induced perturbations of cytokines and immune cell networks. *Biochim. Biophys. Acta.* 2014, 1845(2), 182–201.

[13] Cuenca-López, M.D.; Brea, D.; Galindo, M.F.; Antón-Martínez, D.; Sanz, M.J.; Agulla, J.; Castillo, J.; Jordán, J.. Inflammatory response during ischaemic processes: adhesion molecules and immunomodulation. *Rev. Neurol.* 2010, 51(1), 30–40.

[14] Numakawa, T.; Richards, M.; Nakajima, S.; Adachi, N.; Furuta, M.; Odaka, H.; Kunugi, H. The role of brain-derived neurotrophic factor in comorbid depression: possible linkage with steroid hormones, cytokines, and nutrition. *Front. Psychiatry.* 2014, 5, 136.

[15] Bonaz, B.; Picq, C.; Sinniger, V.; Mayol, J.F.; Clarençon, D. Vagus nerve stimulation: from epilepsy to the cholinergic anti-inflammatory pathway. *Neurogastroenterol Motil.* 2013, 25(3), 208–21.

[16] Santarsieri, M.; Kumar, R.G.; Kochanek, P.M.; Berga, S.; Wagner, A.K. Variable neuroendocrine-immune dysfunction in individuals with unfavorable outcome after severe traumatic brain injury. *Brain Behav. Immun.* 2014. [Epub ahead of print]

[17] Silverman, H.A.; Dancho, M.; Regnier-Golanov, A.; Nasim, M.; Ochani, M.; Olofsson, P.S.; Ahmed, M.; Miller, E.J.; Chavan, S.S.; Golanov, E.; Metz, C.N.; Tracey, K.J.; Pavlov, V.A. . Brain Region-specific Alterations in the Gene Expression of Cytokines, Immune Cell Markers and Cholinergic System Components During Peripheral Endotoxin-induced Inflammation. *Mol. Med.* 2014. [Epub ahead of print]

[18] Suzuki, S.; Tanaka, K.; Suzuki, N. Ambivalent aspects of interleukin-6 in cerebral ischemia: inflammatory versus neurotrophic aspects. *J. Cereb. Blood Flow Metab.* 2009, 29(3), 464–79.

[19] Conroy, S.M.; Nguyen, V.; Quinta, L.A.; Blakely-Gonzales, P.; Ur, C.; Netzeband, J.G.; Prieto, A.L.; Gruol, D.L.. Interleukin-6 produces neuronal loss in developing cerebellar granule neuron cultures. *J. Neuroimmunol.* 2004, 155(1-2), 43-54.

[20] Marz, P.; Cheng, J.G.; Gadient, R.A.; Patterson, P.H.; Stoyan, T.; Otten, U.; Rose-John S. Sympathetic neurons can produce and respond to interle ukin 6. *Proc. Natl. Acad. Sci.* 1998, 95(6), 3251-6

[21] Gadient, R.A.; Otten, U.H. Interleukin-6 (IL-6)--a molecule with both beneficial and destructive potentials. *Prog. Neurobiol.* 1997, 52(5), 379-90.

[22] Leibinger, M.; Andreadaki, A.; Diekmann, H.; Fischer, D. Neuronal STAT3 activation is essential for CNTF- and inflammatory stimulation-induced CNS axon regeneration. *Cell Death Dis*. 2013, 4:e805.

[23] Olah, M.E.; Stiles G.L. Adenosine receptor subtypes: characterization and therapeutic regulation. *Annu. Rev. Pharmacol. Toxicol*. 1995, 581-606

[24] Latini, S.; Pedata, F. Adenosine in the central nervous system: release mechanisms and extracellular concentrations. *J. Neurochem.* 2001, 79(3), 463-84.

[25] Schwaninger, M.; Neher, M.; Viegas, E.; Schneider, A.; Spranger, M. Stimulation of interleukin-6 secretion and gene transcription in primary astrocytes by adenosine. *J. Neurochem.* 1997, 69(3), 1145–50.

[26] Sun, Y.; Wu, F.; Sun, F.; Huang, P. Adenosine promotes IL-6 release in airway epithelia. *J. Immunol.* 2008, 180(6), 4173–81.

[27] Haskó, G.; Linden, J.; Cronstein, B.; Pacher, P. Adenosine receptors: therapeutic aspects for inflammatory and immune diseases. *Nat, Rev, Drug Discov.* 2008, 7(9), 759–70.

[28] Lima, F.O.; Souza, G.R.; Verri, W.A.; Parada, C.A.; Ferreira, S.H.; Cunha, F.Q.; Cunha, T.M.. Direct blockade of inflammatory hypernociception by peripheral A1 adenosine receptors: involvement of the NO/cGMP/PKG/KATP signaling pathway. *Pain*. 2010, 151(2), 506–15.

[29] Phillis, J.W.; Goshgarian, H.G. Adenosine and neurotrauma: therapeutic perspectives. *Neurol. Res.* 2001, 23(2-3), 183–9.

[30] Levy, O.; Coughlin, M.; Cronstein, B.N.; Roy, R.M.; Desai, A.; Wessels, M.R. The adenosine system selectively inhibits TLR-mediated TNF-alpha production in the human newborn. *J. Immunol.* 2006, 177(3), 1956–66.

[31] Tsutsui, S.; Schnermann, J.; Noorbakhsh, F.; Henry, S.; Yong, V.W.; Winston, B.W.; Warren, K.; Power, C. A1 adenosine receptor upregulation and activation attenuates neuroinflammation and demyelination in a model of multiple sclerosis. *J. Neurosci.* 2004, 24(6), 1521–9.

[32] Mayne, M.; Shepel, P.N.; Jiang, Y.; Geiger, J.D.; Power C. Dysregulation of adenosine A1 receptor-mediated cytokine expression in peripheral blood mononuclear cells from multiple sclerosis patients. *Ann. Neurol.* 1999, 45(5), 633–9.

[33] Teng, B.; Smith, J.D.; Rosenfeld, M.E.; Robinet, P.; Davis, M.E.; Morrison, R.R.; Mustafa, S.J.. A_1 adenosine receptor deficiency or inhibition reduces atherosclerotic lesions in apolipoprotein E deficient mice. *Cardiovasc. Res.* 2014, 102(1), 157–65.

[34] Gallos, G.; Ruyle, T.D.; Emala, C.W.; Lee, H.T. A1 adenosine receptor knockout mice exhibit increased mortality, renal dysfunction, and hepatic injury in murine septic peritonitis. *Am. J. Physiol. Renal Physiol.* 2005, 289(2), F369–76.

[35] Biber, K.; Fiebich, B.L.; Gebicke-Härter, P.; van Calker, D. Carbamazepine-induced upregulation of adenosine A1-receptors in astrocyte cultures affects coupling to the phosphoinositol signaling pathway. *Neuropsychopharmacology*. 1999, 20(3), 271–8.

[36] Biber, K.; Lubrich, B.; Fiebich, B.L.; Boddeke, H.W.; van Calker, D. Interleukin-6 enhances expression of adenosine A(1) receptor mRNA and signaling in cultured rat cortical astrocytes and brain slices. *Neuropsychopharmacology*. 2001, 24(1), 86–96.

[37] Biber, K.; Pinto-Duarte, A.; Wittendorp, M.C.; Dolga, A.M.; Fernandes, C.C.; Von Frijtag Drabbe Künzel, J.; Keijser, J.N.; de Vries, R.; Ijzerman, A.P.; Ribeiro, J.A.; Eisel, U.; Sebastião, A.M.; Boddeke, H.W.. Interleukin-6 upregulates neuronal adenosine A1 receptors: implications for neuromodulation and neuroprotection. *Neuropsychopharmacology*. 2008, 33(9), 2237–50.

[38] Mendonça Torre, P.M.; de Araujo, E.G. Interleukin-6 increases the survival of retinal ganglion cells in vitro. *J. Neuroimmuno*. 2001, 117(1-2), 43-50.

[39] Perígolo-Vicente, R.; Ritt, K.; Pereira, M.R.; Torres, P.M.; Paes-de-Carvalho, R.; Giestal-de-Araujo, E. IL-6 treatment increases the survival of retinal ganglion cells in vitro: the role of adenosine A1 receptor. *Biochem. Biophys. Res. Commun*. 2013, 430(2), 512-8.

[40] Perígolo-Vicente, R.; Ritt, K.; Gonçalves-de-Albuquerque, C.F.; Castro-Faria-Neto, H.C.; Paes-de-Carvalho, R.; Giestal-de-Araujo, E. IL-6, A1 and A2aR: a crosstalk that modulates BDNF and induces neuroprotection. *Biochem. Biophys. Res. Commun*. 2014, 449(4), 477-82.

[41] Ohta, A.; Sitkovsky, M. Role of G-protein-coupled adenosine receptors in downregulation of inflammation and protection from tissue damage. *Nature*. 2001, 414(6866), 916–20.

[42] Lappas, C.M.; Rieger, J.M.; Linden, J. A2A adenosine receptor induction inhibits IFN-gamma production in murine CD4+ T cells. *J. Immunol*. 2005, 174(2), 1073–80.

[43] Sevigny, C.P.; Li, L.; Awad, A.S.; Huang, L.; McDuffie, M.; Linden, J.; Lobo, P.I.; Okusa, M.D.. Activation of adenosine 2A receptors attenuates allograft rejection and alloantigen recognition. *J. Immunol*. 2007, 178(7), 4240–9.

[44] Naganuma, M.; Wiznerowicz, E.B.; Lappas, C.M.; Linden, J.; Worthington, M.T.; Ernst, P.B. Cutting edge: Critical role for A2A adenosine receptors in the T cell-mediated regulation of colitis. *J. Immunol*. 2006, 177(5), 2765–9.

[45] Erdmann, A.A.; Gao, Z.G.; Jung, U.; Foley, J.; Borenstein, T.; Jacobson, K.A.; Fowler, D.H.. Activation of Th1 and Tc1 cell adenosine A2A receptors directly inhibits IL-2 secretion in vitro and IL-2-driven expansion in vivo. *Blood*. 2005, 105(12), 4707–14.

[46] Csóka, B.; Németh, Z.H.; Virág, L.; Gergely, P.; Leibovich, S.J.; Pacher, P.; Sun, C.X.; Blackburn, M.R.; Vizi, E.S.; Deitch, E.A.; Haskó, G.. A2A adenosine receptors and

C/EBPbeta are crucially required for IL-10 production by macrophages exposed to Escherichia coli. *Blood.* 2007, 110(7), 2685–95.

[47] McColl, S.R.; St-Onge, M.; Dussault, A.A.; Laflamme, C.; Bouchard, L.; Boulanger, J.; Pouliot, M.. Immunomodulatory impact of the A2A adenosine receptor on the profile of chemokines produced by neutrophils. *FASEB J.* 2006, 20(1), 187–9.

[48] Ritchie, P.K.; Spangelo, B.L.; Krzymowski, D.K.; Rossiter, T.B.; Kurth, E.; Judd, A.M. Adenosine increases interleukin 6 release and decreases tumour necrosis factor release from rat adrenal zona glomerulosa cells, ovarian cells, anterior pituitary cells, and peritoneal macrophages. *Cytokine.* 1997, 9(3), 187–98.

[49] Zhang, L.; Franchini, M.; Wehrli Eser, M.; Dip, R. Enhanced IL-6 transcriptional response to adenosine receptor ligands in horses with lower airway inflammation. *Equine. Vet. J.* 2012, 44(1), 81–7.

[50] Fredholm, B.B.; Irenius, E.; Kull, B.; Schulte, G. Comparison of the potency of adenosine as an agonist at human adenosine receptors expressed in Chinese hamster ovary cells. *Biochem. Pharmacol.* 2001, 61(4), 443–8.

[51] Gyorgy, H.; Linden, J.; Cronstein, B.; Pacher, P. Adenosine receptors: therapeutic aspects for inflammatory and immune diseases. *Nat. Rev. Drug. Discov.* 2008, 7(9), 759–70.

[52] Németh, Z.H.; Lutz, C.S.; Csóka, B.; Deitch, E.A.; Leibovich, S.J.; Gause, W.C.; Tone, M.; Pacher, P.; Vizi, E.S.; Haskó, G.. Adenosine augments IL-10 production by macrophages through an A2B receptor-mediated posttranscriptional mechanism. *J. Immunol.* 2005, 175(12), 8260–70.

[53] Ryzhov, S.; Zaynagetdinov, R.; Goldstein, A.E.; Novitskiy, S.V.; Dikov, M.M.; Blackburn, M.R.; Biaggioni, I.; Feoktistov, I.. Effect of A2B Adenosine Receptor Gene Ablation on Proinflammatory Adenosine Signaling in Mast Cells. *J. Immunol.* 2008, 180(11), 7212–20.

[54] Feoktistov, I.; Biaggioni, I. Adenosine A2b receptors evoke interleukin-8 secretion in human mast cells. An enprofylline-sensitive mechanism with implications for asthma. *J. Clin. Invest.* 1995, 96(4), 1979–86.

[55] Zhong, H.; Belardinelli, L.; Maa, T.; Feoktistov, I.; Biaggioni, I.; Zeng, D. A(2B) adenosine receptors increase cytokine release by bronchial smooth muscle cells. *Am. J. Respir. Cell. Mol. Biol.* 2004, 30(1), 118–25.

[56] Zhong, H.; Wu, Y.; Belardinelli, L.; Zeng, D. A2B adenosine receptors induce IL-19 from bronchial epithelial cells, resulting in TNF-alpha increase. *Am. J. Respir. Cell. Mol. Biol.* 2006, 35(5), 587–92.

[57] Zhong, H.; Belardinelli, L.; Maa, T.; Zeng, D. Synergy between A2B adenosine receptors and hypoxia in activating human lung fibroblasts. *Am. J. Respir. Cell Mol. Biol.* 2005, 32(1), 2–8.

[58] Sitaraman, S.V.; Merlin, D.; Wang, L.; Wong, M.; Gewirtz, A.T.; Si-Tahar, M.; Madara, J.L.. Neutrophil-epithelial crosstalk at the intestinal lumenal surface mediated by reciprocal secretion of adenosine and IL-6. *J. Clin. Invest.* 2001, 107(7), 861–9.

[59] Ren, T.; Qiu, Y.; Wu, W.; Feng, X.; Ye, S.; Wang, Z.; Tian, T.; He, Y.; Yu, C.; Zhou, Y. Activation of adenosine A3 receptor alleviates TNF-α-induced inflammation through inhibition of the NF-κB signaling pathway in human colonic epithelial cells. *Mediators Inflamm.* 2014, 2014, 818251.

[60] Ochaion, A.; Bar-Yehuda, S.; Cohn, S.; Del Valle, L.; Perez-Liz, G.; Madi, L.; Barer, F.; Farbstein, M.; Fishman-Furman, S.; Reitblat, T.; Reitblat, A.; Amital, H.; Levi, Y.; Molad, Y.; Mader, R.; Tishler, M.; Langevitz, P.; Zabutti, A.; Fishman, P. Methotrexate enhances the anti-inflammatory effect of CF101 via up-regulation of the A3 adenosine receptor expression. *Arthritis. Res. Ther.* 2006, 8(6), R169.

[61] Madi, L.; Cohen, S.; Ochayin, A.; Bar-Yehuda, S.; Barer, F.; Fishman, P. Overexpression of A3 adenosine receptor in peripheral blood mononuclear cells in rheumatoid arthritis: involvement of nuclear factor-kappaB in mediating receptor level. *J. Rheumatol.* 2007, 34(1), 20–6.

[62] Szabó, C.; Scott, G.S.; Virág, L.; Egnaczyk, G.; Salzman, A.L.; Shanley, T.P.; Haskó, G. Suppression of macrophage inflammatory protein (MIP)-1alpha production and collagen-induced arthritis by adenosine receptor agonists. *Br. J. Pharmacol.* 1998, 125(2), 379–87.

[63] Lee, H.S.; Chung, H.J.; Lee, H.W.; Jeong, L.S.; Lee, S.K. Suppression of inflammation response by a novel A_3 adenosine receptor agonist thio-Cl-IB-MECA through inhibition of Akt and NF-κB signaling. *Immunobiology*. 2011, 216(9), 997–1003.

[64] Mabley, J.; Soriano, F.; Pacher, P.; Haskó, G.; Marton, A.; Wallace, R.; Salzman, A.; Szabó, C.. The adenosine A3 receptor agonist, N6-(3-iodobenzyl)-adenosine-5'-N-methyluronamide, is protective in two murine models of colitis. *Eur. J. Pharmacol.* 2003, 466(3), 323–9.

[65] Lee, H.T.; Kim, M.; Joo, J.D.; Gallos, G.; Chen, J.F.; Emala, C.W. A3 adenosine receptor activation decreases mortality and renal and hepatic injury in murine septic peritonitis. *Am. J. Physiol. Regul. Integr. Comp. Physiol.* 2006, 291(4), R959–69.

[66] Choi, I-Y; Lee, J-C; Ju, C; Hwang, S; Cho, G-S; Lee, HW; Choi, W.J.; Jeong, L.S.; Kim, W.K.. A3 adenosine receptor agonist reduces brain ischemic injury and inhibits inflammatory cell migration in rats. *Am. J. Pathol.* 2011, 179(4), 2042–52.

[67] Zhong, H.; Shlykov, S.G.; Molina, J.G.; Sanborn, B.M.; Jacobson, M.A.; Tilley, S.L.; Blackburn, M.R. Activation of murine lung mast cells by the adenosine A3 receptor. *J. Immunol.* 2003, 171(1), 338–45.

[68] Forte, G.; Sorrentino, R.; Montinaro, A.; Pinto, A.; Morello, S. Cl-IB-MECA enhances TNF-α release in peritoneal macrophages stimulated with LPS. *Cytokine*. 2011, 54(2), 161–6.

[69] Tsutsui, S.; Schnermann, J.; Noorbakhsh, F.; Henry, S.; Yong, V.W.; Winston, B.W.; Warren, K.; Power, C. A1 adenosine receptor upregulation and activation attenuates neuroinflammation and demyelination in a model of multiple sclerosis. *J. Neurosci.* 2004, 24(6), 1521–9.

[70] King, A.; Balaji, S.; Le, L.D.; Crombleholme, T.M.; Keswani, S.G. Regenerative Wound Healing: The Role of Interleukin-10. *Adv. Wound Care (New Rochelle).* 2014, 3(4), 315–23.

[71] Ngiow, S.F.; Teng, M.W.; Smyth, M.J. A balance of interleukin-12 and -23 in cancer. *Trends Immunol*. 2013, 34(11), 548–55.

[72] Whiteside, T.L.; Mandapathil, M.; Schuler, P. The role of the adenosinergic pathway in immunosuppression mediated by human regulatory T cells (Treg). *Curr. Med. Chem.* 2011, 18(34), 5217–23.

[73] Chen, J.F.; Eltzschig, H.K.; Fredholm, B.B. Adenosine receptors as drug targets--what are the challenges? *Nat. Rev. Drug Discov.* 2013, 12(4), 265–86.

[74] Müller, C.E.; Jacobson, K.A. Recent developments in adenosine receptor ligands and their potential as novel drugs. *Biochim. Biophys. Acta*. 2011, 1808(5), 1290–308.

In: Adenosine Signaling Mechanisms
Editors: V. Ramkumar and R. Paes de Carvalho
ISBN: 978-1-63483-186-4

Chapter 8

ROLE OF ADENOSINE RECEPTORS IN CANCER

Sandeep Sheth, Srinivisan Tupal and Vickram Ramkumar
Department of Pharmacology,
Southern Illinois University School of Medicine,
Springfield, IL, US

ABSTRACT

Adenosine, a ubiquitous nucleoside produced from the metabolism of adenosine triphosphate (ATP), is vital for a number of physiological processes. The physiological functions of adenosine are mediated by its interaction with four distinct G-protein-coupled receptor subtypes, termed adenosine receptors (ARs). These ARs are classified into A_1, A_{2A}, A_{2B} and A_3Rs, each of which shows diverse tissue distribution and functions. Due to the high metabolic rates of cancer cells, high levels of adenosine are produced by these cells. Several studies have demonstrated that modulation of ARs activity can profoundly influence cellular proliferation, survival, apoptosis, angiogenesis or metastasis, which are considered to be critical hallmarks of cancer growth and progression. This chapter highlights intrinsic and extrinsic pathway mediated by ARs in carcinogenesis. We present a detailed account of how ARs in different cancers regulate tumor growth and metastasis and discuss the potential of AR ligands in cancer treatment.

INTRODUCTION

The nucleoside adenosine is vital for normal physiological functions. Various intracellular and extracellular biochemical pathways and transporters are involved in regulating the concentrations of adenosine, as described in previous chapters. Intracellular adenosine is produced from the sequential enzymatic hydrolysis of adenosine triphosphate (ATP) to adenosine monophosphate (AMP) by endonucleotidases and then to adenosine by endo-5'-AMP nucleotidase. Adenosine is metabolized by adenosine deaminase to inosine and hypoxanthine, and further to uric acid by xanthine oxidase. Moreover, intracellular adenosine is also known to be produced from S-adenosylhomocysteine (SAH) by SAH hydrolase.

Likewise, extracellular adenosine is produced by the conversion of ATP into AMP and ultimately to adenosine by CD39 (ecto-ATP apyrase) and CD73 (ecto-5'-AMP nucleotidase), respectively. The extracellular disposition of adenosine is not only regulated by the adenosine deaminase and xanthine oxidase in the milieu but also by bidirectional passive equilibrative and Na^+-dependent active concentrative transporters [1].

The physiological extracellular concentration of adenosine is maintained in the low nanomolar range but could increase to high micromolar levels under metabolically stressful conditions (such as inflammation, tissue hypoxia and tumorogenesis) [2-5]. In the tumor microenvironment, extracellular adenosine mediates its effects on tumor cells by acting on cell surface ARs, namely the A_1, A_{2A}, A_{2B} and A_3 ARs (depending on cell types). Cancer cell killing mediated by ARs involves activation of pro-apoptotic pathways involving caspase-3, poly ADP-ribose polymerase (PARP) and nicotinamide adenine dinucleotide phosphate (NADPH) oxidase activity to promote cytochrome c release.

Alternatively, AR activation could promote direct activation of caspase-8, further downstream in this apoptosis pathway [6-8]. The extrinsic apoptotic pathway involves activation of G_q proteins and subsequently trigger stress-induced mitogen-activated protein kinases (MAPKs), including c-Jun N-terminal kinase (JNK) and p38 [9, 10]. The intrinsic pro-apoptotic pathway mediated by adenosine involves membrane transport as well as excessive intracellular generation of adenosine, which then gets catalytically converted to adenosine mono-, di- and triphosphates and cause up-regulation of p53 and AMP kinase followed by nuclear translocation of apoptosis-inducing factors [11-14]. This book chapter offers comprehensive and up-to-date information pertaining to role of adenosine, its receptors and effector pathways that regulate cancer cell proliferation, angiogenesis, metastasis and host immune response.

A_1 Adenosine Receptors

A_1AR is a G inhibitory (Gi) protein-coupled receptor (GPCR) that mediates many functions of adenosine throughout the body. Activation of the A_1AR inhibits cyclic AMP (cAMP) production, and decreases protein kinase A (PKA) activity and cAMP response element-binding protein (CREB) phosphorylation [15-17], while in some cases it activates phospholipase C (PLC) through a pertussis toxin-sensitive G protein. In the heart, A_1AR can directly couple to and inhibit cardiac K^+ channels and Q-, N- and P-type voltage sensitive Ca^{2+} channels. Activation of the A_1AR can also activate MAPK pathway in cells of the Chinese hamster ovary (CHO) [18, 19] and COS-7 fibroblast-like cells by $\beta\gamma$-subunit of $G_{i/o}$ protein [20].

Role of A_1AR in Brain Cancer

In the brain, A_1ARs are widely expressed in the presynaptic regions of the neurons in cortex, hippocampus and cerebellum [16], where they decrease the probability of the release of neurotransmitters, such as glutamate, acetylcholine, serotonin and GABA [15]. These receptors are also present on astrocytes [21], oligodendrocytes [22] and microglia [23]. Since

A_1AR is expressed throughout the brain, adenosine is implicated in different pathological condition of brain including brain tumors [24]. A_1AR expression has been detected in F98 and C6 rat brain tumors [25] and in rat C6 glioma cells [26]. Studies in F98 glioma-bearing rats showed higher than normal density of A_1AR in the zones surrounding experimental tumors. Immunostaining identified activated astrocytes as the main source of peri-tumoral A_1AR upregulation. This result was confirmed in a patient with recurrent glioblastoma by positron emission tomography (PET) using 8-cyclopentyl-3-(3-18F-fluoropropyl)-1-propylxanthine (18F-CPFPX) [27].

A_1AR expressed in microglia, the native immune cells of the brain, have shown some inhibitory effect on tumor growth. A_1AR-deficient mice as hosts exhibited larger Gl261 glioblastoma tumor volume as compared to wild-type mice, which demonstrated high infiltration of microglia around the tumor [28]. This suggests that adenosine-mediated inhibition of glioblastoma growth is dependent on activation of A_1ARs in microglia. Microglial cells are important source of matrix metalloproteinase (MMP)-2 [29], which increase the invasive behavior of glioma cells [30]. Activation of A_1ARs clearly decreased the activity of MMP-2 in microglia suggesting the anti-invasive effect of these receptors in glioma [28]. Other metalloproteinases, such as MMP-9 and MMP-12, have also been shown to be significantly elevated in A_1AR-deficient mice [31]. Furthermore, A_1AR activation on microglia/macrophages inhibits the production of not only cytokines, such as interleukin-1β, but also MMPs, such as MMP-12 [31].

Unlike glioma cells, activation of A_1AR stimulated chemotaxis in A2058 human melanoma cells. The chemotactic stimulation by adenosine and AMP was pertussis toxin sensitive and was mimicked by agonists and inhibited by antagonists specific for A_1AR. This finding suggests the possibility of anti-metastatic therapies based on inhibition of A_1ARs [32]. Thus, the effect of A_1AR on tumor cell invasion might be dependent on the type of cancer.

Conflicting Role of A_1AR in Carcinogenesis

The role of A_1AR in cancer development is unclear. Few studies have characterized the role of A_1AR to be pro-carcinogenic. Using reverse transcription-polymerase chain reaction (RT-PCR), high amounts of A_1AR mRNA were detected in the human colorectal adenocarcinoma as compared to peri-tumoral normal tissues, indicating a possible role of A_1AR in the carcinogenesis of colorectal cancer [33]. This increase in A_1AR was suggested to be in response to increased adenosine release from the hypoxic cells found in malignant tumor tissues. However, in CW2 human colonic cancer cells, adenosine acting via A_1AR induced apoptosis by activating canonical cell death pathways involving caspase-3, -8 and -9, and this effect was inhibited by A_1AR antagonist [34]. Similarly, A_1AR-mediated activation of caspase-3 and caspase-9 was also involved in the death of astrocytoma cells [35]. In addition to this, in human LoVo metastatic cells [36], human leukemia MOLT-4 cells, human breast cancer (T47D, HS578T and MCF-7) cells, A_1AR inhibited proliferation [3]. On the contrary, upregulation of A_1AR in different human breast cancer cell lines and human breast cancer tissues than in their normal counterparts was found to contribute to tumor cell growth and survival [37]. Knockdown of A_1AR by small interfering RNA (siRNA) in MDA-MB-468 breast cancer cells attenuated cell proliferation and increased cell death. Cell cycle analysis indicated that knockdown of A_1AR in these cells arrested them in G2/M phase due to impairment of G1 checkpoint. Further studies into the molecular mechanisms in HeLa cervical cancer cells showed that siRNA-induced depletion of A_1AR expression increased the expression of cyclin-dependent kinase (CDK) inhibitor, p27, with concomitant reduction of CDK4 and cyclin E proteins resulting in extensive cell death and apoptosis [37].

In another study using estrogen receptor (ER)-α positive breast cancer cells, it was shown that A_1AR gene is located proximally to an ERα binding site and that its expression is induced by ERα activation by estradiol [38]. Interestingly, a selective A_1AR antagonist, 1,3-dipropyl-8-cyclopentylxanthine (DPCPX), reduced proliferation of ERα positive breast cancer cells (MCF-7) by decreasing the expression of ERα [38].

These findings demonstrate the presence of a short feed-forward loop involving ERα and A_1AR that favors breast cancer growth. Thus, further studies are necessary to resolve in detail the molecular basis governing pro- or anti-cancer role of A_1AR in different types of cancers.

A_{2A} Adenosine Receptors

The signal transduction initiated by the activation of $A_{2A}ARs$ is mediate intracellularly via G_{olf} and G_s proteins. The alpha-subunit of the G_s protein physically interacts to activate the enzyme adenylyl cylcase (AC), which results in increased intracellular cAMP. Moreover, in kidney fibroblast-like (COS-7) cells, it was found that $A_{2A}AR$ mediated increase in cAMP via pertussis toxin-insensitive $G_{\alpha 15}$ and $G_{\alpha 16}$ proteins can in turn induce inositol 1,4,5 triphosphate production, causing a surge in intracellular Ca^{2+} release from the endoplasmic reticulum stores via ryanodine receptors and protein kinase C (PKC) activation [16, 52]. These molecular mechanisms initiated by $A_{2A}AR$ activation play a critical role in adenosine-mediated effects on cell proliferation, death and angiogenesis. $A_{2A}Rs$ are overexpressed in various cancer cells including A375 melanoma [39, 40], MCF-7 breast cancer [41], Jurkat T-cell leukemia [42], U937 monocytic lymphoma [43], U87MG human glioblastoma [44], SH-SY5Y neuroblastoma [45], neuroendocrine tumors [46], lung cancer cell lines [47, 48], A431 epidermoid carcinoma [49] and colon cancer cells [50, 51].

Role of $A_{2A}AR$ in Cancer Cell Proliferation and Apoptosis

Molecular events mediated by $A_{2A}AR$ activation results in different outcomes in tumor cell survival, proliferation and apoptosis, depending up on the cancer cell type. In B cell lymphomas and multiple myelomas, several $A_{2A}AR$ agonists showed significant pharmacodynamic synergism in exerting anti-proliferative effect and induction of cell death with phosphodiesterase (particularly PDE4B isoform) inhibitors as well as β-2 adrenergic receptor agonist, salmeterol. This effect was seen both in glucocorticoid-sensitive and -insensitive multiple myeloma and is associated with increase in cAMP. This combination therapy is promising in the treatment of multiple myeloma and diffuse large B-cell lymphoma and helps reduce the dose and their untoward effects [53, 54].

Likewise, in Caco-2 human colon cancer cells, anti-proliferative and pro-apoptotic effect of A_{2A}AR agonist, 2-p-[2-carboxyethyl] phenethyl-amino-5'-N-ethylcarboxamido-adenosine (CGS21680), involved the disruption of the mitochondrial membrane potential and activation of caspase-9 and -3. Moreover, these effects have a clear dose-response relationship that is associated with increase in cAMP, mimicked by forskolin and blocked by adenylyl cyclase inhibitor, 9-(Tetrahydro-2-furanyl)-9*H*-purin-6-amine (SQ22536) [51].

Merighi et al., [39] showed that in human cultured melanoma A375 cells, the anti-cancer effect of adenosine and A_{2A} receptor agonists, 2-hexynyl-5'*N*–ethylcarboxamido adenosine (HENECA) and 7-(2-phenylethyl)2-(2-furyl)pyrazolo[4,3*e*]1,2,4-triazolo[1,5*c*]pyrimidine (SCH58261), were critically mediated by PKC-extracellular signal-regulated kinase (ERK)-1/2 pathway. On the contrary, agonist of A_{2A}AR, CGS21680, exhibited a pro-proliferative effect on MCF-7 breast adenocarcinoma cells by increasing transcription of progesterone receptors [41]. Thus, the effect of activation and inhibition of A_{2A}AR can have different effects on survival and apoptosis depending on the cell type and phenotype of cancer cells.

Role of A_{2A}AR in Cancer Angiogenesis

A_{2A}ARs play a critical role in adenosine-induced angiogenesis in response to tissue injury by increasing the production of angiogenic-growth factor such as vascular endothelial growth factor (VEGF) and promoting proliferation of vascular endothelial cells [55-58]. Angiogenesis is critical for tumor growth and metastasis in melanomas and cancers of breast, lung, colon, and prostate. In some of these cancers (breast and prostate), vessel density has proven to be a significant prognostic indicator. Thus, it has been hypothesized that A_{2A}AR antagonists may exert anti-tumor effects by impairing solid tumor growth and metastasis. However, there is a paucity of evidence demonstrating the role of A_{2A}AR in angiogenesis. Desai et al., [59] have shown, using human microvascular epithelial cells, that CGS-21680 and 2-[2-(4-chlorophenyl) ethoxy]adenosine (MRE-0094), A_{2A}AR agonists, promotes neovascularization by decreasing the expression of anti-angiogenic factors such as thrombospondin-1.

Role of A_{2A}AR in Anti-Tumor Immunity

In addition to the anti-proliferative effect on cancer cells, the effect on survival and proliferation of immune cells would have major implications on the disease prognosis. CD4(+), CD8(+) and LAK cells express A_{2A}ARs and its activation caused suppression of anti-tumor immune cells. A_{2A}AR-mediated PKA activation enhance immunesuppression of T-lymphocytes and lymphokine-activated killer (LAK) cells. Autoimmune rejection of immunogenic melanoma cells occurred in ~60% of A_{2A}AR-deficient mice [3, 52]. As described earlier, agonists of A_{2A}AR synergistically promoted apoptosis with PD4B inhibitors in multiple myeloma and diffuse large B-cell lymphoma [53]. Interestingly, A_{2A}AR activation exerts a differential anti-apoptotic effect via Bcl-2 and pro-apoptotic effect via Bax proteins. This results in a reduction of neutrophil accumulation in normal tissue, while promoting tumor cell death of human A375 melanoma cells and Caco-2 human colon cancer cells [39,

51, 60]. Adenosine also exerts a significant immunosuppression on anti-tumor cytotoxic T lymphocytes via A_{2A}AR [61-64]. Adenosine can reduce tissue damaging inflammatory response by acting on A_{2A}AR and suppressing bone marrow-derived immune cells. Also, A_{2A}AR activation has been shown to suppress other immune cells such as macrophages, T cells and dendritic cells [65-67]. In terms of adaptive immune response, A_{2A}AR activation inhibits both CD4(+) and CD8(+) T-cell function [67]. Interestingly, A_{2A}AR-activation on T cells seems to selectively inhibit pro-inflammatory cytokine expression while sparing anti-inflammatory cytokine expression [67]. Adenosine acting via the A_{2A}AR might partially mediate the suppressive function of regulatory T (Treg) cells, which are crucial for maintaining tolerance to self. Thus, the A_{2A}AR prevents autoimmune diseases [68] by releasing the humoral mediators such as IL-10, TGF-β, interferon (IFN)-γ, galectin-1, and Foxp3 [69]. A_{2A}AR knockout mice exhibit enhanced antitumor immune responses by CD8(+) T cells, as well as a dramatic reduction in the growth of experimental tumors in comparison to wild-type controls. A_{2A}AR signaling has also been implicated in adenosine-mediated inhibition of cytokine production and cytotoxic activity of natural killer (NK) cells [70].

A_{2A}AR antagonists could provide an effective strategy to enhance immunosurveillance of cancer by interfering with the ability of extracellular adenosine (and A_{2A}AR) to induce local immunosuppression in tumors. Indeed, it has recently been reported that adenosine resistant effector T cells were successfully obtained by the exposure of activated T cells to NECA, an A_{2A}AR agonist. These *in vitro* studies form the basis for future attempts to produce anti-tumor T cells that are more conducive for adoptive immunotherapy [71]. In light of the wide distribution of A_{2A}AR, precautions must be taken to ensure that these A_{2A}R blockers do not mitigate immune suppression.

A_{2B} Adenosine Receptors

A_{2B}ARs are coupled to G_s and G_q protein which mediate increased intracellular levels of cAMP and Inositol 1,4,5 triphosphate (IP3) and diacylglycerol (DAG), respectively. Activation of IP3/DAG pathway in turn activates PKC and arachidonic acid pathways [72]. These receptors are highly expressed in spleen, lungs, colon and kidney, primarily in their vasculature. In addition these receptors are expressed in high levels in smooth muscle cells, endothelial cells and macrophages [73]. Owing to its tissue expression, activation of A_{2B}ARs elicits diverse molecular events important for carcinogenesis.

Role of A_{2B}AR in Cancer Cell Proliferation

A_{2B}ARs play different roles in survival and growth or apoptosis of tumors. In the highly invasive estrogen receptor unresponsive MDA-MB-231 breast cancer cells, activation of A_{2B}ARs promote proliferation and growth by activating adenylyl cyclase (to increase cAMP) and increasing intracellular Ca^{2+} concentrations (via PLC) [74]. In colorectal cancers, hypoxia induces the expression of A_{2B}ARs and promotes proliferation of cancer cells [75]. However, activation of A_{2B}AR receptors can promote apoptosis of cancer cells by activating caspase and Puma-dependent down regulation of Bcl-2 proteins [76]. In another recent study, Wei et al., [77] showed that non-selective and selective agonists for A_{2B}ARs, 5′-*N*-ethylcarboxamidoadenosine (NECA) and BAY60-6583, respectively, diminish the release of lactate dehydrogenase (LDH) and reduce tumor necrosis factor (TNF)-α-induced pro-apoptotic caspase-3 in PC-3, DU145 and LNCaP prostate cancer cells. Recently, Limm et al., [78] reported that melanoma cells lack the expression of methylthioadenosine phosphorylase, which results in the accumulation of 5'-methylthioadenosine (MTA). MTA affected cell proliferation of surrounding stroma cells and cell invasiveness via the A_{2B}AR by activating of the activator protein (AP)-1 transcription factor [78]. Therefore, A_{2B}ARs play a prominent role in regulating proliferation and growth of several types of cancers.

Role of A_{2B}AR in Tumor Angiogenesis and Invasion

A_{2B}ARs play a prominent physiological role in promoting angiogenesis. In retinal endothelial cells, A_{2B}ARs promote VEGF-mediated neovascularization [79]. Activation of A_{2B}ARs in human microvascular endothelial cells increases the expression of pro-angiogenic factors, such as VEGF, IL-8 and basic fibroblast growth factor (bFGF) in a PKA-dependent manner [80]. Hypoxia, a key regulator of angiogenesis, increases the expression of A_{2B}AR, by activating hypoxia inducible factor (HIF)-1α on the receptor promoter. Thus, transcriptional activation of A_{2B}AR increases angiogenesis [81]. Hypoxia-mediated angiogenesis in human melanoma cells could be effectively treated with A_{2B}AR antagonists by reducing the release of pro-angiogenic growth factor IL-8 [82]. Similarly, adenosine decreases the release of IL-8 from hypoxic U87MG cells which is dependent on the activation of A_{2B}AR [83]. Breast and colon cancer cell migration and invasion are promoted by A_{2B}AR activation. Breast cancer show increased expression of *CD73,* which is primarily responsible for catalyzing the dephosphorylation of extracellular AMP to produce adenosine [84]. Increase in adenosine concentration promotes cancer cell migration and invasion through activation of A_{2B}AR. Activation of this receptor promotes tumor cell chemotaxis *in vitro* and metastasis *in vivo* [85]. The activation of A_{2B}AR increased invasiveness of deleted in colon cancer (DCC)-deficient human colon cancer cells [86]. This effect was mediated via a PKA-mediated modulation of Rho GTPase and Rho kinase on cytoskeletal proteins.

Role of A_{2B}AR in Anti-Tumor Immunity

A_{2B}ARs have also been investigated for their role in anti-cancer immunity. Using A_{2B}AR knockout mouse model, Ryzhov et al., [87] demonstrated that activation of A_{2B}ARs located on tumor-infiltrating host immune cells, increased intra-tumoral vessel density and promoted Lewis lung carcinoma growth. This response was due to a deficit in VEGF in A_{2B}AR knock-out mice, compared to the tumors growing in wild-type animals [87]. Consistent with this finding, intermittent intra-tumor injection of a non-selective AR antagonist, aminophylline (AMO; theophylline ethylenediamine), and a selective A_{2B}AR antagonist, ATL801, into MB49 bladder and 4T1 breast tumors slowed their growth in syngeneic mice and reduced metastasis of breast cancer cells from mammary fat to lung. This effect was mediated by A_{2B}ARs and associated with IFN-γ and the IFN-inducible chemokine CXCL10 (a ligand for

CXCR3). This resulted in increased infiltration of activated tumor-infiltrating CXCR3$^+$ T cells and a decrease in endothelial cell precursors within tumors, leading to inhibition of tumor growth [88].

In another study, melanoma-bearing mice treated with a selective A_{2B}AR agonist showed increased melanoma growth. This effect was associated with higher levels of immune regulatory mediators, such as interleukin-10 (IL-10) and monocyte chemoattractant protein (MCP)-1, and accumulation of tumor-associated CD11b positive Gr1 positive cells (CD11b$^+$Gr1$^+$) myeloid-derived suppressor cells (MDSCs).

Conversely, pharmacological blockade of A_{2B}AR reversed immune suppression in the tumor microenvironment, leading to a significant delay in melanoma growth [89]. Thus, A_{2B}ARs can effectively shape immune response in regulating cancer growth and metastasis.

A_3 Adenosine Receptors As Tumor Cell Marker

A_3ARs are coupled to $G_{i/o}$ proteins. A_3ARs have shown to regulate the activity of PLC via a pertussis toxin-sensitive G protein in rat basophilic leukemia cells or by direct coupling to G_q protein [1]. A_3ARs are known to be upregulated in cancer cells. Higher A_3AR mRNA and protein expression has been reported in different kinds of cancer cells, such as human melanoma, colon, breast, small-cell lung and pancreatic carcinoma as compared to adjacent non-neoplastic or normal tissues [90]. Similarly, higher A_3AR expression was seen in A375 melanoma cells [40], HL-60 human leukemia [91, 92], Jurkat lymphoma [93], U87MG human glioblastoma [44, 94] and PC-3MM human prostate cancer cells [95], suggesting that A_3AR upregulation is a common phenomenon to different malignancies. Moreover, there is substantial evidence correlating increased A_3AR expression with disease severity [96]. In a study where 73 paired samples of tumor and surrounding peri-tumoral normal mucosa and blood samples from a cohort of 30 patients with colorectal cancer were investigated, it was revealed that the density of A_3ARs was higher in colon carcinomas as compared with normal mucosa originating from the same individuals. In peripheral blood cells, the density of A_3AR was approximately 3-fold higher in cancer patients compared with healthy subjects. Interestingly, large adenomas showed increased binding than small adenomas, which had affinity and density values similar to those of the mucosa of healthy subjects. Moreover, expression of A_3ARs in circulating blood cells returned to normal after surgical resection for colorectal cancer [96]. In a different study, lymph node metastasis expressed higher levels of A_3AR than in the primary tumors, supporting the notion that A_3AR expression levels are directly associated with disease severity [90]. In a comparative study between primary thyroid cancer tissues and normal thyroid tissue samples, it was reported that A_3ARs were expressed in cancerous tissues (as determined by immunohistochemistry), whereas normal tissue did not express A_3AR [97]. Considering the elevated expression of A_3AR in many cancers, several studies have suggested that the A_3ARs could serve as a tumor cell marker and as a target to for cancer treatment.

Role of A_3AR in Cancer Cell Proliferation and Apoptosis

Majority of the studies has reported an anti-cancer role of A_3AR agonists. One of the earliest studies elucidating the tumor suppressive role of adenosine, conducted by Fishman et al., [98], showed that skeletal muscles are resistant to tumor metastases due to the release of adenosine by these cells , which exerted differential effect on tumor vs. normal cell growth. It was later demonstrated that the anti-cancer activity exerted by muscle cell conditioned media was mediated primarily through the A_3ARs [99]. Subsequently, the activation of A_3AR was shown to inhibit the growth of lymphoma [100], melanoma [101], glioblastoma [44], hepatoma [102], mesothlioma [103], leukemia [104], colon cancer [105, 106], breast cancer [107], prostate cancer [95, 108], and lung cancer [109] cells.

The anti-cancer effect mediated by activation of A_3AR involves inhibition of cancer cell proliferation, cell cycle arrest and increased cell apoptosis. In Nb2-11C rat lymphoma cells, adenosine acting through A_3AR induced inhibition of telomerase activity and arrest in the G_0/G_1 phase of cell cycle [100]. A_3AR activation also inhibited the growth of HCT-116 colon carcinoma in mice by decreasing the expression level of PKA while increasing glycogen synthase kinase-3β (GSK-3β) expression. This results in destabilization of β-catenin and the subsequent suppression of expression of cyclin D1 and c-Myc, genes crucial for cell cycle progression [105]. Additionally, A_3AR agonist was found to down-regulate PKB/Akt expression levels, resulting in decrease in the level and DNA-binding capacity of nuclear factor (NF)-κB, which is a well-known transcriptional regulator of cyclin D1 and c-Myc [105, 108]. Both of these pathways led to cell cycle arrest and apoptosis, thereby leading to inhibition of tumor growth. Recently, CF102, an orally active analog of A_3AR agonist, 2-chloro-N6-3-iodobenzyladenosine-5′-N-methyluronamide (Cl-IB-MECA), was shown to protect the liver against inflammation and inhibited the growth of hepatocellular carcinoma. These actions were produced by deregulating NF-κB and Wnt signal transduction pathways, resulting in apoptosis of tumor cells [110]. CF102 has been tested in Phase I/II clinical trials to study the safety and clinical effect in patients with advanced un-resectable hepatocellular carcinoma. The results showed that it was well-tolerated with favorable pharmacokinetic properties, justifying further clinical development of this drug [111].

Role of A_3AR in Cancer Cell Metastasis

The effect of A_3AR activation on cancer cell migration and invasion has been investigated. Studies from our laboratory have demonstrated that A_3AR activation reduces the ability of prostate cancer cells to migrate *in vitro* and metastasize *in vivo*. The anti-metastatic action was due to A_3AR-mediated reduction in high levels of reactive oxygen species (ROS) via NADPH oxidase in these cells. We reported that metastatic prostate cancer cells have high NADPH oxidase activity, knockdown of which reduces invasiveness in these cells. Activation of A_3AR suppressed PKA-mediated stimulation of ERK1/2, leading to lower NADPH oxidase activity and cancer cell invasiveness [95] (Figure 1).

However, in human glioblastoma U87MG cells, A_3AR activation increased MMP-9 levels by phosphorylation of ERK1/2, c-Jun N-teminal kinase/stress-activated protein kinase (pJNK/SAPK), PKB/Akt and finally AP-1. This effect led to increased invasiveness of glioblastoma cells [44].

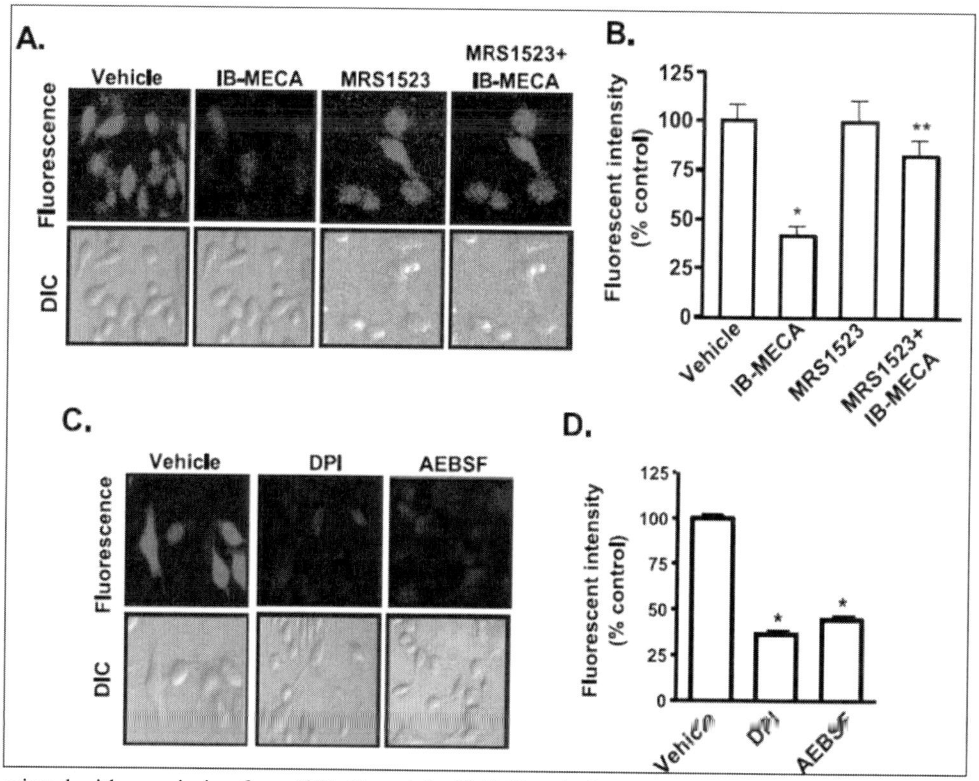

Reprinted with permission from [95], Copyright 2009 Neoplasia Press, Inc.

Figure 1. Activation of A3AR reduces ROS generation in AT6.1 prostate cancer cells: AT6.1 cells were pretreated with either vehicle or MRS-1523, $A_{3A}R$ antagonist, for 30 mins and then treated with IB-MECA, $A_{3A}R$ agonist for 15 mins. ROS production was determined by incubating the cells with H2DCFDA dye using florescence microscopy. IB-MECA-induced suppression of ROS was reversed by MRS-1523, indicating the role of $A_{3A}R$ (**A,B**). Similar effects were obtained with DPI and AEBSF, NADPH oxidase inhibitors (**C,D**). Histograms represent mean ± SEM of 4 different experiments. (*) and (**) indicate $P < 0.05$ vs. vehicle and IB-MECA treatments, respectively.

Therefore, caution should be used if A_3AR agonists are to be used to treat other cancer types.

Role of A_3AR in Anti-Tumor Immunity

In addition to the direct effect of A_3AR on tumor cells, these receptors can also exert anti-cancer activity via an indirect mechanism by increasing anti-tumor immune response. In a murine melanoma model, A_3AR agonist, Cl-IB-MECA, potentiated the infiltration and activation of the NK cells in naïve and tumor bearing mice through the induction of IL-12 production [112]. IL-12 is a known stimulant of NK cells and is also involved in the anti-tumor activity as it exerts cytotoxic effects against the tumor cells [113]. Similarly, A_3AR agonist, CF101, suppressed colon carcinoma growth and liver metastases and exerted myeloprotective effect by upregulating IL-12 production and potentiating NK cell activity

[106]. In another study, Cl-IB-MECA treatment significantly inhibited tumor growth in immune-competent mice by increasing the recruitment of tumor-infiltrating CD8(+) T cells and NK1.1(+) cells. This effect was associated with high levels of TNF-α and IFN-γ in melanoma tissue. Moreover, the killing of melanoma cells was dependent on anti-tumor immune cells, as depletion of either CD8(+) T cells or NK1.1(+) cells completely abrogated the antitumor effect of Cl-IB-MECA [114].

Role of A_3AR in Tumor Angiogenesis

Although activation of A_3ARs has primarily shown to inhibit tumor growth, few studies have shown that A_3AR stimulation facilitates tumor growth by promoting angiogenesis. The tumor microenvironment is severely hypoxic and angiogenic response by the endothelial cells is crucial for the tumor cells to survive under such conditions of oxygen deprivation. HIF-1α is a major transcription factor that is regulated under hypoxic conditions. Stabilization of HIF-1α induces the transcription of pro-angiogenic growth factors such as vascular endothelial growth factors VEGF, which mediates tumor angiogenesis. A_3AR activation led to accumulation of HIF-1α in melanoma [115], glioblastoma [94], osteosarcoma, and carcinomas of the colon, ovary and breast [115]. In hypoxic human glioblastoma A172 and U87MG cells, adenosine acting via A_3AR up-regulated HIF-1α and VEGF, and this response was shown to be inhibited by A_3AR antagonist [94]. Furthermore, it was reported that A_3AR-induced HIF-1α and VEGF required activation of ERK1/2 and p38 MAPKs in hypoxic glioblastoma cells [94], whereas blockade of A_3AR reduced VEGF production in A375 human melanoma cells treated with etoposide and doxorubicin [82]. Adenosine was also shown to stimulate angiogenesis in a mechanism that was independent of VEGF. This action involved modulation of other pro- or anti-angiogenic growth factors. A_3AR agonist, in hypoxic A375 melanoma cells, induced stabilization of HIF-1α which significantly increased the expression of pro-angiogenic factor angiopoietin-2. This response was also regulated via ERK1/2 and p38 MAPK pathways [115]. Thus, these studies have opened up a new possibility for A_3AR antagonists as anti-cancer drugs by blocking angiogenesis. However, these findings could complicate the use of A_3AR agonists as anticancer agents, as these agents could also stimulate tumor growth by increase blood supply to the tumors.

CONCLUSION

Due to the diverse roles of ARs in cancers, they offer a number of potentially novel, selective and effective anti-neoplastic drug targets. As described in this chapter, preclinical studies involving cell line, syngeneic, xenograft, orthotopic and metastatic experimental animal models have shown that ARs can act as distinct cancer targets - A_1AR for inhibiting tumor growth, $A_{2A}AR$ for immune therapy, $A_{2B}AR$ as anti-angiogenic therapy and A_3AR for the prevention of metastasis. A case in point is Cl-IB-MECA (CF102, Can-Fite BioPharma), an A_3AR agonist, which has shown promising results in Phase I/II clinical trials against hepatocellular carcinoma. A major task would be to design drugs which are selective for each

AR subtype, to limit potential side effects. Future studies should better elucidate the importance of subtype-selective AR drugs for treating different cancers.

Acknowledgment

Research described in this chapter was supported, in part, by grants from National Institutes of Health (R15 CA135494, R15DC011412, RO1-DC002396, and R01-CA166907), Deafness Research Foundation, National Organization for Hearing Research and by various grants from SIU School of Medicine.

References

[1] Sheth, S.; Brito, R.; Mukherjea, D.; Rybak, L. P.; Ramkumar, V. Adenosine receptors: expression, function and regulation. *Int. J. Mol. Sci.,* 2014, 15, 2024-52.
[2] Fredholm, B. B. Adenosine receptors as drug targets. *Exp. Cell Res.,* 2010, 316, 1284-8.
[3] Gessi, S.; Merighi, S.; Sacchetto, V.; Simioni, C.; Borea, P. A. Adenosine receptors and cancer. *Biochim. Biophys. Acta*, 2011, 1808, 1400-12.
[4] Schulte, G.; Fredholm, B. B. Signalling from adenosine receptors to mitogen-activated protein kinases. *Cell Signal*, 2003, 15, 813-27.
[5] Stagg, J.; Smyth, M. J. Extracellular adenosine triphosphate and adenosine in cancer. *Oncogene,* 2010, 29, 5346-58.
[6] Ma, Y.; Zhang, J.; Zhang, Q.; Chen, P.; Song, J.; Yu, S.; Liu, H.; Liu, F.; Song, C.; Yang, D.; Liu, J. Adenosine induces apoptosis in human liver cancer cells through ROS production and mitochondrial dysfunction. *Biochem. Biophys. Res. Commun.*, 2014, 448, 8-14.
[7] Yang, D.; Yaguchi, T.; Lim, C. R.; Ishizawa, Y.; Nakano, T.; Nishizaki, T. Tuning of apoptosis-mediator gene transcription in HepG2 human hepatoma cells through an adenosine signal. *Cancer Lett.,* 2010, 291, 225-9.
[8] Yang, D.; Yaguchi, T.; Yamamoto, H.; Nishizaki, T. Intracellularly transported adenosine induces apoptosis in HuH-7 human hepatoma cells by downregulating c-FLIP expression causing caspase-3/-8 activation. *Biochem. Pharmacol.*, 2007, 73, 1665-75.
[9] Vaque, J. P.; Dorsam, R. T.; Feng, X.; Iglesias-Bartolome, R.; Forsthoefel, D. J.; Chen, Q.; Debant, A.; Seeger, M. A.; Ksander, B. R.; Teramoto, H.; Gutkind, J. S. A genome-wide RNAi screen reveals a Trio-regulated Rho GTPase circuitry transducing mitogenic signals initiated by G protein-coupled receptors. *Mol.Cell*, 2013, 49, 94-108.
[10] Young, A.; Mittal, D.; Stagg, J.; Smyth, M. J. Targeting cancer-derived adenosine: new therapeutic approaches. *Cancer Discov.*, 2014, 4, 879-88.
[11] Saitoh, M.; Nagai, K.; Nakagawa, K.; Yamamura, T.; Yamamoto, S.; Nishizaki, T. Adenosine induces apoptosis in the human gastric cancer cells via an intrinsic pathway relevant to activation of AMP-activated protein kinase. *Biochem. Pharmacol.*, 2004, 67, 2005-11.
[12] Seetulsingh-Goorah, S. P. Mechanisms of adenosine-induced cytotoxicity and their clinical and physiological implications. *Biofactors,* 2006, 27, 213-30.

[13] Tsuchiya, A.; Kanno, T.; Saito, M.; Miyoshi, Y.; Gotoh, A.; Nakano, T.; Nishizaki, T. Intracellularly transported adenosine induces apoptosis in [corrected] MCF-7 human breast cancer cells by accumulating AMID in the nucleus. *Cancer Lett.,* 2012, 321, 65-72.

[14] Virtanen, S. S.; Kukkonen-Macchi, A.; Vainio, M.; Elima, K.; Harkonen, P. L.; Jalkanen, S.; Yegutkin, G. G. Adenosine Inhibits Tumor Cell Invasion via Receptor-independent Mechanisms. *Mol. Cancer Res.*, 2014. [Epub ahead of print].

[15] Cunha, R. A. Adenosine as a neuromodulator and as a homeostatic regulator in the nervous system: different roles, different sources and different receptors. *Neurochem. Int.,* 2001, 38, 107-25.

[16] Fredholm, B. B.; AP, I. J.; Jacobson, K. A.; Linden, J.; Muller, C. E. International Union of Basic and Clinical Pharmacology. LXXXI. Nomenclature and classification of adenosine receptors--an update. *Pharmacol. Rev*., 2011, 63, 1-34.

[17] Paes-De-Carvalho, R. Adenosine as a signaling molecule in the retina: biochemical and developmental aspects. *An. Acad. Bras. Cienc*., 2002, 74, 437-51.

[18] Dickenson, J. M.; Blank, J. L.; Hill, S. J. Human adenosine A1 receptor and P2Y2-purinoceptor-mediated activation of the mitogen-activated protein kinase cascade in transfected CHO cells. *Br. J. Pharmacol*., 1998, 124, 1491-9.

[19] Schulte, G.; Fredholm, B. B. Human adenosine A(1), A(2A), A(2B), and A(3) receptors expressed in Chinese hamster ovary cells all mediate the phosphorylation of extracellular-regulated kinase 1/2. *Mol. Pharmacol.,* 2000, 58, 477-82.

[20] Faure, M.; Voyno-Yasenetskaya, T. A.; Bourne, H. R. cAMP and beta gamma subunits of heterotrimeric G proteins stimulate the mitogen-activated protein kinase pathway in COS-7 cells. *J. Biol. Chem*., 1994, 269, 7851-4.

[21] Biber, K.; Klotz, K. N.; Berger, M.; Gebicke-Harter, P. J.; van Calker, D. Adenosine A1 receptor-mediated activation of phospholipase C in cultured astrocytes depends on the level of receptor expression. *J. Neurosci*., 1997, 17, 4956-64.

[22] Othman, T.; Yan, H.; Rivkees, S. A. Oligodendrocytes express functional A1 adenosine receptors that stimulate cellular migration. *Glia,* 2003, 44, 166-72.

[23] Johnston, J. B.; Silva, C.; Gonzalez, G.; Holden, J.; Warren, K. G.; Metz, L. M.; Power, C. Diminished adenosine A1 receptor expression on macrophages in brain and blood of patients with multiple sclerosis. *Ann. Neurol.,* 2001, 49, 650-8.

[24] Dunwiddie, T. V.; Masino, S. A. The role and regulation of adenosine in the central nervous system. *Ann. Rev. Neurosci*., 2001, 24, 31-55.

[25] Dehnhardt, M.; Palm, C.; Vieten, A.; Bauer, A.; Pietrzyk, U. Quantifying the A1AR distribution in peritumoural zones around experimental F98 and C6 rat brain tumours. *J. Neurooncol*., 2007, 85, 49-63.

[26] Castillo, C. A.; Albasanz, J. L.; Fernandez, M.; Martin, M. Endogenous expression of adenosine A1, A2 and A3 receptors in rat C6 glioma cells. *Neurochem. Res*., 2007, 32, 1056-70.

[27] Bauer, A.; Langen, K. J.; Bidmon, H.; Holschbach, M. H.; Weber, S.; Olsson, R. A.; Coenen, H. H.; Zilles, K. 18F-CPFPX PET identifies changes in cerebral A1 adenosine receptor density caused by glioma invasion. *J. Nucl. Medicine*, 2005, 46, 450-4.

[28] Synowitz, M.; Glass, R.; Farber, K.; Markovic, D.; Kronenberg, G.; Herrmann, K.; Schnermann, J.; Nolte, C.; van Rooijen, N.; Kiwit, J.; Kettenmann, H. A1 adenosine

receptors in microglia control glioblastoma-host interaction. *Cancer Res.,* 2006, 66, 8550-7.

[29] Markovic, D. S.; Glass, R.; Synowitz, M.; Rooijen, N.; Kettenmann, H. Microglia stimulate the invasiveness of glioma cells by increasing the activity of metalloprotease-2. *J. Neuropathol. Exp. Neurol.*, 2005, 64, 754-62.

[30] Rao, J. S. Molecular mechanisms of glioma invasiveness: the role of proteases. *Nat. Rev. Cancer*, 2003, 3, 489-501.

[31] Tsutsui, S.; Schnermann, J.; Noorbakhsh, F.; Henry, S.; Yong, V. W.; Winston, B. W.; Warren, K.; Power, C. A1 adenosine receptor upregulation and activation attenuates neuroinflammation and demyelination in a model of multiple sclerosis. *J. Neurosci.*, 2004, 24, 1521-9.

[32] Woodhouse, E. C.; Amanatullah, D. F.; Schetz, J. A.; Liotta, L. A.; Stracke, M. L.; Clair, T. Adenosine receptor mediates motility in human melanoma cells. *Biochem. Biophys. Res. Commmun.*, 1998, 246, 888-94.

[33] Khoo, H. E.; Ho, C. L.; Chhatwal, V. J.; Chan, S. T.; Ngoi, S. S.; Moochhala, S. M. Differential expression of adenosine A1 receptors in colorectal cancer and related mucosa. *Cancer Lett.,* 1996, 106, 17-21.

[34] Saito, M.; Yaguchi, T.; Yasuda, Y.; Nakano, T.; Nishizaki, T. Adenosine suppresses CW2 human colonic cancer growth by inducing apoptosis via A(1) adenosine receptors. *Cancer Lett.,* 2010, 290, 211-5.

[35] Sai, K.; Yang, D.; Yamamoto, H.; Fujikawa, H.; Yamamoto, S.; Nagata, T.; Saito, M.; Yamamura, T.; Nishizaki, T. A(1) adenosine receptor signal and AMPK involving caspase-9/-3 activation are responsible for adenosine-induced RCR-1 astrocytoma cell death. *Neurotoxicol.*, 2006, 27, 458-67.

[36] D'Ancona, S.; Ragazzi, E.; Fassina, G.; Mazzo, M.; Gusella, M.; Berti, T. Effect of dipyridamole, 5'-(N-ethyl)-carboxamidoadenosine and 1,3-dipropyl-8-(2-amino-4-chlorophenyl)-xanthine on LOVO cell growth and morphology. *Anticancer Res.*, 1994, 14, 93-7.

[37] Mirza, A.; Basso, A.; Black, S.; Malkowski, M.; Kwee, L.; Pachter, J. A.; Lachowicz, J. E.; Wang, Y.; Liu, S. RNA interference targeting of A1 receptor-overexpressing breast carcinoma cells leads to diminished rates of cell proliferation and induction of apoptosis. *Cancer Biol. Ther.,* 2005, 4, 1355-60.

[38] Lin, Z.; Reierstad, S.; Huang, C. C.; Bulun, S. E. Novel estrogen receptor-alpha binding sites and estradiol target genes identified by chromatin immunoprecipitation cloning in breast cancer. *Cancer Res.,* 2007, 67, 5017-24.

[39] Merighi, S.; Mirandola, P.; Milani, D.; Varani, K.; Gessi, S.; Klotz, K. N.; Leung, E.; Baraldi, P. G.; Borea, P. A. Adenosine receptors as mediators of both cell proliferation and cell death of cultured human melanoma cells. *J. Invest. Dermatol.*, 2002, 119, 923-33.

[40] Merighi, S.; Varani, K.; Gessi, S.; Cattabriga, E.; Iannotta, V.; Ulouglu, C.; Leung, E.; Borea, P. A. Pharmacological and biochemical characterization of adenosine receptors in the human malignant melanoma A375 cell line. *Br. J. Pharmacol.*, 2001, 134, 1215-26.

[41] Etique, N.; Grillier-Vuissoz, I.; Lecomte, J.; Flament, S. Crosstalk between adenosine receptor (A2A isoform) and ERalpha mediates ethanol action in MCF-7 breast cancer cells. *Oncol. Rep.*, 2009, 21, 977-81.

[42] van der Ploeg, I.; Ahlberg, S.; Parkinson, F. E.; Olsson, R. A.; Fredholm, B. B. Functional characterization of adenosine A2 receptors in Jurkat cells and PC12 cells using adenosine receptor agonists. Naunyn. *Schmiedebergs Arch. Pharmacol.*, 1996, 353, 250-60.

[43] Mayne, M.; Fotheringham, J.; Yan, H. J.; Power, C.; Del Bigio, M. R.; Peeling, J.; Geiger, J. D. Adenosine A2A receptor activation reduces proinflammatory events and decreases cell death following intracerebral hemorrhage. *Ann. Neurol.*, 2001, 49, 727-35.

[44] Gessi, S.; Sacchetto, V.; Fogli, E.; Merighi, S.; Varani, K.; Baraldi, P. G.; Tabrizi, M. A.; Leung, E.; Maclennan, S.; Borea, P. A. Modulation of metalloproteinase-9 in U87MG glioblastoma cells by A3 adenosine receptors. *Biochem. Pharmacol.*, 2010, 79, 1483-95.

[45] Hillion, J.; Canals, M.; Torvinen, M.; Casado, V.; Scott, R.; Terasmaa, A.; Hansson, A.; Watson, S.; Olah, M. E.; Mallol, J.; Canela, E. I.; Zoli, M.; Agnati, L. F.; Ibanez, C. F.; Lluis, C.; Franco, R.; Ferre, S.; Fuxe, K. Coaggregation, cointernalization, and codesensitization of adenosine A2A receptors and dopamine D2 receptors. *J. Biol. Chem.*, 2002, 277, 18091-7.

[46] Kalhan, A.; Gharibi, B.; Vazquez, M.; Jasani, B.; Neal, J.; Kidd, M.; Modlin, I. M.; Pfragner, R.; Rees, D. A.; Ham, J. Adenosine A2A and A2B receptor expression in neuroendocrine tumours: potential targets for therapy. *Purinergic Signal*, 2012, 8, 265-74.

[47] Ahmad, A.; Ahmad, S.; Glover, L.; Miller, S. M.; Shannon, J. M.; Guo, X.; Franklin, W. A.; Bridges, J. P.; Schaack, J. B.; Colgan, S. P.; White, C. W. Adenosine A2A receptor is a unique angiogenic target of HIF-2Alpha in pulmonary endothelial cells. *Proc. Natl. Acad. Sci. U. S. A.*, 2009, 106, 10684-9.

[48] Mediavilla-Varela, M.; Luddy, K.; Noyes, D.; Khalil, F. K.; Neuger, A. M.; Soliman, H.; Antonia, S. J. Antagonism of adenosine A2A receptor expressed by lung adenocarcinoma tumor cells and cancer associated fibroblasts inhibits their growth. *Cancer Biol. Ther.*, 2013, 14, 860-8.

[49] Tey, H. B.; Khoo, H. E.; Tan, C. H. Adenosine modulates cell growth in human epidermoid carcinoma (A431) cells. *Biochem. Biophys. Res. Commun.*, 1992, 187, 1486-92.

[50] Gessi, S.; Merighi, S.; Varani, K.; Cattabriga, E.; Benini, A.; Mirandola, P.; Leung, E.; Mac Lennan, S.; Feo, C.; Baraldi, S.; Borea, P. A. Adenosine receptors in colon carcinoma tissues and colon tumoral cell lines: focus on the A(3) adenosine subtype. *J. Cell. Physiol.*, 2007, 211, 826-36.

[51] Yasuda, Y.; Saito, M.; Yamamura, T.; Yaguchi, T.; Nishizaki, T. Extracellular adenosine induces apoptosis in Caco-2 human colonic cancer cells by activating caspase-9/-3 via A(2A) adenosine receptors. *J. Gastroenterol.*, 2009, 44, 56-65.

[52] Jajoo, S.; Sheth, S.; Mukherjea, D.; Kaur, T.; Rybak, L. P.; Ramkumar, V. Targeting adenosine receptors for the treatment of cancers. *Curr. Top. Biochem. Res.*, 2011, 13, 1.

[53] Rickles, R. J.; Pierce, L. T.; Giordano, T. P., 3rd; Tam, W. F.; McMillin, D. W.; Delmore, J.; Laubach, J. P.; Borisy, A. A.; Richardson, P. G.; Lee, M. S. Adenosine A2A receptor agonists and PDE inhibitors: a synergistic multitarget mechanism discovered through systematic combination screening in B-cell malignancies. *Blood*, 2010, 116, 593-602.

[54] Rickles, R. J.; Tam, W. F.; Giordano, T. P., 3rd; Pierce, L. T.; Farwell, M.; McMillin, D. W.; Necheva, A.; Crowe, D.; Chen, M.; Avery, W.; Kansra, V.; Nawrocki, S. T.; Carew, J. S.; Giles, F. J.; Mitsiades, C. S.; Borisy, A. A.; Anderson, K. C.; Lee, M. S. Adenosine A2A and beta-2 adrenergic receptor agonists: novel selective and synergistic multiple myeloma targets discovered through systematic combination screening. *Mol. Cancer Ther.*, 2012, 11, 1432-42.

[55] Leibovich, S. J.; Chen, J. F.; Pinhal-Enfield, G.; Belem, P. C.; Elson, G.; Rosania, A.; Ramanathan, M.; Montesinos, C.; Jacobson, M.; Schwarzschild, M. A.; Fink, J. S.; Cronstein, B. Synergistic up-regulation of vascular endothelial growth factor expression in murine macrophages by adenosine A(2A) receptor agonists and endotoxin. *Am. J. Pathol.*, 2002, 160, 2231-44.

[56] Lutty, G. A.; McLeod, D. S. Retinal vascular development and oxygen-induced retinopathy: a role for adenosine. *Prog. Retin. Eye Res.*, 2003, 22, 95-111.

[57] Montesinos, M. C.; Desai, A.; Chen, J. F.; Yee, H.; Schwarzschild, M. A.; Fink, J. S.; Cronstein, B. N. Adenosine promotes wound healing and mediates angiogenesis in response to tissue injury via occupancy of A(2A) receptors. *Am. J. Pathol.*, 2002, 160, 2009-18.

[58] Sexl, V.; Mancusi, G.; Holler, C.; Gloria-Maercker, E.; Schutz, W.; Freissmuth, M. Stimulation of the mitogen-activated protein kinase via the A2A-adenosine receptor in primary human endothelial cells. *J. Biol. Chem.*, 1997, 272, 5792-9.

[59] Desai, A.; Victor-Vega, C.; Gadangi, S.; Montesinos, M. C.; Chu, C. C.; Cronstein, B. N. Adenosine A2A receptor stimulation increases angiogenesis by down-regulating production of the antiangiogenic matrix protein thrombospondin 1. *Mol. Pharmacol.*, 2005, 67, 1406-13.

[60] Zhao, Z. Q.; Budde, J. M.; Morris, C.; Wang, N. P.; Velez, D. A.; Muraki, S.; Guyton, R. A.; Vinten-Johansen, J. Adenosine attenuates reperfusion-induced apoptotic cell death by modulating expression of Bcl-2 and Bax proteins. *J. Mol. Cell. Cardiol.*, 2001, 33, 57-68.

[61] Blay, J.; White, T. D.; Hoskin, D. W. The extracellular fluid of solid carcinomas contains immunosuppressive concentrations of adenosine. *Cancer Res.*, 1997, 57, 2602-5.

[62] Hoskin, D. W.; Reynolds, T.; Blay, J. Adenosine as a possible inhibitor of killer T-cell activation in the microenvironment of solid tumours. *Int. J. Cancer*, 1994, 59, 854-5.

[63] Koshiba, M.; Kojima, H.; Huang, S.; Apasov, S.; Sitkovsky, M. V. Memory of extracellular adenosine A2A purinergic receptor-mediated signaling in murine T cells. *J. Biol. Chem.*, 1997, 272, 25881-9.

[64] Ohta, A.; Gorelik, E.; Prasad, S. J.; Ronchese, F.; Lukashev, D.; Wong, M. K.; Huang, X.; Caldwell, S.; Liu, K.; Smith, P.; Chen, J. F.; Jackson, E. K.; Apasov, S.; Abrams, S.; Sitkovsky, M. A2A adenosine receptor protects tumors from antitumor T cells. *Proc. Natl. Acad. Sci. U. S. A.*, 2006, 103, 13132-7.

[65] Erdmann, A. A.; Gao, Z. G.; Jung, U.; Foley, J.; Borenstein, T.; Jacobson, K. A.; Fowler, D. H. Activation of Th1 and Tc1 cell adenosine A2A receptors directly inhibits IL-2 secretion in vitro and IL-2-driven expansion in vivo. *Blood*, 2005, 105, 4707-14.

[66] Huang, S.; Apasov, S.; Koshiba, M.; Sitkovsky, M. Role of A2A extracellular adenosine receptor-mediated signaling in adenosine-mediated inhibition of T-cell activation and expansion. *Blood*, 1997, 90, 1600-10.

[67] Naganuma, M.; Wiznerowicz, E. B.; Lappas, C. M.; Linden, J.; Worthington, M. T.; Ernst, P. B. Cutting edge: Critical role for A2A adenosine receptors in the T cell-mediated regulation of colitis. *J. Immunol.*, 2006, 177, 2765-9.
[68] Deaglio, S.; Dwyer, K. M.; Gao, W.; Friedman, D.; Usheva, A.; Erat, A.; Chen, J. F.; Enjyoji, K.; Linden, J.; Oukka, M.; Kuchroo, V. K.; Strom, T. B.; Robson, S. C. Adenosine generation catalyzed by CD39 and CD73 expressed on regulatory T cells mediates immune suppression. *J. Exp. Med.*, 2007, 204, 1257-65.
[69] Sitkovsky, M. V. T regulatory cells: hypoxia-adenosinergic suppression and re-direction of the immune response. *Trends Immunol.*, 2009, 30, 102-8.
[70] Hoskin, D. W.; Mader, J. S.; Furlong, S. J.; Conrad, D. M.; Blay, J. Inhibition of T cell and natural killer cell function by adenosine and its contribution to immune evasion by tumor cells (Review). *Int. J. Oncol.*, 2008, 32, 527-35.
[71] Ohta, A.; Kjaergaard, J.; Sharma, S.; Mohsin, M.; Goel, N.; Madasu, M.; Fradkov, E.; Ohta, A.; Sitkovsky, M. In vitro induction of T cells that are resistant to A2 adenosine receptor-mediated immunosuppression. *Br. J. Pharmacol.*, 2009, 156, 297-306.
[72] Donoso, M. V.; Lopez, R.; Miranda, R.; Briones, R.; Huidobro-Toro, J. P. A2B adenosine receptor mediates human chorionic vasoconstriction and signals through arachidonic acid cascade. *Am. J. P. Heart Circ. Physiol.*, 2005, 288, H2439-49.
[73] Trincavelli, M. L.; Daniele, S.; Martini, C. Adenosine receptors: what we know and what we are learning. *Curr. Top. Med. Chem.*, 2010, 10, 860-77.
[74] Panjehpour, M.; Castro, M.; Klotz, K. N., Human breast cancer cell line MDA-MB-231 expresses endogenous A2B adenosine receptors mediating a Ca2+ signal. *Br. J. Pharmacol.*, 2005, 145, 211-8.
[75] Ma, D. F.; Kondo, T.; Nakazawa, T.; Niu, D. F.; Mochizuki, K.; Kawasaki, T.; Yamane, T.; Katoh, R. Hypoxia-inducible adenosine A2B receptor modulates proliferation of colon carcinoma cells. *Hum. Pathol.*, 2010, 41, 1550-7.
[76] Long, J. S.; Crighton, D.; O'Prey, J.; Mackay, G.; Zheng, L.; Palmer, T. M.; Gottlieb, E.; Ryan, K. M. Extracellular adenosine sensing-a metabolic cell death priming mechanism downstream of p53. *Mol. Cell,* 2013, 50, 394-406.
[77] Wei, Q.; Costanzi, S.; Balasubramanian, R.; Gao, Z. G.; Jacobson, K. A. A2B adenosine receptor blockade inhibits growth of prostate cancer cells. *Purinergic Signal*, 2013, 9, 271-80.
[78] Limm, K.; Wallner, S.; Milenkovi, V. M.; Wetzel, C. H.; Bosserhoff, A. K. The metabolite 5'-methylthioadenosine signals through the adenosine receptor A2B in melanoma. *Eur. J. Cancer*, 2014, 50, 2714-24.
[79] Grant, M. B.; Tarnuzzer, R. W.; Caballero, S.; Ozeck, M. J.; Davis, M. I.; Spoerri, P. E.; Feoktistov, I.; Biaggioni, I.; Shryock, J. C.; Belardinelli, L. Adenosine receptor activation induces vascular endothelial growth factor in human retinal endothelial cells. *Circ. Res.,* 1999, 85, 699-706.
[80] Feoktistov, I.; Goldstein, A. E.; Ryzhov, S.; Zeng, D.; Belardinelli, L.; Voyno-Yasenetskaya, T.; Biaggioni, I. Differential expression of adenosine receptors in human endothelial cells: role of A2B receptors in angiogenic factor regulation. *Circ. Res.*, 2002, 90, 531-8.
[81] Kong, T.; Westerman, K. A.; Faigle, M.; Eltzschig, H. K.; Colgan, S. P. HIF-dependent induction of adenosine A2B receptor in hypoxia. *FASEB J.*, 2006, 20, 2242-50.

[82] Merighi, S.; Simioni, C.; Gessi, S.; Varani, K.; Mirandola, P.; Tabrizi, M. A.; Baraldi, P. G.; Borea, P. A. A(2B) and A(3) adenosine receptors modulate vascular endothelial growth factor and interleukin-8 expression in human melanoma cells treated with etoposide and doxorubicin. *Neoplasia,* 2009, 11, 1064-73.

[83] Zeng, D.; Maa, T.; Wang, U.; Feoktistov, I.; Biaggioni, I.; Belardinelli, L. Expression and function of A2B adenosine receptors in the U87MG tumor cells. *Drug. Dev. Res.*, 2003, 58, 405–411.

[84] Colgan, S. P.; Eltzschig, H. K.; Eckle, T.; Thompson, L. F. Physiological roles for ecto-5'-nucleotidase (CD73). *Purinergic Signal,* 2006, 2, 351-60.

[85] Stagg, J.; Divisekera, U.; McLaughlin, N.; Sharkey, J.; Pommey, S.; Denoyer, D.; Dwyer, K. M.; Smyth, M. J. Anti-CD73 antibody therapy inhibits breast tumor growth and metastasis. *Proc. Natl. Acad. Sci. U. S. A.,* 2010, 107, 1547-52.

[86] Rodrigues, S.; De Wever, O.; Bruyneel, E.; Rooney, R. J.; Gespach, C. Opposing roles of netrin-1 and the dependence receptor DCC in cancer cell invasion, tumor growth and metastasis. *Oncogene*, 2007, 26, 5615-25.

[87] Ryzhov, S.; Novitskiy, S. V.; Zaynagetdinov, R.; Goldstein, A. E.; Carbone, D. P.; Biaggioni, I.; Dikov, M. M.; Feoktistov, I. Host A(2B) adenosine receptors promote carcinoma growth. *Neoplasia,* 2008, 10, 987-95.

[88] Cekic, C.; Sag, D.; Li, Y.; Theodorescu, D.; Strieter, R. M.; Linden, J. Adenosine A2B receptor blockade slows growth of bladder and breast tumors. *J. Immunol.*, 2012, 188, 198-205.

[89] Iannone, R.; Miele, L.; Maiolino, P.; Pinto, A.; Morello, S. Blockade of A2B adenosine receptor reduces tumor growth and immune suppression mediated by myeloid-derived suppressor cells in a mouse model of melanoma. *Neoplasia*, 2013, 15, 1400-9.

[90] Madi, L.; Ochaion, A.; Rath-Wolfson, L.; Bar-Yehuda, S.; Erlanger, A.; Ohana, G.; Harish, A.; Merimski, O.; Barer, F.; Fishman, P. The A3 adenosine receptor is highly expressed in tumor versus normal cells: potential target for tumor growth inhibition. *Clin. Cancer Res.,* 2004, 10, 4472-9.

[91] Gessi, S.; Varani, K.; Merighi, S.; Cattabriga, E.; Iannotta, V.; Leung, E.; Baraldi, P. G.; Borea, P. A. A(3) adenosine receptors in human neutrophils and promyelocytic HL60 cells: a pharmacological and biochemical study. *Mol. Pharmacol.*, 2002, 61, 415-24.

[92] Kohno, Y.; Sei, Y.; Koshiba, M.; Kim, H. O.; Jacobson, K. A. Induction of apoptosis in HL-60 human promyelocytic leukemia cells by adenosine A(3) receptor agonists. *Biochem. Biophys. Res. Commun.,* 1996, 219, 904-10.

[93] Gessi, S.; Varani, K.; Merighi, S.; Morelli, A.; Ferrari, D.; Leung, E.; Baraldi, P. G.; Spalluto, G.; Borea, P. A. Pharmacological and biochemical characterization of A3 adenosine receptors in Jurkat T cells. *Br. L. Pharmacol.*, 2001, 134, 116-26.

[94] Merighi, S.; Benini, A.; Mirandola, P.; Gessi, S.; Varani, K.; Leung, E.; Maclennan, S.; Borea, P. A. Adenosine modulates vascular endothelial growth factor expression via hypoxia-inducible factor-1 in human glioblastoma cells. *Biochem. Pharmacol.*, 2006, 72, 19-31.

[95] Jajoo, S.; Mukherjea, D.; Watabe, K.; Ramkumar, V. Adenosine A(3) receptor suppresses prostate cancer metastasis by inhibiting NADPH oxidase activity. *Neoplasia,* 2009, 11, 1132-45.

[96] Gessi, S.; Cattabriga, E.; Avitabile, A.; Gafa, R.; Lanza, G.; Cavazzini, L.; Bianchi, N.; Gambari, R.; Feo, C.; Liboni, A.; Gullini, S.; Leung, E.; Mac-Lennan, S.; Borea, P. A. Elevated expression of A3 adenosine receptors in human colorectal cancer is reflected in peripheral blood cells. *Clin. Cancer Res.*, 2004, 10, 5895-901.

[97] Morello, S.; Petrella, A.; Festa, M.; Popolo, A.; Monaco, M.; Vuttariello, E.; Chiappetta, G.; Parente, L.; Pinto, A. Cl-IB-MECA inhibits human thyroid cancer cell proliferation independently of A3 adenosine receptor activation. *Cancer Biol. Ther.*, 2008, 7, 278-84.

[98] Fishman, P.; Bar-Yehuda, S.; Vagman, L., Adenosine and other low molecular weight factors released by muscle cells inhibit tumor cell growth. *Cancer Res.*, 1998, 58, 3181-7.

[99] Bar-Yehuda, S.; Barer, F.; Volfsson, L.; Fishman, P. Resistance of muscle to tumor metastases: a role for a3 adenosine receptor agonists. *Neoplasia*, 2001, 3, 125-31.

[100] Fishman, P.; Bar-Yehuda, S.; Ohana, G.; Pathak, S.; Wasserman, L.; Barer, F.; Multani, A. S. Adenosine acts as an inhibitor of lymphoma cell growth: a major role for the A3 adenosine receptor. *Eur. J. Cancer,* 2000, 36, 1452-8.

[101] Fishman, P.; Bar-Yehuda, S.; Barer, F.; Madi, L.; Multani, A. S.; Pathak, S. The A3 adenosine receptor as a new target for cancer therapy and chemoprotection. *Exp. Cell Res.*, 2001, 269, 230-6.

[102] Wen, L. T.; Knowles, A. F. Extracellular ATP and adenosine induce cell apoptosis of human hepatoma Li-7A cells via the A3 adenosine receptor. *Br. J. Pharmacol.*, 2003, 140, 1009-18.

[103] Varani, K.; Maniero, S.; Vincenzi, F.; Targa, M.; Stefanelli, A.; Maniscalco, P.; Martini, F.; Tognon, M.; Borea, P. A. A(3) receptors are overexpressed in pleura from patients with mesothelioma and reduce cell growth via Akt/nuclear factor-kappaB pathway. *Am. J. Respir.Crit. Care Med.*, 2011, 183, 522-30.

[104] Lee, E. J.; Min, H. Y.; Chung, H. J.; Park, E. J.; Shin, D. H.; Jeong, L. S.; Lee, S. K. A novel adenosine analog, thio-Cl-IB-MECA, induces G0/G1 cell cycle arrest and apoptosis in human promyelocytic leukemia HL-60 cells. *Biochem.Pharmacol.*, 2005, 70, 918-24.

[105] Fishman, P.; Bar-Yehuda, S.; Ohana, G.; Barer, F.; Ochaion, A.; Erlanger, A.; Madi, L. An agonist to the A3 adenosine receptor inhibits colon carcinoma growth in mice via modulation of GSK-3 beta and NF-kappa B. *Oncogene,* 2004, 23, 2465-71.

[106] Ohana, G.; Bar-Yehuda, S.; Arich, A.; Madi, L.; Dreznick, Z.; Rath-Wolfson, L.; Silberman, D.; Slosman, G.; Fishman, P. Inhibition of primary colon carcinoma growth and liver metastasis by the A3 adenosine receptor agonist CF101. *Br. J. Cancer*, 2003, 89, 1552-8.

[107] Panjehpour, M.; Karami-Tehrani, F. An adenosine analog (IB-MECA) inhibits anchorage-dependent cell growth of various human breast cancer cell lines. *Int. J. Biochem. Cell Biol.,* 2004, 36, 1502-9.

[108] Fishman, P.; Bar-Yehuda, S.; Ardon, E.; Rath-Wolfson, L.; Barrer, F.; Ochaion, A.; Madi, L. Targeting the A3 adenosine receptor for cancer therapy: inhibition of prostate carcinoma cell growth by A3AR agonist. *Anticancer Res.,* 2003, 23, 2077-83.

[109] Nakamura, K.; Yoshikawa, N.; Yamaguchi, Y.; Kagota, S.; Shinozuka, K.; Kunitomo, M. Antitumor effect of cordycepin (3'-deoxyadenosine) on mouse melanoma and lung

carcinoma cells involves adenosine A3 receptor stimulation. *Anticancer Res.,* 2006, 26, 43-7.

[110] Cohen, S.; Stemmer, S. M.; Zozulya, G.; Ochaion, A.; Patoka, R.; Barer, F.; Bar-Yehuda, S.; Rath-Wolfson, L.; Jacobson, K. A.; Fishman, P. CF102 an A3 adenosine receptor agonist mediates anti-tumor and anti-inflammatory effects in the liver. *J. Cell. Physiol.*, 2011, 226, 2438-47.

[111] Stemmer, S. M.; Benjaminov, O.; Medalia, G.; Ciuraru, N. B.; Silverman, M. H.; Bar-Yehuda, S.; Fishman, S.; Harpaz, Z.; Farbstein, M.; Cohen, S.; Patoka, R.; Singer, B.; Kerns, W. D.; Fishman, P. CF102 for the treatment of hepatocellular carcinoma: a phase I/II, open-label, dose-escalation study. *Oncologist,* 2013, 18, 25-6.

[112] Harish, A.; Hohana, G.; Fishman, P.; Arnon, O.; Bar-Yehuda, S., A3 adenosine receptor agonist potentiates natural killer cell activity. *Int. J. Oncol.*, 2003, 23, 1245-9.

[113] Mirandola, P.; Ponti, C.; Gobbi, G.; Sponzilli, I.; Vaccarezza, M.; Cocco, L.; Zauli, G.; Secchiero, P.; Manzoli, F. A.; Vitale, M. Activated human NK and CD8+ T cells express both TNF-related apoptosis-inducing ligand (TRAIL) and TRAIL receptors but are resistant to TRAIL-mediated cytotoxicity. *Blood*, 2004, 104, 2418-24.

[114] Morello, S.; Sorrentino, R.; Montinaro, A.; Luciano, A.; Maiolino, P.; Ngkelo, A.; Arra, C.; Adcock, I. M.; Pinto, A. NK1.1 cells and CD8 T cells mediate the antitumor activity of Cl-IB-MECA in a mouse melanoma model. *Neoplasia*, 2011, 13, 365-73.

[115] Merighi, S.; Benini, A.; Mirandola, P.; Gessi, S.; Varani, K.; Leung, E.; MacLennan, S.; Baraldi, P. G.; Borea, P. A. A3 adenosine receptors modulate hypoxia-inducible factor-1alpha expression in human A375 melanoma cells. *Neoplasia*, 2005, 7, 894-903.

In: Adenosine Signaling Mechanisms
Editors: V. Ramkumar and R. Paes de Carvalho
ISBN: 978-1-63483-186-4

Chapter 9

ROLE OF ADENOSINE RECEPTORS IN AUDITORY FUNCTIONS

Vickram Ramkumar*[1,*], *Vikrant Borse*[1], *Sumana Ghosh*[1], *Tejbeer Kaur*[3], *Kelly Sheehan*[2], *Sandeep Sheth*[1], *Asmita Dhukhwa*[1], *Srinivisan Tupal*[1], *Debashree Mukherjea*[2] and *Leonard P. Rybak*[2]

[1]Department of Pharmacology
[2]Department of Surgery (Otolaryngology), Southern Illinois University School of Medicine, Springfield, IL, US
[3]Department of Otolaryngology, Washington University School of Medicine, St. Louis, MO, US

ABSTRACT

Following the initial characterization of adenosine receptors in the cochlea, it was shown that activation of these receptors increased the activity of antioxidant enzymes in the cochlea. The cochlea also expresses various ATP hydrolytic enzymes responsible for the synthesis of adenosine, along with equilibrative and concentrative transporters. Adenosine receptors show distinct distribution patterns in the cochlea. Their ability to increase antioxidant enzymes, prompted future studies to determine the efficacy of adenosine receptor agonists to treat hearing loss induced by chemotherapeutic agents and noise trauma, conditions linked to the generation of reactive oxygen species. To date, agents that activate the A_1 adenosine receptor appear to show efficacy for the treatment of hearing loss in animal studies. It is yet to be shown that these agents are effective for preventing and treating hearing loss in humans. A major problem yet to be addressed is how to administer these drugs for treatment of hearing loss without producing significant side effects.

---

* To whom correspondence should be addressed: E-mail: vramkumar@siumed.edu, Phone: +1(217) 545-2171, Fax: +1(217) 545-0145.

Introduction

Early studies indicated that adenosine could modulate afferent neurotransmission in hair cell of the frog labyrinth system, a model system for studying hair cell [1]. Adenosine inhibited the afferent firing rate in the frog ampullar nerve of the semicircular canal. This effect was antagonized by theophylline, a nonselective antagonist of the adenosine receptors (ARs), and adenosine deaminase, which degrades adenosine to inosine. In addition, the adenosine uptake inhibitor, dipyridamole, decreased afferent firing rate. Furthermore, adenosine could be released from the isolated semicircular canal by electrical stimulation [1]. These findings suggest a functional role of adenosine in the semicircular canal. Dulon et al., [2] demonstrated that the precursor of adenosine, ATP, increases intracellular Ca^{2+} levels in mammalian inner hair cells through the activation of purinergic P_{2Z} receptors. Studies from our laboratory demonstrated the presence of A_1AR in the rat cochlea, using several biochemical techniques [3]. Activation of these receptors did not appear to affect auditory parameters, as measured by auditory brainstem responses and sound-evoked endocochlear potentials [4]. However, these receptors increased the activities of antioxidant enzymes (namely superoxide dismutase and glutathione peroxidase) in the cochlea [4]. This study was the first to link the A_1AR with activation of cochlear antioxidant defense system.

Since oxidative stress is implicated in cisplatin, aminoglycoside and noise-induced hearing loss (NIHL), we determined whether activation of the A_1AR could protect against cisplatin-induced ototoxicity. Activation of these receptors was shown to confer protection against cisplatin-induced hearing loss [5, 6]. This finding has stimulated subsequent interest in understanding the mechanism(s) underlying a protective role of adenosine and ARs in the cochlea, as described below.

Expression of AR Subtypes in the Cochlea

Studies from the laboratory of Srdjan Vlajkovic (University of Auckland, New Zealand) showed expression of three subtypes of the ARs in the cochlea [7, 8]. The A_1AR are localized primarily to Deiters cells, inner hair cells (IHCs) and spiral ganglion neurons (SGNs) while the $A_{2a}AR$ are localized to Deiters cells, IHCs, SGNs, spiral ligament and modiolar blood vessels. A_3AR immunolabeling was observed in Deiters cells, Hensen's cells, Claudius cells, pillar cells, inner and outer sulcus cells, IHCs and outer hair cells (OHCs), and interdental cells of the spiral limbus [7].

Although the A_3AR subtype exhibits a wider distribution in the cochlea than either the A_1 or $A_{2a}AR$, it is the activation of the A_1AR that is linked to otoprotection against cisplatin [6] and noise-induced hearing loss [9]. An even more perplexing finding was that significant protection of OHCs against cisplatin-induced damage is observed by A_1AR activation, even though the OHCs do not express this receptor subtype [7]. One explanation for this finding is that protection is mediated by activation of A_1AR present on supporting cells, such as Deiters cells, which communicate this information to the OHCs through release of paracrine factors into the extracellular fluid. In support of this hypothesis, a recent study showed that Deiters cells could protect hair cells by secreting heat shock protein 70 (Hsp70) into the extracellular space [10]. Another possibility is that protection is mediated by low levels of A_1AR present on OHCs which are not detected by immunohistochemical techniques. Unlike the OHCs, the IHCs express high levels of A_1AR. Interestingly, these cells appear to be relatively resistant to killing by ototoxic drugs or noise. Thus, it is tempting to speculate that the A_1AR on these cells confer, at least in part, resistance to these ototoxic insults.

In addition to the differential expression of AR subtypes, the cochlea expresses a dynamic adenosine generation process. The most important source of adenosine is ATP, which is released by the cochlea in response to stress [11]. ATP is metabolized by various extracellular NTPDases and ecto-nucleotidases, leading to the production of extracellular adenosine [12-14]. Since the intracellular levels of ATP is considerably higher than extracellular adenosine, small changes extracellular ATP could lead to significant increases in extracellular adenosine [15]. This could lead to elevated levels of extracellular adenosine under stressful conditions which result in damage or death of cochlear cells. The cochlea also expresses a variety of both equilibrative (ENT1, ENT2) and concentrative types (CNT1, CNT2) of adenosine transporters [16]. Equilibrative transporters, as the name implies, normalize the levels of adenosine between areas containing high and low concentrations of the nucleoside. Concentrative transporters move adenosine against its concentration gradient, using energy derived from plasma membrane Na^+-K^+-ATPase [17].

Activation of the A1AR Protects Against Cisplatin Ototoxicity

Cisplatin is a widely used chemotherapeutic agent to treat various solid tumors, including ovarian, testicular and head and neck cancers. Dose limiting side effects of cisplatin include hearing loss, nephrotoxicity and peripheral neuropathy [18]. Children and the elderly are especially susceptible to these side effects. Hearing loss in children could affect their speech, cognitive and social development. Various approaches are being studied for the alleviation of cisplatin-induced ototoxicity, the foremost among which are antioxidants [18, 19]. However, there are no clinically approved treatments for cisplatin-induced hearing loss. The generation of reactive oxygen species (ROS) by cochlear NOX3 NADPH oxidase appears to be critical for mediating cisplatin ototoxicity. As such, knockdown of NOX3 by trans-tympanic administration of NOX3 short interfering (si)RNA protected against cisplatin ototoxicity, as demonstrated by preservation of hearing thresholds and inner ear sensory cells [20].

The first study to show the efficacy of A_1AR against cisplatin-induced hearing loss was published by Whitworth et al., [6]. In this study, various agonists were administered by round window application (for 90 min and then the excess fluid was removed) prior to cisplatin. Pretreatment with 1 mM *R*-phenylisopropyladenosine (*R*-PIA) and 100 μM 2-chloro-N6-cyclopentyladenosine (CCPA) to chinchilla led to reduced cisplatin-induced hearing loss, assessed by auditory brainstem responses three days following cisplatin administration. Similarly, *R*-PIA protected against loss of OHCs in the organ of Corti. These beneficial actions were associated with significant reductions in malondialdehyde, a marker of oxidative stress. Blockade of the A_1AR by pretreatment with 8-cyclopentyl-1,3-dipropylxanthine (DPCPX) abrogated the beneficial effects of *R*-PIA and even potentiated cisplatin ototoxicity, suggesting a beneficial effect of endogenous adenosine in hearing loss (Figure 1).

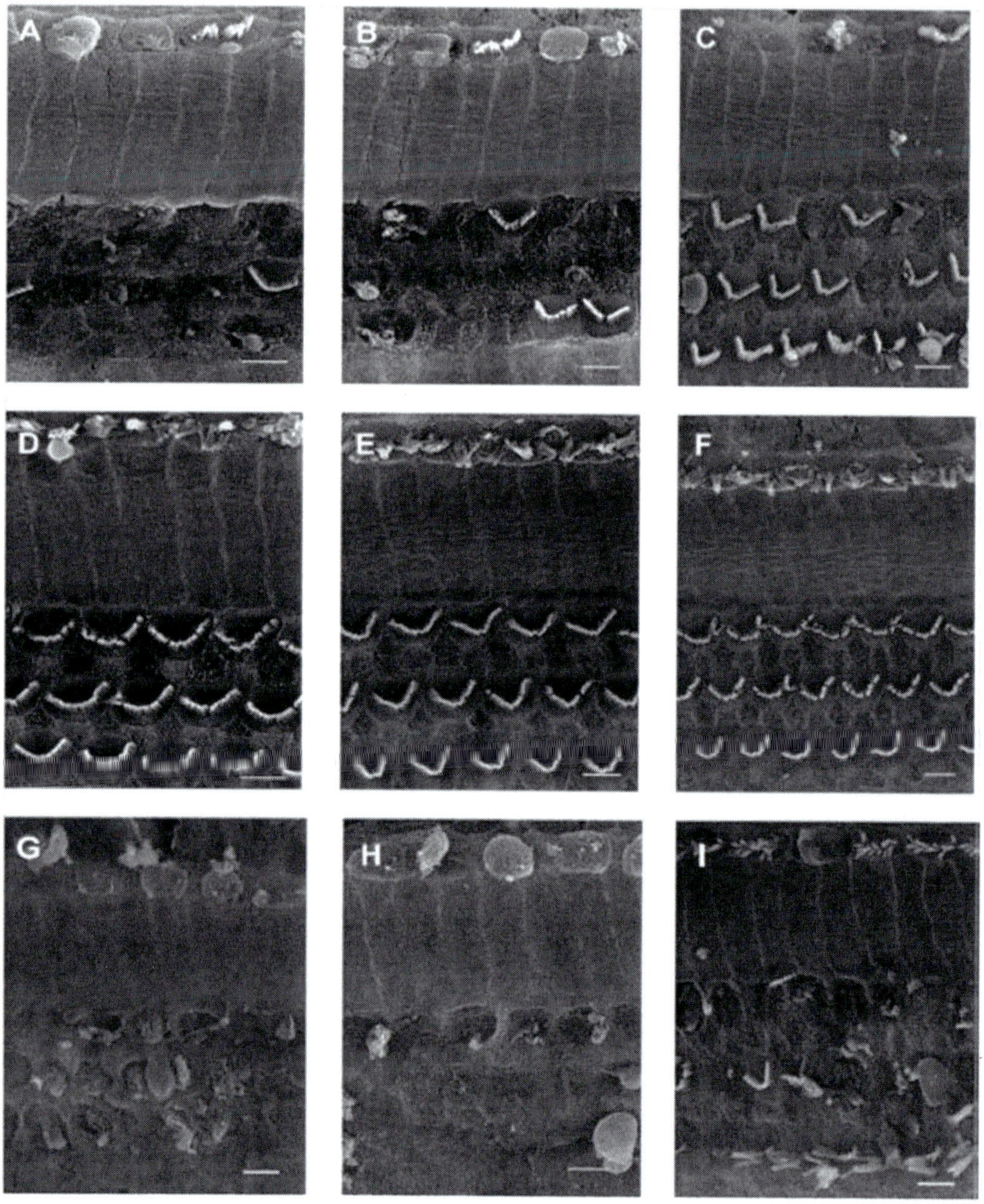

Figure 1. Comparison of organ of Corti morphology between an ear treated with saline + cisplatin, one treated with R-PIA + cisplatin and one treated with DPCPX/R-PIA + cisplatin. Application of saline + cisplatin resulted in almost complete IHC and OHC loss in the hook (A) and basal turn (B) and moderate loss in the middle turn (C). In contrast, the organ of Corti morphology was very well preserved in the hook, basal and middle turns (D, E and F, respectively) of ears treated with R-PIA þ cisplatin. Some cytoplasmic extrusions from the IHCs and disruption of stereocilia were evident in the hook region of the R-PIA + cisplatin-treated ear that displayed the largest threshold changes (D). The administration of DPCPX/R-PIA cocktail prior to cisplatin resulted in extensive hair cell loss in the hook, basal and middle turns (G, H and I, respectively). The morphological differences between the apical turns from the three groups were negligible (not shown). Bars = 5 μm. Figure reprinted from Whitworth et al., [6] with permission.

We also observed that round window application of 2-[4-(2-p-carboxy-ethyl)phenylamino]-5'-N-ethylcarboxamidoadenosine (CGS), an agonist of the A_{2A}AR did not provide any protection against cisplatin-induced ototoxicity. In fact, this agonist appeared to exacerbate cisplatin-induced ototoxicity. It was estimated that the concentrations of these agonists and antagonist achieved in the perilymph would be in the low micromolar range, which show efficacy in *in vitro* studies [6]. Another important finding from this study is that the drugs (namely *R*-PIA and CCPA) are effective over a three day period following a single round window application [6]. Translation of these findings observed in the chinchilla for the treatment of hearing loss in human would require the development of minimally invasive techniques to administer these drugs locally, as systemic administration of these agents is associated with significant bradycardia and sedation. Furthermore, it is still unclear whether systemic administration of these agents could interfere with cisplatin chemotherapeutic efficacy.

An interesting property of the A_1AR in the cochlea is that it could be induced by oxidative stress produced by chemotherapeutic agents (such as cisplatin) [5] and noise [17] (see Figure 2). Induction of A_1AR appears to be dependent on the activation of nuclear factor (NF)-κB by ROS which enhance transcriptional activity of the A_1AR gene [18]. We propose that such feedback regulation could increase the levels of the A_1AR and increase the activity of cochlear antioxidant enzymes. This in turn could decrease the levels of ROS and reduce NF-κB activity, thereby contributing to subsequent negative feedback regulation of this transcription factor.

The role of inflammation in mediating hearing loss is becoming an important area of research [19]. Recent studies indicate that cisplatin can activate cochlear inflammatory processes by activating signal transducer and activator of transcription 1 (STAT1) protein [20, 21]. Activation of STAT1 was mediated by ROS produced by the NOX3 NADPH oxidase system, linking cisplatin-mediated ROS generation to cochlear inflammation. Initial studies from our laboratory indicate that activation of the A_1AR inhibits cisplatin-induced STAT1 activation in the cochlea [22]. Cisplatin-induced activation of mitogen activated protein kinases (MAPKs) and NF-κB pathways are also linked to inflammation in immortalized organ of Corti (HEI-OC1) cells [23]. Interestingly, activation of MAPKs is linked to cisplatin-mediated activation of STAT1 [21].

Role of Adenosine in Noise-Induced Hearing Loss

Exposure to harmful noise is one of the leading causes of noise-induced hearing loss (NIHL) in people of all ages. According to a Central Disease Control and Prevention (CDC) report in 2010, approximately 16% of teens of age 12-19 in the United States show some signs of hearing loss due to harmful exposure to loud noise. Hearing loss may have a great impact on psychosocial life of an individual. Poor psychosocial behavior may contribute to the anxiety, depression, and cognitive degeneration [24]. Although NIHL is preventable, there are currently no clinically effective treatments for this problem.

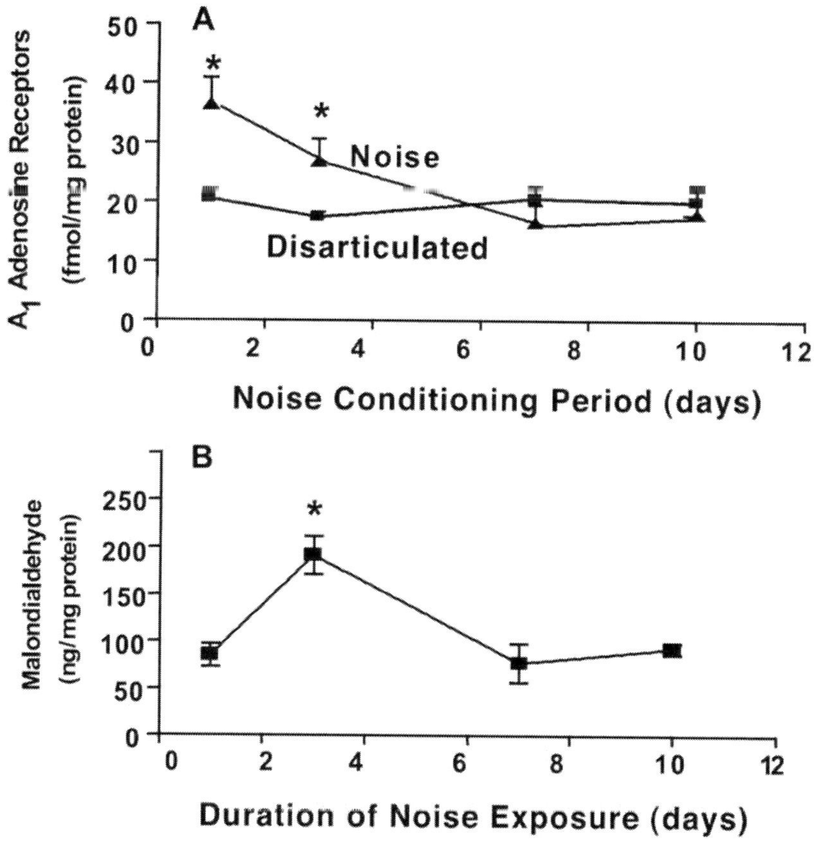

Figure 2. Noise induces A1AR expression and increases MDA. (A) Chinchillas were exposed to 96 dB (4 kHz) noise for a period of 6 h (day 1), or for 6 h daily. The levels of A1AR were determined using a single concentration (~1 nM) of [125I]ABMECA. The group labeled 'disarticulated' had their middle ear bones (malleus, incus) surgically removed, and served as the no noise control. The data are presented as the mean ± S.E.M. of four to six animals from each treatment group and at each time period tested. (B) Malondialdehyde (MDA) levels were determined using samples derived from A. Asterisks indicates statistically significant difference from control (P<0.05). Figure reprinted with permission from Ramkumar et al., [17].

Noise exposure increases metabolic activity and production of ROS in the cochlea [25-28]. Increase in ROS overwhelms the antioxidant defense mechanism in cochlea, promoting cochlear damage and NIHL [29-32]. Adenosine and adenosine receptors have been shown to activate the cochlear antioxidant defense system in the cochlea [4]. Noise exposure increases the levels of ectonucleotidases such as NTPDase1, NTPDase2 and NTPDase3 in the cochlea, which in turn increases the production of adenosine [33, 34]. However, exposure to broadband noise has no effect on the expression patterns of adenosine kinase (ADK), an enzyme involved in adenosine phosphorylation and purine salvage in the cochlea [35]. These data suggest that ATP hydrolysis is the primary source of adenosine production in the cochlea upon noise exposure.

A few studies have demonstrated protection of the cochlea by activating the A_1AR. For example, round window application of R-PIA, prior to noise exposure, significantly decreased hair cell loss and subsequently protects from NIHL [29]. Recently, CCPA and adenosine amine congener (ADAC), a selective A_1 adenosine receptor agonists, have been shown to protect against NIHL [36, 37]. As such, ingestion of caffeine, a nonselective AR antagonist, exacerbated hearing loss in a guinea pig model of acoustic trauma [38]. Inhibition of ADK by ABT-702 did not protect against NIHL induced by traumatic noise exposure [39] but was partially effective against of age-related hearing loss [40].

SUMMARY

This review highlights the importance of the adenosine/A_1AR system in the cochlea in mediating protection against insults. While endogenous adenosine provided a small degree of protection against cisplatin ototoxicity, it is clear that local administration of non-hydrolyzable adenosine analogs provided significant protection against cisplatin and noise-induced hearing loss. A significant hurdle to overcome is to develop efficient methods to deliver these drugs into the cochlea, which would facilitate their use as otoprotective agents and reduce potential side effects. The use of round window application of these drugs in human might be too invasive. The recent demonstration that the A_1AR agonist, ADAC, was effective for treating NIHL when administered by the intraperitoneal route and had less systemic side effects [36, 37] is quite promising. A better understanding of the mechanism(s) underlying the protective action of adenosine in the cochlea would certainly stimulate the development of newer agents to treat different forms of hearing loss.

ACKNOWLEDGMENTS

Research described in this review was supported, in part, by grants from the National Institutes of Health (R15 CA135494, R15 DC011412, R01-DC002396 and R01-CA166907), Deafness Research Foundation, National Organization for Hearing Research and by various grants from SIU School of Medicine.

REFERENCES

[1] Bryant, GM; Barron, SE; Norris, CH; Guth, PS. Adenosine is a modulator of hair cell-afferent neurotransmission. *Hear. Res.* 1987, *30*, 231–237.

[2] Dulon, D; Mollard, P; Aran, JM. Extracellular ATP elevates cytosolic Ca2+ in cochlear inner hair cells. *Neuroreport* 1991, *2*, 69–72.

[3] Ramkumar, V; Ravi, R; Wilson, MC; Gettys, TW; Whitworth, C; Rybak, LP. Identification of A_1 adenosine receptors in rat cochlea coupled to inhibition of adenylyl cyclase. *Am. J. Physiol.* 1994, *267*, C731–C737.

[4] Ford, MS; Maggirwar, SB; Rybak, LP; Whitworth, C; Ramkumar, V. Expression and function of adenosine receptors in the chinchilla cochlea. *Hear. Res.* 1997, 105, 130-140.

[5] Ford, MS; Nie, Z; Whitworth, C; Rybak, LP; Ramkumar, V. Up-regulation of adenosine receptors in the cochlea by cisplatin. *Hear. Res.* 1997, *111*, 143–152.

[6] Whitworth, CA; Ramkumar, V; Jones, B; Tsukasaki, N; Rybak, LP. Protection against cisplatin ototoxicity by adenosine agonists. *Biochem. Pharmacol.* 2004, *67*, 1801–1807.

[7] Vlajkovic, SM; Abi, S; Wang, CJ; Housley, GD; Thorne, PR. Differential distribution of adenosine receptors in rat cochlea. *Cell Tissue Res.* 2007, *328*, 461–471.

[8] Vlajkovic, SM; Housley, GD; Thorne, PR. Adenosine and the auditory system. *Curr. Neuropharmacol.* 2009, *7*, 246–256.

[9] Wong AC; Guo CX; Gupta R; Housley GD; Thorne PR; Vlajkovic SM. Post exposure administration of A_1 adenosine receptor agonists attenuate noise-induced hearing loss. *Hear. Res. 2010,* 260, 81–88.

[10] May, LA; Kramarenko, II; Brandon, CS; Voelkel-Johnson, C; Roy, S; Truong, K; Francis, SP; Monzack, EL; Lee, F-S; Cunningham, LL. Inner ear supporting cells protect hair cells by secreting HSP70. *J. Clin. Invest.* 2012, 123, 3577–3587.

[11] Muñoz, DJB, Kendrick, IS, Rassam, M, Thorne, PR (2001) Vesicular storage of adenosine triphosphate in the guinea-pig cochlear lateral wall and concentrations of ATP in the endolymph during sound exposure and hypoxia. *Acta Otolaryngol.* 2001, 121, 10-15.

[12] Vlajkovic, SM; Thorne, PR; Muñoz, DJB; Housley, GD. Ectonucleotidase activity in the perilymphatic compartment of the guinea pig cochlea. *Hear. Res.* 1996, 99, 31-37.

[13] Vlajkovic, SM; Thorne, PR; Housley, GD; Munoz, DJB; Kendrick, IS. Ecto-nucleotidases terminate purinergic signaling in the cochlear endolymphatic compartment. *Neuroreport.* 1998, 9, 1559-1565.

[14] Vlajkovic, SM; Thorne, PR; Sévigny, J; Robson, SC; Housley, GD. NTPDase1 and NTPDase2 immunolocalization in mouse cochlea: implications for regulation of P2 receptor signalling. *J. Histochem. Cytochem.* 2002, 50, 1435-1442.

[15] Fredholm, BB; Chen, JF; Cunha, RA; Svenningsson, P; Vaugeois, JM. Adenosine and brain function. *Int. Rev. Neurobiol.* 2005, 63, 191-270.

[16] Khan, AF; Thorne, PR; Muñoz, DJ; Wang, CJH; Housley, GD; Vlajkovic, SM. Nucleoside transporter expression and adenosine uptake in the rat cochlea. *Neuroreport.* 2007, **18**, 235-259.

[17] Ramkumar, V; Whitworth, CA; Pingle, SC; Hughes, LF; Rybak, LP. Noise induces A1 adenosine receptor expression in the chinchilla cochlea. Hear. Res., 2004, 188, 47-56.

[18] Rybak, LP; Whitworth, CA; Mukherjea, D; Ramkumar, V. Mechanisms of cisplatin-induced ototoxicity and prevention. *Hear. Res.* 2007, *226*, 157–167.

[19] Abi-Hachem, RN; Zine, A; Van De Water, TR. The injured cochlea as a target for inflammatory processes, initiation of cell death pathways and application of related otoprotectives strategies. *Recent Pat. CNS Drug Discov.* 2010, 5, 147-163.

[20] Schmitt, NC; Rubel, EW; Nathanson, NM. Cisplatin-induced hair cell death requires STAT1 and is attenuated by epigallocatechin gallate. J Neurosci. 2009, 29, 3843-3851.

[21] Kaur, T; Mukherjea, D; Sheehan, K; Jajoo, S; Rybak, LP; Ramkumar, V. Short interfering RNA against STAT1 attenuates cisplatin-induced ototoxicity in the rat by suppressing inflammation. *Cell Death Dis.* 2011, *2*, e180.

[22] Kaur, T; Ramkumar, V. Unpublished data.

[23] Kim, J; Cho, HJ; Sagong, B; Kim, SJ; Lee, JT; So, HS; Lee, IK; Kim, UK; Lee, KY; Choo YS. Alpha-lipoic acid protects against cisplatin-induced ototoxicity via the regulation of MAPKs and proinflammatory cytokines. *Biochem Biophys Res Commun.* 2014, 449, 183-189.

[24] Gates, GA; Mills, JH. Presbycusis. *Lancet* 2005, 366,1111-1120.

[25] Liu, Z. Experimental study on the mechanism of free radical in blast trauma induced hearing loss. *Zhonghua Er Bi Yan Hou Ke Za Zhi* 1992, 27, 24-26.

[26] Ohlemiller, KK; Dugan, LL. Elevation of reactive oxygen species following ischemia-reperfusion in mouse cochlea observed in vivo. *Audiol. Neurootol.* 1999, 4, 219-228.

[27] Yamane, H; Nakai, Y; Takayama, M; Iguchi, H; Nakagawa, T; Kojima, A. Appearance of free radicals in the guinea pig inner ear after noise-induced acoustic trauma. *Eur. Arch. Otorhinolaryngol.* 1995, 252, 504-508.

[28] Yamane, H; Nakai, Y; Takayama, M; Konishi, K; Iguchi, H; Nakagawa, T; Shibata, S; Kato, A; Sunami, K; Kawakatsu, C. The emergence of free radicals after acoustic trauma and strial blood flow. *Acta Otolaryngol. Suppl.* 1995, 519, 87-92.

[29] Hu, BH; Zheng, XY; McFadden, SL; Kopke, RD; Henderson, D. *R*-phenylisopropyl adenosine attenuates noise-induced hearing loss in the chinchilla. *Hear. Res.* 1997, 113, 198-206.

[30] Kopke, RD; Weisskopf, PA; Boone, JL; Jackson, RL; Wester, DC; Hoffer, ME; Lambert, DC; Charon, CC; Ding, DL; McBride, D. Reduction of noise-induced hearing loss using L-NAC and salicylate in the chinchilla. *Hear. Res.* 2000, 149, 138-146.

[31] Seidman, MD; Shivapuja, BG; Quirk, WS. The protective effects of allopurinol and superoxide dismutase on noise-induced cochlear damage. *Otolaryngol. Head Neck Surg.* 1993, 109, 1052-1056.

[32] Yamasoba, T; Harris, C; Shoji, F; Lee, RJ; Nuttall, AL; Miller, JM. Influence of intense sound exposure on glutathione synthesis in the cochlea. *Brain Res.*1998, 804, 72-78.

[33] Vlajkovic, SM; Housley, GD; Muñoz, DJ; Robson, SC; Sévigny, J; Wang, CJ; Thorne, PR. Noise exposure induces up-regulation of ecto-nucleoside triphosphate diphosphohydrolases 1 and 2 in rat cochlea. *Neuroscience* 2004, 126, 763-773.

[34] Vlajkovic, SM; Vinayagamoorthy, A; Thorne, PR; Robson, SC; Wang, CJ; Housley, GD. Noise-induced up-regulation of NTPDase3 expression in the rat cochlea: implications for auditory transmission and cochlear protection. *Brain Res.* 2006, 1104, 55-63.

[35] Vlajkovic, SM; Guo, CX; Dharmawardana, N; Wong, AC; Boison, D; Housley, GD; Thorne, PR. Role of adenosine kinase in cochlear development and response to noise. *J. Neurosci. Res.* 2010, 88, 2598-2609.

[36] Wong, AC; Guo, CX; Gupta, R; Housley, GD; Thorne, PR; Vlajkovic, SM. Post exposure administration of A_1 adenosine receptor agonists attenuate noise-induced hearing loss. *Hear. Res.* 2010, 260, 81–88.

[37] Vlajkovic, SM; Chang, H; Paek, SY; Chi, HH; Sreebhavan, S; Telang, RS; Tingle, M; Housley, GD; Thorne, PR. Adenosine amine congener as a cochlear rescue agent. *Biomed. Res. Int.* 2014, 841-849.

[38] Mujica-Mota, MA; Gasbarrino, K; Rappaport, JM; Shapiro, RS; Daniel, SJ. The effect of caffeine on hearing in a guinea pig model of acoustic trauma. *Am. J. Otolaryngol.* 2014, 35, 99-105.

[39] Vlajkovic, SM; Guo, CX; Dharmawardana, N; Wong, AC; Boison, D; Housley, GD; Thorne, PR. Role of adenosine kinase in cochlear development and response to noise. *J. Neurosci. Res.* 2010, 88, 2598-2609.

[40] Vlajkovic, SM; Guo, CX; Telang, R; Wong, AC; Paramananthasivam, V; Boison, D; Housley, GD; Thorne, PR. Adenosine kinase inhibition in the cochlea delays the onset of age-related hearing loss. *Exp. Gerontol.* 2011, 46, 905-914.

In: Adenosine Signaling Mechanisms
Editors: V. Ramkumar and R. Paes de Carvalho

ISBN: 978-1-63483-186-4

Chapter 10

ADENOSINE RECEPTOR MODULATION OF GABAERGIC TRANSMISSION

Ana M. Sebastião[1,2,*]***, Diogo M. Rombo***[1,2]
and Joaquim A. Ribeiro[1,2]

[1]Instituto de Farmacologia e Neurociências, Faculdade de Medicina, Universidade de Lisboa, Lisboa, Portugal
[2]Unidade de Neurociências, Instituto de Medicina Molecular, Universidade de Lisboa, Lisboa, Portugal

ABSTRACT

Neurotransmission mediated by γ-aminobutyric acid (GABA) is well established, as GABA is the predominant inhibitory transmitter in the adult brain. The multiplicity of inhibitory neurons, differences between phasic and tonic GABAergic signaling, together with the need to discriminate between inhibition of excitatory neurons from inhibition of inhibitory ones, increases the complexity, therefore the interest, of the analysis of neuromodulation of inhibitory transmission. The neuromodulator adenosine, by operating high affinity adenosine receptors, has in addition a double potential: to restrain transmission via A1 receptor (A1R) and to facilitate it via A2A receptor (A2AR). One could therefore anticipate a precise location of these receptors in key points of the network to control the activity of GABAergic neurons.

The present review shall discuss how GABA is regulated by adenosine and how this affects several brain functions, namely hippocampal network stability, retina, cerebellum, basal ganglia functioning and its implications for fine control of movement and of drug addiction, brain areas involved in sleep control and in the control of the autonomic nervous system. This cross-talk between adenosine and GABA is part of brain homeostasis regulation and neuroprotection. Some implications for therapy are also discussed.

---

* Corresponding author: Ana M. Sebastião. E-mail: anaseb@fm.ul.pt, phone: +351-21-7985183.

INTRODUCTION

The predominant inhibitory neurotransmitter in the central nervous system is γ-aminobutyric acid (GABA). Therefore, the control of its activity is a key process not only for overall excitability adjustment, but also for the establishment of patterns of neuronal synchronization/desynchronization. This is crucial for most brain functions where a tight balance between excitation and inhibition is required to adjust network activity to homeostatic needs. Evidence for modulation of GABAergic transmission by adenosine is observed in several brain areas relevant for fine control of movement, drug addition, sleep or central command of autonomic nervous system functioning. At the forebrain, evidence to date point to a predominant action of adenosine on excitatory transmission, in particular in the action of the most abundantly expressed adenosine receptor, the A1R.

Indeed, up to recently A1R have been considered to be mostly devoid of influence upon GABAergic transmission in the hippocampus. We now know that this applies only to phasic GABAergic transmission. As we will detail below, adenosine exerts selected actions in hippocampal interneurons controlling their GABAergic inputs, which adds to its influence upon glutamatergic transmission, therefore allowing a fine control of network oscillations and synchronization of hippocampal output. The evidence for the modulatory action of adenosine upon GABAergic transmission in the other brain areas, as well as the corresponding functional implications, namely fine control of movement, drug addiction, sleep control and regulation of the autonomic nervous system, will also be reviewed.

Before that, we will summarize the main aspects of GABAergic transmission, so that the implications of its modulation by adenosine can be better perceived.

MAJOR CHARACTERISTICS OF GABAERGIC TRANSMISSION

GABA is the main inhibitory neurotransmitter in the adult brain, being primary released by around 20% of brain neurons [1, 2]. When first described in neurons, GABA was shown to produce inhibitory hyperpolarizing responses [3] that were blocked by bicuculline [4]. The fast signaling action of GABA, responsible for inhibitory post-synaptic potentials (IPSP) is mediated by the chloride (Cl^-) permeable ionotropic receptor, the $GABA_A$ receptor ($GABA_AR$) [5]. GABA also operates metabotropic receptors, the $GABA_B$ receptor ($GABA_BR$) [6, 7] coupled, through a G protein, to potassium (K^+) channels at the postsynaptic site, which leads to the late inhibitory postsynaptic potential characteristic of a slow GABA response [8]. $GABA_BR$ at presynaptic nerve terminals mediate inhibition of transmitter release [9]. The ionotropic, Cl^-permeable, GABA receptor at a subpopulation of retinal neurons has pharmacological properties distinct from the most abundant $GABA_ARs$, and was initially referred to as $GABA_C$ [10] but later included in the $GABA_AR$ class, on the recommendations of IUPHAR Nomenclature Committee [11].

The $GABA_ARs$ are heteromeric pentamers with the five subunits arranged around a central pore. Each subunit exists in different isoforms. Current evidence shows that most $GABA_AR$ subtypes are formed from two copies of a single α subunit isoform, two copies of a single β subunit isoform, and one copy of another subunit, such as γ, δ, ε, π or θ [12]. The physiological significance of the structural heterogeneity of $GABA_ARs$ may lie on the provision of functional diversity such as channel kinetics, affinity for GABA, rate of desensitization and susceptibility for transient chemical modification (e.g., phosphorylation) [13]. In all cases, the net flow response that results from the increase in membrane permeability to Cl^- caused by $GABA_AR$ activation depends on the relative concentrations of Cl^- at each side of the membrane and on the membrane potential of the cell. In most mature neurons of the central nervous system the expression of the K^+/Cl^- co-transporter 2 (KCC2) [14, 15], which is a Cl^- extruder, will result in a Cl^- equilibrium potential that is more negative than the resting membrane potential of the neuron [16, 17]. Activation of $GABA_AR$ in mature neurons is thus usually "inhibitory" for two main reasons: (1) the increase in the permeability to Cl^- causes a general increase in membrane input conductance that shunts the ability of excitatory potentials to depolarize the membrane; (2) the Cl^--mediated hyperpolarization of the membrane sums to any depolarizing signal arriving to the neuron, reducing the probability of the cell to fire an action potential [18, 19]. In immature and developing neurons, however, the activation of $GABA_AR$ can lead to membrane depolarization and, in some cases, firing of action potential [20–23]. This results from a higher intracellular concentration of Cl^- due to early developmental expression of $Na^+/K^+/2Cl^-$ co-transporter 1 (NKCC1) [24] and lack of expression of KCC2 [17]. This intracellular accumulation of Cl^- in immature neurons determines a Cl^- equilibrium potential less negative than the resting membrane potential, leading to an outflow of Cl^-, therefore to depolarization upon activation of $GABA_ARs$. Even in mature neurons, neuronal activity, such as epileptiform discharges, can transiently change the reversal potential for GABA and turn $GABA_AR$-mediated currents into depolarizing [25–31].

Information carried out by activation of $GABA_ARs$ can be categorized in two main groups, phasic transmission and tonic transmission. Phasic GABAergic transmission allows a fast and precisely-timed communication between the GABAergic presynaptic terminal and the postsynaptic target. With the arrival of an action potential at the interneuron axonal terminal, a pool of GABA-containing vesicles is synchronously released to the synaptic cleft in a calcium-dependent manner. This will transiently increase local GABA concentration up to about 1.5 to 3.0 mM that lasts for less than 1 ms [32]. Released GABA is rapidly removed from the synapse either by high affinity GABA transporters in presynaptic nerve terminals and surrounding astrocytes or, in a lesser extent, by passive diffusion [33, 34]. $GABA_ARs$ clustered opposite to the releasing site are activated, producing an inhibitory postsynaptic current (IPSC) that in its minimal amplitude may represent one quantum, i.e., the amount of GABA present in one vesicle, which gives rise to a miniature IPSC (mIPSC) [32, 35, 36]. For a mIPSC to occur there is no need of the arrival of the action potential to the nerve terminal, since vesicles can be spontaneously released.

Actually, whenever aiming to quantify solely the release probability or quantal size, action potential generation in the network is blocked by adding a voltage-dependent channel blocker, frequently tetrodotoxin (TTX), to the extracellular solution. Besides the $GABA_ARs$ at the active zone of neurotransmitter release, there are also receptors adjacent to the synaptic bouton that can be recruited. Their activation may result from diffusion of GABA, which reaches $GABA_ARs$ located perisynaptically or located in the nearby synapses [37, 38]. This form of inhibition can also be considered phasic transmission since it is time-locked to presynaptic GABA release and only transiently activates $GABA_ARs$.

Tonic $GABA_AR$-mediated inhibition results from the continuous activation of $GABA_ARs$ by low concentrations of ambient GABA.

Receptors responsible for this form of transmission have an extra- or perisynaptic location, which are in an ideal position to sense ambient levels of GABA continuously present in the extracellular spacebut are less influenced by huge fluctuations of GABA concentrations that occur at the synaptic level [38].

Other important property of these $GABA_A$Rs is their high affinity for GABA, conferring the ability to sense very low concentrations of ambient GABA that range from tens of nanomolar to a few micromolar [39-41]. There are also $GABA_A$Rs that can be activated even in the absence of any ligand and contributing to the tonic current [42].

In addition, $GABA_A$Rs mediating tonic currents are slow desensitizing receptors [43, 44]. The source of ambient GABA mediating tonic inhibition varies in different brain regions, cell types or even synaptic anatomy.

The non-vesicular sources include release from astrocytes [45-47], reversed transport of GABA by its transporter [40], non-vesicular GABA exocytosis [48] and channel-mediated GABA release from glia [49]. Experimentally, $GABA_A$R-mediated tonic transmission can be recorded by exogenously applying a $GABA_A$R antagonist while monitoring the holding current required to voltage-clamp the cell at a given membrane potential [50].

There are clear physiological differences between phasic and tonic neurotransmission, albeit the fact that both control neuronal excitability. In the adult central nervous system, phasic inhibition is mainly involved in suppressing the activity of principal glutamatergic cells and preventing over-excitation of neurons.

Besides this classical role of synaptic GABAergic transmission, fast and precisely timed phasic responses mediated by GABA-releasing interneurons have other important and complex functions in neuronal communication. These include a key role in feedback and feedforward inhibition of principal cells with consequent synchronization of population activity and induction and/or maintenance of rhythmic network oscillations (e.g., gamma and theta frequency oscillations). Tonic transmission, on the other hand, acts on a much larger time window when compared to phasic responses.

A persistent increase in GABA input conductance in a particular neuron will significantly contribute to a phenomena called "shunting effect" [51], leading to a decrease in neuronal excitability [52–54]. The physiological significance of this is that the same excitatory input arriving to a neuron (e.g., glutamatergic input) will lead to a decrease in the output firing rate of the same neuron. Also, there will be a reduction in the duration of the excitatory potentials, narrowing the temporal fidelity of the excitatory input and decreasing the overall gain of the neuronal input-output relationship [54, 55].

Although tonic conductances can be considered as a constant and uninterrupted form of GABAergic transmission, changes in the concentration of ambient GABA or in the number and properties of extrasynaptic $GABA_A$Rs contribute to changes in the magnitude of tonic inhibition, acting to control and fine-tune neuronal excitability [56]. Because tonic and phasic inhibition display distinct functional roles in GABA-mediated actions, selectively modulating these different forms of inhibition also affect the network excitability differently.

Main Brain Areas and Functions Where Adenosine Has Been Shown to Influence GABAergic Function

Cerebral Cortex, Hippocampus and Amygdala

Early evidence for a relevant facilitatory action of A2AR upon GABAergic transmission was provided by John Phillis showing that the A2AR-mediated inhibitory action of GCS 21680 on spontaneous neuronal firing in vivo was prevented upon blockade of $GABA_A$Rs [57], suggesting that in the cerebral cortex A2AR actions on GABAergic transmission predominate over those on excitatory transmission. The same team has also reported that the selective activation of either A1R or A2AR results in an inhibition of ischemia-evoked GABA release in vivo [58]. A2AR-induced inhibition of GABA release from hippocampal slices under ischemic conditions have been also reported in a study using rather high concentrations of selective ligands [59]. Modifications of GABA release without experimental blockade of glutamatergic transmission are hard to interpret since GABA-releasing neurons in the forebrain are interneurons. Therefore, it is hard to distinguish between direct actions on GABAergic nerve terminals from disynaptic changes through modification of glutamatergic inputs to GABAergic neurons. By using isolated nerve terminals of the rat hippocampus it has been shown that a selective concentration of an A2AR agonist enhances depolarization-evoked GABA release, while selective concentrations of an A1R agonist were devoid of effect upon GABA release [60].

It is widely accepted that, in contrast with A2AR, A1R are mostly devoid of effect upon exocytotic GABA release or upon phasic GABAergic transmission at the hippocampus [60–65]. This lack of effect of A1R led to the idea that GABAergic transmission in the hippocampus is insensitive to A1R modulation; as we will detail in this section, there is now solid evidence that other components of GABAergic transmission are under A1R-mediated control. A first report on the absence of influence of A1R agonists (locally perfused in the hippocampus in vivo) upon veratridine- or hypoxia-induced release of GABA, quantified through microdialysis and HPLC, appeared more than 20 years ago [66]. In this study A1R agonists had a marked inhibitory action upon glutamate release. By using hippocampal slices, it was then reported that adenosine, while reducing the K^+-evoked release of glutamate and aspartate, did not affect the evoked release of GABA [61, 67]. Similar findings were reported by using an electrophysiological approach in cultured rat hippocampal explants [62], acute hippocampal slices [63] or cultured neurons [64]. All these studies showed that adenosine did not suppress fast IPSCs, but inhibited excitatory postsynaptic currents (EPSCs). Interestingly, one study noted that adenosine depressed late IPSPs evoked by direct activation of interneurons in the presence of glutamate receptor blockers [63]. Spontaneous quantal release of GABA, assessed by recording mISPCs in hippocampal slices in the presence of glutamate receptor antagonists and sodium channel blockade, is also not affected by A1R activation [65]. Furthermore, the failure to influence phasic GABAergic transmission applies either when recording from pyramidal neurons as well as while recording from interneurons [65].

In spite of not affecting GABA release from GABAergic nerve terminals at the hippocampus, A1R are able to modulate the action of other modulators of GABAergic transmission.

Thus CB1-mediated inhibitory actions upon GABA release are under control of adenosine A1R and this impacts upon forebrain-mediated behavioral actions of cannabinoids [68]. Thus, chronic caffeine intake, while upregulating A1R in the hippocampus, exacerbate memory deficits induced by acute intake of cannabinoids in the absence of caffeine [68]. Acute intake of caffeine or of an A1R antagonist together with cannabinoids also exacerbates memory impairment caused by cannabinoids, providing that the animals arc naïve for caffeine [69]. In animals with A1R upregulated by chronic intake of caffeine, A1R blockade prevented cannabinoid-induced memory impairment [68]. These studies, in spite of highlighting a complex relationship between A1R and cannabinoid CB1 receptors (CB1R), that may involve both GABAergic and glutamagergic transmission, among others, provide clues from behavioraly relevant interaction between A1R and CB1R at hippocampal neurons.

Interestingly, in interneurons A1R actions are present in CB1R-positive interneurons, but are undetectable in CB1R-negative interneurons [65]. Modulatory actions of vasoactive intestinal peptide (VIP) upon GABA release are also under control of A1R, since endogenous adenosine removal with adenosine deaminase (ADA) or blockade of A1R prevents the facilitatory action of VIP upon GABA release, while activation of A1R in the presence of ADA rescues the effect of VIP [70]. A2AR activation is also required for the facilitatory action of VIP upon GABA release from hippocampal nerve terminals [70].

GABA release from immature hippocampal neurons seems sensitive to A1R activation, in contrast to what occurs in mature neuron under similar experimental conditions. Thus, in acutely isolated hippocampal neurons from P12-P15 rats, adenosine A1R activation causes presynaptic inhibition of GABAergic transmission, as assessed by the reduction of the frequency of mIPSCs recorded in the presence of glutamate receptor blockers and tetrodotoxin [71]. The same study reported absence of effect of A1R agonists in P30 neurons, therefore suggesting a developmental dependency of this inhibitory action of A1R [71]. In P5-7 neurons from cortical layer I, A1R blockade enhanced the amplitude of GABA-mediated fast synaptic responses, and decreased paired-pulse response ratio, suggesting a reduction of GABA release probability via presynaptic A1R activation by endogenous adenosine [72]. Interestingly, this action is also absent at P1-2 neurons [72]. Being evident at P5-7, a developmental stage where GABAegic activity can depolarize pyramidal neurons and influence glutamatergic synaptogenesis in the lower cortical layers, A1R may mediate developmental downregulation of the GABAergic excitatory drive when glutamatergic synapses are formed [72]. Delayed migration and insertion of GABAergic neurons into the hippocampal circuitry during the first postnatal week in offspring has been reported to occur when dams are treated with an A2AR antagonist or caffeine [73]. This was associated with increased neuronal network excitability and increased susceptibility to seizures, loss of hippocampal GABAergic neurons, and mild cognitive deficits, suggesting embryonic exposure to caffeine may have adverse consequences upon neuronal maturation [73].

The main output from the entorhinal cortex to the hippocampus is provided by stellate neurons. Interestingly A1R activation inhibits GABAergic inputs to stellate neurons, an action that involves presynaptic mechanisms as assessed by an inhibition of the frequency of mIPSCs [74]. Thus, it appears that while phasic GABAergic transmission within the hippocampus is unaffected by A1R activation [62–65], phasic GABAergic transmission to neurons that project into the hippocampus can be affected by A1R activation [74]. Neuronal responsiveness in the amygdala is largely controlled by inhibitory processes, which are also presynaptically downregulated by adenosine A1R activation [75].

While essentially not directly affecting phasic GABAergic transmission in mature neurons at the hippocampus, adenosine A1R at this brain area selectively modulate tonic inhibition mediated by activation of extrasynaptic $GABA_A$Rs [65]. The A1R-mediated modulation of tonic GABAergic currents is consistent in CA1 pyramidal cells, but present only in a specific population of postsynaptic CA1 GABAergic inhibitory interneurons, the CB1/CCK expressing interneurons [65]. Accordingly, sustained A1R activity results in a decreased expression of $GABA_A$R δ-subunit, a key component of extrasynaptic receptors mediating tonic GABAergic currents [65].

Since tonic GABAergic inhibition is more pronounced in interneurons than in pyramidal cells, the reduction in the disinhibition of interneurons caused by A1R-mediated suppression of tonic GABAergic transmission may increase inhibitory GABAergic output to the hippocampal principal cell population. The selective influence of A1R upon a specific interneuron population, the CB1/CCK expressing interneurons that are involved in synchronous network oscillations [76], may confer to adenosine an important modulatory action on hippocampal network oscillations that are the critical bases for hippocampal-dependent behavior and cognitive processes. A selective action upon tonic inhibition without affecting phasic inhibition may also confer to adenosine A1R a control of neuronal gain without disrupting fidelity of synaptic GABAergic inhibition. Such an action may prove relevant in the context of the anticonvulsant action of adenosine.

In contrast with A1R which, as mentioned above, inhibit tonic inhibitory GABAergic input to a subpopulation GABAergic interneurons in the hippocampus, the CB1R-positive neurons [65], A2ARs located in the nerve terminals of another subpopulation of interneurons (the parvalbumin-positive neurons, responsible for network synchronization) enhance phasic GABAergic inputs into other GABAergic interneurons, leading to disinhibition of pyramidal cells [77]. Interestingly, A2AR do not affect GABAergic inputs to excitatory neurons, as they also do not influence glutamatergic inputs to GABAergic neurons [77], but as expected A2AR enhance glutamatergic inputs to glutamatergic neurons [77]. These highly selective synapse- and cell type specific influence of adenosine A2AR, revealed by optogenetic stimulation of different interneuron subtypes combined with electrical stimulation of glutamatergic synapses, suggest an influence of A2AR upon synchronous pyramidal cell firing in hyperexcitable conditions. Accordingly, an A2AR antagonist was shown to robustly suppress spontaneous interictal like events in an ex vivo model of epilepsy [77].

Ischemia induces adaptive actions upon synaptic transmission and these can be influenced by adenosine that, as very well known, is released by neuronal cells under conditions of low oxygen or glucose supply [78–82]. Adenosine through A1R activation contributes to a post-synaptic enhancement of inhibitory synaptic transmission in CA1 neurons that occur for several hours after forebrain ischemia in vivo [83]. This may contribute to protect synapses, allowing functional recovery, similar to the A1R-mediated depression of glutamatergic synapses in an ex vivo model of hypoxia [84].

Influences of A1R upon the gating properties of $GABA_A$Rs also occur. This was first suggested through a rather indirect way, using a biochemical index that reveals changes in the function of postsynaptic $GABA_A$Rs, through modifications in the binding of butylbicyclophosphorothionate, known to bind to the ion recognition site of the $GABA_A$R as a function of its activity state; by using such approach it was suggested that A1R activation could reduce the function of the GABA-coupled chloride channel in several brain areas, including cortex and hippocampus [85].

The same indirect approaches also lead to the suggestion that caffeine administration in vivo significantly alters the Cl^- transport function of the $GABA_AR$ complex [86]. On the contrary, a positive interaction between adenosine and a $GABA_AR$ agonist, through a mechanism also involving chloride channels in hippocampal neurons, was also suggested to occur [87]. Control of $GABA_AR$ activity may be particularly relevant during seizures, where GABAergic responses transiently switch from hyperpolarizing to depolarizing, therefore excitatory and contributing to seizure progression. Under these conditions A1R-dependent activation of potassium channels increase membrane conductance and has a shunting effect on $GABA_AR$-mediated currents, which could limit seizure activity [88].

Other AR subtypes seem to play a role in the control of GABAergic currents during epilepsy, as antagonists of A2AR, A2BR and A3R (but not of A1R) in different epileptic tissues increase the stability of $GABA_AR$-mediated currents, assessed by a decreased run-down of the currents that occurs after repeated GABA applications [89, 90]. A reduction of the extracellular levels of adenosine through enhanced expression of adenosine kinase also leads to increased stability $GABA_AR$-mediated currents in pyramidal neurons of the hippocampus [91]. How this non-A1R mediated increase in the stability of GABAergic currents during seizures promotes excitatory GABAergic signaling and activity during seizures, is yet unknown. Seizure occurrence is in many cases effectively controlled by inhibitors of GABA transporters (GAT) which enhance extracellular GABA levels, and prolongs its inhibitory action, thus keeping neurons in a more hyperpolarized state. As such, modulation of GAT modulation hashas important implications for the control of epilepsy [34, 92]. GAT activity also determines the levels of ambient GABA available for extrasynaptic receptors and therefore for tonic inhibition. For experimental reasons, studies on modulation of tonic inhibition are performed under conditions of GAT blockade, which allows studying the activity of high affinity/slow desensitizing $GABA_ARs$ [65].

It is however also relevant to directly assess the influence upon GAT activity. Adenosine receptors are also able to modulate this aspect of GABAergic transmission. A2AR activation has been shown to enhance GABA transport into nerve endings [93], while the A1R is devoid of this function in nerve terminals. In astrocytes, A1R activation leads to an inhibition of GAT activity, while A2AR activation leads to its facilitation. This influence is observed either while evaluating GABA transport through GAT1 or GAT3, both present in astrocytes [94]. In addition, A2AR are required to gate the facilitatory action of brain-derived neurotrophic factor upon GAT1 mediated GABA transport into astrocytes [95].

Interestingly, there is a tight interaction between A1R and A2AR in the astrocytes to control GABA transport, so that blockade of one of the receptors prevents the activity of both receptors [94]. Actually, this regulation of GABA transport in astrocytes by adenosine occurs through A1R-A2AR heteromers (A1R-A2AR-A1R-A2AR tetramers) that signal via two different G proteins, G_s and $G_{i/0}$ and either enhances (A2AR) or inhibits (A1R) GABA uptake. Inhibition requires lower adenosine concentrations, but slight increases in concentration are enough to gate A2AR activation and to engage a completely opposite modulation of GABA uptake [94]. The levels of endogenous adenosine at the astrocytic vicinity may vary as a function of network excitability, since adenosine and ATP, a source of adenosine, are released in response to neuronal activity from the pre-synaptic, post-synaptic and astrocytic components [82, 96–98]. The presence or absence of microglia may also determine the levels of extracellular endogenous adenosine to control GABA transport into astrocytes [99].

One may then speculate that A1R-A2AR heteromers in astrocytes behave as adenosine sensors to fine-tune the synaptic levels of the inhibitory transmitter as a function of neuronal status. Thus, A1R-A2AR heteromers in astrocytes may behave as dual amplifiers, facilitating excitation at intense astrocytic to neuronal signalling and increasing inhibition at low neuronal firing rates. The main advantage of heteromerization of A1R and A2AR in astrocytes must be to avoid sudden transitions between inhibition and excitation.

Indeed, overstimulation of just one of the receptor leads to internalization of the whole functional unit [94] and probably avoids a sudden switch from excitation to inhibition or vice-versa as a consequence of desensitization of only one of the receptor subtypes.

In contrast to what occurs in the hippocampus [93, 94], in slices of the rat globus pallidus GAT1 mediated GABA transport is inhibited by adenosine A2AR activation, through a PKA-dependent mechanism [100]. A2AR-mediated facilitation of GAT1 activity in hippocampal nerve endings also involves a cyclic AMP/PKA-dependent mechanism [93], suggesting that similar mechanisms may have opposing actions upon GAT activity, depending upon the brain area involved. Indeed, GAT1 activity in slices from the globus pallidus is inhibited by substances which increase protein kinase A activity, such as forskolin and 8-bromo-cAMP [100], while GAT1 activity in hippocampal synaptosomes is enhanced by forskolin [93]. Differences resulting from the use of different preparations (slices vs isolated nerve endings) cannot be fully discarded, though the data is consistent with GAT1 modulation by A2AR activation in different preparations (isolated nerve endings vs cultured astrocytes) in the same brain area (forebrain).

Endogenous GABA might exert an inhibitory effect over adenosine A1R-mediated responses in the hippocampus [101], which may represent a physiologic regulatory mechanism between the two inhibitory mediators. Conversely, A1R activation may mask $GABA_A$R-mediated responses under certain conditions [102]. Thus, blockade of $GABA_A$Rs induces a potentiation of the inhibitory action of A1R upon synaptic transmission, an effect that involves NO acting through guanylyl cyclase [101].

Also, a GABAergic contribution to the depression of synaptic transmission caused by hypoxia, can only be revealed when A1R are blocked [102]. In conclusion, the data summarized above clearly indicates that adenosine acting on A1R and A2AR at pre-, post- and non-synaptic sites affects GABAergic transmission at the hippocampus.

Retina

Depolarization-evoked GABA release from retinal cells is not affected by A1R activation, though A1R present at the retina are able to mediate inhibition of acethylcholine release [103]. Therefore, the absence of modulation of phasic GABAergic transmission by A1R activationin retinal neurons is similar to observations in hippocampal neurons.

Retinal A1R, however, regulates GABA release mediated by reversal of the GABA transporters. In retinal explants of chick embryo, it was shown that caffeine and a selective A1R antagonist potentiates GAT1 mediated GABA release induced by activation of excitatory amino acid receptors [104], suggestive of tonic inhibition of GAT1 activity by A1Rs. Again, there is a similarity between the influence of adenosine A1R-mediated modulation of GAT activity in the retina and in forebrain areas since A1R activation leads to an inhibition of GAT activity in cultured astrocytes [104].

Cerebellum

Information on the actions of ARs at the cerebellum is scarce. It is known that A1R mediate presynaptic inhibition of GABAergic inputs to cerebellar granule cells [105]. Adenosine release in the rat cerebellum, on the other hand, seems to be under control of $GABA_BR$ activation [106]. Interestingly A1R activation at the cerebellum also inhibits adenosine release [106], suggesting that adenosine receptors can control the release of their ligand. A similar process also occurs at the hippocampus [80] and at chromaffin cells [107], though in these cases, control of adenosine transporters is mediated by A2ARs.

Basal Ganglia and Fine Control of Movement

It is well established that the predominant adenosine receptor at the basal ganglia is the excitatory adenosine A2AR, with a main (but not exclusive) location in the medium spiny GABAergic neurons, where A2AR are located postsynaptically to control afferent dopaminergic transmission (Figure 1). The selective expression of adenosine A2ARs in striatal enkephalin-expressing GABAergic medium spiny neurons, but not in striatal somatostatin-expressing GABAergic cells, suggests that adenosine may play a selective modulatory role in the activity of GABA/enkephalin striatopallidal neurones [108].

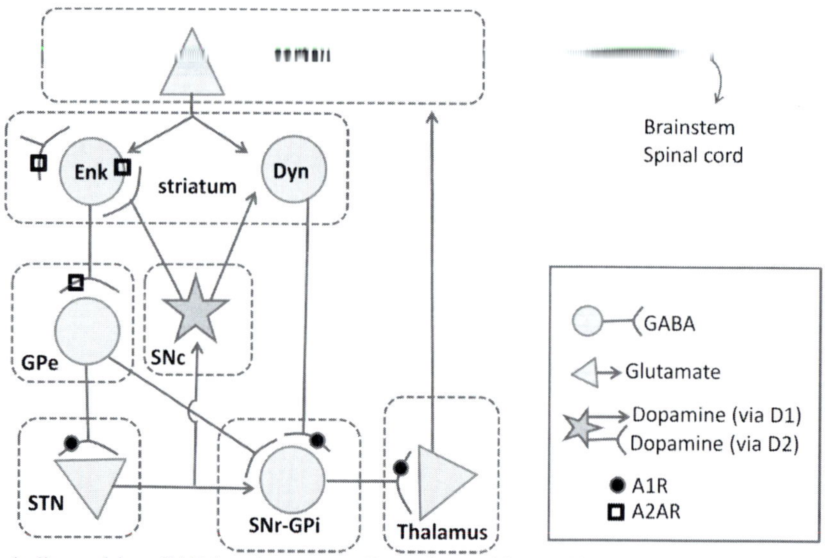

Enk: encephalin positive GABAergic neuron; Dyn: dynorphin positive GABAergic neuron; SNr-Gpi: substancia nigra pars compacta-globus pallidus pars interna complex; SNc: substantia nigra pars compacta; Gpe: globus pallidus pars externa; STN: subthalamic nucleus.

Figure 1. Schematic location of adenosine A1R and A2AR on GABAergic neurons to control basal ganglia circuitry. Presynaptic A2ARs at the basal ganglia tend to inhibit GABA release, whereas postsynaptic A2AR in enkephalin containing neurons inhibit D2R-mediated inhibitory signaling; presynaptic A1R inhibit GABA release (See text for references). The presence of A1R or A2AR in glutamatergic or dopaminergic neurons is not represented, but they are known to inhibit (A1R) or facilitate (A2AR) the release of glutamate at several synapses [112]. A2AR may also be present at dopaminergic nerve terminals to control the action of another neuromodulator (see text for details).

By studying the effect of intrastriatal infusion of adenosine and dopamine receptor ligands and evaluating the release of GABA from the terminals of the striopallidal neuron in awake, freely moving rats, it was also recognized that the GABAergic striopallidal neuron, which is a key component of the indirect striatal efferent pathway, is a main locus for A2AR and dopamine D2 receptor (A2AR-D2R) interactions in the brain and a main target for the actions of A2AR in the basal ganglia [109]. In this study it was shown that while intrastriatal infusion of A2AR agonists did not affect GABA release in the ipsilateral globus pallidus, they markedly antagonized the inhibitory action of D2R agonists upon GABA release in the same brain area [109]. It is now fully established that A2AR in the basal ganglia mostly operate in the indirect pathway, by modifying the D2R-mediated dopaminergic input to GABAergic striatal neurons that project to the globus pallidus pars externa. The implication of this action for fine control of movement has been the focus of several reviews [110, 111].

Evidence that striatal A2AR activation increases spontaneous GABA outflow through a mechanism that relates to the ability of A2AR to counteract the inhibitory influence of D2R-mediated responses has been provided in an early study using a microdialysis approach [113]. It was also that A2AR activation enhances GABA release in slices of striatum containing globus pallidus, and that the negative control exerted by D2R activation is counteracted by A2AR [114]. Using an acute model of Parkinson's disease (PD), it was shown that microinjection of an A2AR agonist into the striatum of non-lesioned animals increased GABA concentrations in the globus pallidus, whereas the injection of an A2AR antagonist reverses the increase in GABA release caused by 6-hydroxydopamine, an action related to the antiparkinsonian effects of A2AR antagonists [115]. Soon after, it was shown that enhancement of striatopallidal GABAergic transmission caused by metabotropic mGLuR5 receptor activation in either the ventral or dorsal striatum is potentiated by co-activation of A2AR, consistent with the idea of a synergistic interaction between mGluR5 and A2AR to facilitate GABAergic transmission in the basal ganglia [116].

As mentioned above, A2AR are predominantly but not exclusively expressed postsynaptically in striatal GABAergic neurons. Ultrastructural localization of adenosine A2AR early suggested multiple cellular sites for modulation of GABAergic neurons in rat striatum [117]. The presence of A2AR in nerve terminals of GABAergic interneurons projecting to the striatopallidal GABAergic neurons has been suggested on the grounds that A2AR activation leads to a pre-synaptic inhibition of inhibitory GABAergic currents at the enkephalin-containing GABAergic medium spiny neurons of the striatum [118, 119]. Pre-synaptic inhibitory actions of A2ARs upon GABA release were also found while using isolated nerve terminals from the striatum [120, 121]. Interestingly, this A2AR-mediated action is apparently not mediated by increases in intracellular cAMP [120]. Inhibition of GABAergic inputs to the striatum will also enhance the activity of striatum-pallidal neurons, thereby decreasing inhibition of subtalamic neurons by GABAergic neurons originated in the globus pallidus. Presynaptic facilitatory actions of A2AR at terminals of striatopallidal medium spiny neurons of the globus pallidus [122–124], will also add to this overall A2AR-mediated disinhibition of glutamatergic subthalamic nucleus activity. This would enhance inhibition of the thalamus by substantia nigra pars reticulate/globus palludus pars interna (SNr/GPi), an output structure of the whole basal ganglia network that projects to the thalamus through inhibitory GABAergic neurons. Oral administration of an A2AR-selective antagonist to an animal model of PD caused a marked and sustained increase of GABA and glutamate levels in the SNr. Selective lesion studies indicate that changes in neurotransmitter release in the SNr brought about by A2AR in the basal ganglia is indeed mostly dependent on their ability to modulate the activity of striatopallidal medium spiny enkephalin-positive neurons that belong to the indirect basal ganglia pathway to control thalamic activity [125]. This facilitatory action of A2AR upon GABA release in the globus pallidus depends, at least in part, from the co-activation of dopamine D2R [126, 127] and is counteracted by co-activation of histamine H3R present in GABAergic terminals of the globus pallidus [128].

The presence of adenosine A2AR in dopaminergic nerve ending was suggested on the grounds of the facilitatory action of glial-derived neurotrophic factor (GDNF) upon dopamine release from isolated striatal nerve endings [129]. GDNF receptors are present in GABAergic nerve terminals of the striatum, though with lower expression than in glutamatergic and dopaminergic nerve terminals [130].

Interestingly the facilitatory action that A2AR exert over GDNF upon dopamine release seems to be under control of GABAergic neurons since in systems where neuronal connections are more preserved, as in striatal slices, the cross-talk between adenosine A2AR and GDNF is markedly affected by drugs that influence GABAergic transmission [129].

It has been shown that adenosine A2AR activation reduce GAT1 mediated GABA uptake in slices of the rat globus pallidus [100]. Interestingly, this is in contrast with what occurs in the forebrain, where A2AR activation facilitates GAT1 mediated GABA uptake into nerve endings [93] as well as GAT1 and GAT3 mediated GABA transport into astrocytes [94]. Since inhibition of GABA transport contributes to an increase in extracellular levels of GABAand enhanced inhibition, one might speculate that at the globus pallidus the action of A2AR upon GAT activity may enhance inhibitory tone of globus pallidus neurons, thereby reinforcing the disinhibition of subthalamic excitatory neurons.

Dopamine D1 receptor (D1R) mediated enhancement of the evoked release of GABA from SNr slices is inhibited by tonic A1R activation [131], suggestive of an A1R/D1R interaction to control the GABAergic inputs to SNr/GPi neurons.

Through this mechanism A1R activation may lead to the disinhibition of SNr/GPi neurons, hence, enhanced thalamic inhibition. Similarly, in ventral pallidum and nucleus accumbens, brain areas belonging to reward circuits and in addictive behavior, stimulation of D1R facilitates GABA release, an action attenuated by concurrent activation of A1R [132]. The facilitation of GABA release by D1Rs localized in the entopeduncular nucleus is also inhibited by A1R [133].

At the level of the thalamus, adenosine, through presynaptic A1Rs, strongly suppresses monosynaptic inhibitory currents both in relay cells of the thalamic ventrobasal complex and in inhibitory neurons of the nucleus reticularis thalami. The ability to presynaptically down-regulate inhibitory postsynaptic responses in thalamus and influence excitatory transmission accounts for the robust antioscillatory effects of A1R activation [134].

Whole-cell current recordings in SNr neurons which receive input from dynorphin containing neurons that, in turn, receive dopaminergic inputs through D1R (direct pathway), revealed that A1R, through a presynaptic mechanism, decreases GABAergic inputs to SNr neurons [135]. This suggests an action of the A1R in the nerve endings of the dynorphin-positive neurons. Consistent with these findings, it was recently shown that adenosine, acting on A1R, blunts D1R enhancement of spontaneous GABAergic synaptic activity in SNr neurons [136]. A decreased inhibition of SNr neurons and consequent increased output from these neurons is known to contribute to the rigidity and bradykinesia in PD [110].

Remarkably, A1R administration in vivo in combination with L-DOPA, reduced the development of abnormal involuntary movements, highlighting the potential benefit of A1R agonists for the treatment of L-DOPA-induced dyskinesia and hyperkinetic disorders [136].

A1Rs also seem to be present in the indirect pathway, as electrophysiological recordings from subthalamic nucleus neurons revealed an A1R-mediated presynaptic inhibition of both GABAergic and glutamatergic inputs to those neurons [137].

Endogenous adenosine, acting on presynaptic A1R, is responsible for early presynaptic depression of IPSCs caused by oxygen/glucose deprivation in striatal slices [138], in clear contrast with the hippocampus where monosynaptic IPSCs are more resistant to hypoxiaand to synaptic modulation by adenosine A1Rs [139].

At the hippocampus, adenosine A1R-mediated inhibition of excitatory synaptic transmission during hypoxia favors recovery after reoxygenation [84]. Whether the early depression of GABA-mediated synaptic inhibition exerted by A1R at the striatum may play a role in the development of ischemic neuronal injury is unknown.

BRAIN AREAS RELATED TO DRUG ADDICTION

The dopaminergic projections from the ventral tegmental area (VTA) into the nucleus accumbens, major players in the reward circuit, are under control of GABAergic neurons that make synapse either at the level of the VTA or at the level of the nucleus accumbens. A1R inhibit GABAergic inputs to nucleus accumbens neurons [140]. Furthermore, the facilitatory action of dopamine D1R upon GABA release at the nucleus accumbens and ventral pallidum is attenuated by concurrent activation of adenosine A1R [132]. In contrast, A2AR may facilitate GABA release in these areas since an A2AR antagonist, KW6002, decreases GABA release while increasing dopamine release from the nucleus accumbens [141].

Patch clamp recordings of GABAergic currents in dopaminergic neurons of the VTA provided evidence that adenosine, through presynaptic adenosine A1R, decreases the release of GABA and attenuated $GABA_B$R-mediated inhibition of dopaminergic neurons [142], therefore providing their disinhibition. Relevant interactions between adenosine A1R and type 2 receptors of corticotrophin releasing factor (CRF-R2) at GABAergic nerve terminals have been recently shown in a cocaine addiction and reinstatement model, which helps to understand the cellular basis for relapse-mediated changes in the activity of VTA dopaminergic neurons [143]. Thus, evidence has been obtained consistent with the interpretation that reinstatement leads to a CRF-R2 dependent enhanced activity of adenosine A1R on VTA GABAergic nerve terminals. This causes a decreasein GABAergic tone and a lower activation of $GABA_B$Rs at the glutamatergic nerve terminalsand increased glutamatergic inputs to VTA dopaminergic neurons [143]. While A1R in VTA may have a facilitatory influence upon reinstatement of cocaine abuse, A2AR in the nucleus accumbens seems to reduce reward associated with cocaine self-administration, an action that also involves changes in the GABAergic tone. This has been suggested based on findings that activation of A2AR attenuates the increase in dopamine levels in the nucleus accumbens caused by cocaine self-administration, while simultaneously enhance GABA levels in the same brain area [144]. This action may be related to A2AR-induced attenuation of inhibitory actions of D2R upon GABAergic neurons in the accumbens, therefore leading to reduced cocaine-induced reward [144]. Also indicative of the key role of adenosine/GABA interplay during different steps of addictive behavior is a study showing that hyperexcitability during morphine withdrawal is retarded with a concomitant increase in endogenous A1R activation by adenosine [145]. Thus, in chronically morphine treated mice both the rate of mIPSCs and the amplitude of evoked IPSCs during naloxone-precipitated withdrawal was profoundly enhanced in the presence of an adenosine A1R antagonist [145].

Brain Areas Involved in Sleep Control

Both A1R and A2AR influence sleep control, by acting in several areas and by operating different mechanisms [112, 146–148]. These actions that involve GABAergic transmission in brain areas more directly related to sleep regulation are summarized below.

Adenosine acts presynaptically to decrease the probability of spontaneous GABA inputs onto tuberomammillary nucleus (TMN) neurons [149], which have important roles in the homeostatic control of sleep. GABAergic innervation from the ventrolateral preoptic nucleus (VLPO) to the TMN also plays pivotal roles in the regulation of sleep-wakefulness [150]. Microinjections of adenosine into the medial preoptic area induces sleep and the early finding that this action is prevented by the non-competitive $GABA_AR$ antagonist flumazenil constituted an initial evidence that the hypnotic properties of adenosine could be mediated by a direct or indirect action on the $GABA_A$-benzodiazepine receptor complex [151]. Activation of presynaptic A1R decreases spontaneous GABAergic transmission onto TMN neurons via the modulation of G-protein coupled inward rectifying potassium (GIRK) channels as well as the adenylate cyclase/cyclic AMP/protein kinase A signal transduction pathway. This adenosine A1R-mediated modulation of GABAergic transmission onto TMN neurons may play an important role in the fine modulation of the excitability of TMN histaminergic neurons as well as the regulation of sleep-wakefulness [149]. Furthermore, adenosine, via the activation of A2AR, stimulates a subpopulation of VLPO GABAergic neurons [152].

GABAergic inputs to hypocretin neurons in the lateral hypothalamus, a wakefulness promoting center, are presynaptically inhibited by A1R located in GABAergic nerve terminals, a process that may contribute to fine regulation of the excitability of these neurons as well as eventually to modulate the sleep–wakefulness cycle [153]. Disinhibition of sleep promotion neurons of the ventrolateral preoptic area of the hypothalamus through inhibitory presynaptic A1R located in GABAergic neurons has also been reported [154].

In freely moving rats, application of an A2AR agonist, CGS 21680, to the subarachnoid space underlying the rostral basal forebrain significantly promoted sleep and inhibited histamine release in the frontal cortex [155]. While inhibiting histamine release in both the frontal cortex and medial pre-optic area, GCS 21680 increased GABA release specifically in the histaminergic TNM neurons but not in the frontal cortex [155].

Moreover, the CGS 21680-induced inhibition of histamine release was antagonized by perfusion of the TMN with a GABAAR antagonist, picrotoxin [155]. Altogether, these data suggest that the A2AR activation can induce sleep by inhibiting the histaminergic system through an increase in GABA release in the TMN [155].

In cultured neurons of the arcuate nucleus and of the suprachiasmatic nucleus of the hypothalamus, the control center of mammalian circadian rhythms, adenosine inhibits synaptic GABA release, an action that may involve not only A1R but also A2AR [156].

CIRCUITS INVOLVED IN PAIN CONTROL

The existence of an interaction between adenosine-mediated mechanisms and GABAergic mechanisms to control pain was initially suggested on the grounds that a $GABA_AR$ and $GABA_BR$ antagonists prevented antinociceptive action of A1R agonists, and that a $GABA_BR$ agonist potentiate the antinociceptive action of a A1R agonist [157].

Adenosine suppresses GABAergic and glycinergic transmission in substantia gelatinosa neurons by activating presynaptic A1R, an action that may contribute to the modulation of pain transmission [158]. Also in the spinal cord, but at the sacral dorsal commissural nucleus, an area involved in central processing of pelvic visceral information, adenosine, by operating post-synaptic A1R, suppresses GABA-induced Cl^- currents, suggestive of an involvement of adenosine in visceral pain control [159]. This action of adenosine involves a Ca^{2+}-insensitive protein kinase C-dependent mechanism [159]. Similar post-synaptic actions of A1R have been detected in the superficial laminae (laminae I and II) of the rat spinal dorsal horn [160]. In addition, convergent and activity-dependent inhibition of GABA release by A1 and GABAB autoreceptors have been shown to occur in dorsal horn neurons of the spinal cord, suggestive of a modulation of the integrative properties of these neurons under physiological conditions and/or during the development of pathological pain states [161].

In isolated rat dorsal root ganglion neurons adenosine A1R inhibit inhibitory currents mediated by $GABA_ARs$ by shifting the GABA concentration-response curve downward without affecting the apparent affinity of GABA [162]. This inhibitory effect of adenosine A1R upon GABAergic currents requires protein kinase C activity and is mimicked by phorbol esters, suggesting involvement of the phospholipase C/protein kinase C transducing pathway [162]. Since GABA is an important neurotransmitter in the generation of presynaptic inhibition in the dorsal horn to modulate sensory inputs such as pain, this inhibitory action of adenosine upon an inhibitory neurotransmitter may impact upon pain processing and synaptic plasticity during overstimulation, namely in wind up phenomena observed in certain spinal sensory pathways [162].

BRAIN AREAS INVOLVED IN THE CONTROL OF AUTONOMIC NERVOUS SYSTEM

Magnocellular neurosecretory cells synthesize vasopressin, known to play a key role in body fluid homeostasis. The cell bodies of magnocellular cells are located at the hypothalamus and project their axons to the neurohypophysis, controlling hormone release. Thus, GABAergic inputs to the cell body of magnocellular neurons impact hormone secretion and fluid homeostasis. Adenosine, via A1R, inhibits synaptic GABAergic inputs to the cell body of magnocellular cells located at supraoptic nucleus of the hypothalamus [163].

The pre-sympathetic neurons in the hypothalamic paraventricular nucleus play an important role in regulating arterial blood pressure and sympathetic outflow through projections to the spinal cord and brainstem. Adenosine A1R expressed on spinally projecting hypothalamic paraventricular nucleus neurons hyperpolarize these neurons and decrease their firing activity [164]. In addition, the GABAergic and glutamatergic inputs to these neurons are also inhibited by adenosine, indicative of the presence of these receptors in nerve terminals that synapse with hypothalamic paraventricular neurons [164]. A presynaptic inhibitory action of adenosine A1R upon spontaneous and evoked GABAergic inputs to neurons of the paraventricular nucleus, as assessed by a decrease in the frequency of mIPSC and an increase paired-pulse ratio of evoked IPSCs, was also reported [165]. These data reinforce previous evidence that adenosine, through control of GABAergic transmission, acts as an important neuromodulator of the central nervous system mediated cardiovascular regulation [166]. Adenosine A1R have been shown to presynaptically inhibit GABAergic inputs to the midbrain periaqueductal grey neurons, which also play a key role in the cardiovascular responses, in particular those related to stress or pain [167].

GABA and adenosine contribute to respiratory inhibition in early postnatal life. Thus, during early development, adenosine contributes to the occurrence of respiratory depression and recurrent apneas, an action that results from an interaction with GABAergic circuits since it is abolished upon blockade of GABAergic transmission [168]. A2AR mRNAs was detected in GABAergic neurons of brainstem areas involved in the central control of respiration, including bulbospinal GABAergic neurons projecting to the phrenic motor nuclei [169], reinforcing the idea that adenosine A2AR can induce respiratory depression through a facilitatory action over GABAergic neurons. In infant (14-16 days after birth), but not in adult rats, the intracerebroventricular injection of an A2AR agonist decreases the respiratory drive, an action prevented by prior GABAAR blockade, also suggestive of an action mediated via GABAergic inputs to the inspiratory timing neural circuitry [170]. A2AR-mediated amplification of GABAergic mechanisms in the nucleus tractus solitarius with implications for reflex respiratory control has also been demonstrated [171]. Interestingly, adenosine, through excitatory A2AR and A2BR located at the carotid body, mediate an increase in carotid sinus nerve firing frequency and an increase in the respiratory drive [172–175], but any relationship between these facilitatory actions and modulation of GABAergic activity within the carotid body neurons is as yet unknown.

Conclusion

As pointed out by Sebastião and Ribeiro [112], synchronization or desynchronization of neurons is essential for most brain functions. Dysfunctions of desynchronization, resulting in hypersynchronism at the temporal lobe, trigger epileptic seizures, and adenosine, as GABA, are known as antiepileptic agents. Indeed, the homeostasis of the brain depends primarily on regulatory molecules, such as GABA and adenosine, which modulate neuronal excitability and control the predominance of the activity at certain brain areas over others. Classical examples are brain oscillations put in evidence while coding specific information, setting and modulating brain attentional states, and assuring the communication between neuronal populations, which constitutes the material core of cognitive functions [176].

Evidence summarized in this review highlights the multiple points where adenosine can act at the tripartite synapse to control different aspects of GABAergic transmission. A role of adenosine upon tonic GABAergic inhibition, exerted either by modulation of GABA transporters at nerve endings and astrocytes, or by modulating high affinity/slow desensitizing $GABA_ARs$, was demonstrated in detail in the hippocampus, basal ganglia and retina. This influence upon tonic inhibition, together with synapse specific actions upon phasic inhibition, will certainly contribute to adjust network activity in different brain areas. Such a function could underlie the adenosine 'destiny' - the homeostatic control of brain function.

ACKNOWLEDGMENTS

The authors Unit has been in receipt of Fundação para a Ciência e Tecnologia (FCT) and European Union (Cost B10 and B30) research grants. DMR has been in receipt of a PhD fellowship from FCT (SFRH/BD/60386/ 2009).

REFERENCES

[1] Beaulieu, C.; Kisvarday, Z.; Somogyi, P.; Cynader, M.; Cowey, A. Quantitative distribution of GABA-immunopositive and -immunonegative neurons and synapses in the monkey striate cortex (area 17). *Cereb. Cortex* 1992, 2, 295–309.

[2] Somogyi, P.; Tamás, G.; Lujan, R.; Buhl, E. H. Salient features of synaptic organisation in the cerebral cortex. *Brain Res. Rev.* 1998, 26, 113–135.

[3] Krnjević, K.; Schwartz, S. The action of gamma-aminobutyric acid on cortical neurones. *Exp. Brain Res.* 1967, 3, 320–336.

[4] Curtis, D. R.; Duggan, A. W.; Felix, D.; Johnston, G. A. GABA, bicuculline and central inhibition. *Nature* 1970, 226, 1222–1224.

[5] Schofield, P. R.; Darlison, M. G.; Fujita, N.; Burt, D. R.; Stephenson, F. A.; Rodriguez, H.; Rhee, L. M.; Ramachandran, J.; Reale, V.; Glencorse, T. A. Sequence and functional expression of the GABA A receptor shows a ligand-gated receptor super-family. *Nature* 1987, 328, 221–227.

[6] Bowery, N. G.; Hill, D. R.; Hudson, A. L.; Doble, A.; Middlemiss, D. N.; Shaw, J.; Turnbull, M. (-)Baclofen decreases neurotransmitter release in the mammalian CNS by an action at a novel GABA receptor. *Nature* 1980, 283, 92–94.

[7] Kerr, D. I.; Ong, J. GABAB receptors. *Pharmacol. Ther.* 1995, 67, 187–246.

[8] Newberry, N. R.; Nicoll, R. A. Comparison of the action of baclofen with gamma-aminobutyric acid on rat hippocampal pyramidal cells in vitro. *J. Physiol.* 1985, 360, 161–185.

[9] Dunlap, K.; Fischbach, G. D. Neurotransmitters decrease the calcium conductance activated by depolarization of embryonic chick sensory neurones. *J. Physiol.* 1981, 317, 519–535.

[10] Drew, C. A.; Johnston, G. A. R.; Weatherby, R. P. Bicuculline-insensitive GABA receptors: Studies on the binding of (-)-baclofen to rat cerebellar membranes. *Neurosci. Lett.* 1984, 52, 317–321.

[11] Barnard, E. A.; Skolnick, P.; Olsen, R. W.; Mohler, H.; Sieghart, W.; Biggio, G.; Braestrup, C.; Bateson, A. N.; Langer, S. Z. International Union of Pharmacology. XV. Subtypes of gamma-aminobutyric acidA receptors: classification on the basis of subunit structure and receptor function. *Pharmacol. Rev.* 1998, 50, 291–313.

[12] McKernan, R. M.; Whiting, P. J. Which GABAA-receptor subtypes really occur in the brain? *Trends Neurosci.* 1996, 19, 139–143.

[13] Macdonald, R. L.; Olsen, R. W. GABAA receptor channels. *Annu. Rev. Neurosci.* 1994, 17, 569–602.

[14] Payne, J. A.; Rivera, C.; Voipio, J.; Kaila, K. Cation-chloride co-transporters in neuronal communication, development and trauma. *Trends Neurosci.* 2003, 26, 199–206.

[15] Rivera, C.; Voipio, J.; Kaila, K. Two developmental switches in GABAergic signalling: the K+-Cl- cotransporter KCC2 and carbonic anhydrase CAVII. *J. Physiol.* 2005, 562, 27–36.

[16] Thompson, S. M.; Gähwiler, B. H. Activity-dependent disinhibition. II. Effects of extracellular potassium, furosemide, and membrane potential on ECl- in hippocampal CA3 neurons. *J. Neurophysiol.* 1989, 61, 512–523.

[17] Rivera, C.; Voipio, J.; Payne, J. A.; Ruusuvuori, E.; Lahtinen, H.; Lamsa, K.; Pirvola, U.; Saarma, M.; Kaila, K. The K+/Cl- co-transporter KCC2 renders GABA hyperpolarizing during neuronal maturation. *Nature* 1999, 397, 251–255.

[18] Kuffler, S. W. Excitation and inhibition in single nerve cells. *Harvey Lect.* 1960, 54, 176–218.

[19] McCormick, D. A. GABA as an inhibitory neurotransmitter in human cerebral cortex. *J. Neurophysiol.* 1989, 62, 1018–1027.

[20] Ben-Ari, Y.; Cherubini, E.; Corradetti, R.; Gaiarsa, J. L. Giant synaptic potentials in immature rat CA3 hippocampal neurones. *J. Physiol.* 1989, 416, 303–325.

[21] Chen, G.; Trombley, P. Q.; van den Pol, A. N. Excitatory actions of GABA in developing rat hypothalamic neurones. *J. Physiol.* 1996, 494, 451–464.

[22] Owens, D. F.; Boyce, L. H.; Davis, M. B.; Kriegstein, A. R. Excitatory GABA responses in embryonic and neonatal cortical slices demonstrated by gramicidin perforated-patch recordings and calcium imaging. *J. Neurosci.* 1996, 16, 6414–6423.

[23] Wang, Y. F.; Gao, X. B.; van den Pol, A. N. Membrane properties underlying patterns of GABA-dependent action potentials in developing mouse hypothalamic neurons. *J. Neurophysiol.* 2001, 86, 1252–1265.

[24] Delpire, E. Cation-Chloride Cotransporters in Neuronal Communication. *News Physiol. Sci.* 2000, 15, 309–312.

[25] Alger, B. E.; Nicoll, R. A. Feed-forward dendritic inhibition in rat hippocampal pyramidal cells studied in vitro. *J. Physiol.* 1982, 328, 105–123.

[26] Huguenard, J. R.; Alger, B. E. Whole-cell voltage-clamp study of the fading of GABA-activated currents in acutely dissociated hippocampal neurons. *J. Neurophysiol.* 1986, 56, 1–18.

[27] Perreault, P.; Avoli, M. 4-aminopyridine-induced epileptiform activity and a GABA-mediated long-lasting depolarization in the rat hippocampus. *J. Neurosci.* 1992, 12, 104–115.

[28] Thompson, S. M.; Gähwiler, B. H. Activity-dependent disinhibition. I. Repetitive stimulation reduces IPSP driving force and conductance in the hippocampus in vitro. *J. Neurophysiol.* 1989, 61, 501–511.
[29] Michelson, H. B.; Wong, R. K. Excitatory synaptic responses mediated by GABAA receptors in the hippocampus. *Science* 1991, 253, 1420–1423.
[30] Staley, K. J.; Soldo, B. L.; Proctor, W. R. Ionic mechanisms of neuronal excitation by inhibitory GABAA receptors. *Science* 1995, 269, 977–981.
[31] Kaila, K.; Lamsa, K.; Smirnov, S.; Taira, T.; Voipio, J. Long-lasting GABA-mediated depolarization evoked by high-frequency stimulation in pyramidal neurons of rat hippocampal slice is attributable to a network-driven, bicarbonate-dependent K+ transient. *J. Neurosci.* 1997, 17, 7662–7672.
[32] Mody, I.; De Koninck, Y.; Otis, T. S.; Soltesz, I. Bridging the cleft at GABA synapses in the brain. *Trends Neurosci.* 1994, 17, 517–525.
[33] Iversen, L. L.; Neal, M. J. The uptake of [3H]GABA by slices of rat cerebral cortex. *J. Neurochem.* 1968, 15, 1141–1149.
[34] Conti, F.; Minelli, A.; Melone, M. GABA transporters in the mammalian cerebral cortex: Localization, development and pathological implications. *Brain Res. Rev.* 2004, 45, 196–212.
[35] Edwards, F. A.; Konnerth, A.; Sakmann, B. Quantal analysis of inhibitory synaptic transmission in the dentate gyrus of rat hippocampal slices: a patch-clamp study. *J. Physiol.* 1990, 430, 213–249.
[36] Nusser, Z.; Cull-Candy, S.; Farrant, M. Differences in synaptic GABA(A) receptor number underlie variation in GABA mini amplitude. *Neuron* 1997, 19, 697–709.
[37] Nusser, Z.; Sieghart, W.; Somogyi, P. Segregation of different GABAA receptors to synaptic and extrasynaptic membranes of cerebellar granule cells. *J. Neurosci.* 1998, 18, 1693–1703.
[38] Wei, W.; Zhang, N.; Peng, Z.; Houser, C. R.; Mody, I. Perisynaptic localization of delta subunit-containing GABA(A) receptors and their activation by GABA spillover in the mouse dentate gyrus. *J. Neurosci.* 2003, 23, 10650–10661.
[39] Lerma, J.; Herranz, A. S.; Herreras, O.; Abraira, V.; Martín del Río, R. In vivo determination of extracellular concentration of amino acids in the rat hippocampus. A method based on brain dialysis and computerized analysis. *Brain Res.* 1986, 384, 145–155.
[40] Attwell, D.; Barbour, B.; Szatkowski, M. Nonvesicular release of neurotransmitter. *Neuron* 1993, 11, 401–407.
[41] Kennedy, R. T.; Thompson, J. E.; Vickroy, T. W. In vivo monitoring of amino acids by direct sampling of brain extracellular fluid at ultralow flow rates and capillary electrophoresis. *J. Neurosci. Methods* 2002, 114, 39–49.
[42] McCartney, M. R.; Deeb, T. Z.; Henderson, T. N.; Hales, T. G. Tonically active GABAA receptors in hippocampal pyramidal neurons exhibit constitutive GABA-independent gating. *Mol. Pharmacol.* 2007, 71, 539–548.
[43] Farrant, M.; Nusser, Z. Variations on an inhibitory theme: phasic and tonic activation of GABA(A) receptors. *Nat. Rev. Neurosci.* 2005, 6, 215–229.
[44] Glykys, J.; Mody, I. Activation of GABAA receptors: views from outside the synaptic cleft. *Neuron* 2007, 56, 763–770.

[45] Kimelberg, H. K.; Goderie, S. K.; Higman, S.; Pang, S.; Waniewski, R. A. Swelling-induced release of glutamate, aspartate, and taurine from astrocyte cultures. *J. Neurosci.* 1990, 10, 1583–1591.

[46] Liu, Q. Y.; Schaffner, A. E.; Chang, Y. H.; Maric, D.; Barker, J. L. Persistent activation of GABA(A) receptor/Cl(-) channels by astrocyte-derived GABA in cultured embryonic rat hippocampal neurons. *J. Neurophysiol.* 2000, 84, 1392–1403.

[47] Wang, C. M.; Chang, Y. Y.; Kuo, J. S.; Sun, S. H. Activation of P2x7 receptors induced [3H]GABA release from the RBA-2 type-2 astrocyte cell line through a Cl-/HCO3 - -dependent mechanism. *Glia* 2002, 37, 8–18.

[48] Rossi, D. J.; Hamann, M.; Attwell, D. Multiple modes of GABAergic inhibition of rat cerebellar granule cells. *J. Physiol.* 2003, 548, 97–110.

[49] Lee, S.; Yoon, B.-E.; Berglund, K.; Oh, S.-J.; Park, H.; Shin, H.-S.; Augustine, G. J.; Lee, C. J. Channel-mediated tonic GABA release from glia. *Science* 2010, 330, 790–796.

[50] Glykys, J.; Mody, I. The main source of ambient GABA responsible for tonic inhibition in the mouse hippocampus. *J. Physiol.* 2007, 582, 1163–1178.

[51] Semyanov, A.; Walker, M. C.; Kullmann, D. M.; Silver, R. A. Tonically active GABA A receptors: modulating gain and maintaining the tone. *Trends Neurosci.* 2004, 27, 262–269.

[52] Brickley, S. G.; Cull-Candy, S. G.; Farrant, M. Development of a tonic form of synaptic inhibition in rat cerebellar granule cells resulting from persistent activation of GABAA receptors. *J. Physiol.* 1996, 497, 753–759.

[53] Holt, G. R.; Koch, C. Shunting inhibition does not have a divisive effect on firing rates. *Neural Comput.* 1997, 9, 1001–1013.

[54] Mitchell, S. J.; Silver, R. A. Shunting inhibition modulates neuronal gain during synaptic excitation. *Neuron* 2003, 38, 433–445.

[55] Chance, F. S.; Abbott, L. F.; Reyes, A. D. Gain modulation from background synaptic input. *Neuron* 2002, 35, 773–782.

[56] Mody, I.; Pearce, R. A. Diversity of inhibitory neurotransmission through GABA(A) receptors. *Trends Neurosci.* 2004, 27, 569–575.

[57] Phillis, J. W. Inhibitory action of CGS 21680 on cerebral cortical neurons is antagonized by bicuculline and picrotoxin - Is GABA involved? *Brain Res.* 1998, 807, 193–198.

[58] O'Regan, M. H.; Simpson, R. E.; Perkins, L. M.; Phillis, J. W. Adenosine receptor agonists inhibit the release of gamma-aminobutyric acid (GABA) from the ischemic rat cerebral cortex. *Brain Res.* 1992, 582, 22–26.

[59] Saransaari, P.; Oja, S. S. GABA release modified by adenosine receptors in mouse hippocampal slices under normal and ischemic conditions. *Neurochem. Res.* 2005, 30, 467–473.

[60] Cunha, R. A.; Ribeiro, J. A. Purinergic modulation of [3H]GABA release from rat hippocampal nerve terminals. *Neuropharmacology* 2000, 39, 1156–1167.

[61] Dolphin, A. C.; Archer, E. R. An adenosine agonist inhibits and a cyclic AMP analogue enhances the release of glutamate but not GABA from slices of rat dentate gyrus. *Neurosci. Lett.* 1983, 43, 49–54.

[62] Kamiya, H. Some pharmacological differences between hippocampal excitatory and inhibitory synapses in transmitter release: an in vitro study. *Synapse* 1991, 8, 229–235.

[63] Lambert, N. A.; Teyler, T. J. Adenosine depresses excitatory but not fast inhibitory synaptic transmission in area CA1 of the rat hippocampus. *Neurosci. Lett.* 1991, 122, 50–52.

[64] Yoon, K. W.; Rothman, S. M. Adenosine inhibits excitatory but not inhibitory synaptic transmission in the hippocampus. *J. Neurosci.* 1991, 11, 1375–1380.

[65] Rombo, D. M.; Dias, R. B.; Duarte, S. T.; Ribeiro, J. A.; Lamsa, K. P.; Sebastiao, A. M. Adenosine A1 Receptor Suppresses Tonic GABAA Receptor Currents in Hippocampal Pyramidal Cells and in a Defined Subpopulation of Interneurons. *Cereb. Cortex* 2014, Dec. 1. pii: bhu288. [Epub. ahead of print].

[66] Heron, A.; Lasbennes, F.; Seylaz, J. Adenosine modulation of amino acid release in rat hippocampus during ischemia and veratridine depolarization. *Brain Res.* 1993, 608, 27–32.

[67] Burke, S. P.; Nadler, J. V. Regulation of glutamate and aspartate release from slices of the hippocampal CA1 area: effects of adenosine and baclofen. *J. Neurochem.* 1988, 51, 1541–1551.

[68] Sousa, V. C.; Assaife-Lopes, N.; Ribeiro, J. A.; Pratt, J. A.; Brett, R. R.; Sebastião, A. M. Regulation of hippocampal cannabinoid CB1 receptor actions by adenosine A1 receptors and chronic caffeine administration: implications for the effects of Δ9-tetrahydrocannabinol on spatial memory. *Neuropsychopharmacology* 2011, 36, 472–487.

[69] Panlilio, L. V.; Ferré, S.; Yasar, S.; Thorndike, E. B.; Schindler, C. W.; Goldberg, S. R. Combined effects of THC and caffeine on working memory in rats. *Br. J. Pharmacol.* 2012, 165, 2529–2538.

[70] Cunha-Reis, D.; Ribeiro, J. A.; Sebastião, A. M. A1 and A2A receptor activation by endogenous adenosine is required for VIP enhancement of K+ -evoked [3H]-GABA release from rat hippocampal nerve terminals. *Neurosci. Lett.* 2008, 430, 207–212.

[71] Jeong, H.-J.; Jang, I.-S.; Nabekura, J.; Akaike, N. Adenosine A1 receptor-mediated presynaptic inhibition of GABAergic transmission in immature rat hippocampal CA1 neurons. *J. Neurophysiol.* 2003, 89, 1214–1222.

[72] Kirmse, K.; Dvorzhak, A.; Grantyn, R.; Kirischuk, S. Developmental downregulation of excitatory GABAergic transmission in neocortical layer I via presynaptic adenosine A1 receptors. *Cereb. Cortex* 2008, 18, 424–432.

[73] Silva, C. G.; Métin, C.; Fazeli, W.; Machado, N. J.; Darmopil, S.; Launay, P.-S.; Ghestem, A.; Nesa, M.-P.; Bassot, E.; Szabó, E.; Baqi, Y.; Müller, C. E.; Tomé, A. R.; Ivanov, A.; Isbrandt, D.; Zilberter, Y.; Cunha, R. A.; Esclapez, M.; Bernard, C. Adenosine receptor antagonists including caffeine alter fetal brain development in mice. *Sci. Transl. Med.* 2013, 5, 197ra104.

[74] Li, Y.; Fan, S.; Yan, J.; Li, B.; Chen, F.; Xia, J.; Yu, Z.; Hu, Z. Adenosine modulates the excitability of layer II stellate neurons in entorhinal cortex through A1 receptors. *Hippocampus* 2011, 21, 265–280.

[75] Heinbockel, T.; Pape, H. C. Modulatory effects of adenosine on inhibitory postsynaptic potentials in the lateral amygdala of the rat. *Br. J. Pharmacol.* 1999, 128, 190–196.

[76] Klausberger, T.; Marton, L. F.; O'Neill, J.; Huck, J. H. J.; Dalezios, Y.; Fuentealba, P.; Suen, W. Y.; Papp, E.; Kaneko, T.; Watanabe, M.; Csicsvari, J.; Somogyi, P. Complementary roles of cholecystokinin- and parvalbumin-expressing GABAergic neurons in hippocampal network oscillations. *J. Neurosci.* 2005, 25, 9782–9793.

[77] Rombo, D. M.; Newton, K.; Nissen, W.; Badurek, S.; Horn, J.; Minichiello, L.; Jefferys, J. G. R.; Sebastiao, A. M.; Lamsa, K. Synaptic mechanisms of adenosine A2A receptor-mediated hyperexcitability in the hippocampus. *Hippocampus* 2014, Nov. 17. doi: 10.1002/hipo. 22392 [Epub. ahead of print].

[78] Latini, S.; Pedata, F. Adenosine in the central nervous system: release mechanisms and extracellular concentrations. *J. Neurochem.* 2001, 79, 463–484.

[79] Pearson, T.; Nuritova, F.; Caldwell, D.; Dale, N.; Frenguelli, B. G. A depletable pool of adenosine in area CA1 of the rat hippocampus. *J. Neurosci.* 2001, 21, 2298–2307.

[80] Pinto-Duarte, A.; Coelho, J. E.; Cunha, R. A.; Ribeiro, J. A.; Sebastião, A. M. Adenosine A2A receptors control the extracellular levels of adenosine through modulation of nucleoside transporters activity in the rat hippocampus. *J. Neurochem.* 2005, 93, 595–604.

[81] Burnstock, G.; Krügel, U.; Abbracchio, M. P.; Illes, P. Purinergic signalling: From normal behaviour to pathological brain function. *Prog. Neurobiol.* 2011, 95, 229–274.

[82] Dias, R. B.; Rombo, D. M.; Ribeiro, J. A.; Henley, J. M.; Sebastião, A. M. Adenosine: setting the stage for plasticity. *Trends Neurosci.* 2013, 36, 248–257.

[83] Liang, R.; Pang, Z. P.; Deng, P.; Xu, Z. C. Transient enhancement of inhibitory synaptic transmission in hippocampal CA1 pyramidal neurons after cerebral ischemia. *Neuroscience* 2009, 160, 412–418.

[84] Sebastião, A. M.; de Mendonça, A.; Moreira, T.; Ribeiro, J. A. Activation of synaptic NMDA receptors by action potential-dependent release of transmitter during hypoxia impairs recovery of synaptic transmission on reoxygenation. *J. Neurosci.* 2001, 21, 6561–6571.

[85] Concas, A.; Santoro, G.; Mascia, M. P.; Maciocco, E.; Dazzi, L.; Ongini, E.; Biggio, G. Anticonvulsant doses of 2-chloro-N6-cyclopentyladenosine, an adenosine A1 receptor agonist, reduce GABAergic transmission in different areas of the mouse brain. *J. Pharmacol. Exp. Ther.* 1993, 267, 844–851.

[86] Lopez, F.; Miller, L. G.; Greenblatt, D. J.; Kaplan, G. B.; Shader, R. I. Interaction of caffeine with the GABAA receptor complex: alterations in receptor function but not ligand binding. *Eur. J. Pharmacol.* 1989, 172, 453–459.

[87] Akhondzadeh, S.; Stone, T. W. Interaction between adenosine and GABAA receptors on hippocampal neurones. *Brain Res.* 1994, 665, 229–236.

[88] Ilie, A.; Raimondo, J. V; Akerman, C. J. Adenosine release during seizures attenuates GABAA receptor-mediated depolarization. *J. Neurosci.* 2012, 32, 5321–5332.

[89] Roseti, C.; Martinello, K.; Fucile, S.; Piccari, V.; Mascia, A.; Di Gennaro, G.; Quarato, P. P.; Manfredi, M.; Esposito, V.; Cantore, G.; Arcella, A.; Simonato, M.; Fredholm, B. B.; Limatola, C.; Miledi, R.; Eusebi, F. Adenosine receptor antagonists alter the stability of human epileptic GABAA receptors. *Proc. Natl. Acad. Sci. US* 2008, 105, 15118–15123.

[90] Roseti, C.; Palma, E.; Martinello, K.; Fucile, S.; Morace, R.; Esposito, V.; Cantore, G.; Arcella, A.; Giangaspero, F.; Aronica, E.; Mascia, A.; Di Gennaro, G.; Quarato, P. P.; Manfredi, M.; Cristalli, G.; Lambertucci, C.; Marucci, G.; Volpini, R.; Limatola, C.; Eusebi, F. Blockage of A2A and A3 adenosine receptors decreases the desensitization of human GABA(A) receptors microtransplanted to Xenopus oocytes. *Proc. Natl. Acad. Sci. US* 2009, 106, 15927–15931.

[91] Diógenes, M. J.; Neves-Tomé, R.; Fucile, S.; Martinello, K.; Scianni, M.; Theofilas, P.; Lopatár, J.; Ribeiro, J. A.; Maggi, L.; Frenguelli, B. G.; Limatola, C.; Boison, D.; Sebastião, A. M. Homeostatic control of synaptic activity by endogenous adenosine is mediated by adenosine kinase. *Cereb. Cortex* 2014, 24, 67–80.

[92] Schousboe, A.; Sarup, A.; Larsson, O. M.; White, H. S. GABA transporters as drug targets for modulation of GABAergic activity. *Biochem. Pharmacol.* 2004, 68, 1557–1563.

[93] Cristóvão-Ferreira, S.; Vaz, S. H.; Ribeiro, J. A.; Sebastião, A. M. Adenosine A2A receptors enhance GABA transport into nerve terminals by restraining PKC inhibition of GAT-1. *J. Neurochem.* 2009, 109, 336–347.

[94] Cristóvão-Ferreira, S.; Navarro, G.; Brugarolas, M.; Pérez-Capote, K.; Vaz, S. H.; Fattorini, G.; Conti, F.; Lluis, C.; Ribeiro, J. A.; McCormick, P. J.; Casadó, V.; Franco, R.; Sebastião, A. M. A1R-A2AR heteromers coupled to Gs and G i/0 proteins modulate GABA transport into astrocytes. *Purinergic Signal.* 2013, 9, 433–449.

[95] Vaz, S. H.; Jørgensen, T. N.; Cristóvão-Ferreira, S.; Duflot, S.; Ribeiro, J. A.; Gether, U.; Sebastião, A. M., Brain-derived neurotrophic factor (BDNF) enhances gaba transport by modulating the trafficking of GABA transporter-1 (GAT-1) from the plasma membrane of rat cortical astrocytes. *J. Biol. Chem.* 2011, 286, 40464–40476.

[96] Cunha, R. A.; Vizi, E. S.; Ribeiro, J. A.; Sebastiao, A. M. Preferential release of ATP and its extracellular catabolism as a source of adenosine upon high- but not low-frequency stimulation of rat hippocampal slices. *J. Neurochem.* 1996, 67, 2180–2187.

[97] Lovatt, D.; Xu, Q.; Liu, W.; Takano, T.; Smith, N. A.; Schnermann, J.; Tieu, K.; Nedergaard, M. Neuronal adenosine release, and not astrocytic ATP release, mediates feedback inhibition of excitatory activity. *Proc. Natl. Acad. Sci. US* 2012, 109, 6265–6270.

[98] Lee, H. U.; Yamazaki, Y.; Tanaka, K. F.; Furuya, K.; Sokabe, M.; Hida, H.; Takao, K.; Miyakawa, T.; Fujii, S.; Ikenaka, K. Increased astrocytic ATP release results in enhanced excitability of the hippocampus. *Glia* 2013, 61, 210–224.

[99] Jacob, P. F.; Vaz, S. H.; Ribeiro, J. A.; Sebastião, A. M. P2Y1 receptor inhibits GABA transport through a calcium signalling-dependent mechanism in rat cortical astrocytes. *Glia* 2014, 62, 1211–1226.

[100] Gonzalez, B.; Paz, F.; Florán, L.; Aceves, J.; Erlij, D.; Florán, B. Adenosine A2A receptor stimulation decreases GAT-1-mediated GABA uptake in the globus pallidus of the rat. *Neuropharmacology* 2006, 51, 154–159.

[101] Fragata, I. R.; Ribeiro, J. A.; Sebastião, A. M. Nitric oxide mediates interactions between GABAA receptors and adenosine A1 receptors in the rat hippocampus. *Eur. J. Pharmacol.* 2006, 543, 32–39.

[102] Lucchi, R.; Latini, S.; de Mendonça, A.; Sebastião, A. M.; Ribeiro, J. A. Adenosine by activating A1 receptors prevents GABA(A)-mediated actions during hypoxia in the rat hippocampus. *Brain Res.* 1996, 732, 261–266.

[103] Santos, P. F.; Caramelo, O. L.; Carvalho, A. P.; Duarte, C. B. Adenosine A1 receptors inhibit Ca2+ channels coupled to the release of ACh, but not of GABA, in cultured retina cells. *Brain Res.* 2000, 852, 10–15.

[104] Ferreira, D. D. P.; Stutz, B.; de Mello, F. G.; Reis, R. A. M.; Kubrusly, R. C. C. Caffeine potentiates the release of GABA mediated by NMDA receptor activation: Involvement of A1 adenosine receptors. *Neuroscience* 2014, 281, 208–215.

[105] Courjaret, R.; Tröger, M.; Deitmer, J. W. Suppression of GABA input by A1 adenosine receptor activation in rat cerebellar granule cells. *Neuroscience* 2009, 162, 946–958.

[106] Klyuch, B. P.; Dale, N.; Wall, M. J. Receptor-mediated modulation of activity-dependent adenosine release in rat cerebellum. *Neuropharmacology* 2012, 62, 815–824.

[107] Delicado, E. G.; Rodrigues, A.; Sen, R. P.; Sebastiao, A. M.; Ribeiro, J. A.; Miras-Portugal, M. T. Effect of 5'-(N-ethylcarboxamido)adenosine on adenosine transport in cultured chromaffin cells. *J. Neurochem.* 1990, 54, 1941–1946.

[108] Augood, S. J.; Emson, P. C. Adenosine A2a receptor mRNA is expressed by enkephalin cells but not by somatostatin cells in rat striatum: a co-expression study. *Brain Res. Mol. Brain Res.* 1994, 22, 204–210.

[109] Ferré, S.; O'Connor, W. T.; Fuxe, K.; Ungerstedt, U. The striopallidal neuron: a main locus for adenosine-dopamine interactions in the brain. *J. Neurosci.* 1993, 13, 5402–5406.

[110] Schwarzschild, M. A.; Agnati, L.; Fuxe, K.; Chen, J.-F.; Morelli, M. Targeting adenosine A2A receptors in Parkinson's disease. *Trends Neurosci.* 2006, 29, 647–654.

[111] Calabresi, P.; Di Filippo, M.; Gallina, A.; Wang, Y.; Stankowski, J. N.; Picconi, B.; Dawson, V. L.; Dawson, T. M. New synaptic and molecular targets for neuroprotection in Parkinson's disease. *Mov. Disord.* 2013, 28, 51–60.

[112] Sebastião, A. M.; Ribeiro, J. A. Adenosine receptors and the central nervous system. *Handb. Exp. Pharmacol.* 2009, 193, 471–534.

[113] Corsi, C.; Melani, A.; Bianchi, L.; Pepeu, G.; Pedata, F. Effect of adenosine A2A receptor stimulation on GABA release from the striatum of young and aged rats in vivo. *Neuroreport* 1999, 10, 3933–3937.

[114] Mayfield, R. D.; Larson, G.; Orona, R. A.; Zahniser, N. R. Opposing actions of adenosine A2a and dopamine D2 receptor activation on GABA release in the basal ganglia: evidence for an A2a/D2 receptor interaction in globus pallidus. *Synapse* 1996, 22, 132–138.

[115] Ochi, M.; Koga, K.; Kurokawa, M.; Kase, H.; Nakamura, J.; Kuwana, Y. Systemic administration of adenosine A(2A) receptor antagonist reverses increased GABA release in the globus pallidus of unilateral 6-hydroxydopamine-lesioned rats: A microdialysis study. *Neuroscience* 2000, 100, 53–62.

[116] Díaz-Cabiale, Z.; Vivó, M.; Del Arco, A.; O'Connor, W. T.; Harte, M. K.; Müller, C. E.; Martínez, E.; Popoli, P.; Fuxe, K.; Ferré, S. Metabotropic glutamate mGlu5 receptor-mediated modulation of the ventral striopallidal GABA pathway in rats. Interactions with adenosine A2A and dopamine D2 receptors. *Neurosci. Lett.* 2002, 324, 154–158.

[117] Hettinger, B. D.; Lee, A.; Linden, J.; Rosin, D. L. Ultrastructural localization of adenosine A2A receptors suggests multiple cellular sites for modulation of GABAergic neurons in rat striatum. *J. Comp. Neurol.* 2001, 431, 331–346.

[118] Mori, A.; Shindou, T.; Ichimura, M.; Nonaka, H.; Kase, H. The role of adenosine A2a receptors in regulating GABAergic synaptic transmission in striatal medium spiny neurons. *J. Neurosci.* 1996, 16, 605–611.

[119] Wirkner, K.; Gerevich, Z.; Krause, T.; Günther, A.; Köles, L.; Schneider, D.; Nörenberg, W.; Illes, P. Adenosine A2A receptor induced inhibition of NMDA and GABAA receptor-mediated synaptic currents in a subpopulation of rat striatal neurons. *Neuropharmacology* 2004, 46, 994–1007.

[120] Kirk, I. P.; Richardson, P. J. Adenosine A2a receptor-mediated modulation of striatal [3H]GABA and [3H]acetylcholine release. *J. Neurochem.* 1994, 62, 960–966.

[121] Kurokawa, M.; Kirk, I. P.; Kirkpatrick, K. A.; Kase, H.; Richardson, P. J. Inhibition by KF17837 of adenosine A2A receptor-mediated modulation of striatal GABA and ACh release. *Br. J. Pharmacol.* 1994, 113, 43–48.

[122] Mayfield, R. D.; Suzuki, F.; Zahniser, N. R. Adenosine A2a receptor modulation of electrically evoked endogenous GABA release from slices of rat globus pallidus. *J. Neurochem.* 1993, 60, 2334–2337.

[123] Shindou, T.; Mori, A.; Kase, H.; Ichimura, M. Adenosine A(2A) receptor enhances GABA(A)-mediated IPSCs in the rat globus pallidus. *J. Physiol.* 2001, 532, 423–434.

[124] Shindou, T.; Richardson, P. J.; Mori, A.; Kase, H.; Ichimura, M. Adenosine modulates the striatal GABAergic inputs to the globus pallidus via adenosine A2A receptors in rats. *Neurosci. Lett.* 2003, 352, 167–170.

[125] Ochi, M.; Shiozaki, S.; Kase, H. Adenosine A2A receptor-mediated modulation of GABA and glutamate release in the output regions of the basal ganglia in a rodent model of Parkinson's disease. *Neuroscience* 2004, 127, 223–231.

[126] Zahniser, N. R.; Simosky, J. K.; Mayfield, R. D.; Negri, C. A.; Hanania, T.; Larson, G. A.; Kelly, M. A.; Grandy, D. K.; Rubinstein, M.; Low, M. J.; Fredholm, B. B. Functional uncoupling of adenosine A(2A) receptors and reduced responseto caffeine in mice lacking dopamine D2 receptors. *J. Neurosci.* 2000, 20, 5949–5957.

[127] Florán, B.; Gonzalez, B.; Florán, L.; Erlij, D.; Aceves, J. Interactions between adenosine A2a and dopamine D2 receptors in the control of [3H]GABA release in the globus pallidus of the rat. *Eur. J. Pharmacol.* 2005, 520, 43–50.

[128] Morales-Figueroa, G.; Márquez-Gómez, R.; González-Pantoja, R.; Escamilla-Sánchez, J.; Arias-Montaño, J. Histamine H3 receptor activation counteracts adenosine A2A receptor-mediated enhancement of depolarization-evoked [3H]-GABA release from rat globus pallidus synaptosomes. *ACS Chem. Neurosci.* 2014, 5, 637–645.

[129] Gomes, C. A. R. V.; Vaz, S. H.; Ribeiro, J. A.; Sebastião, A. M. Glial cell line-derived neurotrophic factor (GDNF) enhances dopamine release from striatal nerve endings in an adenosine A2A receptor-dependent manner. *Brain Res.* 2006, 1113, 129–136.

[130] Gomes, C. A. R. V.; Simões, P. F.; Canas, P. M.; Quiroz, C.; Sebastião, A. M.; Ferré, S.; Cunha, R. A.; Ribeiro, J. A. GDNF control of the glutamatergic cortico-striatal pathway requires tonic activation of adenosine A2A receptors. *J. Neurochem.* 2009, 108, 1208–1219.

[131] Florán, B.; Barajas, C.; Florán, L.; Erlij, D.; Aceves, J. Adenosine A1 receptors control dopamine D1-dependent [3H]GABA release in slices of substantia nigra pars reticulata and motor behavior in the rat. *Neuroscience* 2002, 115, 743–751.

[132] Mayfield, R. D.; Jones, B. A.; Miller, H. A.; Simosky, J. K.; Larson, G. A.; Zahniser, N. R. Modulation of endogenous GABA release by an antagonistic adenosine A1/ dopamine D1 receptor interaction in rat brain limbic regions but not basal ganglia. *Synapse* 1999, 33, 274–281.

[133] Ferre, S.; O'Connor, W. T.; Svenningsson, P.; Bjorklund, L.; Lindberg, J.; Tinner, B.; Stromberg, I.; Goldstein, M.; Ogren, S. O.; Ungerstedt, U.; Fredholm, B. B.; Fuxe, K. Dopamine D1 receptor-mediated facilitation of GABAergic neurotransmission in the rat strioentopenduncular pathway and its modulation by adenosine A1 receptor-mediated mechanisms. *Eur. J. Neurosci.* 1996, 8, 1545–1553.

[134] Ulrich, D.; Huguenard, J. R. Purinergic inhibition of GABA and glutamate release in the thalamus: implications for thalamic network activity. *Neuron* 1995, 15, 909–918.

[135] Shen, K. Z.; Johnson, S. W. Presynaptic GABAB and adenosine A1 receptors regulate synaptic transmission to rat substantia nigra reticulata neurones. *J. Physiol.* 1997, 505, 153–163.

[136] Mango, D.; Bonito-Oliva, A.; Ledonne, A.; Cappellacci, L.; Petrelli, R.; Nisticò, R.; Berretta, N.; Fisone, G.; Mercuri, N. B. Adenosine A1 receptor stimulation reduces D1 receptor-mediated GABAergic transmission from striato-nigral terminals and attenuates l-DOPA-induced dyskinesia in dopamine-denervated mice. *Exp. Neurol.* 2014, 261, 733–743.

[137] Shen, K. Z.; Johnson, S. W. Presynaptic inhibition of synaptic transmission by adenosine in rat subthalamic nucleus in vitro. *Neuroscience* 2003, 116, 99–106.

[138] Centonze, D.; Saulle, E.; Pisani, A.; Bernardi, G.; Calabresi, P. Adenosine-mediated inhibition of striatal GABAergic synaptic transmission during in vitro ischaemia. *Brain* 2001, 124, 1855–1865.

[139] Goda, H.; Ooboshi, H.; Nakane, H.; Ibayashi, S.; Sadoshima, S.; Fujishima, M. Modulation of ischemia-evoked release of excitatory and inhibitory amino acids by adenosine A1 receptor agonist. *Eur. J. Pharmacol.* 1998, 357, 149–155.

[140] Uchimura, N.; North, R. A. Baclofen and adenosine inhibit synaptic potentials mediated by gamma-aminobutyric acid and glutamate release in rat nucleus accumbens. *J. Pharmacol. Exp. Ther.* 1991, 258, 663–668.

[141] Harper, L. K.; Beckett, S. R.; Marsden, C. A.; McCreary, A. C.; Alexander, S. P. H. Effects of the A 2A adenosine receptor antagonist KW6002 in the nucleus accumbens in vitro and in vivo. *Pharmacol. Biochem. Behav.* 2006, 83, 114–121.

[142] Wu, Y. N.; Mercuri, N. B.; Johnson, S. W. Presynaptic inhibition of gamma-aminobutyric acidB-mediated synaptic current by adenosine recorded in vitro in midbrain dopamine neurons. *J. Pharmacol. Exp. Ther.* 1995, 273, 576–81.

[143] Williams, C. L.; Buchta, W. C.; Riegel, A. C. CRF-R2 and the Heterosynaptic Regulation of VTA Glutamate during Reinstatement of Cocaine Seeking. *J. Neurosci.* 2014, 34, 10402–10414.

[144] Wydra, K.; Gołembiowska, K.; Suder, A.; Kamińska, K.; Fuxe, K.; Filip, M. On the role of adenosine (A)2A receptors in cocaine-induced reward: a pharmacological and neurochemical analysis in rats. *Psychopharmacology* (Berl.). 2015, 232, 421–435.

[145] Hack, S. P.; Vaughan, C. W.; Christie, M. J. Modulation of GABA release during morphine withdrawal in midbrain neurons in vitro. *Neuropharmacology* 2003, 45, 575–584.

[146] Huang, Z. L.; Urade, Y.; Hayaishi, O. The role of adenosine in the regulation of sleep. *Curr. Top* 2011, 11, 1047–1057.

[147] Porkka-Heiskanen, T. Methylxanthines and sleep. *Handb. Exp. Pharmacol.* 2011, 200, 331–348.

[148] Lazarus, M.; Chen, J.-F.; Urade, Y.; Huang, Z.-L. Role of the basal ganglia in the control of sleep and wakefulness. *Curr. Opin. Neurobiol.* 2013, 23, 780–785.

[149] Yum, D.-S.; Cho, J.-H.; Choi, I.-S.; Nakamura, M.; Lee, J.-J.; Lee, M.-G.; Choi, B.-J.; Choi, J.-K.; Jang, I.-S. Adenosine A1 receptors inhibit GABAergic transmission in rat tuberomammillary nucleus neurons. *J. Neurochem.* 2008, 106, 361–371.

[150] Sherin, J. E.; Elmquist, J. K.; Torrealba, F.; Saper, C. B. Innervation of histaminergic tuberomammillary neurons by GABAergic and galaninergic neurons in the ventrolateral preoptic nucleus of the rat. *J. Neurosci.* 1998, 18, 4705–4721.

[151] Mendelson, W. B. Sleep-inducing effects of adenosine microinjections into the medial preoptic area are blocked by flumazenil. *Brain Res.* 2000, 852, 479–481.

[152] Gallopin, T.; Luppi, P. H.; Cauli, B.; Urade, Y.; Rossier, J.; Hayaishi, O.; Lambolez, B.; Fort, P. The endogenous somnogen adenosine excites a subset of sleep-promoting neurons via A2A receptors in the ventrolateral preoptic nucleus. *Neuroscience* 2005, 134, 1377–1390.

[153] Xia, J. X.; Xiong, J. X.; Wang, H. K.; Duan, S. M.; Ye, J. N.; Hu, Z. A. Presynaptic inhibition of GABAergic synaptic transmission by adenosine in mouse hypothalamic hypocretin neurons. *Neuroscience* 2012, 201, 46–56.

[154] Morairty, S.; Rainnie, D.; McCarley, R.; Greene, R. Disinhibition of ventrolateral preoptic area sleep-active neurons by adenosine: A new mechanism for sleep promotion. *Neuroscience* 2004, 123, 451–457.

[155] Hong, Z. Y.; Huang, Z. L.; Qu, W. M.; Eguchi, N.; Urade, Y.; Hayaishi, O. An adenosine A2A receptor agonist induces sleep by increasing GABA release in the tuberomammillary nucleus to inhibit histaminergic systems in rats. *J. Neurochem.* 2005, 92, 1542–1549.

[156] Chen, G.; van den Pol, A. N. Adenosine modulation of calcium currents and presynaptic inhibition of GABA release in suprachiasmatic and arcuate nucleus neurons. *J. Neurophysiol.* 1997, 77, 3035–3047.

[157] Sabetkasai, M.; Zarrindast, M. R. Antinociception: interaction between adenosine and GABA systems. *Arch. Int. Pharmacodyn. Ther.* 1993, 322, 14–22.

[158] Yang, K.; Fujita, T.; Kumamoto, E. Adenosine inhibits GABAergic and glycinergic transmission in adult rat substantia gelatinosa neurons. *J. Neurophysiol.* 2004, 92, 2867–2877.

[159] Li, H.; Wu, L.; Li, Y. Q. Adenosine suppresses GABAA receptor-mediated responses in rat sacral dorsal commissural neurons. *Auton. Neurosci. Basic Clin.* 2004, 111, 71–79.

[160] Wu, L.; Li, H.; Li, Y. Q. Adenosine suppresses the response of neurons to gaba in the superficial laminae of the rat spinal dorsal horn. *Neuroscience* 2003, 119, 145–154.

[161] Hugel, S.; Schlichter, R. Convergent control of synaptic GABA release from rat dorsal horn neurones by adenosine and GABA autoreceptors. *J. Physiol.* 2003, 551, 479–489.

[162] Hu, H. Z.; Li, Z. W. Modulation by adenosine of GABA-activated current in rat dorsal root ganglion neurons. *J. Physiol.* 1997, 501, 67–75.

[163] Oliet, S. H. R.; Poulain, D. A. Adenosine-induced presynaptic inhibition of IPSCs and EPSCs in rat hypothalamic supraoptic nucleus neurones. *J. Physiol.* 1999, 520, 815–825.

[164] Li, D. P.; Chen, S. R.; Pan, H. L. Adenosine inhibits paraventricular pre-sympathetic neurons through ATP-dependent potassium channels. *J. Neurochem.* 2010, 113, 530–542.

[165] Han, T. H.; Jang, S. H.; Lee, S. Y.; Ryu, P. D. Adenosine reduces GABAergic IPSC frequency via presynaptic A_1 receptors in hypothalamic paraventricular neurons projecting to rostral ventrolateral medulla. *Neurosci. Lett.* 2011, 490, 63–67.

[166] Thomas, T.; Spyer, K. M. A novel influence of adenosine on ongoing activity in rat rostral ventrolateral medulla. *Neuroscience* 1999, 88, 1213–1223.

[167] Bagley, E. E.; Vaughan, C. W.; Christie, M. J. Inhibition by adenosine receptor agonists of synaptic transmission in rat periaqueductal grey neurons. *J. Physiol.* 1999, 516, 219–225.

[168] Wilson, C. G.; Martin, R. J.; Jaber, M.; Abu-Shaweesh, J.; Jafri, A.; Haxhiu, M. A.; Zaidi, S. Adenosine A2A receptors interact with GABAergic pathways to modulate respiration in neonatal piglets. *Respir. Physiol. Neurobiol.* 2004, 141, 201–211.

[169] Zaidi, S. I. A.; Jafri, A.; Martin, R. J.; Haxhiu, M. A. Adenosine A2A receptors are expressed by GABAergic neurons of medulla oblongata in developing rat. *Brain Res.* 2006, 1071, 42–53.

[170] Mayer, C. a; Haxhiu, M. a; Martin, R. J.; Wilson, C. G. Adenosine A2A receptors mediate GABAergic inhibition of respiration in immature rats. *J. Appl. Physiol.* 2006, 100, 91–97.

[171] Duy, P. M.; Xia, L.; Bartlett, D.; Leiter, J. C. An adenosine A(2A) agonist injected in the nucleus of the solitary tract prolongs the laryngeal chemoreflex by a GABAergic mechanism in decerebrate piglets. *Exp. Physiol.* 2010, 95, 774–787.

[172] McQueen, D. S.; Ribeiro, J. A. On the specificity and type of receptor involved in carotid body chemoreceptor activation by adenosine in the cat. *Br. J. Pharmacol.* 1983, 80, 347–354.

[173] McQueen, D. S.; Ribeiro, J. A. Pharmacological characterization of the receptor involved in chemoexcitation induced by adenosine. *Br. J. Pharmacol.* 1986, 88, 615–620.

[174] Conde, S. V.; Obeso, A.; Vicario, I.; Rigual, R.; Rocher, A.; Gonzalez, C. Caffeine inhibition of rat carotid body chemoreceptors is mediated by A2A and A2B adenosine receptors. *J. Neurochem.* 2006, 98, 616–628.

[175] Livermore, S.; Nurse, C. A. Enhanced adenosine A2b receptor signaling facilitates stimulus-induced catecholamine secretion in chronically hypoxic carotid body type I cells. *Am. J. Physiol. Cell Physiol.* 2013, 305, C739–750.

[176] Lopes da Silva, F. EEG and MEG: Relevance to neuroscience. *Neuron* 2013, 80, 1112–1128.

In: Adenosine Signaling Mechanisms
Editors: V. Ramkumar and R. Paes de Carvalho
ISBN: 978-1-63483-186-4

Chapter 11

ADENOSINE RECEPTORS, OLIGOMERIZATION AND PARKINSON'S DISEASE PHARMACOTHERAPY

Francisco Ciruela[1,*], Leandra C. Constantino[2], Debora Lanzaster[2], Tharine Dal-Cim[2], Carla I. Tasca[2,3,4] and Víctor Fernández-Dueñas[1]

[1]Unitat de Farmacologia, Departament de Patologia i Terapèutica Experimental, Facultat de Medicina, IDIBELL, Universitat de Barcelona, L'Hospitalet de Llobregat, Barcelona, Spain

[2]Departamento de Bioquímica, Centro de Ciências Biológicas, Universidade Federal de Santa Catarina, UFSC, Florianópolis, SC, Brazil

[3]Programa de Pós-graduação em Bioquímica, Centro de Ciências Biológicas, Universidade Federal de Santa Catarina, UFSC, Florianópolis, SC, Brazil

[4]Programa de Pós-graduação em Neurociências, Centro de Ciências Biológicas, Universidade Federal de Santa Catarina, UFSC, Florianópolis, SC, Brazil

ABSTRACT

Adenosine, a well-known neuromodulator in the central nervous system (CNS), exerts its actions through G protein-coupled adenosine receptors (ARs), which are also expressed in a large variety of cells and tissues. Interestingly, AR oligomerization has been shown in the CNS, constituting a molecular mechanism by which extracellular adenosine may modulate other neurotransmitter systems (i.e., dopaminergic and glutamatergic transmission). These AR-containing complexes have been largely studied, since alterations on their functioning are thought to be involved in a number of disorders (i. e., Parkinson's disease).

* Corresponding author: Francisco Ciruela. Unitat de Farmacologia, Departament de Patologia i Terapèutica Experimental, Facultat de Medicina, IDIBELL, Universitat de Barcelona, L'Hospitalet de Llobregat, 08907 Barcelona, Spain. E-mail: fciruela@ub.edu.

Accordingly, they may also represent relevant targets in the search of effective treatments for these pathologies. Here, we focus on how AR oligomerization may impact in Parkinson's disease pharmacotherapy.

Abbreviations

AC	adenylyl cyclase
A1R	adenosine A1 receptor
A2AR	adenosine A2A receptor
ATP	adenosine triphosphate
ADP	adenosine diphosphate
AMP	adenosine monophosphate
cAMP	cyclic adenosine monophosphate
CNS	central nervous system
CPu	caudate-putamen
CYP3A4	cytochrome P450 3A4
D1R	dopamine D1 receptor
D2R	dopamine D2 receptor
GABA	γ-aminobutyric acid
GIRK	G protein-dependent inwardly rectifying K^+ channel
GPCR	G protein-coupled receptor
GPi	globus pallidus pars interna
L-DOPA	L-3,4-dihydroxyphenylalanine
LID	L-DOPA induced dyskinesias
mGlu1α	metabotropic glutamate type 1α receptor
mGlu5	metabotropic glutamate type 5 receptor
MAPK	mitogen-activated protein kinase
MSN	GABAergic medium spiny neurons
NAc	nucleus accumbens
PD	Parkinson's disease
PLA2	phospholipase A2
PLC	phospholipase C
PLD	phospholipase D
SAH	S-Adenosyl-L-homocysteine
SN	substantia nigra
SNc	substantia nigra pars compacta
SNr	substantia nigra pars reticulate

Introduction

In 1929 Drury and Szent-Györgyi described the profound hypotension, bradycardia and kidney regulation produced by adenosine in mammals [1]

Since then, the physiological roles of adenosine have been extensively studied [2]. Adenosine is a nucleoside mostly generated by ATP degradation both at the intracellular and extracellular level (Figure 1). In addition, small amounts of adenosine can be generated intracellularly by S-adenosyl-L-homocysteine (SAH) catabolism (Figure 1). Once produced, intracellular adenosine can be: (*i*) secreted outside the cell, (*ii*) phosphorylated forming AMP, (*iii*) converted to SAH by reacting with L-homocysteine and (*iv*) deaminated to form inosine by the action of adenosine deaminase (Figure 1).

On the other hand, extracellular adenosine can be also deaminated by the action of ecto-adenosine deaminase (ADA, Figure 1).

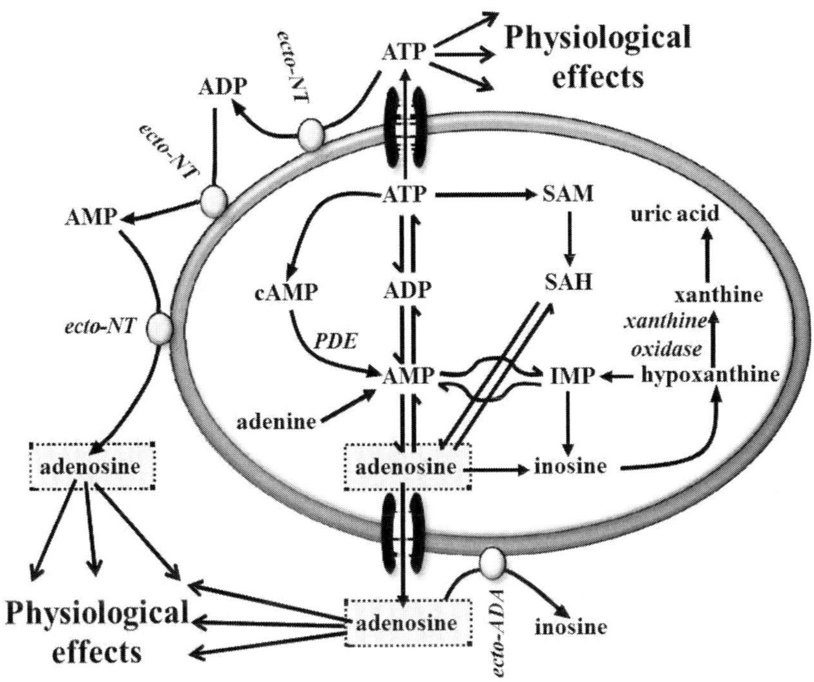

ATP is degraded to hypoxanthine, which can be salvaged by hypoxanthine– guanine– phosphoribosyl–transferase (HGPRT) or further metabolized to xanthine and uric acid by xanthine oxidase. Enzymes are in italics. Guanine- based purines (GTP to guanosine, guanine and xanthine) are not illustrated. Allopurinol inhibits the enzyme xanthine oxidase, possibly favoring purine salvage. The figure illustrates purinergic system and the sources of extracellular adenosine. ATP stored in presynaptic vesicles is released after depolarization, acting on P2 receptors and being dephosphorylated by ecto-NTPDases and ecto-5′nucleotidase (ecto-5′NT) into adenosine (ADO). ADO acts mainly on A1R and A2AR and can be converted into inosine extracellularly by ecto-adenosine deaminase (ecto-ADA) or uptaken by nucleoside transporters. Inside the cell, ADO preferentially forms AMP by adenosine kinase (AK) but also may be deaminated by ADA (not shown). The sources of extracellular adenosine are ATP, adenosine released as such by nucleoside transporters (in situation of energetic imbalance, such as brain damage), and cAMP, which is converted into AMP by an ecto- phosphodiesterase and later into ADO. A1R and A2ARs are present both pre- and post-synaptically. P2 are receptors for ATP.

Figure 1. Schematic representation of purine metabolism.

Extracellular adenosine, either ubiquitously released by means of Na^+-dependent transporters and/or formed extracellularly by the breakdown of ATP, is considered a "retaliatory metabolite," thus mediating the increase of oxygen supply and decrease of oxygen consumption [3]. In addition, adenosine is involved in modulating a large array of physiological processes: respiratory function [4], neural activity [5], platelet aggregation [6], neutrophil function [7], lymphocyte differentiation [8] and vascular tone [9]. Furthermore, adenosine is able to provoke both coronary artery dilatation and kidney blood vessels contraction, thus reducing renal filtration [10]. Also, it exerts a negative chronotropic and dromotropic effect on the heart [11], as well as mediates the inhibition of neurotransmitters release [12] and lipolysis [13]. Overall, this purine nucleoside, postulated to be a mediator of metabolic distress, has a considerable impact on a large variety of physiological cellular functions.

ADENOSINE RECEPTORS IN THE CENTRAL NERVOUS SYSTEM

In the central nervous system (CNS) adenosine plays a regulatory role by acting as a presynaptic, postsynaptic and/or non-synaptic neuromodulator [14]. Interestingly, extracellular adenosine concentration is at the high nM range upon basal conditions and it is related to the intracellular concentration of adenosine and nucleotides (i.e., ATP, AMP and cAMP) [15]. Indeed, intracellular adenosine concentration depends on the ATP metabolism rate [15], thus once adenosine is synthesized as a result of ATP breakdown it is readily released to the extracellular milieu where it acts as a neuromodulator [16] in response to an increased metabolic demand [17]. Interestingly, it has been proposed that in the striatum the main source of extracellular adenosine is the intracellular cAMP accumulation [18]. Thus, upon cAMP generation it is metabolized to AMP by means of phosphodiesterases (PDEs) and then to adenosine, which is secreted (Figure 1). Of note, since cAMP is generated by adenylyl cyclase (AC) activation, it seems likely that striatal extracellular adenosine may correlate well with the activity of receptors that are positively coupled to AC, thus it is clear the regulatory role of adenosine finely sensing intracellular changes. Interestingly, early in the seventies it was demonstrated that the electrical stimulation of brain slices promoted adenosine release [19], which mirrored intracellular cAMP accumulation, a phenomenon that was precluded by methylxanthines (i.e., caffeine and theophylline) [20], and that was also observed in other tissues (i.e., heart) [21]. Indeed, these experimental observations constituted the first piece of evidence suggesting that extracellular adenosine exerted its effects via specific plasma membrane receptors. Next, it was shown that the adenosine-mediated antilipolytic effect on fat cells was mediated by cAMP reduction [22]. Interestingly, this dual effect on cAMP formation was further corroborated when the adenosine-mediated inhibition or stimulation of AC activity was demonstrated. Collectively, these observations ended with the first sub classification of adenosine receptors into Ri and Ra [23], or alternatively, A1 and A2 adenosine receptors [24]. Currently, it is well assumed that adenosine mediates its actions throughout specific G protein-coupled receptors (GPCRs), for which four subtypes (A1R, A2AR, A2BR and A3R) have been identified so far. Interestingly, these receptors have distinctive pharmacological profile, tissue distribution and effector coupling [25], and its functioning have been largely assessed in the CNS (Table 1).

Table 1. Adenosine Receptors

Receptor	Adenosine affinity	G protein	Transduction mechanisms[¥]	Physiological actions
A_1	~ 70 nM	$G_{i/o}$[*] $G_{q/11}$ G_s	•Inhibits: AC[*], •Activates: PLC, AC	Vasoconstriction; [76] hypothermia and sedation; [77] analgesia; [78] neurotransmitter release; [79, 80] chemotaxis; [81] Neuroprotection [82].
A_{2A}	~ 150 nM	G_s[*] G_{olf} $G_{15,16}$[§]	•Activates: AC[*], PLC •Inhibits: Ca^{2+} channels	Platelet aggregation inhibition; [83] vasodilation; [84] neurotransmitter release; [85] regulation of sensorimotor integration in basal ganglia; [86] sleep promotion [87].
A_{2B}	~ 5000 nM	G_s[*] $G_{q/11}$	•Activates: AC[*], PLC	Vasodilation; [88] vasoconstriction; [89] cytokine production; [90] inhibition of cell proliferation [91].
A_3	~ 6500 nM	$G_{i/o}$[*]	•Inhibits: AC[*] •Activates: PLC	Mast cell activation; [92] preconditioning; [93] coronary vasodilation; [94] regulation of intraocular pressure; [95] hypotension [96].

[*]Main mechanism of coupling.

[¥]AC, adenylyl cyclase; PLC, phospholipase C; PLA2, phospholipase A2; PLD, phospholipase D; GIRKs, G protein-dependent inwardly rectifying K^+ channels.

Adenosine receptors belong to the class A of GPCRs, also known as rhodopsin-like [26], which indeed share some common molecular features, namely the presence of the conserved NPxxY(x)5,6F and the DRY motifs [27, 28], within others. Accordingly, adenosine-mediated conformational rearrangement of adenosine receptors determines the binding and activation of specific G proteins (Table 1), which are responsible for activation of different intracellular signaling pathways associated to adenosine function (Table 1). A1Rs and A2ARs are mainly responsible for the central effects of adenosine (Table 1) [29].

The most abundant and homogeneously distributed adenosine receptor within the brain is the A1R. This adenosine receptor couples to pertussis toxin-sensitive G proteins (G_{i1}, G_{i2}, G_{i3} and G_o), whose activation regulates several intracellular effector molecules (i.e., AC, Ca^{2+} channels, K^+ channels and phospholipase C, Table 1) [30].

On the other hand, the A2AR is expressed at high levels only in some restricted brain regions, for instance striatum, olfactory tubercle and nucleus accumbens [18, 31]. This adenosine receptor mainly couples to Gs/Golf proteins [32], thus prompting AC activation and cAMP accumulation (Table 1). In addition, A2AR may also signal through a G-protein independent pathway eventually associated to mitogen-activated protein kinase (MAPK) signalling cascade activation [33]. Next, the A2BR is positively coupled to adenylyl cyclase and PLC through a Gs and Gq protein, respectively (Table 1) [2]. This adenosine receptor is thought to be fairly ubiquitous in the brain, and its study has been limited by the lack of specific pharmacological tools [34].

Finally, the A3R is basically expressed peripherally and involved in immune and inflammatory processes. It has been shown to be coupled to Gi/o proteins, thus inhibiting AC and also stimulating PLC (Table 1) [2].

Adenosine Receptor Containing Oligomers

The concept that GPCRs may be interacting at the plasma membrane level forming receptor oligomers was initially postulated by Luigi F. Agnati and Kjell Fuxe early in the eighties. Accordingly, these authors proposed that an intramembrane neuropeptide and monoamine receptor-receptor interaction was behind the observed functional cross- talk between these two neurotransmitter systems [35, 36]. Initially, evidences indicating GPCR oligomerization exclusively relied on the interpretation of radioligand binding experiments, which permitted to propose the occurrence of oligomerization for several pair of receptors (for review see [37]).

Subsequently, the GPCR oligomerization concept was supported by photo-affinity labeling and radiation inactivation experiments, and also by hydrodynamic and cross-linking analysis [38-41]. However, the straight evidence demonstrating a direct receptor-receptor interaction between GPCRs in native tissue (e.g., the presence of oxytocin receptor dimers and/or oligomers in mammary glands) has not been reported until a few years ago [42].

Interestingly, in the last years several publications have corroborated the existence of adenosine receptors containing oligomers (for review see [43]). For instance, the A1R was found to heteromerize with the dopamine D1 receptor (D1R), a fact shown to be essential for desensitization and receptor trafficking [44]. Similarly, while the heterodimer composed by the A1R and the metabotropic glutamate type 1α (mGlu1α) receptor seems to play a key role in preventing glutamate excitotoxicity [45], the physiological impact of A_1R and purinergic $P2Y_1$ receptor ($P2Y_1R$) heteromerization is still under study [46]. More recently, the oligomerization of A1R with the thromboxane A_2 [47] and the β1- and β2-adrenergic receptors [48] has been described. Conversely, the A2AR is able to oligomerize with the metabotropic glutamate type 5 (mGlu5) receptor [33], thus a synergistic functional interaction has been documented both in biochemical and behavioral experiments [49]. Finally, the most studied adenosine receptor contained oligomer is the one formed by the A2AR with the dopamine D2 receptor (D2R) [50–52]. Interestingly, this A2AR/D2R oligomer may underline the molecular mechanism behind the antagonistic adenosine-dopamine interactions that regulate the function of the GABAergic enkephalinergic neurons, thus constituting a key target in the pharmacotherapy of certain pathologies, namely Parkinson's disease [53–55]. This last issue will be further discussed in the following section.

Parkinson's Disease and Adenosine Receptor Oligomers: Impact on Pharmacotherapy

Parkinson's disease (PD) is a progressive neurodegenerative disorder associated with the loss of dopaminergic nigrostriatal neurons [56]. It was named in this manner after the English physician James Parkinson described the pathology as the shaking palsy in 1817.

Currently, PD is recognized as one of the most common neurological disorders, affecting approximately 1% of individuals older than 60 years. Associated features and symptoms include resting tremor, rigidity, bradykinesia, and postural instability. Unfortunately, as a neurodegenerative disease, the motor phase of this pathology continues to ascend late in the disease to affect the cortex and patients may then exhibit cognitive dysfunction and dementia [57]. The main neuropathologic PD hallmark is the loss of pigmented dopaminergic neurons in the substantia nigra (SN) and the presence of Lewy bodies (LB), which is also common to other types of Parkinsonism. The loss of dopaminergic neurons occurs most prominently in the ventral lateral SN. Unfortunately, 60-80% of dopaminergic neurons are lost before the motor signs of PD emerge. This fact points out the relevance of finding a biomarker for the pathology, especially for early stages of the disease.

The basal ganglia [58], where the SN belongs to, modulates the cortical output necessary for normal movement. Inhibitory output to the thalamocortical motor circuit is directed from the globus pallidus pars interna (GPi) and the substantia nigra pars reticulata (SNr) in the basal ganglia, thus suppressing movement. Within the basal ganglia, two pathways exist and they are referred as to the direct and the indirect pathways (Figure 2). Dopamine is released by the dopaminergic neurons of the substantia nigra pars compacta (SNc) and acts on the striatum via activation of either D1R or D2R. The striatal GABAergic neurons containing D1R constitute the direct pathway and project to the GPi/SNr. On the other hand, the striatal GABAergic neurons containing D2R are part of the indirect pathway and project to the globus pallidus externa (GPe), which tonically inhibits the subthalamic nucleus (STN). The STN, in turn, projects excitatory inputs to the SNr-GPi complex (which inhibits the thalamus). The end-result of this circuitry consists of the inhibition of the thalamus and, therefore, decreased stimulation of the motor cortex by the thalamus. In non-pathological conditions, dopamine released by the SNc activates the direct pathway and inhibits the indirect pathway resulting in the inhibition of the GPi/SNr.

In PD, because of the loss of dopaminergic neurons in the SNc, striatal dopamine is reduced and the inhibition of the GPi/SNr decreases. By diminishing the inhibition of the GPi/SNr, the GPi/SNr inhibitory output to the thalamus is increased and it is augmented the inhibition of the thalamocortical pathway that suppresses movement (Figure 2). As a result, the central role that the striatum plays in processes such as motor activity control, motor learning and some forms of associative and visual learning is altered [59].

Treatment of PD mainly comprises pharmacotherapy [59], although surgical interventions (i.e., deep brain stimulation) are also in use [60]. However, all surgical procedures have associated risks and surgery is not suitable for everyone. Research around gene therapy and stem cell therapy is also promising but it still remains underdeveloped [61]. Accordingly, the best management for most people suffering PD consists of pharmacological treatment. Since the seventies, the most effective therapy for the management of PD is based on the use of drugs that mimic dopamine, initially the precursor L-DOPA (Levodopa) [62] and later on other dopaminergic agents [63], trying to correct the loss of this neurotransmitter.

Interestingly, although L-DOPA has been proved very efficacious in the treatment of PD symptoms, long term treatments tend to lose efficacy and also to induce severe collateral motor effects (dyskinesia and rigidity) and psychiatric problems [64]. Thus, as PD progresses, fewer dopamine neurons are available to store and release L-DOPA-derived dopamine, and the patient's clinical status begins to fluctuate more and more closely in accordance with plasma L- DOPA levels.

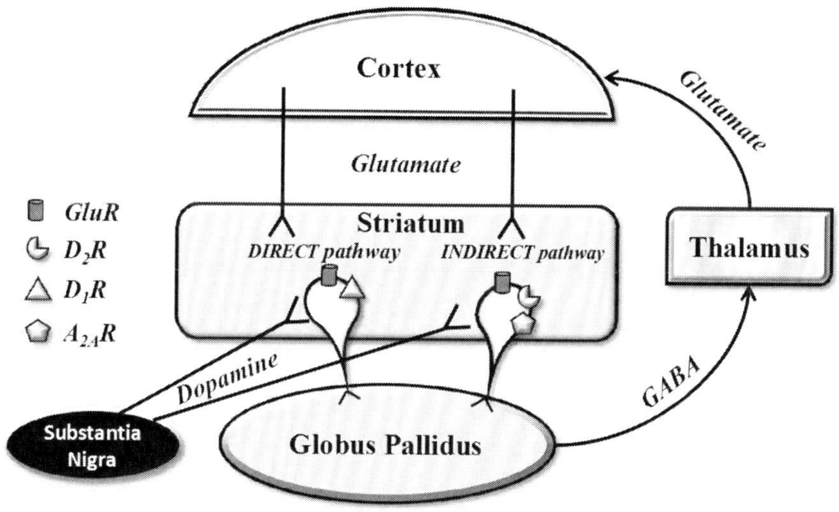

Figure 2. Basal ganglia circuitry and its connection to the thalamocortical circuit. Dopaminergic neurons from the substantia nigra control the direct and indirect pathways within the basal ganglia via activation of D1 or D2 dopamine receptors. Glutamate and adenosine receptors expressed in the striatum have also been postulated as therapeutic targets in PD. The co-localization of receptors for dopamine, glutamate and adenosine in the same striatal neurons puts into relevance the potential role of oligomers formed by these GPCRs.

Exposing striatal dopamine receptors to fluctuating dopamine concentrations may cause hypersensitivity, which is expressed clinically as peak-dose dyskinesia (twisting, turning movements). Finally, fluctuating L-DOPA-derived dopamine concentrations in association with the advancement of the disease may be responsible for development of motor fluctuations and dyskinesia [64]. As a consequence, research is nowadays focused on finding better pharmacological therapies that provide an efficacious treatment of the symptoms and lower risks of dyskinesias, particularly for younger patients where the development of L-DOPA induced dyskinesias (LID) is more likely. On the other hand, another important area of research is based on neuroprotecting dopaminergic neurons from death, thus aiming to slow down the progression of the disease. Unfortunately, there currently is no effective neuroprotective therapy against PD. Therefore, there is a pressing need for an early diagnosis, ideally before the degeneration of dopaminergic neurons, and therapeutic approaches that not only ameliorate the disease symptoms but also the underlying pathophysiology.

Attempts to discover new medications have traditionally operated under the assumption that GPCRs are monomers and that a specific drug activates one single receptor coupled to one single signal transduction pathway. However, it is now well- accepted that in the neuronal membrane, GPCRs may be arranged as homo- and hetero- oligomers [65], although, as mentioned earlier, the direct demonstration of GPCR oligomerization in native tissue is still limited [42]. Nevertheless, adenosine receptors (i.e., A1R and A2AR) located at striatal glutamatergic terminals have been proposed to oligomerize not only with other GPCRs (for review see [43]) but also to heterodimerize forming an A1R/A2AR oligomer [66]. Interestingly, the A1R/A2AR oligomer would function as an adenosine concentration switch that can fine tune glutamate release [67]. Also, dopamine receptors (i.e., D1R and D2R) have been postulated to homo- and heteromerize (for review see [68]). Indeed, as commented above, in striatopallidal GABAergic neurons, D2R/A2AR heterodimerization would perfectly explain the observed reciprocal inhibition between both receptors [69, 70]. Importantly, we have demonstrated D2R/A2AR oligomerization (i.e., direct receptor-receptor interaction) in native tissue (i.e., striatum) and its molecular unbalance in experimental Parkinsonism [71]. It has been postulated that the cross-talk within the D2R/A2AR heteromer might help to modulate postsynaptic plasticity at the glutamatergic synapse. Thus, it seems likely that drugs targeted to modulate this receptor-receptor interaction would have an impact in pathologies (i.e., PD) in which the area where these oligomers are expressed is affected.

In theory, multimodal therapies targeting oligomeric receptors may offer different therapeutic outputs to that expected for singlehanded dugs. In this sense, the increased knowledge about D2R/A2AR oligomerization has propelled the development of A2AR antagonists for the treatment of PD [72]. Indeed, A2AR antagonists have been shown to possess antiparkinsonian properties in human PD patients with LID [73]. This antiparkinsonian effect would probably be due to the action of the A2AR antagonist within the A2AR/D2R oligomer, thereby reducing the A2AR-mediated tonic inhibition of D2R functioning. Furthermore, the status of some oligomers might be altered in pathological situations such as PD or LID, as demonstrated in our preclinical studies [71].

In line with this, the appearance of LID has been suggested to be associated to the prevalence of A2AR/A2AR homomers over A2AR/D2R heteromers and D2R/D2R homomers, thus providing an abnormal increased A2AR signaling [74]. Overall, the demonstration of adenosine receptor-containing oligomers in striatal neurons adds further complexity to the understanding of striatopallidal neurochemistry, which may hold the promise of potentially new therapeutic targets.

Conclusion

Increased knowledge about GPCR oligomerization in general and on adenosine receptor heteromerization in particular may help scientists to design and develop specific A2AR antagonists for PD management [75]. Interestingly, preclinical data suggests that these A2AR antagonists might have major applications in the treatment of PD by: (*i*) providing motor benefit as monotherapy, (*ii*) potentiating the benefit of dopamine-based drugs (*iii*) preventing the development of LID, and (*iv*) preserving the motor response with less dyskinesia while using a lower dose of L-DOPA. On the other hand, clinical data shows that A2AR antagonists possess anti-parkinsonian properties in human PD patients with LID [73], thus several clinical trials are currently on the way [72]. Overall, A2AR antagonists are nowadays considered, based on its action through A2AR/D2R oligomers, the most attractive potential non-dopaminergic drugs for the management of PD.

Acknowledgments

This work was supported by grants SAF2011-24779, Consolider-Ingenio CSD2008-00005 and PCIN-2013-019-C03-03 from Ministerio de Economía y Competitividad and ICREA Academia-2010 from the Catalan Institution for Research and Advanced Studies to F.C. Grants IBN-Net # 01.06.0842-00 from CNPq, Excitotoxicity and Neuroprotection from

INCT, NENASC/PRONEX from FAPESC to C.I.T. Also, support comes from the "Programa Pesquisador Visitante Especial-Ciência sem Fronteiras" by CAPES (Brazil) to FC. VF-D and FC belong to the "Neuropharmacology and Pain" accredited research group (Generalitat de Catalunya, 2014 SGR 1251). C.I.T. is recipient of CNPq productivity fellowship.

REFERENCES

[1] Drury, A. N.; Szent-Gyorgyi, A. The physiological activity of adenine compounds with especial reference to their action upon the mammalian heart. *J. Physiol.* 1929, 68, 213–237.

[2] Jacobson, K. A.; Gao, Z. G. Adenosine receptors as therapeutic targets. *Nat. Rev. Discov.* 2006, 5, 247–264.

[3] Newby, A. C. Adenosine and the concept of "retaliatory metabolites". *Trends Biochem.* 1984, 9, 42–44.

[4] Brown, R. A.; Spina, D.; Page, C. P. Adenosine receptors and asthma. *Br. J. Pharmacol.* 2008, 153 Suppl., S446–56.

[5] Burnstock, G.; Fredholm, B. B.; Verkhratsky, A. Adenosine and ATP receptors in the brain. *Curr. Top. Med. Chem.* 2011, 11, 973–1011.

[6] Johnston-Cox, H. A.; Ravid, K. Adenosine and blood platelets. *Purinergic Signal.* 2011, 7, 357–65.

[7] Barletta, K. E.; Ley, K.; Mehrad, B. Regulation of neutrophil function by adenosine. *Arterioscler. Thromb. Vasc. Biol.* 2012, 32, 856–64.

[8] Ernst, P. B.; Garrison, J. C.; Thompson, L. F. Much ado about adenosine: adenosine synthesis and function in regulatory T cell biology. *J. Immunol.* 2010, 185, 1993–8.

[9] Burnstock, G. Purinergic regulation of vascular tone and remodelling. *Auton. Autacoid Pharmacol.* 2009, 29, 63–72.

[10] Vallon, V.; Mühlbauer, B.; Osswald, H. Adenosine and kidney function. *Physiol. Rev.* 2006, 86, 901–40.

[11] Fredholm, B. B.; Sollevi, A. Cardiovascular effects of adenosine. *Clin. Physiol.* 1986, 6, 1–21.

[12] Sebastião, A. M.; Ribeiro, J. A. Tuning and fine-tuning of synapses with adenosine. *Curr. Neuropharmacol.* 2009, 7, 180–94.

[13] Fredholm, B. B. Adenosine and lipolysis. *Int. J. Obes.* 1981, 5, 643–9.

[14] Sebastiao, A. M.; Ribeiro, J. A. Fine-tuning neuromodulation by adenosine. *Trends Pharmacol. Sci.* 2000, 21, 341–346.

[15] Latini, S.; Pedata, F. Adenosine in the central nervous system: release mechanisms and extracellular concentrations. *J. Neurochem.* 2001, 79, 463–484.

[16] Snyder, S. H. Adenosine as a neuromodulator. *Annu. Rev. Neurosci.* 1985, 8, 103–124.

[17] Ferre, S.; Fuxe, K. Adenosine as a volume transmission signal. A feedback detector of neuronal activation. *Prog. Brain Res.* 2000, 125, 353–361.

[18] Fredholm, B. B. Purinoceptors in the nervous system. *Pharmacol. Toxicol.* 1995, 76, 228–239.

[19] Pull, I.; McIlwain, H. Adenine derivatives as neurohumoral agents in the brain. The quantities liberated on excitation of superfused cerebral tissues. *Biochem. J.* 1972, 130, 975–981.

[20] Sattin, A.; Rall, T. W. The effect of adenosine and adenine nucleotides on the cyclic adenosine 3', 5'-phosphate content of guinea pig cerebral cortex slices. *Mol. Pharmacol.* 1970, 6, 13–23.

[21] Degubareff, T.; Sleator Jr, W. Effects of Caffeine on Mammalian Atrial Muscle, and its Interaction with Adenosine and Calcium. *J. Pharmacol. Exp. Ther.* 1965, 148, 202–214.

[22] Trost, T.; Stock, K. Effects of adenosine derivatives on cAMP accumulation and lipolysis in rat adipocytes and on adenylate cyclase in adipocyte plasma membranes. *Naunyn. Schmiedebergs. Arch. Pharmacol.* 1977, 299, 33–40.

[23] Londos, C.; Cooper, D. M.; Wolff, J. Subclasses of external adenosine receptors. *Proc. Natl. Acad. Sci. US* 1980, 77, 2551–2554.

[24] Van Calker, D.; Muller, M.; Hamprecht, B. Adenosine regulates via two different types of receptors, the accumulation of cyclic AMP in cultured brain cells. *J. Neurochem.* 1979, 33, 999–1005.

[25] Olah, M. E.; Stiles, G. L. Adenosine receptor subtypes: characterization and therapeutic regulation. *Annu. Rev. Pharmacol. Toxicol.* 1995, 35, 581–606.

[26] Kolakowski Jr, L. F. GCRDb: a G-protein-coupled receptor database. *Receptors Channels.* 1994, 2, 1–7.

[27] Fritze, O.; Filipek, S.; Kuksa, V.; Palczewski, K.; Hofmann, K. P.; Ernst, O. P. Role of the conserved NPxxY(x)5,6F motif in the rhodopsin ground state and during activation. *Proc. Natl. Acad. Sci. US* 2003, 100, 2290–2295.

[28] Rovati, G. E.; Capra, V.; Neubig, R. R. The highly conserved DRY motif of class A G protein-coupled receptors: beyond the ground state. *Mol. Pharmacol.* 2007, 71, 959–964.

[29] Dunwiddie, T. V; Masino, S. A. The role and regulation of adenosine in the central nervous system. *Annu. Rev. Neurosci.* 2001, 24, 31–55.

[30] Palmer, T. M.; Stiles, G. L. Adenosine receptors. *Neuropharmacology.* 1995, 34, 683–694.

[31] Rosin, D. L.; Hettinger, B. D.; Lee, A.; Linden, J. Anatomy of adenosine A2A receptors in brain: morphological substrates for integration of striatal function. *Neurology.* 2003, 61, S12–8.

[32] Marala, R. B.; Mustafa, S. J. Direct evidence for the coupling of A2-adenosine receptor to stimulatory guanine nucleotide-binding-protein in bovine brain striatum. *J. Pharmacol. Exp. Ther.* 1993, 266, 294–300.

[33] Ferre, S.; Karcz-Kubicha, M.; Hope, B. T.; Popoli, P.; Burgueno, J.; Gutierrez, M. A.; Casado, V.; Fuxe, K.; Goldberg, S. R.; Lluis, C.; Franco, R.; Ciruela, F. Synergistic interaction between adenosine A2A and glutamate mGlu5 receptors: implications for striatal neuronal function. *Proc. Natl. Acad. Sci. US* 2002, 99, 11940–11945.

[34] Feoktistov, I.; Biaggioni, I. Adenosine A2B receptors. *Pharmacol. Rev.* 1997, 49, 381–402.

[35] Agnati, L. F.; Fuxe, K.; Zini, I.; Lenzi, P.; Hokfelt, T. Aspects on receptor regulation and isoreceptor identification. *Med. Biol.* 1980, 58, 182–187.

[36] Fuxe, K.; Agnati, L. F.; Benfenati, F.; Celani, M.; Zini, I.; Zoli, M.; Mutt, V. Evidence for the existence of receptor--receptor interactions in the central nervous system. Studies on the regulation of monoamine receptors by neuropeptides. *J. Neural Transm.* 1983, 18, 165–179.

[37] Agnati, L. F.; Ferre, S.; Lluis, C.; Franco, R.; Fuxe, K. Molecular mechanisms and therapeutical implications of intramembrane receptor/receptor interactions among heptahelical receptors with examples from the striatopallidal GABA neurons. *Pharmacol. Rev.* 2003, 55, 509–550.

[38] Avissar, S.; Amitai, G.; Sokolovsky, M. Oligomeric structure of muscarinic receptors is shown by photoaffinity labeling: subunit assembly may explain high- and low-affinity agonist states. *Proc. Natl. Acad. Sci. US* 1983, 80, 156–159.

[39] Fraser, C. M.; Venter, J. C. The size of the mammalian lung beta 2-adrenergic receptor as determined by target size analysis and immunoaffinity chromatography. *Biochem. Biophys. Res. Commun.* 1982, 109, 21–29.

[40] Peterson, G. L.; Rosenbaum, L. C.; Broderick, D. J.; Schimerlik, M. I. Physical properties of the purified cardiac muscarinic acetylcholine receptor. *Biochemistry.* 1986, 25, 3189–3202.

[41] Herberg, J. T.; Codina, J.; Rich, K. A.; Rojas, F. J.; Iyengar, R. The hepatic glucagon receptor. Solubilization, characterization, and development of an affinity adsorption assay for the soluble receptor. *J. Biol. Chem.* 1984, 259, 9285–9294.

[42] Albizu, L.; Cottet, M.; Kralikova, M.; Stoev, S.; Seyer, R.; Brabet, I.; Roux, T.; Bazin, H.; Bourrier, E.; Lamarque, L.; Breton, C.; Rives, M. L.; Newman, A.; Javitch, J.; Trinquet, E.; Manning, M.; Pin, J. P.; Mouillac, B.; Durroux, T. Time-resolved FRET between GPCR ligands reveals oligomers in native tissues. *Nat. Chem. Biol.* 2010, 6, 587–594.

[43] Ciruela, F.; Gomez-Soler, M.; Guidolin, D.; Borroto-Escuela, D. O.; Agnati, L. F.; Fuxe, K.; Fernandez-Duenas, V. Adenosine receptor containing oligomers: their role in the control of dopamine and glutamate neurotransmission in the brain. *Biochim. Biophys. Acta.* 2011, 1808, 1245–1255.

[44] Torvinen, M.; Gines, S.; Hillion, J.; Latini, S.; Canals, M.; Ciruela, F.; Bordoni, F.; Staines, W.; Pedata, F.; Agnati, L. F.; Lluis, C.; Franco, R.; Ferre, S.; Fuxe, K. Interactions among adenosine deaminase, adenosine A(1) receptors and dopamine D(1) receptors in stably cotransfected fibroblast cells and neurons. *Neuroscience.* 2002, 113, 709–719.

[45] Ciruela, F.; Escriche, M.; Burgueno, J.; Angulo, E.; Casado, V.; Soloviev, M. M.; Canela, E. I.; Mallol, J.; Chan, W. Y.; Lluis, C.; McIlhinney, R. A.; Franco, R. Metabotropic glutamate 1alpha and adenosine A1 receptors assemble into functionally interacting complexes. *J. Biol. Chem.* 2001, 276, 18345–18351.

[46] Nakata, H.; Yoshioka, K.; Kamiya, T.; Tsuga, H.; Oyanagi, K. Functions of heteromeric association between adenosine and P2Y receptors. *J. Mol. Neurosci.* 2005, 26, 233–238.

[47] Mizuno, N.; Suzuki, T.; Hirasawa, N.; Nakahata, N. Hetero-oligomerization between adenosine A1 and thromboxane A2 receptors and cellular signal transduction on stimulation with high and low concentrations of agonists for both receptors. *Eur. J. Pharmacol.* 2012, 677, 5–14.

[48] Chandrasekera, P. C.; Wan, T. C.; Gizewski, E. T.; Auchampach, J. A.; Lasley, R. D. Adenosine A1 receptors heterodimerize with β1- and β2-adrenergic receptors creating novel receptor complexes with altered G protein coupling and signaling. *Cell. Signal.* 2013, 25, 736–42.

[49] Ferre, S.; Agnati, L. F.; Ciruela, F.; Lluis, C.; Woods, A. S.; Fuxe, K.; Franco, R. Neurotransmitter receptor heteromers and their integrative role in "local modules": the striatal spine module. *Brain Res. Rev.* 2007, 55, 55–67.

[50] Hillion, J.; Canals, M.; Torvinen, M.; Casado, V.; Scott, R.; Terasmaa, A.; Hansson, A.; Watson, S.; Olah, M. E.; Mallol, J.; Canela, E. I.; Zoli, M.; Agnati, L. F.; Ibanez, C. F.; Lluis, C.; Franco, R.; Ferre, S.; Fuxe, K. Coaggregation, cointernalization, and codesensitization of adenosine A2A receptors and dopamine D2 receptors. *J. Biol. Chem.* 2002, 277, 18091–18097.

[51] Ciruela, F.; Burgueno, J.; Casado, V.; Canals, M.; Marcellino, D.; Goldberg, S. R.; Bader, M.; Fuxe, K.; Agnati, L. F.; Lluis, C.; Franco, R.; Ferre, S.; Woods, A. S. Combining mass spectrometry and pull-down techniques for the study of receptor heteromerization. Direct epitope-epitope electrostatic interactions between adenosine A2A and dopamine D2 receptors. *Anal. Chem.* 2004, 76, 5354–5363.

[52] Canals, M.; Marcellino, D.; Fanelli, F.; Ciruela, F.; de Benedetti, P.; Goldberg, S. R.; Neve, K.; Fuxe, K.; Agnati, L. F.; Woods, A. S.; Ferre, S.; Lluis, C.; Bouvier, M.; Franco, R. Adenosine A2A-dopamine D2 receptor-receptor heteromerization: qualitative and quantitative assessment by fluorescence and bioluminescence energy transfer. *J. Biol. Chem.* 2003, 278, 46741–46749.

[53] Fuxe, K.; Agnati, L. F.; Jacobsen, K.; Hillion, J.; Canals, M.; Torvinen, M.; Tinner-Staines, B.; Staines, W.; Rosin, D.; Terasmaa, A.; Popoli, P.; Leo, G.; Vergoni, V.; Lluis, C.; Ciruela, F.; Franco, R.; Ferre, S. Receptor heteromerization in adenosine A2A receptor signaling: relevance for striatal function and Parkinson's disease. *Neurology.* 2003, 61, S19–23.

[54] Fuxe, K.; Marcellino, D.; Borroto-Escuela, D. O.; Guescini, M.; Fernandez-Duenas, V.; Tanganelli, S.; Rivera, A.; Ciruela, F.; Agnati, L. F. Adenosine- dopamine interactions in the pathophysiology and treatment of CNS disorders. *CNS Neurosci. Ther.* 2010, 16, e18–42.

[55] Fuxe, K.; Borroto-Escuela, D. O.; Romero-Fernandez, W.; Palkovits, M.; Tarakanov, A. O.; Ciruela, F.; Agnati, L. F. Moonlighting proteins and protein- protein interactions as neurotherapeutic targets in the g protein-coupled receptor field. *Neuropsychopharmacology.* 2014, 39, 131–55.

[56] Lang, A. E.; Lozano, A. M. Parkinson's disease. First of two parts. *N. Engl. J. Med.* 1998, 339, 1044–1053.

[57] Poewe, W.; Mahlknecht, P. The clinical progression of Parkinson's disease. *Parkinsonism Relat. Disord.* 2009, 15 Suppl. 4, S28–32.

[58] Gerfen, C. R. In: *The Rat Nervous System*; Paxinos, G., Ed.; Elsevier Academic Press: Amsterdam, 2004; 445–508.

[59] Meissner, W. G.; Frasier, M.; Gasser, T.; Goetz, C. G.; Lozano, A.; Piccini, P.; Obeso, J. A.; Rascol, O.; Schapira, A.; Voon, V.; Weiner, D. M.; Tison, F.; Bezard, E. Priorities in Parkinson's disease research. *Nat. Rev. Discov.* 2011, 10, 377–393.

[60] Volkmann, J. Deep brain stimulation for the treatment of Parkinson's disease. *J. Clin. Neurophysiol.* 21, 6–17.

[61] Mochizuki, H.; Choong, C.-J.; Yasuda, T. The promises of stem cells: stem cell therapy for movement disorders. *Parkinsonism Relat. Disord.* 2014, 20 Suppl. 1, S128–31.

[62] Cotzias, G. C.; Van Woert, M. H.; Schiffer, L. M. Aromatic amino acids and modification of parkinsonism. *N. Engl. J. Med.* 1967, 276, 374–379.

[63] Romrell, J.; Fernandez, H. H.; Okun, M. S. Rationale for current therapies in Parkinson's disease. *Expert Opin. Pharmacother.* 2003, 4, 1747–61.

[64] Huot, P.; Johnston, T. H.; Koprich, J. B.; Fox, S. H.; Brotchie, J. M. The pharmacology of L-DOPA-induced dyskinesia in Parkinson's disease. *Pharmacol. Rev.* 2013, 65, 171–222.

[65] Ciruela, F.; Vallano, A.; Arnau, J. M.; Sanchez, S.; Borroto-Escuela, D. O.; Agnati, L. F.; Fuxe, K.; Fernandez-Duenas, V. G protein-coupled receptor oligomerization for what? *J. Recept. Signal Transduct. Res.* 2010, 30, 322–330.

[66] Ciruela, F.; Casado, V.; Rodrigues, R. J.; Lujan, R.; Burgueno, J.; Canals, M.; Borycz, J.; Rebola, N.; Goldberg, S. R.; Mallol, J.; Cortes, A.; Canela, E. I.; Lopez-Gimenez, J. F.; Milligan, G.; Lluis, C.; Cunha, R. A.; Ferre, S.; Franco, R. Presynaptic control of striatal glutamatergic neurotransmission by adenosine A1-A2A receptor heteromers. *J. Neurosci.* 2006, 26, 2080–2087.

[67] Ciruela, F.; Ferre, S.; Casado, V.; Cortes, A.; Cunha, R. A.; Lluis, C.; Franco, R. Heterodimeric adenosine receptors: a device to regulate neurotransmitter release. *Cell. Mol. Life Sci.* 2006, 63, 2427–2431.

[68] Ciruela, F.; Fernandez-Duenas, V.; Llorente, J.; Borroto-Escuela, D.; Cuffi, M. L.; Carbonell, L.; Sanchez, S.; Agnati, L. F.; Fuxe, K.; Tasca, C. I. G protein- coupled receptor oligomerization and brain integration: Focus on adenosinergic transmission. *Brain Res.* 2012, 1476, 86–95.

[69] Ferre, S.; Quiroz, C.; Woods, A. S.; Cunha, R.; Popoli, P.; Ciruela, F.; Lluis, C.; Franco, R.; Azdad, K.; Schiffmann, S. N. An update on adenosine A2A-dopamine D2 receptor interactions: implications for the function of G protein-coupled receptors. *Curr. Pharm. Des.* 2008, 14, 1468–1474.

[70] Ferre, S.; Ciruela, F.; Woods, A. S.; Lluis, C.; Franco, R. Functional relevance of neurotransmitter receptor heteromers in the central nervous system. *Trends Neurosci.* 2007, 30, 440–446.

[71] Fernández-Dueñas, V.; Taura, J. J.; Cottet, M.; Gómez-Soler, M.; López-Cano, M.; Ledent, C.; Watanabe, M.; Trinquet, E.; Pin, J. P.; Luján, R.; Durroux, T.; Ciruela, F. Untangling dopamine-adenosine receptor assembly in experimental parkinsonism. *Dis. Model. Mech.* In press.

[72] Vallano, A.; Fernandez-Duenas, V.; Pedros, C.; Arnau, J. M.; Ciruela, F. An update on adenosine A2A receptors as drug target in Parkinson's disease. *CNS Neurol. Disord. Drug Targets.* 2011, 10, 659–669.

[73] LeWitt, P. A.; Guttman, M.; Tetrud, J. W.; Tuite, P. J.; Mori, A.; Chaikin, P.; Sussman, N. M.; Group, 6002-US-005 Study Adenosine A2A receptor antagonist istradefylline (KW-6002) reduces "off" time in Parkinson's disease: a double- blind, randomized, multicenter clinical trial (6002-US-005). *Ann. Neurol.* 2008, 63, 295–302.

[74] Antonelli, T.; Fuxe, K.; Agnati, L.; Mazzoni, E.; Tanganelli, S.; Tomasini, M. C.; Ferraro, L. Experimental studies and theoretical aspects on A2A/D2 receptor interactions in a model of Parkinson's disease. Relevance for L-dopa induced dyskinesias. *J. Neurol. Sci.* 2006, 248, 16–22.

[75] Schwarzschild, M. A.; Agnati, L.; Fuxe, K.; Chen, J. F.; Morelli, M. Targeting adenosine A2A receptors in Parkinson's disease. *Trends Neurosci.* 2006, 29, 647–654.

[76] Murray, R. D.; Churchill, P. C. Effects of adenosine receptor agonists in the isolated, perfused rat kidney. *Am. J. Physiol.* 1984, 247, H343–8.

[77] Anderson, R.; Sheehan, M. J.; Strong, P. Characterization of the adenosine receptors mediating hypothermia in the conscious mouse. *Br. J. Pharmacol.* 1994, 113, 1386–90.

[78] Yamamoto, S.; Nakanishi, O.; Matsui, T.; Shinohara, N.; Kinoshita, H.; Lambert, C.; Ishikawa, T. Intrathecal adenosine A1 receptor agonist attenuates hyperalgesia without inhibiting spinal glutamate release in the rat. *Cell. Mol. Neurobiol.* 2003, 23, 175–85.

[79] De Lorenzo, S.; Veggetti, M.; Muchnik, S.; Losavio, A. Presynaptic inhibition of spontaneous acetylcholine release induced by adenosine at the mouse neuromuscular junction. *Br. J. Pharmacol.* 2004, 142, 113–24.

[80] Scholz, K. P.; Miller, R. J. Inhibition of quantal transmitter release in the absence of calcium influx by a G protein-linked adenosine receptor at hippocampal synapses. *Neuron.* 1992, 8, 1139–50.

[81] Schnurr, M.; Toy, T.; Shin, A.; Hartmann, G.; Rothenfusser, S.; Soellner, J.; Davis, I. D.; Cebon, J.; Maraskovsky, E. Role of adenosine receptors in regulating chemotaxis and cytokine production of plasmacytoid dendritic cells. *Blood.* 2004, 103, 1391–7.

[82] MacGregor, D. G.; Miller, W. J.; Stone, T. W. Mediation of the neuroprotective action of R-phenylisopropyl-adenosine through a centrally located adenosine A1 receptor. *Br. J. Pharmacol.* 1993, 110, 470–6.

[83] Varani, K.; Portaluppi, F.; Gessi, S.; Merighi, S.; Ongini, E.; Belardinelli, L.; Borea, P. A. Dose and time effects of caffeine intake on human platelet adenosine A(2A) receptors: functional and biochemical aspects. *Circulation.* 2000, 102, 285–9.

[84] Carroll, M. A.; Doumad, A. B.; Li, J.; Cheng, M. K.; Falck, J. R.; McGiff, J. C. Adenosine2A receptor vasodilation of rat preglomerular microvessels is mediated by EETs that activate the cAMP/PKA pathway. *Am. J. Physiol. Renal Physiol.* 2006, 291, F155–61.

[85] Popoli, P.; Betto, P.; Reggio, R.; Ricciarello, G. Adenosine A2A receptor stimulation enhances striatal extracellular glutamate levels in rats. *Eur. J. Pharmacol.* 1995, 287, 215–7.

[86] Nagel, J.; Schladebach, H.; Koch, M.; Schwienbacher, I.; Müller, C. E.; Hauber, W. Effects of an adenosine A2A receptor blockade in the nucleus accumbens on locomotion, feeding, and prepulse inhibition in rats. *Synapse.* 2003, 49, 279–86.

[87] Scammell, T. E.; Gerashchenko, D. Y.; Mochizuki, T.; McCarthy, M. T.; Estabrooke, I. V; Sears, C. A.; Saper, C. B.; Urade, Y.; Hayaishi, O. An adenosine A2a agonist increases sleep and induces Fos in ventrolateral preoptic neurons. *Neuroscience.* 2001, 107, 653–63.

[88] Kemp, B. K.; Cocks, T. M. Adenosine mediates relaxation of human small resistance-like coronary arteries via A2B receptors. *Br. J. Pharmacol.* 1999, 126, 1796–800.

[89] Donoso, M. V.; López, R.; Miranda, R.; Briones, R.; Huidobro-Toro, J. P. A2B adenosine receptor mediates human chorionic vasoconstriction and signals through arachidonic acid cascade. *Am. J. Physiol. Heart Circ. Physiol.* 2005, 288, H2439–49.

[90] Zhong, H.; Belardinelli, L.; Maa, T.; Feoktistov, I.; Biaggioni, I.; Zeng, D. A(2B) adenosine receptors increase cytokine release by bronchial smooth muscle cells. *Am. J. Respir. Cell Mol. Biol.* 2004, 30, 118–25.

[91] Dubey, R. K.; Gillespie, D. G.; Mi, Z.; Jackson, E. K. Adenosine inhibits PDGF-induced growth of human glomerular mesangial cells via A(2B) receptors. *Hypertension.* 2005, 46, 628–34.

[92] Zhong, H.; Shlykov, S. G.; Molina, J. G.; Sanborn, B. M.; Jacobson, M. A.; Tilley, S. L.; Blackburn, M. R. Activation of murine lung mast cells by the adenosine A3 receptor. *J. Immunol.* 2003, 171, 338–45.

[93] Das, S.; Cordis, G. A.; Maulik, N.; Das, D. K. Pharmacological preconditioning with resveratrol: role of CREB-dependent Bcl-2 signaling via adenosine A3 receptor activation. *Am. J. Physiol. Heart Circ. Physiol.* 2005, 288, H328–35.

[94] Hinschen, A. K.; Rose'Meyer, R. B.; Headrick, J. P. Adenosine receptor subtypes mediating coronary vasodilation in rat hearts. *J. Cardiovasc. Pharmacol.* 2003, 41, 73–80.

[95] Avila, M. Y.; Stone, R. A.; Civan, M. M. Knockout of A3 adenosine receptors reduces mouse intraocular pressure. *Invest. Ophthalmol. Vis. Sci.* 2002, 43, 3021–6.

[96] Stella, L.; de Novellis, V.; Marabese, I.; Berrino, L.; Maione, S.; Filippelli, A.; Rossi, F. The role of A3 adenosine receptors in central regulation of arterial blood pressure. *Br. J. Pharmacol.* 1998, 125, 437–40.

In: Adenosine Signaling Mechanisms
Editors: V. Ramkumar and R. Paes de Carvalho

ISBN: 978-1-63483-186-4

Chapter 12

ADENOSINE RECEPTORS IN NEUROPROTECTION EVOKED BY PRE AND POSTCONDITIONING

***Leandra C. Constantino*[2]*, Tharine Dal-Cim*[3]*, Simone Molz*[4]*, Francisco Ciruela*[5]*, Carina R. Boeck*[6] *and Carla I. Tasca*[1,2,3,*]**

[1]Departamento de Bioquímica, [2]Programa de Pós-graduação em Bioquímica,
[3]Programa de Pós-graduação em Neurociências, Centro de Ciências Biológicas,
Universidade Federal de Santa Catarina,
UFSC, Campus Trindade, Florianópolis, SC, Brazil
[4]Curso de Farmácia, Universidade do Contestado - UnC, Canoinhas, SC, Brazil
[5]Unitat de Farmacologia, Departament de Patologia i Terapèutica Experimental,
Facultat de Medicina, IDIBELL – Universitat de Barcelona, L'Hospitalet de Llobregat,
Barcelona, Spain
[6]Programa de Pós-graduação em Nanociências, Centro Universitário Franciscano,
Santa Maria, RS, Brazil

ABSTRACT

Cellular tolerance or resistance can be achieved by interventions before or after injury through potential toxic agents used in low stimulus or dose. Brain tolerance evoked by pre- or postconditioning is a neuroprotective strategy that can afford attenuation of motor, cognitive, emotional, or memory deficits aroused following a brain damage. Preconditioning is a well-established experimental and clinical translational strategy with great beneficial effects, but limited applications. Postconditioning has recently emerged as a new neuroprotective strategy, which has shown interesting results when applied after a stroke event. Adenosine receptors have been reported as a fundamental endogenous neuroprotective system in the adjacent cellular mechanisms contributing to the development of cardiac or brain tolerance. In the present chapter, tissue tolerance achieved by pre- and postconditioning is discussed as a protective paradigm and the role

* Corresponding author: Carla I. Tasca, PhD Departamento de Bioquímica, CCB, UFSC Trindade, 88040-900 Florianópolis, SC, Brazil. Telephone number: + 55 48 3721-5046. FAX number: + 55 48 3721-9672. E-mail: carla.tasca@ufsc.br.

of adenosine receptors in the mechanism of neuroprotection evoked by tissue conditioning is focused.

Introduction

Brain Tolerance Evoked by Pre or Postconditioning

Preconditioning is a natural adaptive process that is considered a mechanism of increasing tissue tolerance and conferring cellular protection in different organs such as heart, kidney and brain [1]. The terms “preconditioning” and “tolerance” were introduced by Janoff [2], from a study conducted in lysosomes in a model of shock. However, it is generally accepted that preconditioning was first described in the canine heart when occlusion of the left circumflex coronary artery for 5 minutes applied four times produced tolerance against a 40 minutes occlusion. These brief periods of ischemia could reduce or prevent substantial damage caused by a subsequent, more prolonged ischemic episode [3]. Thus, one could define that the general principle of preconditioning is to induce an adaptive process of cellular, tissue or whole organism protection through exposure to sub-lethal insults that therefore grant certain tolerance to a subsequent lethal insult [2, 4].

Brain is a very sensitive organ to injury. Neurons are rarely replaced once they have died, so the damage in certain regions may be permanent. In this scenario, studying neuroprotective strategies is a theme of great relevance. Among these neuroprotective strategies, preconditioning attracts substantial attention as a novel therapeutic approach for neuroprotection. The first description of preconditioning in the central nervous system (CNS) demonstrated that the neuroprotection promoted by a brief period of cerebral ischemia produced a status of protection against a subsequent lethal ischemic event in several regions of the brain, e.g., the CA_1 and CA_3 areas in the hippocampus, considered more vulnerable to injuries [5].

The mechanism of brain tolerance represents a transient resistance of the cerebral tissue to a lethal insult, which is established by preconditioning with a mild insult of short/defined duration. Increased brain tolerance can be achieved by stressor mechanisms, e.g., by chemical, electrical or anoxic stimulus [6-8]. Ischemic preconditioning has been the target of most studies in this research field [9, 10]. The idea is that exposing the brain to short periods of non-damaging ischemia affords protection, or tolerance to, a later ischemic event that would typically be devastating to the brain. Beyond ischemic preconditioning, a large variety of chemical compounds has been proposed as chemical agents inducing preconditioning, resulting in cardio-, renal- or cerebral protective effects. Among them are anesthetics, ethanol, selective ligands of the GABAergic (GABA, gamma amino butyric acid) system, opioids and glutamate receptors, and potassium channels activators. Ischemic and chemical preconditioning will be discussed latter in this chapter.

The induction of brain tolerance by preconditioning involves modulation of signaling pathways that activate effector mechanisms related to neuroprotection. Brain preconditioning may be due, at least in part, to the upregulation of cellular defense systems, such as upregulation of cellular survival pathways [11-13]. Researches have succeeded in demonstrating that the mechanisms of endogenous neuroprotection triggered by preconditioning involves: antiinflammatory actions, reduction of apoptotic markers, induction of antioxidant defenses and decreasing excitotoxicity response, promoting metabolic and neurovascular protection [14, 11, 15-19].

The brain tolerance status can be maintained through two different ways. After application of sub-lethal stimuli, preconditioning can take effect within hours, described as rapid or early tolerance [20], or days, described as delayed or classical tolerance [21]. Cellular protection induced by rapid, early tolerance occurs within hours, independent of new gene transcription. Conversely, classical/delayed tolerance develops over 1–3 days, involves the transcription of new genes and *de novo* protein synthesis and neuroprotective effect continues for several days.

The clinical approach of preconditioning has obvious limitations, e.g., the inability to predict the onset of the injury. Therefore, tissue conditioning is also evoked by postconditioning that is a neuroprotective strategy that has been well studied for ischemic events, where the reperfusion is considered as an entity apart from ischemic injury, subjected to control or reduction [22]. Pharmacological postconditioning, as a protective strategy against delayed neuronal death, has been intensively studied in recent years [23-25]. Cellular protective pathways activated by postconditioning are, at least in part, similar to those activated in preconditioning, e.g., adenosine receptors, antioxidant defenses and survival pathways [26]. However, postconditioning has the advantage of being able to be applied after the insult. Several studies discuss basic approaches to postconditioning and show two main methods of how this can be achieved: i) rapid postconditioning, where the interruption of reperfusion occurs between minutes and hours after the injury [27, 28], and ii) delayed postconditioning, where treatment is applied between hours and days after the incident [29]. The signaling pathways and the mechanisms activated in cells which are triggered by preconditioning or postconditioning are not yet completely understood. However, several evidences demonstrated the involvement of adenosine receptors in neuroprotection elicited by pre- or postconditioning.

Therefore, in the next sections of this chapter, the role of adenosine receptors in the neuroprotective mechanisms and the relationship between pre- and postconditioning strategies - e.g., ischemia, anesthetics, ethanol, selective ligands to the GABAergic system, opioids, potassium channel activators, glutamate receptors agonist, N-methyl-D-aspartate (NMDA) and the involvement of distinct adenosine receptors will be discussed.

Adenosine Receptors and Neuroprotection in Neurological and Psychiatric Disorders

Epidemiological, electrophysiological and pharmacological studies emphasize the importance of adenosine receptors in a variety of neurological diseases such as stroke, epilepsy, Alzheimer's, Parkinson's and Huntington's diseases as well as in psychiatric disorders like depression, anxiety and schizophrenia [30-32].

Neuronal and glial cells express all four types of adenosine receptors, namely A_1, A_{2A}, A_{2B} and A_3 adenosine receptors (A_1R, $A_{2A}R$, $A_{2B}R$ and A_3R). However, due to the highest density of A_1R and $A_{2A}R$ in comparison to $A_{2B}R$ and A_3R, most data on adenosine effects on brain function and dysfunction are focused at A_1R and $A_{2A}R$ and the fine tuning between these two receptors [33].

A_1R are particularly important at hippocampal neurons, where they are located presynapticaly, postsynapticaly and nonsynapticaly [34] modulating synaptic transmission [35]. Tonic activation of presynaptic A_1R limits calcium entry in the nerve terminals thus inhibiting the release of neurotransmitters such as glutamate, acetylcholine and serotonin [36]. At the postsynaptic density, A_1R modulates glutamate receptors [35], decreases NMDA receptors currents [37, 38], inhibits voltage sensitive calcium and potassium channels [36] contributing to neuronal excitability reduction at postsynaptic moieties. By the other hand, activation of $A_{2A}R$ increase glutamate presynaptic release and are involved on facilitation of synaptic plasticity [33].

Excessive glutamate release and excitotoxicity occurs during acute brain insults such as hypoxia/ischemia, traumatic brain injury and epilepsy. Since A_1R activation suppresses [39, 40] and $A_{2A}R$ activation enhances synaptic transmission [33], one could expect that A_1R stimulation as well as $A_{2A}R$ blockade could afford neuroprotection (reviewed in [30, 41]). Indeed, hypoxic/ischemic conditions as well as status epilepticus increase adenosine levels in animal models of ischemia and epilepsy [42, 43]. In line with this view, adenosine and adenosine-potentiating agents (e.g., inhibitors of adenosine kinase) generally decrease the amount of brain damage in animal models of hypoxic/ischemia and in different models of epilepsy [44-46]. On the other hand, A_1R blockade is related to exacerbation of brain injury [47, 48] and mice lacking A_1R has a decreased threshold for seizures [49, 42]. Taken together, these findings support the idea that activation of the A_1R could be beneficial to stroke and epileptic patients [50, 51].

However, pharmacological manipulation of A_1R are only effective if made before or shortly after the ischemic insult, whereas chronic manipulation of A_1R function caused paradoxical effects [44, 30]. Thus, probably acute stimulation of synaptic A_1R provides protection by decreasing synaptic transmission, but during recovery, other mechanisms may also play a role [33]. Additionally, in genetic knockout models, the absence of A_1R had no effect on the level of ischemia/hypoxia-induced damage [52]. Differences in A_1R manipulation also take place in immature versus adult brain. In immature brain, pharmacologic A_1R stimulation has been shown to exacerbate rather than attenuate ischemic injury [53]. Consistent with this, genetic deletion of A_1R was neuroprotective in immature mice raised under hypoxic conditions [54]. These data highlight the potential differences in outcomes generated by pharmacologic versus genetic models and suggest that the protective effects of A_1R activation are temporally regulated [32].

$A_{2A}R$ are likely to be activated under stressful situations, when both ATP and adenosine reach high concentrations in the extracellular space [55, 56]. Given that activation of $A_{2A}R$ facilitates glutamate release, one possible explanation for the afforded neuroprotection with the blockade of these receptors in brain acute injury [41] is that it prevents glutamate overflow [57-59]. In line with this view, activation of these receptors inhibited glutamate uptake by astrocytes both *in vitro* [60] and *in vivo* [61].

While A_1R play a critical function in controlling seizures, the role of $A_{2A}R$ in seizures development are less clear. Pharmacological studies provide evidences that $A_{2A}R$ activation can either suppress or induce seizures [62, 63]. Data show that genetic deletion of facilitatory $A_{2A}R$ might also attenuate convulsions [64] and kindled seizures induced by pentilenetetrazol [65]. Moreover, after a convulsive behavior there is a long-term decrease of A_1R accompanied by an increased density of $A_{2A}R$, suggesting that $A_{2A}R$ antagonists rather than A_1R agonists may be more promising anticonvulsive drugs [66]. Genetic variation in the adenosine $A_{2A}R$ (ADORA2A), whose activation is involved in excitotoxicity, may be a predisposing factor of encephalopathy with febrile *status epilepticus* [67].

In comparison to A_1R and $A_{2A}R$, there are few studies in the literature concerning the role of A_3R in the pathophysiology of cerebral ischemia and they are rather contradictory. A_3R knockout mice are more sensitive to hypoxic neurodegeneration [68] and administration of A_3R agonists before or immediately after ischemic insults was shown to significantly protect brain tissues in rodent ischemic models [69, 70]. Consistent with these reports, it was demonstrated in rat cortical neurons that the selective activation of A_3R during hypoxia is involved in the inhibition of excitatory neurotransmission, indicating that A_3R also contribute to neuroprotective action of adenosine [71]. A_1R and A_3R activation protect astrocytes against hypoxic damage [72] and also reduce the lipopolisacharide-mediated hypoxia inducible factor (HIF)-1α accumulation in murine astrocytes [73]. Thus, these data point to the possibility of A_3R might exert A_1R-like protective effects and may be a new target to relieve cerebral ischemic injury. On the other hand, acute A_3R stimulation has been shown to exacerbate the damage caused by a concomitant ischemic episode *in vivo* [74]. Selective antagonism of A_3R facilitates the recovery of synaptic activity induced by ischemic preconditioning in rat hippocampal slices [75] and a harmful role of A_3R during *in vitro* ischemia was confirmed by the observation that blocking the A_3R protects from the irreversible disruption of excitatory neurotransmission caused by a severe ischemic episode [76]. Based on these observations, one could suggest A_3R stimulated by high levels of adenosine released during brief ischemia might exert A_1R-like protective effects on neurotransmission. However, severe ischemia would transform the A_3R-mediated effects from protective to injurious [77]. Regarding A_3R and epilepsy, it seems that antagonists also resulted in a reduction in seizure duration and intensity [62]. Facilitation of the onset of epileptiform discharge has been observed in the presence of a selective A_3R agonist and a reduction in such epileptic activity was observed when this receptor, activated by endogenously released adenosine during seizures, were blocked by the selective antagonist [78].

The involvement of $A_{2B}R$ is even less explored, and astrocytes are the main cellular target of $A_{2B}R$-mediated effects (revised in [79]) where they participated in adenosine-induced expression of glial cell line-derived neurotrophic factor (GDNF) [80] and in adenosine-induced glycogen synthesis [81]. $A_{2B}R$-mediated release of leukemia inhibitory factor from astrocytes also protected cortical neurons from excitotoxicity [82]. The selective $A_{2B}R$ antagonist MRS1706 regulated the stability of $GABA_A$ receptors from epileptic human brains transplanted into Xenopus oocytes [83].

Regarding Alzheimer's disease, epidemiological and animal studies have shown caffeine consumption has cognitive as well as protective effects [30]. Caffeine, a non-selective adenosine receptors antagonist, also reverses cognitive impairment and brain levels of amyloid-β peptide in mice [84-87], prevents amyloid-β-peptide-induced toxicity in cerebellar granular neurons [88] and has beneficial effects in a model of Alzheimer disease-like tau pathology [89]. The aforementioned benefits of caffeine against cognitive deficits in Alzheimer disease animal models as well as the neuroprotective actions against this disease hallmarks (e.g., amyloid-β peptide toxicity and tau phosphorylation) are not mimicked by A_1R antagonists but rather by antagonizing $A_{2A}R$ [90, 33, 87] indicating that caffeine-induced neuroprotection is mainly due to $A_{2A}R$ rather than A_1R blockade. The mechanisms of neuroprotection afforded by $A_{2A}R$ antagonism against amyloid-β peptide-induced toxicity still remain to be fully characterized, but $A_{2A}R$ can control the neurochemical consequences of enhanced production of free radicals [91], glutamate excitotoxicity [92] and astrocyte glutamate uptake [93]. In line with the epidemiological and pharmacological evidences of $A_{2A}R$ as potential targets to alleviate cognitive and neurochemical dysfunction in Alzheimer disease, the intracerebral administration of amyloid-β peptide did not cause cognitive deficits or toxicity in $A_{2A}R$ knockout mice [90].

Clinical and epidemiological evidence also has led to increased interest in $A_{2A}R$ as novel, non-dopaminergic therapeutic targets for Parkinson's disease [94, 95]. Several $A_{2A}R$ specific antagonists developed over the last decade exhibit the anti-parkinsonian property of enhancing motor activity in animal models of Parkinson disease and more recently, in Parkinson disease patients [96]. Regarding the possible mechanisms involved in $A_{2A}R$-induced neuroprotection against Parkinson disease, it seems that blockade of these receptors reduces dopaminergic degeneration as well as mitochondrial bioenergetics defects in animal models of the disease (reviewed in [94]).

Considering adenosine and psychiatric disorders, indirect evidences suggest that hypofunction of adenosinergic system is involved in schizophrenia. Adenosine receptor agonist display antipsychotic-like profile and blockade of adenosine receptors with caffeine may exacerbate psychotic behavior [97]. Moreover, activation of $A_{2A}R$ reduces the effectiveness of dopamine D_2 receptors signaling, and thus being the probable antipsychotic-like profile of adenosine agonist. Animal studies also demonstrate that adenosine is also involved in the regulation of mood related behavior [98]. A_1R and $A_{2A}R$ are expressed in the limbic system and have profound pre- and postsynaptic modulatory actions on both glutamatergic and monoaminergic transmission [99], but adenosine neuromodulatory system in depression is complex, with depressive [100] as well as antidepressive-like effects [101]. Genetic depletion of $A_{2A}R$ results in antidepressant-like effect whereas overexpression of $A_{2A}R$ induces depressive-like behavior [102]. Conversely, antidepressant effects of sleep deprivation require astrocyte-dependent adenosine A_1R-mediated signaling [103]. Thus, as far as the underlying receptor subtype is concerned, the involvement of both A_1 and $A_{2A}R$ seems likely [104].

In summary, increased adenosine levels and thus A_1R activation is becoming a more consolidated strategy in order to manage early damage induced by ischemic injury or epilepsy. Conversely, $A_{2A}R$ play a more prominent role in chronic neurodegenerative diseases (e.g., Alzheimer and Parkinson Disease) as well as in neuropsychiatric disorders such as depression and schizophrenia. Besides these evidences, there are several major open questions regarding adenosine receptors function in neuroprotection that remain to be tackled.

Adenosine Receptors in Preconditioning

Adenosine Receptors in Ischemic Preconditioning

Cerebral ischemia is one of the most common causes of death and the main cause of disability. Ischemic stroke arises in humans as a consequence of cardiac arrest or the stoppage of blood flow to the brain, which can be due to embolic or thrombic occlusion of arteries. Both global and focal ischemic events are very severe pathogenic processes with multiple, parallel, and sequential pathogenesis. When blood flow decreases during focal ischemia, the area known as "penumbra" (surrounding necrotic core of ischemia) is perfused by collateral vessels and the ischemia leads to selective cell death of vulnerable regions such as the hippocampus and cerebral cortex [105, 106].

In the last decades it was demonstrated interest in defining the evolutionarily conserved endogenous protective mechanisms of the brain, which occurs in ischemic tolerance or after ischemic preconditioning. In the context, the ischemic preconditioning was first described by Murry and collaborators in 1986. They proposed that brief ischemic episodes protect the heart of dogs from a subsequent sustained ischemic insult [3]. The first study reporting ischemic tolerance phenomenon in the brain, showed neuroprotection promoted by a brief ischemic episode against subsequent lethal ischemia in various brain areas, such as the CA_1 and CA_3 regions of the hippocampus [5]. Thus, the phenomenon of ischemic preconditioning has been extensively investigated as a neuroprotective strategy in experimental models of cerebral ischemia [9].

Possible neurotransmitter systems involved in ischemic preconditioning include adenosine receptors [107]. The ischemic preconditioning increases adenosine levels and initiates a series of intracellular signaling by which induces neuronal tolerance through a signal transduction pathway that consists of the activation of A_1R which leads to activation of protein kinase C (PKC), followed by activation of potassium channel sensitive to ATP (K_{ATP} channels) and inhibits presynaptic calcium fluxes. The response to calcium influx is also important for the development of protection against ischemia, because calcium influx has been linked to the production of reactive oxygen species (ROS) and the initiation of a number of signaling cascades leading to cell damage [108-111] including the inhibition of glutamate release [36]. Confirming this, many studies demonstrate that the inhibitory effects of A_1R result in neuroprotection, both *in vitro* and *in vivo*, particularly in models of hypoxia/ischemia and seizures [112, 113].

While the CNS protective role of adenosine has been attributed to the A_1R, an additional role of the A_3R has not been ruled out [74]. In contrast to the A_1R and A_3R, activation of the $A_{2A}R$ in the brain appears to aggravate ischemia [114]. A_3R stimulation by released adenosine during ischemia may exert A_1R-like protective effects, favoring depression of synaptic transmission [115]. However, A_3R activation during severe ischemia delays recovery of synaptic transmission [115], a process that is probably related to internalization and degradation of α-amino-3-hydroxy-5-methyl-4-isoxazolepropionic acid (AMPA) subtype of glutamate receptors [116], which by itself has some neuroprotective potential since it may prevent delayed cell death due to excessive activation of calcium-permeable AMPA receptors.

Hence, adenosine exerts influence on the cerebral ischemic events and participates in ischemic preconditioning, mainly through the A_1R.

Adenosine Receptors in Chemical Preconditioning

A large variety of chemical compounds have been proposed as chemical agents inducing preconditioning, resulting in cardio- and neuroprotective effects. Among them are anesthetics, opioids, selective ligands to the GABAergic system, ethanol, K_{ATP} channel activators and NMDA receptor ligands.

Anesthetics and Opioids-evoked Preconditioning

The anesthetics as well as opioids demonstrate some protective effects toward several organs including the heart and brain [117]. As volatile anesthetic agents have been proven safe in clinical treatments, they may be applied as a more medically viable approach for preventing ischemic insults in the future. However, the exact molecular and cellular mechanisms underlying the protective effects of preconditioning remain unclear.

In the 1960s, first evidences that general anesthetics can induce tolerance against cerebral ischemia have appeared [118]. Goldstein and colleagues showed that pentobarbital can offer tolerance to cerebral anoxia [119]. Several studies have been carried out illustrating that intravenous and volatile anesthetics can decrease the neuronal injury during periods of insult. Anesthetic-induced preconditioning has many fundamental steps, including activation of G-protein-coupled receptors, multiple protein kinases and K_{ATP} channels. Many characteristics of preconditioning by volatile anesthetics are similar to those of ischemic preconditioning. These involve activation of A_1R, PKC and K_{ATP} channels [120].

Similarly, opioids activate δ and κ-opioid receptors, and this also leads to PKC activation. Activated PKC acts as an amplifier of the preconditioning stimulus and stabilizes, by phosphorylation, the open state of the mitochondrial ATP sensitive potassium channel (mK_{ATP}) - the main end-effector in anesthetic preconditioning - and the sarcolemmal K_{ATP} channel [120]. Also, other protein kinases are activated as a result of PKC activation, somehow promoting signaling from adenosine A_2R early in reperfusion [121]. Recently, studies demonstrated that functionally interaction of δ and κ opioid receptors with adenosine A_1R mediates cardioprotection by remote ischemic preconditioning in Human Embryonic Kidney 293 cells (HEK-293) [122].

Changes in GABA function following cerebral ischemia and possible protective benefits of GABAergic drugs have been assessed. Volatile anesthetics can produce an increase in $GABA_A$ receptor activity and release of GABA during ischemia can lead to the suppression of inhibitory synaptic transmission by stimulating adenosine receptors [123]. Furthermore, synergism may occur between GABA and adenosine receptors. $GABA_B$ and A_1R/A_2R were able to act synergistically to reduce an *in vitro* hypoxic insult to an isolated optic nerve. This interaction appeared to be at the level of PKC and may involve the down-regulation of the sodium/calcium exchanger [124].

Ethanol-evoked Preconditioning

Ethanol consumption has been historically known due to its harmful effects on the health of humans, however in the last two decades it has become evident that moderate ethanol drinking may also have beneficial therapeutic effects on human cardiovascular system [125]. Several evidence suggest that chronic ingestion of small quantities of ethanol may protect myocardium from ischemic injury and cerebral ischemia by activating many of the endogenous signal transduction elements that have been implicated during other forms of preconditioning [126 130]. Ethanol drinking has been show to promote anti-inflammatory processes, involving A_1R, α1-adrenoceptors, PKC, nitric oxide synthase (NOS), heat shock proteins (HSP) and K_{ATP} channels [131].

The involvement of A_1R in ethanol-induced preconditioning is not completely understood, although it has been hypothesized that ethanol-induced preconditioning is dependent on these adenosine receptors via PKC signaling pathway. Ethanol at low doses results in acute cardioprotection by activating εPKC. On activation, εPKC translocates to the mitochondria, where it phosphorylates cytoprotective targets such as the mitochondrial permeability transition pore (mPTP), cytochrome C oxidase and the mK_{ATP}. At high doses, ethanol activates δPKC and is not acutely cardioprotective, but is cardioprotective if the onset of ischemia is delayed to 60 min post-ethanol treatment. This may be due to δPKC-mediated increase in extracellular adenosine (via increased activity of the equilibrative nucleoside transporter 1, ENT1), resulting in A_1R occupancy and subsequent activation of εPKC. It remains to be determined whether ethanol exerts additional cardioprotective effects in the mitochondria or elsewhere in the cell that are different from those mediated by εPKC activation [125].

Also, brain functional comparison between older moderate alcohol consumers and non-drinkers has received more recent epidemiological study. The moderate alcohol consumption has been associated with a lower risk of cognitive loss and dementia [132]. Experimental evidence of preconditioning with moderate alcohol levels in studies *in vivo* and in rat brain organotypic cultures showed protection against ischemia, endotoxin, amyloid-β, or glycoprotein 120 (gp120), the neuroinflammatory HIV-1 (Human Immunodeficiency Virus) envelope protein [126].

Potassium Channels-evoked Preconditioning

Potassium channels are involved in almost all main cell activities, such as: excitability of neurons and muscle cells, shaping of action potentials, coupling of many chemical and/or mechanical stimuli with given intracellular events, function of secretive cells, regulation of mitochondrial calcium homeostasis. Thus, drugs activating potassium channels have been considered to be interesting tools for the potential treatment of several pathological conditions. Only in the last few years, it was unravel an undoubted role of K_{ATP} channels, in particular of the mitochondrial K_{ATP} channels (mK_{ATP}), in ischemic preconditioning, suggesting an exciting scenario for K_{ATP} activators as innovative cardioprotective anti-ischemic drugs. The opening of K_{ATP} channels may be promoted through receptor-dependent mechanisms (such as adenosine and opioids), which involve the translocation and the activation of PKC and may be favored by the receptor-independent mechanisms of NO, ischemic preconditioning or chemical preconditioning, which prevent calcium overload [133].

NMDA-Evoked Preconditioning

Among different chemical agents responsible for evoking preconditioning, our group has focused in evaluating the mechanisms related to preconditioning evoked by the agonist of the main ionotropic glutamate receptor, NMDA, a ligand-operated calcium channel. The activation of NMDA receptors is known to have dual effects, promoting neuronal survival or death, depending on the activity level and the subunit composition of the receptor [134], which has been discussed with respect to its activity at synaptic or extra-synaptic sites.

NMDA receptors containing GluN2B subunits are mainly located at extra-synaptic processes and are involved in excitotoxicity, while NMDA receptors containing GluN2A subunit and expressed at synaptic membranes and are involved in trophic effects of glutamate receptors that are responsible for neuroprotection [135].

High levels of adenosine in the extracellular space are observed in situations of brain tissue injury, as well as when NMDA receptors are activated [136-138]. This increase in extracellular levels of adenosine may result from its release through its transportation system or through the hydrolysis of extracellular nucleotides, since the enzymatic cascade of ecto-nucleotidases responsible for producing adenosine from released nucleotides is activated by glutamatergic excitotoxicity [139, 140, 136]. For example, it is known that NMDA receptor activation in cultured cerebellar neurons stimulates the activation of ecto-5'-nucleotidase [139], which in turn degrades adenosine-5'-monophosphate (AMP) to adenosine.

Despite many studies showed the involvement of NMDA receptors activation in neuronal loss after ischemia, relevant evidence on the role of NMDA receptors as a major factor in the induction of preconditioning was clearly established by administration of NMDA receptor antagonists such as MK-801 [(+) - 5-methyl-10,11-dihydro-5H-dibenzo [a, d] cycloheptene-5,10-imine maleate)] or ketamine [141-144]. Thus, it is now widely accepted that a moderate activation of NMDA receptors during a preconditioning state induction is considered as a possible effective clinical strategy. Subtoxic concentrations of NMDA are able to prevent neuronal death induced by glutamate, NMDA [11, 145, 146], or oxygen/glucose deprivation in cultured cells [147, 148]. *In vivo*, the intraperitoneal administration of subtoxic doses of NMDA has also been recognized as a preconditioning model against several lethal posterior stimuli such as kainate-induced toxicity [149], ischemia [150] and seizures induced by quinolinic acid, an agonist of NMDA receptors [151].

The neuroprotective effect of NMDA preconditioning has been identified, although the neural mechanisms involved in NMDA preconditioning are not completely understood [12, 152-154]. NMDA-mediated neuroprotection evokes an alteration on antioxidant enzymes activity [18] and depends on the activation of A_1R, because NMDA preconditioning could not be achieved when NMDA receptors or A_1R was blocked with selective antagonists [151]. Blocking NMDA receptors with MK-801 neutralized the neuroprotective effects of NMDA preconditioning against behavioral seizures and the hippocampal cellular damage evoked by quinolinic acid. However, the blockade of A_1R with the selective antagonist 8-cyclopentyl-1,3-dimethylxanthine (CPT) also eliminated the neuroprotection against seizures, but it did not alter the hippocampal protection promoted by NMDA preconditioning. Therefore, it seems that activation of adenosine receptors may be involved in an early mechanism of NMDA preconditioning of preventing behavioral seizures evoked by quinolinic acid, whereas they are not involved in the delayed response of NMDA preconditioned which is observed as a protection of hippocampal cellular damage. These mechanisms remain to be further investigated.

An *in vitro* evaluation of the role of adenosine receptors in the mechanism of NMDA preconditioning in cerebellar granule neurons, revealed that NMDA preconditioning is also abolished by an A_1R antagonist. However, NMDA preconditioning did not alter the functionality of A_1R, but it facilitates a desensitization of the $A_{2A}R$ response. The resulting reduction of $A_{2A}R$-evoked cyclic AMP accumulation might favor the activation of A_1R [11] and then contribute to NMDA-mediated preconditioning. The antagonistic effect of adenosine receptors activation, mainly A_1R and $A_{2A}R$, is well understood and discussed previously in this chapter. Thus, it has been widely described that adenosine interaction with their receptors may be related to decreased (via A_1R activation) or facilitated (via $A_{2A}R$ activation) glutamatergic transmission [155]. Moreover, a balance between A_1R and $A_{2A}R$ function and/or oligomerization state may also be related to glutamatergic transmission [156, 157], a mechanism that might be involved on NMDA preconditioning and deserves further investigation.

We have recently evaluated putative changes on A_1R properties and functionality following NMDA preconditioning induction. A slight increased affinity of A_1R to a selective ligand in the hippocampus of NMDA preconditioned mice was observed. However, the most surprisingly result was that post-activation of A_1R abolished NMDA preconditioned-evoked responses. The moderate activation of NMDA receptors by preconditioning induced an antinociceptive effect in mice and the A_1R selective agonist 2-chloro-N6-cyclopentyl-adenosine (CCPA) in an inactive dose *per se*, reversed the antinociceptive effect triggered by NMDA preconditioning, showing that post-activation of A_1R impairs NMDA preconditioning-mediated responses. Moreover, NMDA preconditioning promotes glutamate clearance from the extracellular space. Accordingly, post-activation of A_1R reversed the NMDA preconditioning-mediated glutamate uptake in hippocampal slices (Constantino et al., unpublished data). The exact mechanism of this cross-talk is not yet understood, although there is hardly any doubt that this glutamatergic/adenosinergic interaction is crucial for the neuroprotective response evoked by NMDA preconditioning. Further in-depth investigations are required in order to reveal the mechanisms by which NMDA preconditioning induce neuroprotection after cerebral injury.

Chemical preconditioning represents a novel therapeutic strategy and additional studies are necessary to increase the knowledge about ligand molecules, receptors and intracellular signaling pathways involved in the neuroprotective mechanisms triggered by preconditioning.

Adenosine Receptors in Postconditioning

Adenosine Receptors in Ischemic Postconditioning

Could we predict a tissue injury induced by and endogenous (or exogenous) event? Preconditioning is a beneficial method in clinical trials but it has hard limitations in translating these protective strategies into clinical practice. Besides, the ethical conduct will be questionable in a 'predictive' manipulation, thus preconditioning is an interesting prevention against cellular damage, but has limitation at its tolerance opportunity window. Even so, while preconditioning event is well known to activate endogenous protective mechanisms in cells, postconditioning strategy is a less adjustable event, because usually its protection window is near to the end of damage period.

Postconditioning approach was initially discovered against cellular damage in myocardia using the principle that the reperfusion is predicted [158]. The strategy was applying brief intermittent cycles of reperfusion alternating with ischemia during first minutes of reperfusion in open-chest canine model of left anterior descending artery occlusion and reperfusion. Zhao and co-works, and other followers demonstrated that the cardioprotection is possible after ischemia-postconditioning similarly to that observed after preconditioning-ischemia in animal models [159, 160]. In translational studies, mechanical postconditioning is induced by four cycles of occlusion-reperfusion via the inflation and deflation of a balloon, reducing the size of microvascular obstruction-evoked infarcts in human patients suffering from acute ST-elevation myocardial infarction [161, 162]. These results indicate that postconditioning could potentially be a safe method to induce neuroprotection, although preclinical and clinical trials remain necessary in order to confirm this hypothesis efficiently.

Adenosinergic system is essentially activated during an ischemic postconditioning period. The participation of specific subtypes of the adenosine receptors in ischemic postconditioning is investigated with administration of agonists or antagonists after onset of reperfusion. In heart of rabbits *in situ*, both non-selective adenosine receptor blocker 8-p-(sulfophenyl) theophylline (8-SPT) or the selective $A_{2B}R$ antagonist, MRS1754, blocked postconditioning-induced cardioprotection [163]. Also in kidney of rats, 8-SPT blocked cardioprotection in a model of remote postconditioning [164] and ischemia-reperfusion postconditioning showing the fundamental participation of $A_{2A}R$ with subsequent PKC activation [165]. In heart of mice, PKC activation occurs after $A_{2B}R$ activation during postconditioning [166]. Besides 8-SPT, ZM241385 (selective $A_{2A}R$ antagonist) or MRS1523 (selective A_3R antagonist) also abolished ischemia-reperfusion postconditioning evoked-protection [167]. The intracellular signaling after activation of adenosine A_2R and/or A_3R has as mediator the phosphatidylinositol-3 kinase/protein kinase B (PI3k/Akt) pathway, opening the mitochondrial K_{ATP} channel, inhibiting the opening of the mitochondrial permeability transition pore and reduction of superoxide anion generation [168]. Thus, there are some controversial data regarding adenosine receptors subtypes involved with ischemia postconditioning, mainly about A_1R and $A_{2A}R$, but certainly $A_{2B}R$ are associated to postconditioning [169, 163].

In brain, postconditioning also represents encouraging strategy to evoke brain tolerance to a damage event [170, 171]. Cerebral ischemic postconditioning induces powerful neuroprotective effects [28] through reduction of pro-inflammatory signals [160] and increasing the content of pro-survival protein Bcl-2 accompanied with reduction of pro-apoptotic protein caspase-3 [15, 172], such as observed for preconditioning strategies [15, 173]. Also, the subunit GluN2A of NMDA receptors is required to neuroprotection in hippocampus and its downstream pro-survival signaling proteins, as calcium/calmodulin-dependent protein kinase II (CaMKII), extracellular-regulated MAP kinase (ERK), cAMP-responsive binding element (CREB), brain-derived neurotrophic factor (BDNF) and the anti-apoptotic protein Bcl2 [174]. In a preconditioning situation, NMDA receptors have been shown to participate effectively, thus it should be expected an association with adenosine receptors also in a postconditioning situation, but studies should be performed in order to endorse this open question.

Adenosine Receptors in Chemical Postconditioning

Chemical preconditioning is a good strategy to study the mechanisms activated (and inhibited) during cellular tolerance. Ischemia, hypoxia and hypothermia preconditioning share common pathways of protection to induce tolerance with several protein targets, differently from chemical preconditioning that has effects through more restrict mechanisms, what could be more accurate when a clinical focus is required. In this sense, chemical postconditioning may be a better therapeutic strategy for reducing reperfusion injury during surgery or following ischemia.

In heart, kidney, liver and brain, chemical postconditioning is an adjuvant strategy to improve the reduced tissue damage size after reperfusion therapy. Chemical postconditioning is evoked by administration of drug during the first minutes of the reoxygenation period [175]. The molecular mechanisms involved in the chemical postconditioning after myocardial ischemia is activation of adenosine, opioid and bradikynin receptors, with ROS production [176, 177]. In mechanisms of cardioprotection, adenosine appears to be a more promising effector [178]. However, the adenosine receptors appear to be differently involved in pre- and postconditioning, mainly because in postconditioning the crucial activation occurs during reperfusion period [167]. Nevertheless, there are the same controversy data concerning the subtypes of adenosine receptors activated [179, 180, 167]. Studies have used adenosine receptor antagonists to block the chemical postconditioning induced-cardioprotection in human clinical trials as in basic research with rodents. The non-selective adenosine receptors antagonist, 8-SPT abolishes postconditioning effect induced by desflurane in human myocardium [177] and morphine in rats [175]. Also, adenosine and its analogues are used as postconditioning effectors. In this way, $A_{2B}R$ is more evident adjuvant in the protection process when PI3k/Akt pathway is required during reperfusion period to inhibit disruptive mitochondrial activity [169]. Regarding A_1R and $A_{2A}R$ there are contradictory data, probably because experimental conditions of drugs administration and postconditioning event should be better evaluated. Postconditioning evoked in isolated mice heart is abolished in presence of the adenosine receptors antagonist 8-SPT, ZM241385, or MRS1523, but not by DPCPX (selective A_1R receptors) [167]. Ischemia-induced hepatocyte death *in vitro* and *in vivo* is reduced when postconditioning is induced by administration of CGS21680 (selective $A_{2A}R$ agonist) or the phosphatase and tensin homologue deleted from chromosome 10 (PTEN) inhibitor dipotassium bisperoxo-(5-hydroxypyridine-2-carboxyl)-oxovanadate [bpV(HOpic)], both via PI3k/Akt pathway [176]. Alternatively, in the rat embryonic cardiomyoblast-derived cell line H9c2, postconditioning with NS1619, an opener of large-conductance calcium-activated potassium channels (BK_{Ca} channels), induces protection via A_1R [180]. It is possible that an activation of multiple adenosine receptor subtypes might be required, because the generation of adenosine via ecto-5'-nucleotidase is crucial, such as the co-activation of $A_{2A}R$ and $A_{2B}R$ during postconditioning in mice [166]. Corroborating with this idea, the extracellular adenosine production by ecto-5′-nucleotidase is suggested to preferentially activate adenosine $A_{2A}R$ [55].

In brain, fewer studies have investigated the participation of adenosine receptors in chemical postconditioning. Additionally, most studies in post conditioned brain focused on the use of volatile anesthetics such as isoflurane, sevoflurane and propofol with similar neuroprotective mechanisms via PKC, PI3k/Akt pathway and K_{ATP} channel [181-184]. McMurtrey and Zuo, in 2010, have demonstrated an increased neuroprotection in cultured hippocampal neurons evoked by isoflurane postconditioning against oxygen/glucose deprivation [185]. The mechanism of neuroprotection probably depends on adenosine $A_{2A}R$ because the agonist CGS21680 increases the isoflurane postconditioning evoked-protection in neurons.

In conclusion, adenosine is released in response to pre- or postconditioning displaying cytoprotective effects against damage. Selective ligands must be administered in order to unravel the participation of adenosine receptors in the 'adenosinergic mechanism' of cytoprotection.

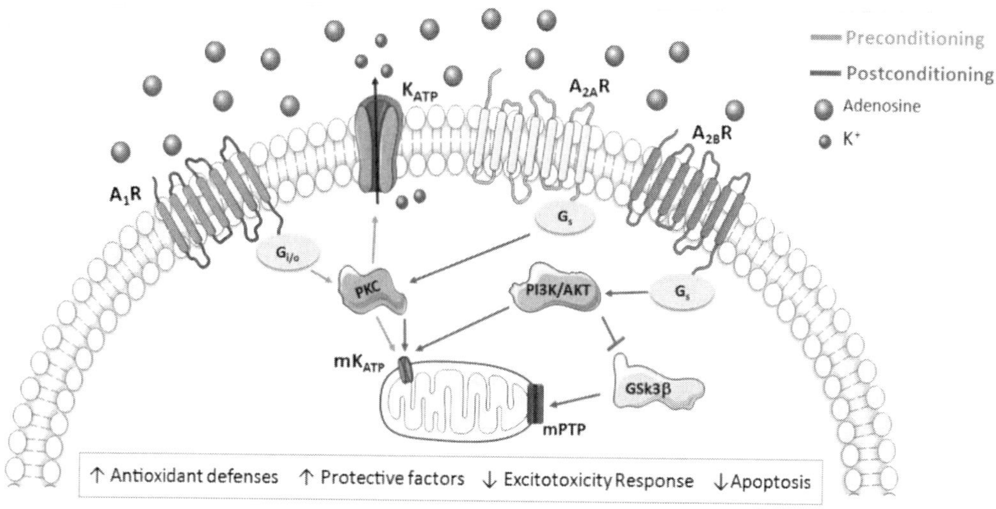

Figure 1. Tissue tolerance evoked by pre and post-conditioning involves the modulation of denosine receptors. Preconditioning (orange arrows) increases adenosine levels and initiates neuroprotective signaling through A_1R (and probably A_3R) leading to PKC activation followed by modulation of cellular membrane mitochondrial K_{ATP} channels (mK_{ATP}). Preconditioning also facilitates the desensitization of the $A_{2A}R$ response contributing to induction of brain tolerance. On the other hand, the mechanism of neuroprotection elicited by postconditioning (blue arrows) involves the activation of $A_{2B}R$ and PI3K/Akt pathway that also inhibit GSK3β activation, promoting the blockade of the opening of mitochondrial permeability transition pore (mPTP). Cellular protective pathways activated by pre- and postconditioning through adenosine receptors result in increased antioxidant defenses and reduction of exicitotoxicity and apoptosis. The figure was designed by T. Dal-Cim and L. C. Constantino, using images from www.servier.com/Powerpoint-image-bank.

Conclusion

In this chapter pre- and postconditioning were presented and discussed as protective strategies and the participation of adenosine receptors is focused. From a translational perspective, preconditioning may have clinical applications in prophylactic situations, as inducing neuroprotection in patients who are undergoing brain surgery or subarachnoid hemorrhages and are exposed to an elevated risk of immediate brain injury. Differently, postconditioning is also an interesting therapeutic approach, since the window of application lies after the diagnosis of a stroke or trauma. There is consensus in the scientific literature that an intervention against cerebral or cardiac damage could be better controlled using pharmacological strategies. In this sense, synthetic adenosine receptors antagonists have potential therapeutic applications even in pre- or postconditioning (Figure 1). At the moment, is accepted that during the period of tolerance achievement adenosine induces: i) cardioprotection through acting on A_1R and A_3R at preconditioning, and on $A_{2B}R$ at postconditioning strategy; ii) cerebroprotection through acting on A_1R and A_3R; iii) anti-inflammatory effects through acting on $A_{2A}R$ and A_3R. However, little is known about chemical pre- and postconditioning, theirs mechanisms and effects. The pursuit of basic research into conditioning study will endow a better comprehension of the underlying mechanisms and signaling pathways associated with the development of tissue tolerance, and ultimately provide the foundation for future clinical applications.

ACKNOWLEDGMENTS

This study was supported by grants from CNPq, Excitotoxicity and Neuroprotection from INCT, NENASC/PRONEX from FAPESC, CAPES/PVE 052/2012 to C. I. T. Also supported by grants from SAF2011-24779, Consolider-Ingenio CSD2008-00005 and PCIN-2013-019-C03-03 from Ministerio de Economía y Competitividad and ICREA Academia-2010 from the Catalan Institution for Research and Advanced Studies to F. C. C. I. T. is recipient of CNPq productivity fellowship and F. C. belongs to the 'Neuropharmacology and Pain' accredited research group (Generalitat de Catalunya, 2014 SGR 1251).

REFERENCES

[1] Obrenovitch, T. P. Molecular physiology of preconditioning-induced brain tolerance to ischemia. *Physiol. Rev.* 2008, 88, 211-247.

[2] Janoff, A. Alterations in Lysosomes (Intracellular Enzymes) during Shock; Effects of Preconditioning (Tolerance) and Protective Drugs. *International Anesthesiology Clinics* 1964, 2, 251-269.

[3] Murry, C. E, Jennings, R. B, Reimer, K. A. Preconditioning with ischemia: a delay of lethal cell injury in ischemic myocardium. *Circulation* 1986, 74, 1124-1136.

[4] Zhang, M. J., Franklin, S., Li, Y., Wang, S., Ru, X., Mitchell-Jordan, S. A., Mano, H., Stefani, E., Ping, P., Vondriska, T. M. Stress signaling by Tec tyrosine kinase in the ischemic myocardium. *Am. J. Physiol.* 2010, 299, H713-H722.

[5] Kitagawa, K., Matsumoto, M., Tagaya, M., Hata, R., Ueda, H., Niinobe, M., Handa, N., Fukunaga, R., Kimura, K., Mikoshiba, K., et al., 'Ischemic tolerance' phenomenon found in the brain. *Brain Res.* 1990, 528, 21-24.

[6] Costa, T., Constantino, L. C., Mendonca, B. P., Pereira, J. G., Herculano, B., Tasca, C. I., Boeck, C. R. N-methyl-D-aspartate preconditioning improves short-term motor deficits outcome after mild traumatic brain injury in mice. *J. Neurosci. Res.* 2010, 88, 1329-1337.

[7] Soriano, F. X., Papadia, S., Hofmann, F., Hardingham, N. R., Bading, H., Hardingham, G. E. Preconditioning doses of NMDA promote neuroprotection by enhancing neuronal excitability. *J. Neurosci.* 2006, 26, 4509-4518.

[8] Youssef, F. F., Addae, J. I., McRae, A., Stone, T. W. Long-term potentiation protects rat hippocampal slices from the effects of acute hypoxia. *Brain Res.* 2001, 907, 144-150.

[9] Dirnagl, U., Simon, R. P., Hallenbeck, J. M. Ischemic tolerance and endogenous neuroprotection. *Trends Neurosci.* 2003, 26, 248-254.

[10] Narayanan, S. V., Dave, K. R., Perez-Pinzon, M. A. Ischemic preconditioning and clinical scenarios. *Curr. Opin. Neurol.* 2013, 26, 1-7.

[11] Boeck, C. R, Kroth, E. H., Bronzatto, M. J., Vendite, D. Adenosine receptors co-operate with NMDA preconditioning to protect cerebellar granule cells against glutamate neurotoxicity. *Neuropharmacology.* 2005, 49, 17-24.

[12] de Araujo Herculano, B., Vandresen-Filho, S., Martins, W. C., Boeck, C. R., Tasca, C. I. NMDA preconditioning protects against quinolinic acid-induced seizures via PKA, PI3K and MAPK/ERK signaling pathways. *Behav. Brain Res.* 2011, 219, 92-97.

[13] Mio, Y., Shim, Y. H., Richards, E., Bosnjak, Z. J., Pagel, P. S., Bienengraeber, M. Xenon preconditioning: the role of prosurvival signaling, mitochondrial permeability transition and bioenergetics in rats. *Anesth. Anal.* 2009, 108, 858-866.

[14] Alfieri, A., Srivastava, S., Siow, R. C., Cash, D., Modo, M., Duchen, M. R., Fraser P. A., Williams, S. C., Mann, G. E. Sulforaphane preconditioning of the Nrf2/HO-1 defense pathway protects the cerebral vasculature against blood-brain barrier disruption and neurological deficits in stroke. *Free Radic. Biol. Med.* 2013, 65, 1012-1022.

[15] Ding, Z. M., Wu, B., Zhang, W. Q., Lu, X. J., Lin, Y. C., Geng, Y. J., Miao, Y. F. Neuroprotective Effects of Ischemic Preconditioning and Postconditioning on Global Brain Ischemia in Rats through the Same Effect on Inhibition of Apoptosis. *Internat. J. Mol. Sci.* 2012, 13, 6089-6101.

[16] Ditz, C., Bahlmann, L., Klaus, S., Keck, A., Onken, N., Gliemroth, J. Cerebral metabolism during experimental endotoxin shock and after preconditioning with monophosphoryl lipid A. *Clin. Neurol. Neurosurg.* 2014, 126C, 115-122.

[17] Kunz, A., Park, L., Abe, T., Gallo, E. F., Anrather, J., Zhou, P., Iadecola, C. Neurovascular protection by ischemic tolerance: role of nitric oxide and reactive oxygen species. *J. Neurosci.* 2007, 27, 7083-7093.

[18] Vandresen-Filho, S., de Araujo Herculano, B., Franco, J. L., Boeck, C. R., Dafre, A. L., Tasca, C. I. Evaluation of glutathione metabolism in NMDA preconditioning against quinolinic acid-induced seizures in mice cerebral cortex and hippocampus. *Brain Res.* 2007, 1184, 38-45.

[19] Zhu, X. C., Jiang, T., Zhang, Q. Q., Cao, L., Tan, M. S., Wang, H. F., Ding, Z. Z., Tan, L., Yu, J. T. Chronic metformin preconditioning provides neuroprotection via suppression of NF-kappaB-mediated inflammatory pathway in rats with permanent cerebral ischemia. *Mol. Neurobiol.* 2014.

[20] Orio, M., Kunz, A., Kawano, T., Anrather, J., Zhou, P., Iadecola, C. Lipopolysaccharide induces early tolerance to excitotoxicity via nitric oxide and cGMP. *Stroke.* 2007, 38, 2812-2817.

[21] Gidday, J. M. Cerebral preconditioning and ischaemic tolerance. *Nat. Rev. Neurosci.* 2006, 7, 437-448.

[22] Tsang, A., Hausenloy, D. J., Yellon, D. M. Myocardial postconditioning: reperfusion injury revisited. *Am. J. Physiol. Heart Circ. Physiol.* 2005, 289, H2-7.

[23] Danielisova, V., Gottlieb, M., Nemethova, M., Kravcukova, P., Domorakova, I., Mechirova, E., Burda, J. Bradykinin postconditioning protects pyramidal CA_1 neurons against delayed neuronal death in rat hippocampus. *Cell. Mol. Neurobiol.* 2009, 29, 871-878.

[24] Nagy, D., Kocsis, K., Fuzik, J., Marosi, M., Kis, Z., Teichberg, V. I., Toldi, J., Farkas, T. Kainate postconditioning restores LTP in ischemic hippocampal CA_1: onset-

dependent second pathophysiological stress. *Neuropharmacology*. 2011, 61, 1026-1032.

[25] Zhang, W., Miao, Y., Zhou, S., Jiang, J., Luo, Q., Qiu, Y. Neuroprotective effects of ischemic postconditioning on global brain ischemia in rats through upregulation of hippocampal glutamine synthetase. *J. Clin. Neurosci.* 2011, 18, 685-689.

[26] Gulati, P., Singh, N. Evolving possible link between PI3K and NO pathways in neuroprotective mechanism of ischemic postconditioning in mice. *Mol. Cell. Biochem.* 2014, 397, 255-265.

[27] Pignataro, G., Meller, R., Inoue, K., Ordonezm A. N., Ashleym, M. D., Xiong, Z., Gala, R., Simon, R. P. *In vivo* and *in vitro* characterization of a novel neuroprotective strategy for stroke: ischemic postconditioning. *J. Cereb. Blood Flow. Metab.* 2008, 28, 232-241.

[28] Zhao, H., Sapolsky, R. M., Steinberg, G. K. Interrupting reperfusion as a stroke therapy: ischemic postconditioning reduces infarct size after focal ischemia in rats. *J. Cereb. Blood Flow Metab.* 2006, 26, 1114-1121.

[29] Ren, C., Gao, X., Niu, G., Yan, Z., Chen, X., Zhao, H. Delayed postconditioning protects against focal ischemic brain injury in rats. *PloS One*. 2008, 3, e3851.

[30] Gomes, C. V., Kaster, M. P., Tome, A. R., Agostinho, P. M., Cunha, R. A. Adenosine receptors and brain diseases: neuroprotection and neurodegeneration. *Biochim. Biophys. Acta*. 2011, 1808, 1380-1399.

[31] Rivera-Oliver, M., Diaz-Rios, M. Using caffeine and other adenosine receptor antagonists and agonists as therapeutic tools against neurodegenerative diseases: a review. *Life Sci*. 2014, 101, 1-9.

[32] Wei, C. J., Li, W., Chen, J. F. Normal and abnormal functions of adenosine receptors in the central nervous system revealed by genetic knockout studies. *Biochim. Biophys. Acta*. 2011, 1808, 1358-1379.

[33] Cunha, R. A. Different cellular sources and different roles of adenosine: A_1 receptor-mediated inhibition through astrocytic-driven volume transmission and synapse-restricted A_{2A} receptor-mediated facilitation of plasticity. *Neurochem. Int.* 2008, 52, 65-72.

[34] Fredholm, B. B., Chen, J. F., Cunha, R. A., Svenningsson, P., Vaugeois, J. M. Adenosine and brain function. *Int. Rev. Neurobiol.* 2005, 63, 191-270.

[35] Rebola, N., Pinheiro, P. C., Oliveira, C. R., Malva, J. O., Cunha, R. A. Subcellular localization of adenosine A(1) receptors in nerve terminals and synapses of the rat hippocampus. *Brain Res*. 2003, 987, 49-58.

[36] Dunwiddie, T. V., Masino, S. A. The role and regulation of adenosine in the central nervous system. *Annu. Rev. Neurosci.* 2001, 24, 31-55.

[37] de Mendonca, A., Sebastiao, A. M., Ribeiro, J. A. Inhibition of NMDA receptor-mediated currents in isolated rat hippocampal neurones by adenosine A_1 receptor activation. *Neuroreport.* 1995, 6, 1097-1100.

[38] Deng, Q., Terunuma, M., Fellin, T., Moss, S. J., Haydon, P. G. Astrocytic activation of A_1 receptors regulates the surface expression of NMDA receptors through a Src kinase dependent pathway. *Glia.* 2011, 59, 1084-1093.

[39] Lovatt, D., Xu, Q., Liu, W., Takano, T., Smith, N. A., Schnermann, J., Tieu, K., Nedergaard, M. Neuronal adenosine release, and not astrocytic ATP release, mediates feedback inhibition of excitatory activity. *Proc. Natl. Acad. Sci. USA*. 2012, 109, 6265-6270.

[40] Pearson, T., Damian, K., Lynas, R. E., Frenguelli, B. G. Sustained elevation of extracellular adenosine and activation of A_1 receptors underlie the post-ischaemic inhibition of neuronal function in rat hippocampus *in vitro*. *J. Neurochem*. 2006, 97, 1357-1368.

[41] Pedata, F., Pugliese, A. M., Coppi, E., Dettori, I., Maraula, G., Cellai, L., Melani, A. Adenosine A receptors modulate acute injury and neuroinflammation in brain ischemia. *Mediators Inflamm*. 2014, 2014, 805198.

[42] Gouder, N., Scheurer, L., Fritschy, J. M., Boison, D. Overexpression of adenosine kinase in epileptic hippocampus contributes to epileptogenesis. *J. Neurosci*. 2004, 24, 692-701.

[43] Van Gompel, J. J., Bower, M. R., Worrell, G. A., Stead, M., Chang, S. Y., Goerss, S. J., Kim, I., Bennet, K. E., Meyer, F. B., Marsh, W. R., Blaha, C. D., Lee, K. H. Increased cortical extracellular adenosine correlates with seizure termination. *Epilepsia*. 2014, 55, 233-244.

[44] Boison, D. Modulators of nucleoside metabolism in the therapy of brain diseases. *Curr. Top. Med. Chem*. 2011, 11, 1068-1086.

[45] Chu, S., Xiong, W., Zhang, D., Soylu, H., Sun, C., Albensi, B. C., Parkinson, F. E. Regulation of adenosine levels during cerebral ischemia. *Acta Pharmacol. Sin*. 2013, 34, 60-66.

[46] Fedele, D. E., Gouder, N., Guttinger, M., Gabernet, L., Scheurer, L., Rulicke, T., Crestani, F., Boison, D. Astrogliosis in epilepsy leads to overexpression of adenosine kinase, resulting in seizure aggravation. *Brain*. 2005, 128, 2383-2395.

[47] Fukuda, M., Suzuki, Y., Hino, H., Ishii, E. Over-activation of adenosine A(2A) receptors and sudden unexpected death in epilepsy. *Epilepsy Behav*. 2012. 387-398.

[48] Hu, S., Dong, H., Zhang, H., Wang, S., Hou, L., Chen, S., Zhang, J., Xiong, L. Noninvasive limb remote ischemic preconditioning contributes neuroprotective effects via activation of adenosine A_1 receptor and redox status after transient focal cerebral ischemia in rats. *Brain Res*. 2012, 1459, 81-90.

[49] Fedele, D. E., Li, T., Lan, J. Q., Fredholm, B. B., Boison, D. Adenosine A_1 receptors are crucial in keeping an epileptic focus localized. *Exp. Neurol*. 2006, 200, 184-190.

[50] Boison, D. Role of adenosine in status epilepticus: a potential new target? *Epilepsia* 2013, 54 Suppl 6, 20-22.

[51] Williams-Karnesky, R. L., Stenzel-Poore, M. P. Adenosine and stroke: maximizing the therapeutic potential of adenosine as a prophylactic and acute neuroprotectant. *Curr. Neuropharmacol*. 2009, 7, 217-227.

[52] Olsson, T., Cronberg, T., Rytter, A., Asztely, F., Fredholm, B. B., Smith, M. L., Wieloch, T. Deletion of the adenosine A_1 receptor gene does not alter neuronal damage following ischaemia *in vivo* or *in vitro*. *Eur. J. Neurosci*. 2004, 20, 1197-2204.

[53] Turner, C. P., Yan, H., Schwartz, M., Othman, T., Rivkees, S. A. A_1 adenosine receptor activation induces ventriculomegaly and white matter loss. *Neuroreport*. 2002, 13, 1199-2204.

[54] Turner, C. P., Seli, M., Ment, L., Stewart, W., Yan, H., Johansson, B., Fredholm, B. B., Blackburn, M., Rivkees, S. A. A_1 adenosine receptors mediate hypoxia-induced ventriculomegaly. *Proc. Natl. Acad. Sci. USA*. 2003, 100, 11718 11722.

[55] Cunha, R. A., Correia-de-Sa, P., Sebastiao, A. M., Ribeiro, J. A. Preferential activation of excitatory adenosine receptors at rat hippocampal and neuromuscular synapses by

adenosine formed from released adenine nucleotides. *Br. J. Pharmacol.* 1996, 119, 253-260.

[56] Pinto-Duarte, A., Coelho, J. E., Cunha, R. A., Ribeiro, J. A., Sebastiao, A. M. Adenosine A_{2A} receptors control the extracellular levels of adenosine through modulation of nucleoside transporters activity in the rat hippocampus. *J. Neurochem.* 2005, 93, 595-604.

[57] Lopes, L. V., Cunha, R. A., Kull, B., Fredholm, B. B., Ribeiro, J. A. Adenosine A(2A) receptor facilitation of hippocampal synaptic transmission is dependent on tonic A(1) receptor inhibition. *Neuroscience.* 2002, 112, 319-329.

[58] Marchi, M., Raiteri, L., Risso, F., Vallarino, A., Bonfanti, A., Monopoli, A., Ongini, E., Raiteri, M. Effects of adenosine A_1 and A_{2A} receptor activation on the evoked release of glutamate from rat cerebrocortical synaptosomes. *Br. J. Pharmacol.* 2002, 136, 434-440.

[59] Rodrigues, R. J., Alfaro, T. M., Rebola, N., Oliveira, C. R., Cunha, R. A. Co-localization and functional interaction between adenosine A(2A) and metabotropic group 5 receptors in glutamatergic nerve terminals of the rat striatum. *J. Neurochem.* 2005, 92, 433-441.

[60] Nishizaki, T., Nagai, K., Nomura, T., Tada, H., Kanno, T., Tozaki, H., Li, X. X., Kondoh, T., Kodama, N., Takahashi, E., Sakai, N., Tanaka, K., Saito, N. A new neuromodulatory pathway with a glial contribution mediated via A(2a) adenosine receptors. *Glia.* 2002, 39, 133-147.

[61] Pintor, A., Galluzzo, M., Grieco, R., Pezzola, A., Reggio, R., Popoli, P. Adenosine A_{2A} receptor antagonists prevent the increase in striatal glutamate levels induced by glutamate uptake inhibitors. *J. Neurochem.* 2004, 89, 152-156.

[62] Etherington, L. A., Frenguelli, B. G. Endogenous adenosine modulates epileptiform activity in rat hippocampus in a receptor subtype-dependent manner. *Eur. J. Neurosci.* 2004, 19, 2539-2550.

[63] Rosim, F. E., Persike, D. S., Nehlig, A., Amorim, R. P., de Oliveira, D. M., Fernandes, M. J. Differential neuroprotection by A(1) receptor activation and A(2A) receptor inhibition following pilocarpine-induced status epilepticus. *Epilepsy Behav.* 2011, 22, 207-213.

[64] Rebola, N., Rodrigues, R. J., Oliveira, C. R., Cunha, R. A. Different roles of adenosine A_1, A_{2A} and A_3 receptors in controlling kainate-induced toxicity in cortical cultured neurons. *Neurochem. Int.* 2005, 47, 317-325.

[65] El Yacoubi, M., Ledent, C., Parmentier, M., Costentin, J., Vaugeois, J. M. Evidence for the involvement of the adenosine A(2A) receptor in the lowered susceptibility to pentylenetetrazol-induced seizures produced in mice by long-term treatment with caffeine. *Neuropharmacology.* 2008, 55, 35-40.

[66] Rebola, N., Porciuncula, L. O., Lopes, L. V., Oliveira, C. R., Soares-da-Silva, P., Cunha, R. A. Long-term effect of convulsive behavior on the density of adenosine A_1 and A_{2A} receptors in the rat cerebral cortex. *Epilepsia.* 2005, 46 Suppl 5, 159-165.

[67] Shinohara, M., Saitoh, M., Nishizawa, D., Ikeda, K., Hirose, S., Takanashi, J., Takita, J., Kikuchi, K., Kubota, M., Yamanaka, G., Shiihara, T., Kumakura, A., Kikuchi, M., Toyoshima, M., Goto, T., Yamanouchi, H., Mizuguchi, M. $ADORA_{2A}$ polymorphism predisposes children to encephalopathy with febrile status epilepticus. *Neurology.* 2013, 80, 1571-1576.

[68] Fedorova, I. M., Jacobson, M. A., Basile, A., Jacobson, K.A. Behavioral characterization of mice lacking the A_3 adenosine receptor: sensitivity to hypoxic neurodegeneration. *Cell. Mol. Neurobiol.* 2003, 23, 431-447.

[69] Chen, G. J., Harvey, B. K., Shen, H., Chou, J., Victor, A., Wang, Y. Activation of adenosine A_3 receptors reduces ischemic brain injury in rodents. *J. Neurosci. Res.* 2006, 84, 1848-1855.

[70] Choi, I. Y., Lee, J. C., Ju, C., Hwang, S., Cho, G. S., Lee, H. W., Choi, W. J., Jeong, L. S., Kim, W. K. A_3 adenosine receptor agonist reduces brain ischemic injury and inhibits inflammatory cell migration in rats. *Am. J. Pathol.* 2011, 179, 2042-2052.

[71] Hentschel, S., Lewerenz, A., Nieber, K. Activation of A(3) receptors by endogenous adenosine inhibits synaptic transmission during hypoxia in rat cortical neurons. *Restor. Neurol. Neurosci.* 2003, 21, 55-63.

[72] Bjorklund, O., Shang, M., Tonazzini, I., Dare, E., Fredholm, B. B. Adenosine A_1 and A_3 receptors protect astrocytes from hypoxic damage. *Eur. J. Pharmacol.* 2008, 596, 6-13.

[73] Gessi, S., Merighi, S., Stefanelli, A., Fazzi, D., Varani, K., Borea, P. A. A(1) and A(3) adenosine receptors inhibit LPS-induced hypoxia-inducible factor-1 accumulation in murine astrocytes. *Pharmacol. Res.* 2013, 76, 157-170.

[74] Von Lubitz, D. K., Lin, R. C., Melman, N., Ji, X. D., Carter, M. F., Jacobson, K. A. Chronic administration of selective adenosine A_1 receptor agonist or antagonist in cerebral ischemia. *Eur. J. Pharmacol.* 1994, 256, 161-167.

[75] Pugliese, A. M., Latini, S., Corradetti, R., Pedata, F. Brief, repeated, oxygen-glucose deprivation episodes protect neurotransmission from a longer ischemic episode in the *in vitro* hippocampus: role of adenosine receptors. *Br. J. Pharmacol.* 2003, 140, 305-314.

[76] Pugliese, A. M., Coppi, E., Spalluto, G., Corradetti, R., Pedata, F. A_3 adenosine receptor antagonists delay irreversible synaptic failure caused by oxygen and glucose deprivation in the rat CA_1 hippocampus *in vitro*. *Br. J. Pharmacol.* 2006, 147, 524-532.

[77] Pugliese, A. M., Coppi, E., Volpini, R., Cristalli, G., Corradetti, R., Jeong, L. S., Jacobson, K. A., Pedata, F. Role of adenosine A_3 receptors on CA_1 hippocampal neurotransmission during oxygen-glucose deprivation episodes of different duration. *Biochem. Pharmacol.* 2007, 74, 768-779.

[78] Laudadio, M. A., Psarropoulou, C. The A_3 adenosine receptor agonist 2-Cl-IB-MECA facilitates epileptiform discharges in the CA_3 area of immature rat hippocampal slices. *Epilepsy Res.* 2004, 59, 83-94.

[79] Popoli, P., Pepponi, R. Potential therapeutic relevance of adenosine A2B and A_{2A} receptors in the central nervous system. *CNS Neurol. Disord. Drug Targets.* 2012, 11, 664-674.

[80] Yamagata, K., Hakata, K., Maeda, A., Mochizuki, C., Matsufuji, H., Chino, M., Yamori, Y. Adenosine induces expression of glial cell line-derived neurotrophic factor (GDNF) in primary rat astrocytes. *Neurosci. Res.* 2007, 59, 467-474.

[81] Allaman, I., Lengacher, S., Magistretti, P. J., Pellerin, L. A2B receptor activation promotes glycogen synthesis in astrocytes through modulation of gene expression. *Am. J. Physiol. Cell. Physiol.* 2003, 284, C696-704.

[82] Moidunny, S., Vinet, J., Wesseling, E., Bijzet, J., Shieh, C. H., van Ijzendoorn, S. C., Bezzi, P., Boddeke, H. W., Biber, K. Adenosine A2B receptor-mediated leukemia

inhibitory factor release from astrocytes protects cortical neurons against excitotoxicity. *J. Neuroinflamm.* 2012, 9, 198.

[83] Roseti, C., Martinello, K., Fucile, S., Piccari, V., Mascia, A., Di Gennaro, G., Quarato, P. P., Manfredi, M., Esposito, V., Cantore, G., Arcella, A., Simonato, M., Fredholm, B. B., Limatola, C., Miledi, R., Eusebi, F. Adenosine receptor antagonists alter the stability of human epileptic $GABA_A$ receptors. *Proc Natl Acad Sci USA*. 2008, 105, 15118-15123.

[84] Arendash, G. W., Schleif, W., Rezai-Zadeh, K., Jackson, E. K., Zacharia, L. C., Cracchiolo, J. R., Shippy, D., Tan, J. Caffeine protects Alzheimer's mice against cognitive impairment and reduces brain beta-amyloid production. *Neuroscience*. 2006, 142, 941-952.

[85] Cao, C., Cirrito, J. R., Lin, X., Wang, L., Verges, D. K., Dickson, A., Mamcarz, M., Zhang, C., Mori, T., Arendash, G. W., Holtzman, D. M., Potter, H. Caffeine suppresses amyloid-beta levels in plasma and brain of Alzheimer's disease transgenic mice. *J Alzheimers Dis*. 2009, 17, 681-697.

[86] Chu, Y. F., Chang, W. H., Black, R. M., Liu, J. R., Sompol, P., Chen, Y., Wei, H., Zhao, Q., Cheng, I. H. Crude caffeine reduces memory impairment and amyloid β(1-42) levels in an Alzheimer's mouse model. *Food Chem.* 2012, 135, 2095-2102.

[87] Espinosa, J., Rocha, A., Nunes, F., Costa, M. S., Schein, V., Kazlauckas, V., Kalinine, E., Souza, D. O., Cunha, R. A., Porciuncula, L. O. Caffeine consumption prevents memory impairment, neuronal damage, and adenosine A_{2A} receptors upregulation in the hippocampus of a rat model of sporadic dementia. *J. Alzheimers Dis*. 2013, 34, 509-518.

[88] Dall'Igna, O. P., Porciuncula, L. O., Souza, D. O., Cunha, R. A., Lara, D. R. Neuroprotection by caffeine and adenosine A_{2A} receptor blockade of beta-amyloid neurotoxicity. *Br. J. Pharmacol.* 2003, 138, 1207-1219.

[89] Laurent, C., Eddarkaoui, S., Derisbourg, M., Leboucher, A., Demeyer, D., Carrier, S., Schneider, M., Hamdane, M., Muller, C. E., Buee, L., Blum, D. Beneficial effects of caffeine in a transgenic model of Alzheimer's disease-like tau pathology. *Neurobiol. Aging*. 2014, 35, 2079-2090.

[90] Canas, P. M., Porciuncula, L. O., Cunha, G. M., Silva, C. G., Machado, N. J., Oliveira, J. M., Oliveira, C. R., Cunha, R. A. Adenosine A_{2A} receptor blockade prevents synaptotoxicity and memory dysfunction caused by beta-amyloid peptides via p38 mitogen-activated protein kinase pathway. *J. Neurosci.* 2009, 29, 14741-14751.

[91] Almeida, C. G., de Mendonca, A., Cunha, R. A., Ribeiro, J. A. Adenosine promotes neuronal recovery from reactive oxygen species induced lesion in rat hippocampal slices. *Neurosci. Lett.* 2003, 339, 127-130.

[92] Cunha, R. A. Neuroprotection by adenosine in the brain: From A(1) receptor activation to A (2A) receptor blockade. *Purinergic Signal.* 2005, 1, 111-134.

[93] Matos, M., Augusto, E., Machado, N. J., dos Santos-Rodrigues, A., Cunha, R. A., Agostinho, P. Astrocytic adenosine A_{2A} receptors control the amyloid-beta peptide-induced decrease of glutamate uptake. *J. Alzheimers Dis.* 2012, 31, 555-567.

[94] Jenner, P., Mori, A., Hauser, R., Morelli, M., Fredholm, B. B., Chen, J. F. Adenosine, adenosine A_{2A} antagonists, and Parkinson's disease. *Parkinsonism Relat. Disord.* 2009, 15, 406-413.

[95] Shook, B. C., Rassnick, S., Osborne, M. C., Davis, S., Westover, L., Boulet, J., Hall, D., Rupert, K. C., Heintzelman, G. R., Hansen, K., Chakravarty, D., Bullington, J. L., Russell, R., Branum, S., Wells, K. M,. Damon, S., Youells, S., Li, X., Beauchamp, D. A., Palmer, D., Reyes, M., Demarest, K., Tang, Y., Rhodes, K., Jackson, P.F. *In vivo* characterization of a dual adenosine A_{2A}/A_1 receptor antagonist in animal models of Parkinson's disease. *J. Med. Chem.* 2010, 53, 8104-8115.

[96] Mizuno, Y., Kondo, T. Adenosine A_{2A} receptor antagonist istradefylline reduces daily off time in Parkinson's disease. *Mov. Disord.* 2013, 28, 1138-1141.

[97] Wardas, J. Potential role of adenosine A_{2A} receptors in the treatment of schizophrenia. *Front. Biosci.* 2008, 13, 4071-4096.

[98] Sperlagh, B., Csolle, C., Ando, R. D., Goloncser, F., Kittel, A., Baranyi, M. The role of purinergic signaling in depressive disorders. *Neuropsychopharmacol. Hung.* 2012, 14, 231-238.

[99] Burnstock, G., Krügel, U., Abbracchio, M. P., Illes, P. Purinergic signalling: from normal behaviour to pathological brain function. *Prog. Neurobiol.* 2011, 95, 229-274.

[100] Woodson, J. C., Minor, T. R., Job, R. F. Inhibition of adenosine deaminase by erythro-9-(2-hydroxy-3-nonyl)adenine (EHNA) mimics the effect of inescapable shock on escape learning in rats. *Behav. Neurosci.* 1998, 112, 399-409.

[101] El Yacoubi, M., Costentin, J., Vaugeois, J. M. Adenosine A_{2A} receptors and depression. *Neurology.* 2003, 61, S82-S87.

[102] Coelho, J. E., Rebola, N., Fragata, I., Ribeiro, J. A., de Mendonca, A., Cunha, R. A. Hypoxia-induced desensitization and internalization of adenosine A_1 receptors in the rat hippocampus. *Neuroscience.* 2006, 138, 1195-1203.

[103] Hines, D. J., Schmitt, L. I., Hines, R. M., Moss, S. J., Haydon, P. G. Antidepressant effects of sleep deprivation require astrocyte-dependent adenosine mediated signaling. *Transl. Psych.* 2013, 3, e212.

[104] Kaster, M. P., Rosa, A. O., Rosso, M. M., Goulart, E. C., Santos, A. R., Rodrigues, A. L. Adenosine administration produces an antidepressant-like effect in mice: evidence for the involvement of A_1 and A_{2A} receptors. *Neurosci. Lett.* 2004, 355, 21-24.

[105] Endres, M., Engelhardt, B., Koistinaho, J., Lindvall, O., Meairs, S., Mohr, J. P., Planas, A., Rothwell, N., Schwaninger, M., Schwab, M. E., Vivien, D., Wieloch, T., Dirnagl, U. Improving outcome after stroke: overcoming the translational roadblock. *Cerebrovasc. Dis.* 2008, 25, 268-278.

[106] Yue, X., Mehmet, H., Penrice, J., Cooper, C., Cady, E., Wyatt, J. S., Reynolds, E. O., Edwards, A. D., Squier, M. V. Apoptosis and necrosis in the newborn piglet brain following transient cerebral hypoxia-ischaemia. *Neuropathol. Appl. Neurobiol.* 1997, 23, 16-25.

[107] Perez-Pinzon, M. A. Mechanisms of neuroprotection during ischemic preconditioning: lessons from anoxic tolerance. *Comp. Biochem. Physiol.* 2007, 147, 291-299.

[108] Di-Capua, N., Sperling, O., Zoref-Shani, E. Protein kinase C-epsilon is involved in the adenosine-activated signal transduction pathway conferring protection against ischemia-reperfusion injury in primary rat neuronal cultures. *J. Neurochem.* 2003, 84, 409-412.

[109] Heurteaux, C., Lauritzen, I., Widmann, C., Lazdunski, M. Essential role of adenosine, adenosine A_1 receptors, and ATP-sensitive K^+ channels in cerebral ischemic preconditioning. *Proc. Natl. Acad. Sci. USA.* 1995, 92, 4666-4670.

[110] Liu, X. Q., Sheng, R., Qin, Z. H. The neuroprotective mechanism of brain ischemic preconditioning. *Acta Pharmacol. Sin.* 2009, 30, 1071-1080.

[111] Reshef, A., Sperling, O., Zoref-Shani, E. Role of K(ATP) channels in the induction of ischemic tolerance by the 'adenosine mechanism' in neuronal cultures. *Adv. Exper. Med. Biol.* 2000, 486, 217-221.

[112] Rudolphi, K. A., Schubert, P., Parkinson, F. E., Fredholm, B. B. Adenosine and brain ischemia. *Cerebrovasc. Brain Metabol. Rev.* 1992, 4, 346-369.

[113] Von Lubitz, D. K., Lin, R. C., Popik, P., Carter, M. F., Jacobson, K. A. Adenosine A_3 receptor stimulation and cerebral ischemia. *Eur. J. Pharmacol.* 1994, 263, 59-67.

[114] Chen, J. F., Huang, Z., Ma, J., Zhu, J., Moratalla, R., Standaert, D., Moskowitz, M. A., Fink, J. S., Schwarzschild, M. A. A(2A) adenosine receptor deficiency attenuates brain injury induced by transient focal ischemia in mice. *J. Neurosci.* 1999, 19, 9192-9200.

[115] Sebastiao, A. M., Ribeiro, F. F., Ribeiro, J. A. From A_1 to A_3 en passant through A(2A) receptors in the hippocampus: pharmacological implications. *CNS Neurol. Disord. Drug Targ.* 2012, 11, 652-663.

[116] Dennis, S. H., Jaafari, N., Cimarosti, H., Hanley, J. G., Henley, J. M., Mellor, J. R. Oxygen/glucose deprivation induces a reduction in synaptic AMPA receptors on hippocampal CA_3 neurons mediated by mGluR1 and adenosine A_3 receptors. *J. Neurosci.* 2011, 31, 11941-11952.

[117] Kinoshita, H. Editorial: novel insights into the role of anesthetics and opioids in organ or tissue protection. *Curr. Pharm. Des.* 2014, 20, 5671-5682.

[118] Wells, B. A., Keats, A. S., Cooley, D. A. Increased tolerance to cerebral ischemia produced by general anesthesia during temporary carotid occlusion. *Surgery.* 1963, 54, 216-223.

[119] Goldstein, A., Jr., Wells, B. A., Keats, A. S. Increased tolerance to cerebral anoxia by pentobarbital. *Arch. Internat. Pharm. Ther.* 1966, 161, 138-143.

[120] Zaugg, M., Lucchinetti, E., Uecker, M., Pasch, T., Schaub, M. C. Anaesthetics and cardiac preconditioning. Part I. Signalling and cytoprotective mechanisms. *Br. J. Anaesth.* 2003, 91, 551-565.

[121] Downey, J. M., Davis, A. M., Cohen, M. V. Signaling pathways in ischemic preconditioning. *Heart Fail. Rev.* 2007, 12, 181-188.

[122] Surendra, H., Diaz, R. J., Harvey, K., Tropak, M., Callahan, J., Hinek, A., Hossain, T., Redington, A., Wilson, G. J. Interaction of delta and kappa opioid receptors with adenosine A_1 receptors mediates cardioprotection by remote ischemic preconditioning. *J. Mol. Cell. Cardiol.* 2013, 60, 142-150.

[123] Centonze, D., Saulle, E., Pisani, A., Bernardi, G., Calabresi, P. Adenosine-mediated inhibition of striatal GABAergic synaptic transmission during *in vitro* ischaemia. *Brain.* 2001, 124, 1855-1865.

[124] MacGregor, D. G., Graham, D. I., Stone, T. W. The attenuation of kainate-induced neurotoxicity by chlormethiazole and its enhancement by dizocilpine, muscimol, and adenosine receptor agonists. *Exp. Neurol.* 1997, 148, 110-123.

[125] Churchill, E. N., Disatnik, M. H., Budas, G. R., Mochly-Rosen, D. Ethanol for cardiac ischemia: the role of protein kinase c. *Therap. Adv. Cardiovasc. Dis.* 2008, 2, 469-483.

[126] Collins, M. A., Neafsey, E. J., Mukamal, K. J., Gray, M. O., Parks, D. A., Das, D. K., Korthuis, R. J. Alcohol in moderation, cardioprotection, and neuroprotection:

epidemiological considerations and mechanistic studies. *Alcohol. Clin. Exp. Res.* 2009, 33, 206-219.

[127] Klatsky, A. L., Armstrong, M. A., Friedman, G. D. Risk of cardiovascular mortality in alcohol drinkers, ex-drinkers and nondrinkers. *Am. J. Cardiol.* 1990, 66, 1237-1242.

[128] Rimm, E. B., Giovannucci, E. L., Willett, W. C., Colditz, G. A., Ascherio, A., Rosner, B., Stampfer, M. J. Prospective study of alcohol consumption and risk of coronary disease in men. *Lancet.* 1991, 338, 464-468.

[129] Sesso, H. D., Stampfer, M. J., Rosner, B., Hennekens, C. H., Manson, J. E., Gaziano, J. M. Seven-year changes in alcohol consumption and subsequent risk of cardiovascular disease in men. *Arch. Intern. Med.* 2000, 160, 2605-2612.

[130] Thun, M. J., Peto, R., Lopez, A. D., Monaco, J. H., Henley, S. J., Heath, C. W. Jr., Doll, R. Alcohol consumption and mortality among middle-aged and elderly U.S. adults. *N. Engl. J. Med.* 1997, 337, 1705-1714.

[131] Pagel, P. S., Kersten, J. R., Warltier, D. C. Mechanisms of myocardial protection produced by chronic ethanol consumption. *Pathophysiology* 2004, 10, 121-129.

[132] Mukamal, K. J., Chung, H., Jenny, N. S., Kuller, L. H., Longstreth, W. T. Jr., Mittleman, M. A., Burke, G. L., Cushman, M., Beauchamp, N. J. Jr., Siscovick, D. S. Alcohol use and risk of ischemic stroke among older adults: the cardiovascular health study. *Stroke.* 2005, 36, 1830-1834.

[133] Testai, L., Rapposelli, S., Calderone, V. Cardiac ATP-sensitive potassium channels: a potential target for an anti-ischaemic pharmacological strategy. *Cardiovasc. Hematol. Agents Med. Chem.* 2007, 5, 79-90.

[134] Lai, T. W., Zhang, S., Wang, Y. T. Excitotoxicity and stroke: identifying novel targets for neuroprotection. *Prog. Neurobiol.* 2014, 115, 157-188.

[135] Vizi, E. S., Kisfali, M., Lorincz, T. Role of nonsynaptic GluN2B-containing NMDA receptors in excitotoxicity: evidence that fluoxetine selectively inhibits these receptors and may have neuroprotective effects. *Brain Res. Bull.* 2013, 93, 32-38.

[136] Craig, C. G., White, T. D. N-methyl-D-aspartate- and non-N-methyl-D-aspartate-evoked adenosine release from rat cortical slices: distinct purinergic sources and mechanisms of release. *J. Neurochem.* 1993, 60, 1073-1080.

[137] Hoehn, K., White, T. D. Role of excitatory amino acid receptors in K^+- and glutamate-evoked release of endogenous adenosine from rat cortical slices. *J. Neurochem.* 1990, 54, 256-265.

[138] Melani, A., Pantoni, L., Corsi, C., Bianchi, L., Monopoli, A., Bertorelli, R., Pepeu, G., Pedata, F. Striatal outflow of adenosine, excitatory amino acids, gamma-aminobutyric acid, and taurine in awake freely moving rats after middle cerebral artery occlusion: correlations with neurological deficit and histopathological damage. *Stroke.* 1999, 30, 2448-2454.

[139] Boeck, C. R., Bronzatto, M. J., Souza, D. G., Sarkis, J. J., Vendite, D. The modulation of ecto-nucleotidase activities by glutamate in cultured cerebellar granule cells. *Neuroreport.* 2000, 11, 709-712.

[140] Bruno, A. N., Bonan, C. D., Wofchuk, S. T., Sarkis, J. J., Battastini, A. M. ATP diphosphohydrolase (NTPDase 1) in rat hippocampal slices and effect of glutamate on the enzyme activity in different phases of development. *Life Sci.* 2002, 71, 215-225.

[141] Bond, A., Lodge, D., Hicks, C. A., Ward, M. A., O'Neill, M. J. NMDA receptor antagonism, but not AMPA receptor antagonism attenuates induced ischaemic tolerance in the gerbil hippocampus. *Eur. J. Pharmacol.* 1999, 380, 91-99.

[142] Rejdak, R., Rejdak, K., Sieklucka-Dziuba, M., Stelmasiak, Z., Grieb, P. Brain tolerance and preconditioning. *Pol. J. Pharmacol.* 2001, 53, 73-79.

[143] Samoilov, M. O., Lazarevich, E. V., Semenov, D. G., Mokrushin, A. A., Tyul'kova, E. I., Romanovskii, D. Y., Milyakova, E. A., Dudkin, K. N. The adaptive effects of hypoxic preconditioning of brain neurons. *Neurosci. Behav. Physiol.* 2003, 33, 1-11.

[144] Schaller, B., Graf, R. Cerebral ischemic preconditioning. An experimental phenomenon or a clinical important entity of stroke prevention? *J. Neurol.* 2002, 249, 1503-1511.

[145] Chuang, D. M., Gao, X. M., Paul, S. M. N-methyl-D-aspartate exposure blocks glutamate toxicity in cultured cerebellar granule cells. *Mol. Pharmacol.* 1992, 42, 210-216.

[146] Dickie, B. G., Holmes, C., Greenfield, S. A. Neurotoxic and neurotrophic effects of chronic N-methyl-D-aspartate exposure upon mesencephalic dopaminergic neurons in organotypic culture. *Neuroscience.* 1996, 72, 731-741.

[147] Pringle, A. K., Thomas, S. J., Signorelli, F., Iannotti, F. Ischaemic pre-conditioning in organotypic hippocampal slice cultures is inversely correlated to the induction of the 72 kDa heat shock protein (HSP72). *Brain Res.* 1999, 845, 152-164.

[148] Valentim, L. M., Rodnight, R., Geyer, A. B., Horn, A. P., Tavares, A., Cimarosti, H., Netto, C. A., Salbego, C. G. Changes in heat shock protein 27 phosphorylation and immunocontent in response to preconditioning to oxygen and glucose deprivation in organotypic hippocampal cultures. *Neuroscience.* 2003, 118, 379-386.

[149] Ogita, K., Okuda, H., Yamamoto, Y., Nishiyama, N., Yoneda, Y. *In vivo* neuroprotective role of NMDA receptors against kainate-induced excitotoxicity in murine hippocampal pyramidal neurons. *J. Neurochem.* 2003, 85, 1336-1346.

[150] Miao, B., Yin, X. H., Pei, D. S., Zhang, Q. G., Zhang, G. Y. Neuroprotective effects of preconditioning ischemia on ischemic brain injury through down-regulating activation of JNK1/2 via N-methyl-D-aspartate receptor-mediated Akt1 activation. *J. Biol. Chem.* 2005, 280, 21693-21699.

[151] Boeck, C. R., Ganzella, M., Lottermann, A., Vendite, D. NMDA preconditioning protects against seizures and hippocampal neurotoxicity induced by quinolinic acid in mice. *Epilepsia.* 2004, 45, 745-750.

[152] do Amaral e Silva Muller, G., Vandresen-Filho, S., Tavares, C. P., Menegatti, A. C., Terenzi, H., Tasca, C. I., Severino, P. C. Proteomic analysis of the mice hippocampus after preconditioning induced by N-methyl-D-aspartate (NMDA). *J. Mol. Neurosci.* 2013, 50, 154-164.

[153] Vandresen-Filho, S., Hoeller, A. A., Herculano, B. A., Duzzioni, M., Duarte, F. S., Piermartiri, T. C., Boeck, C. R., de Lima, T. C., Marino-Neto, J., Tasca, C. I. NMDA preconditioning attenuates cortical and hippocampal seizures induced by intracerebroventricular quinolinic acid infusion. *Neurotox. Res.* 2013, 24, 55-62.

[154] Vandresen-Filho, S., Severino, P.C., Constantino, L. C., Martins, W. C., Molz, S., Dal-Cim T Bertoldo, D. B., Silva, F. R., Tasca, C. I. N-methyl-D-aspartate preconditioning prevents quinolinic acid-induced deregulation of glutamate and calcium homeostasis in mice hippocampus. *Neurotox. Res.* 2015, 27, 118-128.

[155] Wu, L. G., Saggau, P. Presynaptic inhibition of elicited neurotransmitter release. *Trends Neurosci*. 1997, 20, 204-212.

[156] Ciruela, F., Casado, V., Rodrigues, R. J., Lujan, R., Burgueno, J., Canals, M., Borycz, J., Rebola, N., Goldberg, S. R., Mallol, J., Cortes, A., Canela, E. I., Lopez-Gimenez, J. F., Milligan, G., Lluis, C., Cunha, R. A., Ferre, S., Franco, R. Presynaptic control of striatal glutamatergic neurotransmission by adenosine A_1-A_{2A} receptor heteromers. *J. Neurosci*. 2006, 26, 2080-2087.

[157] Ciruela, F., Fernandez-Duenas, V., Llorente, J., Borroto-Escuela, D., Cuffi, M. L., Carbonell, L., Sanchez, S., Agnati, L. F., Fuxe, K., Tasca, C. I. G protein-coupled receptor oligomerization and brain integration: focus on adenosinergic transmission. *Brain Res*. 2012, 1476, 86-95.

[158] Zhao, Z. Q., Corvera, J. S., Halkos, M. E., Kerendi, F., Wang, N. P., Guyton, R. A.,Vinten-Johansen, J. Inhibition of myocardial injury by ischemic postconditioning during reperfusion: comparison with ischemic preconditioning. *Am. J. Physiol*. 2003, 285, H579-H588.

[159] Roubille, F., Franck-Miclo, A., Covinhes, A., Lafont, C., Cransac, F., Combes, S., Vincent, A., Fontanaud, P., Sportouch-Dukhan, C., Redt-Clouet, C., Nargeot, J., Piot, C., Barrere-Lemaire, S. Delayed postconditioning in the mouse heart *in vivo*. *Circulation*. 2011, 124, 1330-1336.

[160] Xing, B., Chen, H., Zhang, M., Zhao, D., Jiang, R., Liu, X., Zhang, S. Ischemic postconditioning inhibits apoptosis after focal cerebral ischemia/reperfusion injury in the rat. *Stroke*. 2008, 39, 2362-2369.

[161] Laskey, W. K., Yoon, S., Calzada, N., Ricciardi, M. J. Concordant improvements in coronary flow reserve and ST-segment resolution during percutaneous coronary intervention for acute myocardial infarction: a benefit of postconditioning. *Cath. Cardiovasc. Intervent*. 2008, 72, 212-220.

[162] Staat, P., Rioufol, G., Piot, C., Cottin, Y., Cung, T. T., L'Huillier, I., Aupetit, J. F., Bonnefoy, E., Finet, G., Andre-Fouet, X., Ovize, M. Postconditioning the human heart. *Circulation*. 2005, 112, 2143-2148.

[163] Philipp, S., Yang, X. M., Cui, L., Davis, A. M., Downey, J. M., Cohen, M.V. Postconditioning protects rabbit hearts through a protein kinase C-adenosine A2b receptor cascade. *Cardiovasc. Res*. 2006, 70, 308-314.

[164] Kerendi, F., Kin, H., Halkos, M. E., Jiang, R., Zatta, A. J., Zhao, Z. Q., Guyton, R. A., Vinten-Johansen. J. Remote postconditioning. Brief renal ischemia and reperfusion applied before coronary artery reperfusion reduces myocardial infarct size via endogenous activation of adenosine receptors. *Basic Res. Cardiol*. 2005, 100, 404-412.

[165] Eldaif, S. M., Deneve, J. A., Wang, N. P., Jiang, R., Mosunjac, M., Mutrie, C. J., Guyton, R. A., Zhao, Z. Q., Vinten-Johansen, J. Attenuation of renal ischemia-reperfusion injury by postconditioning involves adenosine receptor and protein kinase C activation. *Transplant Internat*. 2010, 23, 217-226.

[166] Methner, C., Schmidt, K., Cohen, M. V., Downey, J. M., Krieg, T. Both A_{2A} and A_{2B} adenosine receptors at reperfusion are necessary to reduce infarct size in mouse hearts. *Am. J. Physiol*. 2010, 299, H1262-H1264.

[167] Kin, H., Zatta, A. J., Lofye, M. T., Amerson, B. S., Halkos, M. E., Kerendi, F., Zhao, Z. Q., Guyton, R. A., Headrick, J. P., Vinten-Johansen, J. Postconditioning reduces infarct

size via adenosine receptor activation by endogenous adenosine. *Cardiovasc. Res.* 2005, 67,124-133.

[168] Vinten-Johansen, J., Zhao, Z. Q., Jiang R, Zatta A. J., Dobson G. P. Preconditioning and postconditioning: innate cardioprotection from ischemia-reperfusion injury. *J. Appl. Physiol.* 2007, 103, 1441-1448.

[169] Cohen, M. V., Downey, J. M. Adenosine: trigger and mediator of cardioprotection. *Basic Res. Cardiol.* 2008, 103, 203-215.

[170] Leconte, C., Tixier, E., Freret, T., Toutain, J., Saulnier, R., Boulouard, M., Roussel, S., Schumann-Bard, P., Bernaudin, M. Delayed hypoxic postconditioning protects against cerebral ischemia in the mouse. *Stroke.* 2009, 40, 3349-3355.

[171] Pignataro, G., Scorziello, A., Di Renzo, G., Annunziato, L. Post-ischemic brain damage: effect of ischemic preconditioning and postconditioning and identification of potential candidates for stroke therapy. *FEBS J.* 2009, 276, 46-57.

[172] Zhang, W., Wang, B., Zhou, S., Qiu, Y. The effect of ischemic post-conditioning on hippocampal cell apoptosis following global brain ischemia in rats. *J. Clin. Neurosci.* 2012, 19, 570-573.

[173] Tanaka, H., Yokota, H., Jover, T., Cappuccio, I., Calderone, A., Simionescu, M., Bennett, M. V., Zukin, R. S. Ischemic preconditioning: neuronal survival in the face of caspase-3 activation. *J. Neurosci.* 2004, 24, 2750-2759.

[174] Zhang, X., Zhang, Q., Tu, J., Zhu, Y., Yang, F., Liu, B., Brann, D., Wang, R. Prosurvival NMDA 2A receptor signaling mediates postconditioning neuroprotection in the hippocampus. *Hippocampus.* 2015, 25, 286-296.

[175] Ling Ling, J., Wong, G. T., Yao, L., Xia, Z., Irwin, M. G. Remote pharmacological post-conditioning by intrathecal morphine: cardiac protection from spinal opioid receptor activation. *Acta Anaesth. Scand.* 2010, 54, 1097-1104.

[176] Dal Ponte, C., Alchera, E., Follenzi, A., Imarisio, C., Prat, M., Albano, E., Carini, R. Pharmacological postconditioning protects against hepatic ischemia/reperfusion injury. *Liver Transplant.* 2011, 17, 474-482.

[177] Lemoine, S., Buleon, C., Rouet, R., Ivascau, C., Babatasi, G., Massetti, M., Gerard, J. L., Hanouz, J. L. Bradykinin and adenosine receptors mediate desflurane induced postconditioning in human myocardium: role of reactive oxygen species. *BMC Anesthesiol.* 2010, 10, 12.

[178] Gerczuk, P. Z., Kloner, R. A. An update on cardioprotection: a review of the latest adjunctive therapies to limit myocardial infarction size in clinical trials. *J. Amer. Coll. Cardiol.* 2012, 59, 969-978.

[179] Bibli, S. I., Iliodromitis, E. K., Lambertucci, C., Zoga, A., Lougiakis, N., Dagres, N., Volpini, R., Dal Ben, D., Kremastinos, D.T., Tsantili Kakoulidou, A., Cristalli, G., Andreadou, I. Pharmacological postconditioning of the rabbit heart with non-selective, A_1, A_{2A} and A_3 adenosine receptor agonists. *J. Pharm. Pharmacol.* 2014, 66, 1140-1149.

[180] Fretwell, L., Dickenson, J. M. Role of large-conductance $Ca(2)^+$-activated K^+ channels in adenosine A(1) receptor-mediated pharmacological postconditioning in H9c2 cells. *Can. J. Physiol. Pharmacol.* 2011, 89, 24-30.

[181] Kim, J. A., Li, L., Zuo, Z. Isoflurane induces a postconditioning effect on bovine pulmonary arterial endothelial cells exposed to oxygen-glucose deprivation. *Eur. J. Pharmacol.* 2009, 615, 144-149.

[182] Li, L., Zuo, Z. Isoflurane postconditioning induces neuroprotection via Akt activation and attenuation of increased mitochondrial membrane permeability. *Neuroscience*. 2011, 199, 44-50.

[183] Wang, H. Y., Wang, G. L., Yu, Y. H., Wang, Y. The role of phosphoinositide-3-kinase/Akt pathway in propofol-induced post-conditioning against focal cerebral ischemia-reperfusion injury in rats. *Brain Res*. 2009, 1297, 177-184.

[184] Yang, Z., Chen, Y., Zhang, Y., Jiang, Y., Fang, X., Xu, J. Sevoflurane postconditioning against cerebral ischemic neuronal injury is abolished in diet-induced obesity: role of brain mitochondrial K_{ATP} channels. *Mol. Med. Rep*. 2014, 9, 843-850.

[185] McMurtrey, R. J., Zuo, Z. Isoflurane preconditioning and postconditioning in rat hippocampal neurons. *Brain Res*. 2010, 1358, 184-190.

In: Adenosine Signaling Mechanisms
Editors: V. Ramkumar and R. Paes de Carvalho
ISBN: 978-1-63483-186-4

Chapter 13

FUNCTIONAL INTERACTIONS OF ADENOSINE RECEPTORS AND THEIR POSSIBLE IMPLICATIONS IN CENTRAL NERVOUS SYSTEM DISEASES

Patrizia Popoli*[*], *Maria Rosaria Domenici, Alberto Martire and Maria Teresa Tebano
Section of Central Nervous System Pharmacology, Department of Therapeutic Research and Medicines Evaluation, Istituto Superiore di Sanità, Rome, Italy

ABSTRACT

Adenosine is an ubiquitous endogenous nucleoside. In the central nervous system (CNS), it plays a pivotal role in the control of neuronal excitability and in the modulation of both physiological and pathological processes. Adenosine activities are mediated by the binding to four distinct G-protein-coupled receptors, namely the A_1, A_{2A}, A_{2B}, and A_3 adenosine receptor subtypes. Multiple and converging evidence demonstrated that adenosine receptors physically and functionally interact with other receptors in the modulation of CNS functions. The aim of this chapter is to describe and critically discuss the most relevant interactions involving the high affinity A_1 and A_{2A} adenosine receptors, and their possible implications for neurological and psychiatric diseases.

INTRODUCTION

Adenosine is an endogenous nucleoside widely distributed in mammal tissues where it is involved in the regulation of several functions. In the central nervous system (CNS), it acts as a neuromodulator, controlling neuron excitability and modulating the activity of other cell types, including astrocytes and microglia. Unlike neurotransmitters, adenosine is not stored in vesicles and released from neurons following their activation, but it can be released from any

[*] Correspondence to: Dr. Patrizia Popoli, Istituto Superiore di Sanità, Viale Regina Elena, 299, Rome, Italy. E-mail: patrizia.popoli@iss.it.

cell when its intracellular concentration increases, through bidirectional equilibrative nucleoside transporters. Alternatively, extracellular adenosine can originate from nucleotidase-mediated hydrolysis of locally released adenosine triphosphate (ATP). Usually, extracellular adenosine levels are kept in the range of tens nanomolar by the activity of adenosine deaminase and adenosine kinase, the two main enzymes involved in the clearance of adenosine. In normal conditions, adenosine activates its high affinity receptors, namely A_1 and A_{2A} subtypes (see below), while in situations in which adenosine extracellular levels are higher, such as during hypoxia, epileptic seizures or in presence of an increased demand and decreased availability of energy, the low affinity A_{2B} and A_3 receptor subtypes are also stimulated. All the adenosine receptors belong to the G-protein coupled receptor (GPCR) family, membrane receptors with seven transmembrane domains that activate different signal transduction pathways through the coupling with different G proteins [1, 2]. Given the scarce expression in the CNS and the paucity of studies on the role of the low affinity A_{2B} and A_3 receptors in CNS diseases, we will focus on A_1 and A_{2A} receptors and their interactions with other receptor systems, discussing the possible implications for CNS diseases.

Function, Expression and Signaling of Adenosine A_1 and A_{2A} Receptors in the CNS

The adenosine A_1 receptor (A_1R) is widely distributed in the CNS with the highest peak of expression in the cortex, hippocampus, cerebellum and spinal cord [3]. Classically, A_1Rs couple with inhibitory G protein (G_i or G_o) to inhibit the adenylate cyclase (AC)-cyclic adenosine monophosphate (cAMP)-protein kinase A (PKA) signaling pathway, although there is evidence that in some circumstances they can couple with different G proteins [4, 5]. As for the subcellular compartments, the A_1R is present both at pre- and postsynaptic level and its stimulation inhibits the presynaptic release of neurotransmitters and (by inhibiting Ca^{2+}-conductances and activating K^+ currents) hyperpolarizes postsynaptic neurons [6], with the final result of reducing neuronal activity. Therefore adenosine, through stimulation of A_1Rs, can be considered as an endogenous neuroprotective agent with antiepileptic activity.

The adenosine A_{2A} receptor ($A_{2A}R$) is also quite ubiquitous in the CNS, although its expression levels vary considerably in the different regions. The dorsal and ventral striatum show the highest levels of $A_{2A}Rs$, localized on medium sized spiny neurons (MSNs), but appreciable levels of expression are also found in other brain areas [7, 8]. At the subcellular level, the $A_{2A}R$ shows a prevalent postsynaptic expression in the caudate putamen and a main presynaptic localization in the hippocampus. Specifically, in the caudate-putamen it is strongly expressed on the MSNs of the indirect pathway that project to the globus pallidus, while hippocampal $A_{2A}Rs$ are mostly located in nerve terminals and particularly in the presynaptic active zone [9, 10]. However, even though at low level, a presynaptic localization of $A_{2A}Rs$ has been demonstrated also in the striatum and specifically in cortical glutamatergic terminals that contact striatal neurons of the direct pathway (neurons that project to the substantia nigra) where they modulate corticostriatal neurotransmission [11]. Originally, $A_{2A}R$ stimulation was demonstrated to activate the AC-cAMP-PKA pathway through coupling with G_s protein [12], although in the striatum the involvement of G_{olf} has been clearly identified. Now it is clear that in addition to signal through a PKA-dependent pathway, $A_{2A}R$-mediated effects can also require the involvement of protein kinase C (PKC,

[13, 14]). Moreover, when expressed in Chinese hamster ovary (CHO) cells, all adenosine receptors are able to induce phosphorylation of extracellular signal-regulated kinases (ERK) [15] and several studies have documented the ability of A_{2A}Rs to control mitogen-activated protein kinases (MAPK) pathways in a cAMP-independent manner [16-19].

Once released from neurons or glial cells, adenosine is able to influence the actions of different neurotransmitters or neuromodulators, through an intracellular interaction occurring at the signaling transducing systems [20]. In addition to this, in the last years receptor-receptor interactions occurring at plasma membrane level have received particular attention and several studies demonstrated that adenosine receptors are able to physically and functionally interact with other receptors. Functional and physical interactions have been demonstrated with neurotransmitter receptors such as dopamine and metabotropic glutamate receptors but also with receptors for endocannabinoids, neuromodulators widely distributed in the CNS that are released 'on demand' to control neuronal activity [21]. In addition, interactions between different adenosine receptors, such as between A_1 and A_{2A} receptors have also been documented. In this chapter, the most relevant (and the most documented) receptor-receptor interactions will be reported and their possible implications for CNS diseases discussed.

Interactions between Adenosine A_{2A} Receptors and Dopamine D_2 Receptors

The interaction of A_{2A} with dopamine D_2 receptors (D_2Rs) has been extensively studied, and has been the object of many reviews, especially concerning the potential relevance of such an interaction for Parkinson's disease [22-24]. For this reason, the whole matter will be not extensively re-reviewed, and only some specific issues will be mentioned in this chapter.

Another important premise concerns the overt (and intentional) reduction of discussing the A_{2A}/D_2 receptor interaction as the basis of the effects elicited by A_{2A}R manipulation in Parkinson's disease, schizophrenia, and other conditions. This is, exactly, a simplification, since the role of A_{2A}Rs in those conditions involves receptors others than D_2, relies on mechanisms different from the inhibition of D_2Rs and is, in a few words, much more complex than that depicted here. Obviously, the same caution applies to all the other interactions reported in this chapter.

Functional Role of A_{2A}-D_2 Receptor Heteromerization

Dopamine receptors (D_1-D_5 Rs) are GPCRs divided into the D_1-like (D_1 and D_5) and D_2-like (D_2, D_3, and D_4) subgroups. Unlike D_1-like, the D_2-like family is inhibitory and couples to Gi/o [25, 26]. In the last years, the study of the interaction between A_{2A} and D_2 receptors has been focused on the occurrence of functional heteromeric receptor complexes. A receptor heteromer is defined as a macromolecular complex composed of at least two (functional) receptor units with biochemical properties that are demonstrably different from those of its individual components [27]. The existence of A_{2A}-D_2 receptor heteromers has been demonstrated in co-immunoprecipitation studies showing that these complexes undergo coaggregation and co-internalization, and that they co-desensitize after combined agonist treatment [28]. Further evidence was provided by biophysical techniques, such as resonance energy transfer (bioluminescence and fluorescence resonance energy transfer: BRET and FRET, respectively) [29, 30].

The adenosine A_{2A}-D_2 receptor heteromerization represents an example of a functionally relevant electrostatic interaction [27]. When such an interaction is disrupted, the quaternary structure of the heteromer is destabilized [31], and functional effects such as the A_{2A}R-dependent inhibition of depression of striatal neuronal firing induced by D_2Rs is lost [32]. Thus, the heteromer represents a major molecular basis for the antagonistic A_{2A}/D_2 receptor interactions.

A_{2A}/D_2 Receptor Interaction and Parkinson's Disease

The interactions between A_{2A} and D_2 receptors are very relevant not only for the regulation of basal ganglia physiology, but also for diseases in which D_2Rs have been classically considered as the major pathophysiological players (and, thus, the major therapeutic targets). The most important of the abovementioned diseases is Parkinson's disease (PD), which is characterized by degeneration of dopaminergic neurons, a consequent reduction of striatal dopamine levels, and thus a severe motor impairment. According to the models of basal ganglia functioning, the indirect and the direct pathways project to the external segment of the globus pallidus and to the internal segment of the globus pallidus/substantia nigra pars reticulata, respectively [33, 34]. As mentioned before, A_{2A}Rs are mainly expressed by the striatopallidal neurons, which are the first station of the indirect pathway and which express D_2Rs as well. Dopamine arising from the substantia nigra pars compacta, by acting on D_2Rs, inhibits the indirect pathway which, in turn, has a repressive influence on movement execution. In PD, the inhibitory activity exerted by dopamine through the activation of D_2Rs is lost.

Since the early '90s, it has been shown that A_{2A}Rs antagonistically interact with D_2Rs at the membrane level. In particular, Ferré et al., [35] showed that the stimulation of A_{2A}Rs potently reduces the affinity of D_2R agonist binding sites in striatal membranes, and hypothesized that such a receptor-receptor interaction could explain the motor stimulating effects of adenosine antagonists. In agreement, the activation of A_{2A}Rs was found to reduce spontaneous motility and to inhibit the hyperlocomotion induced by D_2Rs agonists [36-38].

The neurochemical and behavioral evidence suggesting that A_{2A}Rs functionally oppose the actions of D_2Rs on striatopallidal neurons paved the way for the investigation of the antiparkinsonian effects of A_{2A}R antagonists. Indeed, A_{2A}R antagonists have been successfully tested in different models of PD (for review see [39-43]). Just to mention some of the results, such compounds have been reported to reverse catalepsy in rodents [44, 45] and to increase motor activity in animals rendered parkinsonian by the administration of neurotoxins 6-hydroxydopamine (6-OHDA) in rodents [46, 47] and 1-methyl-4-phenyl-1,2,3,6-tetrahydropyridine (MPTP) in rodents and non-human primates [48-50]. These data represent a strong preclinical signal for the potential of A_{2A}R antagonists in the management of PD patients. Indeed, many of these drugs have been tested in clinical trials [51 53]. Although several positive results have been obtained, in some cases efficacy was not consistently proven. Anyway, while levodopa remains the most effective agent to treat motor symptoms in PD, $A_{2A}R$ antagonists can still be regarded as a good approach to reduce the doses of levodopa, to target symptoms non-levodopa-responsive and to prevent some levodopa associated drawbacks. Furthermore, these drugs not only have the advantage of a non-dopaminergic approach, but do have a clear neuroprotective potential. Thus, the discovery and the clinical development of new, efficacious and safe $A_{2A}R$ antagonists remains a priority for the treatment of PD.

A_{2A}/D_2 Receptor Interaction and Schizophrenia

Schizophrenia is a debilitating, incurable mental disorder presenting (among others) with both 'positive' (delusions, hallucinations) and 'negative' (reduced effectiveness, anhedonia, social isolation) symptoms [54]. Positive symptoms are classically ascribed to a hyperdopaminergic state related to high dopamine D_2R activity [55, 56]. In agreement, D_2Rs of the ventral striatopallidal neurons appear to be involved in the antipsychotic effects of neuroleptics. Besides the dorsal striatum, a powerful antagonistic A_{2A}-D_2 receptor interaction was demonstrated in the ventral striatopallidal system, thus suggesting that $A_{2A}R$ agonists can be considered as a potential treatment for schizophrenia [57]. In agreement, agonists of $A_{2A}Rs$ showed an antipsychotic profile in animal models of psychosis [36, 58, 59]. For reviews on the role of adenosine and adenosine receptors in schizophrenia see [60-62].

Recently, the relevance of the A_{2A}/D_2 heteroreceptor complexes within the ventral striatum for the treatment of schizophrenia has been critically re-examined [63]. According to the Authors, the antipsychotic properties of $A_{2A}R$ agonists may depend, at least in part, on the increase in β-arrestin 2 signaling over the D_2 protomer in the A_{2A}-D_2 heteromer.

A_{2A}/D_2 Receptor Interaction and Huntington's Disease

Huntington's disease (HD) is a dominantly inherited neurodegenerative disorder characterized by progressively worsening chorea, psychiatric disturbances and cognitive impairment [64]. Interestingly, the striatopallidal neurons are the most vulnerable in HD [65] and their early dysfunction is responsible for the development of chorea.

The (very) complex and controversial role of $A_{2A}Rs$ in the pathophysiology of HD has been the object of several reviews [66-68] and will not be discussed here. In this chapter we will only consider some issues specifically linked to the potential relevance of the A_{2A}/D_2 receptor interaction in a hyperkinetic condition such as HD.

Unfortunately, the most widely used transgenic model of HD (R6/2 mice) [69] fails to reproduce the pivotal symptom of the human disease, i.e., the hyperkinetic syndrome. Rather, R6/2 mice are a hypokinetic model, and thus they are not suitable to study the potential anti-hyperkinetic effects of a drug. To test the influence of an adenosine $A_{2A}R$ agonist, we used the quinolinic acid (QA) model of HD in rats [70]. Interestingly, the intrastriatal injection of the $A_{2A}R$ agonist CGS 21680 completely prevented the motor hyper-responsiveness to d-amphetamine typical of lesioned rats. Although the involvement of D_2Rs in such an effect was not directly tested, it seems very likely, since the motor depressant activity of $A_{2A}R$ agonists mainly relies on the inhibition of D_2R-mediated signaling.

A_{2A}/D_2 Receptor Interaction and Dystonias

Dystonias are a complex class of movement disorders, including at least 15 syndromes referred to as DYT1 to DYT15 [71]. DYT1 dystonia, also known as early-onset torsion dystonia or Oppenheim's disease, is the most severe form of hereditary dystonia [72]. Although the molecular and cellular events underlying dystonia still need to be clarified, an abnormal dopaminergic neurotransmission has been suggested to play a role [73-75]. According to the hypothesis that dystonia occurs after a preferential decrease in D_2R-mediated inhibition of the indirect basal ganglia pathway [76], a blockade of A_{2A}Rs was considered as a potential approach to increase D_2R activity. In agreement, Napolitano et al., [77] found that the selective A_{2A}R antagonist SCH 58621 fully restored the impairment of synaptic plasticity observed in hMT mice (a transgenic model of DYT1 dystonia), thus indicating A_{2A}Rs as a potential target for this disease.

Interactions between Adenosine A_{2A} and Metabotropic mGlu$_5$ Receptors

Group I metabotropic glutamate receptors (mGluRs) are highly expressed in the striatum where they contribute to synaptic plasticity [78, 79] and to excitotoxic injury [80, 81]. Metabotropic glutamate receptor 5 (mGlu$_5$R) is particularly abundant in the striatal GABAergic efferent neurons [82], where, as already mentioned, A_{2A}Rs are abundantly expressed. As for the hippocampus, in spite of their low levels of expression, A_{2A}Rs influence normal synaptic transmission and modulate excitotoxic processes [83]. In this area, mGlu$_5$Rs are markedly expressed and, by setting the tone of N-methyl-D-aspartate (NMDA) receptors [79-81], mediate both physiological and pathological processes such as learning and memory, epilepsy, excitotoxicity [84-87]. In the striatum, both A_{2A} and mGlu$_5$ receptors are mainly localized postsynaptically [9, 88] but they have been found to co-localize in glutamatergic nerve terminals [89]. In the hippocampus, A_{2A} and mGlu$_5$ receptors are co-localized in glutamatergic synapses and they are present both in the presynaptic active zone and at the postsynaptic site, although with different subsynaptic distribution [90]. The co-localization of A_{2A} and mGlu$_5$ receptors and their proximity state on synapses in different brain areas [9, 90], together with the demonstrated ability of mGluRs to form constitutive homo- or heteromers [91], are in line with the existence of physical and functional interactions between these two receptors.

The existence of A_{2A}/mGlu$_5$ receptor heteromeric complexes and the occurrence of a functional synergistic interaction have been brightly demonstrated by Ferré et al., [92]. Since these receptors synergistically act on ERK1/2 phosphorylation and c-fos induction, and given the key role of ERK1/2 and c-*fos* in the coupling of early neuronal responses to long-term adaptive changes [93-95], the A_{2A}/mGlu$_5$ receptor interactions seem to be very important in corticostriatal plasticity [96, 97] and in dopamine-independent c-*fos* expression in the striatopallidal neurons associated with sensitization to psychostimulants [98]. Other functional interactions between striatal mGlu$_5$ and A_{2A} receptors have been reported in the modulation of excitatory neurotransmission [illegible] and in the phosphorylation of dopamine and cAMP-regulated phosphoprotein (DARPP-32) [100]. Moreover, in the striatum both A_{2A}

and $mGlu_5$ receptors modulate presynaptic release of glutamate [101-103]. In addition, in the striatum and in the hippocampus, the state of activation of A_{2A}Rs influences the ability of $mGlu_5$Rs to potentiate NMDA effects and toxicity [90, 99]; in the hippocampus this happens mainly at the postsynaptic level, where all the three receptors are expressed [90]. Then, the neuroprotective effects of A_{2A}R antagonists towards neurodegeneration and excitotoxicity [104] could be mediated not only by their ability to modulate glutamate outflow, but also by the removal of the permissive role of A_{2A}Rs on $mGlu_5$R-mediated toxicity.

A_{2A}/$mGlu_5$ Receptor Interaction in the Modulation of Excitotoxicity

Given the influence of both A_{2A} and $mGlu_5$ receptors on glutamate release and NMDA functioning, excitotoxicity (defined as cell death ensuing from the toxic action of glutamate through excessive activation of glutamate receptors), represents a condition in which the interaction between these two receptors plays a primary role. A_{2A}R blockade has been consistently shown to be protective in various neuropathological models involving excitotoxicity [68, 105, 106]. Oppositely, the A_{2A}R agonist CGS21680 was shown able to significantly increase extracellular glutamate in the rat striatum [101, 107]. As already mentioned, the activation of $mGlu_5$Rs contributes to set the tone of NMDA receptors [108] and to enhance their effects (reviewed in [84, 109]). Since, as previously reported, A_{2A}Rs exert a strong permissive/facilitatory activity on $mGlu_5$R-induced effects it can be argued that the state of activation of A_{2A}Rs regulates the effects exerted by $mGlu_5$Rs in terms of excitotoxicity [110]. Among the neurodegenerative disorders involving primarily the striatum, the altered glutamatergic neurotransmission is one of the main determinants in the pathophysiology of HD [111]. In the QA rat model of HD, both A_{2A}R antagonists and group I mGluR antagonists exert a protective action [112, 113]. The A_{2A}/$mGlu_5$ receptor interaction thus deserves to be further investigated as a potential therapeutic approach to this disease.

By using mice lacking A_{2A}Rs, Tebano et al., [90] demonstrated that also hippocampal A_{2A}Rs exert a permissive role on $mGlu_5$R-mediated potentiation of NMDA responses. A_{2A}Rs might thus represent an interesting target for the development of new therapeutic strategies for disorders involving changes in NMDA receptor signaling, such as Alzheimer's disease, epilepsy and schizophrenia, where they might exert a neuroprotective effect similar to that recognized for striatal diseases [113].

A_{2A}/$mGlu_5$ Receptor Interaction and Schizophrenia

Schizophrenia is a disorder in which NMDA receptor dysfunction seems to be implicated [114-117]. Antagonists of NMDA receptors exacerbate pre-existing symptoms in patients with schizophrenia [118-120], and produce schizophrenia-like symptoms in healthy individuals [121, 122]. Thus, NMDA receptor hypofunction has been implicated in the pathophysiology of this disease [123], suggesting that augmenting such activity may be a plausible therapeutic strategy for the treatment of some symptoms of schizophrenia [117, 124, 125]. It has been suggested that group I mGluRs, particularly $mGlu_5$R, may also contribute to schizophrenia [126, 127]. Given the role of NMDA in this disease, and on the basis of the existing interaction between $mGlu_5$ and NMDA receptors (see above), the modulation of

$mGlu_5Rs$ could represent a complementary therapeutic approach to schizophrenia [115, 117, 128, 129], and the $A_{2A}/mGlu_5$ receptor interaction might play a role as well. Interestingly, indeed, after their central co-administration, A_{2A} and $mGlu_5$ receptors agonists synergistically counteracted phencyclidine-induced motor activation [92], which represents a model of schizophrenia [121].

$A_{2A}/mGlu_5$ Receptor Interaction and Parkinson's Disease

It is now recognized that the control of the responsiveness of the medium spiny neurons of the indirect pathway depends not only on D_2 and A_{2A} receptors but also on $mGlu_5Rs$ [130]. At the membrane level, while the A_{2A} and the D_2 receptors interact antagonistically (see above), $mGlu_5Rs$ and $A_{2A}Rs$ act synergistically to counteract the D_2R-mediated signaling. A synergistic interaction between A_{2A} and $mGlu_5$ receptors was demonstrated at the level of MAPK, DARPP-32 phosphorylation and early gene expression. The existence of a trimeric complex between D_2, A_{2A} and $mGlu_5$ receptors has been also demonstrated [130]. As occurs with drugs blocking $A_{2A}Rs$ [131], antagonists of $mGlu_5Rs$ also display anti-parkinsonian properties [132-135] and reduced L-DOPA-induced dyskinesias (LID) in parkinsonian monkeys [136]. Interestingly, the latter effect was associated with a normalization of glutamate [137] and dopamine neurotransmission [138]. The antidyskinetic activity of $mGlu_5Rs$ antagonists may be caused by desensitization of the $A_{2A}/mGlu_5$ receptor heteromeric complex, leading to removal of the blockade of D_2R-induced signaling effect [23]. The striatal A_{2A} D_2 $mGlu_5$ multimeric receptors complexes are involved in the striatal plasticity and seem to be relevant for the management of PD and LID [92, 23, 139]. Accordingly, A_{2A} and $mGlu_5$ receptors antagonists showed synergistic antiparkinsonian effects in 6-OH-dopamine lesioned rats [140, 141]. It could be hypothesised that the antiparkinsonian effects of A_{2A} or $mGlu_5$ receptors antagonists might not only depend on their ability to control the responsiveness of neurons of the indirect pathway, which is hyperfunctional in PD [142], but also on their ability to control the release of glutamate that is actually the trigger of striatal pathways [78].

$A_{2A}/mGlu_5$ Receptor Interaction and Fragile X

Fragile X Syndrome (FXS) is the most common heritable form of mental retardation and is accompanied by neuropsychiatric problems [143]. The disease is due to a mutation of the *fmr1* gene and to the resulting absence of the fragile X mental retardation protein (FMRP). FMRP normally inhibits the translation of mRNA at synaptic sites [144, 145] which, oppositely, is stimulated by mGluRs. The absence of FMRP leads to increased protein synthesis and to defects in synaptic plasticity including an exacerbated $mGlu_5R$-dependent long-term depression (LTD) found in Fmr1-KO mice (a mouse model of FXS) [146]. Several $mGlu_5R$ inhibitors are under clinical examination but, unfortunately, a clear beneficial effect has not been observed. Alternatively to the direct inhibition of $mGlu_5R$, which is important for synaptic activity, the research of other therapeutic strategies is requested. As we demonstrated that $A_{2A}Rs$ exert a strong permissive role on $mGlu_5R$-mediated synaptic effects [99, 90], the inhibition of $A_{2A}R$ signaling could lead to a reduced $mGlu_5R$ functioning.

INTERACTION BETWEEN ADENOSINE AND ENDOCANNABINOID RECEPTORS

The endocannabinoids, mainly represented by 2-arachidonoylglycerol (2-AG) and N-arachidonoylethanolamide (anandamide), are neuromodulators that exert their action in different regions of the CNS, having a role on several functions such as movement control, mechanisms of reward and addiction, memory and cognition, pain and others. They bind two subtypes of receptors, CB_1 and CB_2 receptors. The CB_1 subtype is the one predominantly expressed in the adult CNS and it is considered the most abundant GPCR in the brain [147, 148]. Typically, CB_1R binding sites are located on presynaptic nerve terminals where, by coupling to Gi proteins control the release of neurotransmitters. CB_1Rs are activated by the retrograde action of endocannabinoids (synthetized and released upon postsynaptic depolarization) and their activation inhibits neurotransmitter release [149, 150], thus influencing synaptic transmission and synaptic plasticity. Although an exclusive presynaptic expression of striatal CB_1Rs has been suggested by some authors [151, 152], in the striatum a postsynaptic localization has been reported by Köfalvi et al., [153], who identified CB_1Rs in the somatodendritic area of MSNs.

The interaction between adenosine and cannabinoid receptors can occur at the level of signal transduction pathways, involving the recruitment of G proteins, as probably occurs between A_1 and CB_1 receptors, or through receptor heteromerization, as it has been demonstrated for A_{2A} and CB_1 receptors.

A_1/CB_1 Receptor Interaction

A_1 and CB_1 receptors are both negatively associated with AC to reduce cAMP levels and mainly presynaptically localized to control the release of neurotransmitters and synaptic transmission. A first demonstration of a functional interaction between A_1 and CB_1 receptors comes from the study of Hoffman et al., [154], which demonstrated in the hippocampus that the activation of A_1R by endogenous adenosine reduces CB_1R-mediated inhibition of glutamate release. Following, an A_1-CB_1 receptor interaction has been demonstrated to influence both GABA and glutamate release in the hippocampus, in a way that A_1R activation attenuated the CB_1R-mediated inhibition of GABA and glutamate release [155].

A_1/CB_1 Receptor Interaction and Cognitive Deficits

Alterations of synaptic transmission can be crucial for the development of cognitive decline typical of aging or neurodegenerative disease, or as a consequence of the use of drugs of abuse such as marijuana. That the endogenous activation of either A_1 or CB_1 receptors is involved in spatial learning is demonstrated by the facilitatory effect of selective A_1R antagonists on learning and memory [156-159], and by the reported increased memory performance in mice subjected to a systemic administration of cannabinoid antagonists [160-163]. The co-administration of CB_1 and A_1 receptors antagonists synergistically improves the acquisition of spatial memory in mice, and this seems to be related to the modulation of glutamate release and NMDA receptor activation in the hippocampus [164]. These findings suggest that this interaction can be relevant and exploited for the treatment of cognitive dysfunctions. Importantly, since both receptors are widely expressed in the CNS, the use of combined low doses of the antagonists might avoid a strong receptor blockade and thus the occurrence of widespread side effects.

A_{2A}/CB_1 Receptor Interaction

A functional interaction between A_{2A} and CB_1 receptors was first reported in the hypolocomotor and rewarding effects of Δ-9-tetrahydrocannabinol (THC) [165, 166]. Then, the existence of a physical interaction has been demonstrated, with the identification of heteromeric complexes between A_{2A} and CB_1 receptors in co-transfected HEK-293T cells and in the rat striatum [167]. This interaction has been shown to play a significant role in the motor depressant effects of CB_1R agonists, in a way that CB_1R signaling controlling motor activity depends on $A_{2A}R$ activation [167]. The A_{2A}/CB_1 receptors interaction is relevant also in the modulation of striatal synaptic transmission, suggesting that an endogenous activation of $A_{2A}Rs$ is generally required to allow a proper CB_1R functioning in the striatum [168].

However, several evidence suggest that $A_{2A}Rs$ may also negatively control CB_1R-dependent effects: i) in the striatum chronic $A_{2A}R$ blockade by caffeine increases the presynaptic effects of CB_1R stimulation at the GABAergic synapses [169]; ii) the blockade or activation of $A_{2A}Rs$ increases or decreases, respectively, CB_1-dependent LTD in the dorsal striatum [170, 171]; iii) striatal $A_{2A}R$ activation inhibited CB_1R-mediated depression of synaptic transmission and CB_1R-mediated inhibition of 4-aminopyridine-evoked glutamate release [172]. The inhibitory effect of $A_{2A}R$ activation on CB_1R-mediated inhibition of glutamate release has been further corroborated in purified striatal nerve terminals, where a precise presynaptic localization of A_{2A}/CB_1 receptor heteromers has been demonstrated [173]. On the contrary, even though a postsynaptic localization cannot be excluded, a clear demonstration of the presence of A_{2A}/CB_1 receptor heteromers at the postsynaptic level is still lacking.

All these data clearly demonstrate the existence of a functional and physical interaction between A_{2A} and CB_1 receptors, which could be relevant in CNS functions. Although the real impact of this interaction in CNS disease is far to be understood, we can identify some pathological condition where this interaction seems to be involved and for which we can anticipate a therapeutic application.

A_{2A}/CB_1 Receptor Interaction and Drug Addiction

The first evidence that some of the effects of THC, the main psychoactive constituent of marijuana, could be controlled by $A_{2A}Rs$ was the finding that in $A_{2A}R$ knock-out mice the rewarding and aversive responses produced by THC were abolished and the severity of THC withdrawal attenuated [166]. Subsequently, it was shown that $A_{2A}R$ blockade prevents synergy between μ opiate and cannabinoid CB_1Rs and eliminates heroin-seeking behavior in addicted rats [174]. The use of $A_{2A}R$ antagonist for the treatment of cannabis abuse was suggested by the study of Justinová et al., [175], which demonstrated that $A_{2A}R$ antagonists, at low doses, might selectively reduce reinforcing effects of cannabinoids. Later it was demonstrated that the preferentially presynaptic blockade of A_{2A}Rs reduces reinforcing effects of THC in squirrel monkeys, while the treatment with a preferentially postsynaptic A_{2A}R antagonist (KW-6002) shifts the dose-response curves for THC to the left, consistently with the potentiation of the reinforcing effects of THC [176]. Thus, striatal presynaptic A_{2A}Rs can be viewed as a new target for the treatment of cannabis abuse. Whether this interaction takes place at the level of the A_{2A}/CB_1 receptor heteromers or involves the signal transduction mechanisms is still not clear.

A_{2A}/CB_1 Receptor Interaction and Parkinson's Disease

As already mentioned, the motor depressions produced by CB_1R agonists are reduced by A_{2A}R inactivation or blockade [165, 167]. These findings, together with the biochemical evidence of the existence of striatal heteromeric complexes, indicate that some functional effects of striatal CB_1Rs depend on their physical and functional interactions with A_{2A}Rs. Recently, the existence of receptor heteromeric complexes has been demonstrated in rodent and primate models of PD [177, 178]. Interestingly, the expression of heteromers formed by A_{2A}, CB_1 and/or D_2 receptors is dramatically diminished in the striatum of parkinsonian monkeys receiving chronic L-DOPA treatment, providing the first evidence of altered heteromer expression during the course of a human disease [177]. The functional cross-talk existing between the adenosine and cannabinoid system could help to explain the finding that, while A_{2A} and CB_1 receptors antagonists have neuroprotective properties against 6-OHDA-induced nigrostriatal degeneration when given alone, their effects disappear when the two drugs are given together [179]. These findings could be related to the permissive/facilitatory role of A_{2A}Rs on CB_1R functions, probably by direct interaction within A_{2A}R-CB_1R heteromers [180]. In addition, since the blockade of CB_1R reduced the locomotor-activating effects of A_{2A}R antagonists [170], the opposite modulation may also occur and explain why in the presence of a CB_1R antagonist, the neuroprotective and anti-inflammatory effects of the A_{2A}R antagonist are blunted [179].

Interactions between Adenosine A_1 and A_{2A} Receptors

As previously reported, endogenous adenosine activates both A_1 and A_{2A} high affinity receptors which, operating through opposing signal transduction pathways, play inhibitory and facilitatory effects, respectively, on neuronal activity. Since A_1Rs are more abundant and have a slightly higher affinity for adenosine, as observed in functional studies [181], their inhibitory effects prevail in basal conditions, whereas the A_{2A}R-dependend stimulatory effects predominate in case of prolonged stimulation [182]. Nevertheless, these two receptors subtypes cannot be considered as entities independent one from another, and when co-expressed in the same cells such as in glutamatergic neurons of the hippocampus [183] and striatum [184], they may establish a molecular and/or functional interaction [103, 185]. The existence of a functional cross talk between A_1 and A_{2A} receptors was clearly documented in the hippocampus in young adult rats where A_{2A}R agonists decreased the inhibitory effect of

A_1R agonists on synaptic transmission [186, 187] and the binding affinity of the A_1R for the agonist [188]. Furthermore, the $A_{2A}R$-induced facilitation of synaptic transmission in the CA1 area of rat hippocampal slices is dependent on tonic A_1R inhibition [103]. These results suggested that $A_{2A}R$ agonists facilitate hippocampal synaptic transmission by attenuating the tonic effect of inhibitory presynaptic A_1Rs located in glutamatergic nerve terminals. Desensitization of A_1R binding site by the $A_{2A}R$ selective agonist CGS 21680 could contribute to this effect [3]. This might represent a fine-tuning role for adenosine $A_{2A}Rs$ to allow frequency-dependent plasticity phenomena without compromising the A_1R-mediated neuroprotective role of adenosine [189, 190]. The finding that A_1 and A_{2A} receptors are co-localized in approximately 25% of glutamatergic terminals of rat hippocampus [183] has provided an anatomical basis for such a functional interaction.

As previously mentioned, the basal ganglia represent the cerebral area with the highest expression of $A_{2A}Rs$. Using the *in vivo* microdialysis technique in freely moving rats, Quarta et al., [191] found that both $A_{2A}R$ agonists and A_1R antagonists induced a dose dependent increase in striatal glutamate release. The existence of A_1-A_{2A} receptor heteromers on the cell surface of co-transfected cells and the co-localization of such receptors in the same striatal glutamatergic nerve terminals was subsequently demonstrated [192]. The main biochemical characteristic of the A_1-A_{2A} receptor heteromer is the ability of the $A_{2A}R$ activation to reduce the affinity of the A_1R for its agonist, and this interaction is unidirectional, namely the $A_{2A}R$ can control the A_1R functionality, but not *vice versa*.

As previously reported, under normal conditions adenosine preferentially stimulates A_1Rs inhibiting glutamatergic neurotransmission, while under conditions of stronger adenosine release, $A_{2A}R$ activation stimulates glutamate release, either directly or by blocking A_1R-mediated effects in the A_1-A_{2A} receptor heteromer. Thus, the A_1-A_{2A} receptor heteromer provides a "concentration-dependent switch" mechanism by which low and high concentrations of synaptic adenosine produce opposite effects, stimulating or inhibiting the release of glutamate [193, 194].

Even though astrocytic A_1-A_{2A} receptor heteromers have been reported to control extracellular gamma-aminobutyric acid (GABA) levels [195], the main result of the interaction between A_1 and A_{2A} receptors is represented by the control of excitatory transmission. Since excitotoxicity is involved in the aetiology of different brain disorders [196], the A_1/A_{2A} receptor interaction might thus constitute a novel target to control the demise of different neurological and psychiatric disorders [197]. Considering the key role of the concentration-dependent switch in the regulation of CNS excitability, the different mechanisms and sources of adenosine need to be briefly summarized (reviewed in [2, 16, 198, 199]).

Adenosine may be derived through two mechanisms: external transport (via nucleoside transporter) of intracellularly produced adenosine and formation in the extracellular space from ectonucleotidase-mediated degradation of adenine nucleotides (ATP and ADP) released by neurons and glia (for review [83, 198, 200]). ATP degradation often occurs in presence of a brain damage [201], probably to attempt preserving cell viability, and leading to a disproportionally higher formation of adenosine (reviewed in [16, 198, 199]). The first mechanism (external transport of adenosine), seems to be prevalent at lower frequencies of stimulation and favors the activation of A_1Rs [202]. Conversely, the contribution of ATP derived adenosine increases at high frequencies of nerve stimulation leading to a preferential activation of $A_{2A}Rs$ (see [83] and references herein). In particular, ecto-5-nucleotidase

(CD73), the only enzyme able to dephosphorylate extracellular AMP into adenosine in the brain [203] is involved in the activation of A_{2A}Rs [204-208]. Manipulations of CD73 activity might thus regulate the state of activation of A_{2A}Rs and, thus, the final effect achieved by targeting the A_1/A_{2A} receptor heteromer. In the following sections, two examples of pathological conditions characterized by increased adenosine levels and in which the interaction A_1/A_{2A} receptor can be involved will be examined.

A_{2A}/A_1 Receptor Interaction and Ischemia

The extracellular adenosine concentration increases dramatically during brain ischemia changing from 30-200 nM in basal conditions [209], up to approximately 30 μM [2, 210]. Such an increase may be considered as an endogenous neuroprotective response [201], thus indicating adenosine receptors as potential therapeutic targets in the treatment of stroke [105]. Although hampered by the occurrence of peripheral side effects and receptor desensitization [211], the stimulation of A_1R has been consistently shown to be neuroprotective against ischemia and excitotoxicity (for review, [6, 16, 83, 105, 190, 198, 200, 211, 212]). The role of A_{2A}Rs is conversely much more controversial [213, 214]. In particular, A_{2A}R antagonists provide an early protection via centrally mediated control of excessive glutamate release, whereas in the hours and days after brain ischemia protection is rather induced by A_{2A}R agonists, due to their peripheral immune-suppressive effects that reduce infiltration and inflammatory responses ([215]; for review [213, 216]). The functional interaction between A_1 and A_{2A} receptors can play a role in the early protective effect of A_{2A}R antagonists. One implication of this interaction is that, when the extracellular level of adenosine reaches levels sufficient to activate A_{2A}R, as in ischemic condition, it may inhibit A_1R function. The blockade of A_{2A}R with the antagonist can thus act, at least in part, by removing the A_{2A}R suppression of A_1R [210, 217, 218]. Interestingly, the use of A_{2A}R antagonist might overcome the side effects of A_1R agonists since they work only when the receptors have been activated by an abnormally increased amount of endogenous adenosine and, thus, these effects are limited to the ischemic brain areas.

A_{2A}/A_1 Receptor Interaction and Epilepsy

Evidence that seizures increase extracellular adenosine levels was obtained both in animals and humans [219]. This observation suggested that adenosine could be considered a sort of endogenous antiepileptic drug. Since A_1Rs are highly expressed in epilepsy-prone regions such as the hippocampus [181] and their activation induces anticonvulsive effects in different animals models [212, 220, 221], this subtype was considered an interesting therapeutic target [222]. However, a reduction in the expression of A_1Rs has been reported both in animal models [223, 224, 225] and in human temporal lobes of patients undergoing epilepsy surgery [226], whereas, interestingly, opposite changes were seen in the expression of A_{2A}Rs [227, 228], as well as in density and activity of ecto-5-nucleotidase [229]. These data demonstrate that there is a dysregulation in the adenosine system in epilepsy, and the resulting imbalance between A_1 and A_{2A} receptors can be regarded as a possible therapeutic target. While A_1R-operated inhibitory system works to avoid the initiation and the spread of epileptogenic activity [230-232], under high frequency stimulation conditions, such as those present in the epilepticus circuits, the activation of facilitatory A_{2A}Rs from released ATP [202] can downregulate A_1R responses [224, 227]. Then, again, the functional interaction between A_1 and A_{2A} receptors could play a role in the protective effect of A_{2A}R antagonists.

Conclusion

Adenosine receptors (A_1 and A_{2A} subtypes in particular) exert a variety of effects on the CNS, and have been regarded as potential therapeutic targets for CNS diseases. Considering that such receptors physically and functionally interact with receptors for other neurotransmitters and neuromodulators, their potential therapeutic relevance should be re-evaluated by a more comprehensive (and much more complex) approach. In this chapter, some of the more therapeutically relevant interactions have been discussed. Some of them (such as the A_{2A}/D_2 interaction in PD) are very well documented, while for others only a theoretical basis exists at the moment. Should these interactions confirm to have a real impact in CNS disease, they could represent alternative/integrative targets for the development of more efficacious and/or less toxic drugs.

References

[1] Fredholm B. B. Adenosine, an endogenous distress signal, modulates tissue damage and repair. *Cell Death Differ.* 2007, 14, 1315-23.

[2] Fredholm, B. B., IJzerman, A. P., Jacobson, K. A., Linden, J., Müller C. E. International Union of Basic and Clinical Pharmacology. LXXXI. Nomenclature and classification of adenosine receptors--an update. *Pharmacol. Rev.* 2011, 63, 1-34.

[3] Dixon, A. K., Widdowson, L., Richardson, P. J. Desensitisation of the adenosine A_1 receptor by the A_{2A} receptor in the rat striatum. *J. Neurochem.* 1997, 69, 315-21.

[4] Cordeaux, Y., Briddon, S. J., Megson, A. E., McDonnell, J., Dickenson, J. M., Hill, S. J. Influence of receptor number on functional responses elicited by agonists acting at the human adenosine A_1 receptor: evidence for signaling pathway dependent changes in agonist potency and relative intrinsic activity. *Mol. Pharmacol.* 2000, 58, 1075-84.

[5] Cordeaux, Y., Ijzerman, A. P., Hill, S. J. Coupling of the human A_1 Adenosine receptor to different heterotrimeric G-proteins, evidence for agonist-specific G-protein activation. *Br. J. Pharmacol.* 2004, 143, 705-14.

[6] Haas, H. L., Selbach, O. Functions of neuronal adenosine receptors. *Naunyn Schmiedebergs Arch. Pharmacol.* 2000, 362, 375-81.

[7] Svenningsson, P., Hall, H., Sedvall, G., Fredholm, B. B. Distribution of adenosine receptors in the postmortem human brain: an extended autoradiographic study. *Synapse* 1997, 27, 322-35.

[8] Rosin, D. L., Hettinger, B. D., Lee, A., Linden, J. Anatomy of adenosine A_{2A} receptors in brain: morphological substrates for integration of striatal function. *Neurology* 2003, 61, S12-S8.

[9] Hettinger, B. D., Lee, A., Linden, J., Rosin, D. L Ultrastructural localization of adenosine A_{2A} receptors suggests multiple cellular sites for modulation of GABAergic neurons in rat striatum. *J. Comp. Neurol.* 2001, 431, 331-46.

[10] Rebola, N., Canas, P. M., Oliveira, C. R., Cunha R. A. Different synaptic and subsynaptic localization of adenosine A_{2A} receptors in the hippocampus and striatum of the rat. *Neuroscience* 2005, 132, 893-903.

[11] Quiroz, C., Luján, R., Uchigashima, M., Simoes, A. P., Lerner, T. N., Borycz, J., Kachroo, A., Canas, P. M., Orru, M., Schwarzschild, M. A., Rosin, D. L., Kreitzer, A. C., Cunha, R. A., Watanabe, M., Ferré S. Key modulatory role of presynaptic adenosine A_{2A} receptors in cortical neurotransmission to the striatal direct pathway. *ScientificWorldJournal* 2009, 9, 1321-44.

[12] Fredholm, B. B., Arslan, G., Halldner, L., Kull, B., Schulte, G., Wasserman, W. Structure and function of adenosine receptors and their genes. *Naunyn-Schmiedebergs Arch. Pharmacol.* 2000, 362, 364-74.

[13] Cunha, R., Ribeiro, J. A. Adenosine A_{2A} receptor facilitation of synaptic transmission in the CA_1 area of the rat hippocampus requires protein kinase C but not protein kinase A activation. *Neurosci. Lett.* 2000, 289, 127-30.

[14] Canals, M., Angulo, E., Casadó, V., Canela, E. I., Mallol, J., Viñals, F., Staines, W., Tinner, B., Hillion, J., Agnati, L., Fuxe, K., Ferré, S., Lluis, C., Franco, R. Molecular mechanisms involved in the adenosine A and A receptor-induced neuronal differentiation in neuroblastoma cells and striatal primary cultures. *J. Neurochem.* 2005, 92, 337-48.

[15] Gudermann, T., Grosse, R., Schultz, G. Contribution of receptor/G protein signaling to cell growth and transformation. *Naunyn Schmiedebergs Arch. Pharmacol.* 2000, 361, 345-62.

[16] Fredholm, B. B., Chen, J. F., Cunha R. A., Svenningsson P., Vaugeois J. M. Adenosine and brain function. *Int. Rev. Neurobiol.* 2005, 63, 191-270.

[17] Gsandtner, I., Charalambous, C., Stefan, E., Ogris, E., Freissmuth, M., Zezula, J. Heterotrimeric G protein-independent signaling of a G protein coupled receptor. Direct binding of ARNO/cytohesin-2 to the carboxyl terminus of the A_{2A} adenosine receptor is necessary for sustained activation of the ERK/MAP kinase pathway. *J. Biol. Chem.* 2005, 280, 31898-905.

[18] Canas, P. M., Porciúncula, L. O., Cunha, G. M., Silva, C. G., Machado, N. J., Oliveira, J. M., Oliveira, C. R., Cunha, R. A. Adenosine A_{2A} receptor blockade prevents synaptotoxicity and memory dysfunction caused by beta-amyloid peptides via p38 mitogen-activated protein kinase pathway. *J. Neurosci.* 2009, 29, 14741-51.

[19] Visentin, S., De Nuccio, C., Bernardo, A., Pepponi, R., Ferrante, A., Minghetti, L., Popoli, P. The stimulation of adenosine A_{2A} receptors ameliorates the pathological phenotype of fibroblasts from Niemann-Pick type C patients. *J. Neurosci.* 2013, 33, 15388-93.

[20] Sebastião, A. M., Ribeiro, J. A. Tuning and fine-tuning of synapses with adenosine. *Curr. Neuropharmacol.* 2009, 7, 180-94.

[21] Alger, B. E. Retrograde signaling in the regulation of synaptic transmission: focus on endocannabinoids. *Prog. Neurobiol.* 2002, 68, 247-86.

[22] Ferré, S., Fredholm, B. B., Morelli, M., Popoli, P., Fuxe, K. Adenosine-dopamine receptor-receptor interactions as an integrative mechanism in the basal ganglia. *Trends Neurosci.* 1997, 20, 482-7.

[23] Fuxe, K., Agnati, L. F., Jacobsen, K., Hillion, J., Canals, M., Torvinen, M., Tinner-Staines, B., Staines, W., Rosin, D., Terasmaa, A., Popoli, P., Leo, G., Vergoni, V., Lluis, C., Ciruela, F., Franco, R., Ferré, S. Receptor heteromerization in adenosine A_{2A} receptor signaling: relevance for striatal function and Parkinson's disease. *Neurology* 2003, 61, S19-23.

[24] Ferré, S., Quiroz, C., Woods, A. S., Cunha, R., Popoli, P., Ciruela, F., Lluis, C., Franco, R., Azdad, K., Schiffmann, S. N. An update on adenosine A_{2A}-dopamine D_2 receptor interactions: implications for the function of G protein-coupled receptors. *Curr. Pharm. Des.* 2008, 14, 1468-74.

[25] Neve, K. A., Seamans, J. K., Trantham-Davidson, H. Dopamine receptor signaling. *J. Recept. Signal. Transduct. Res.* 2004, 24, 165-205.

[26] Beaulieu, J. M., Gainetdinov, R. R. The physiology, signaling, and pharmacology of dopamine receptors. *Pharmacol. Rev.* 2011, 63, 182-217.

[27] Ferré, S., Casadó, V., Devi, L. A., Filizola, M., Jockers, R., Lohse, M. J., Milligan, G., Pin, J. P., Guitart, X. G protein-coupled receptor oligomerization revisited: functional and pharmacological perspectives. *Pharmacol. Rev.* 2014, 66, 413-34.

[28] Hillion, J., Canals, M., Torvinen, M., Casado, V., Scott, R., Terasmaa, A., Hansson, A., Watson, S., Olah, M. E., Mallol, J., Canela, E. I., Zoli, M., Agnati, L. F., Ibanez, C. F., Lluis, C., Franco. R., Ferre, S., Fuxe, K. Coaggregation, cointernalization, and codesensitization of adenosine A_{2A} receptors and dopamine D_2 receptors. *J. Biol. Chem.* 2002, 277, 18091-7.

[29] Canals, M., Marcellino, D., Fanelli, F., Ciruela, F., de Benedetti, P., Goldberg, S. R., Neve, K., Fuxe, K., Agnati, L. F., Woods, A. S., Ferré, S., Lluis, C., Bouvier M., Franco, R. Adenosine A_{2A}-dopamine D_2 receptor-receptor heteromerization: qualitative and quantitative assessment by fluorescence and bioluminescence energy transfer. *J.Biol. Chem.* 2003, 278, 46741-9.

[30] Kamiya, T., Saitoh, O., Yoshioka, K., Nakata, H. Oligomerization of adenosine A_{2A} and dopamine D_2 receptors in living cells. *Biochem. Biophys. Res. Commun.* 2003, 306, 544-9.

[31] Navarro, G., Ferré, S., Cordomi, A., Moreno, E., Mallol, J., Casadó, V., Cortés, A., Hoffmann, H., Ortiz, J., Canela, E. I., Lluís, C., Pardo, L., Franco, R., Woods, A.S. Interactions between intracellular domains as key determinants of the quaternary structure and function of receptor heteromers. *J. Biol. Chem.* 2010, 285, 27346-59.

[32] Azdad, K., Gall, D., Woods, A. S., Ledent, C., Ferré, S., Schiffmann, S. N. Dopamine D_2 and adenosine A_{2A} receptors regulate NMDA-mediated excitation in accumbens neurons through A_{2A}-D_2 receptor heteromerization. *Neuropsychopharmacol.* 2009, 34, 972-86.

[33] Alexander, G. R., Crutcher, M. D. Functional architecture and basal ganglia circuits: neural substrates of parallel processing. *Trends Neurosci.* 1990, 13, 266-71.

[34] Gerfen, C. R., Engber, T. M., Mahan, L., Susel, Z., Chase, T. N., Monsma, F. J., Sibley, D. R. D_1 and D_2 dopamine receptor-regulated gene expression of strionigral and striopallidal neurons. *Science* 1990, 250, 1429-32.

[35] Ferré, S., von Euler, G., Johansson, B., Fredholm, B. B., Fuxe, K. Stimulation of high-affinity adenosine A_2 receptors decreases the affinity of dopamine D_2 receptors in rat striatal membranes. *Proc. Natl. Acad. Sci. U.S.A.* 1991, 88, 7238-41.

[36] Rimondini, R., Ferré, S., Ogren, S. O., Fuxe, K. Adenosine A_{2A} agonists: a potential new type of atypical antipsychotic. *Neuropsychopharmacol.* 1997, 17, 82-91.

[37] Ferré, S., Popoli, P., Rimondini, R., Reggio, R., Kehr, J., Fuxe, K. Adenosine A_{2A} and group I metabotropic glutamate receptors synergistically modulate the binding characteristics of dopamine D_2 receptors in the rat striatum. *Neuropharmacol.* 1999, 38, 129-140.

[38] Popoli, P., Reggio, R., Pèzzola, A. Effects of SCH 58261, an adenosine A(2A) receptor antagonist, on quinpirole-induced turning in 6-hydroxydopamine-lesioned rats. Lack of tolerance after chronic caffeine intake. *Neuropsychopharmacol.* 2000, 22, 522-9.

[39] Fuxe, K., Strömberg, I., Popoli, P., Rimondini-Giorgini, R., Torvinen, M., Ogren, S. O., Franco, R., Agnati, L. F., Ferré, S. Adenosine receptors and Parkinson's disease. Relevance of antagonistic adenosine and dopamine receptor interactions in the striatum. *Adv. Neurol.* 2001, 86, 45-53.

[40] Chen, J.F. The adenosine A(2A) receptor as an attractive target for Parkinson's disease treatment. *Drug News Perspect.* 2003, 16, 597-604.

[41] Schwarzschild, M.A., Agnati, L., Fuxe, K., Chen, J. F., Morelli, M. Targeting adenosine A_{2A} receptors in Parkinson's disease. *Trends Neurosci.* 2006, 29, 647-54.

[42] Simola, N., Morelli, M., Pinna, A. Adenosine A_{2A} receptor antagonists and Parkinson's disease: state of the art and future directions. *Curr. Pharm. Des.* 2008, 14, 1475-89.

[43] Morelli, M., Carta, A. R., Jenner, P. Adenosine A_{2A} receptors and Parkinson's disease. *Handb Exp. Pharmacol.* 2009, 193, 589-615.

[44] Kanda, T., Shiozaki, S., Shimada, J., Suzuki, F., Nakamura, J. KF17837: a novel selective adenosine A_{2A} receptor antagonist with anticataleptic activity. *Eur. J. Pharmacol.* 1994, 256, 263-8.

[45] Pinna, A., Volpini, R., Cristalli, G., Morelli, M. New adenosine A_{2A} receptor antagonists: actions on Parkinson's disease models. *Eur. J. Pharmacol.* 2005, 512, 157-64.

[46] Fenu, S., Pinna, A., Ongini, E., Morelli, M. Adenosine A_{2A} receptor antagonism potentiates L-DOPA-induced turning behaviour and c-fos expression in 6-hydroxydopamine-lesioned rats. *Eur. J. Pharmacol.* 1997, 321, 143-7.

[47] Strömberg, I., Popoli, P., Müller, C. E., Ferré, S., Fuxe, K. Electrophysiological and behavioural evidence for an antagonistic modulatory role of adenosine A_{2A} receptors in dopamine D_2 receptor regulation in the rat dopamine-denervated striatum. *Eur. J. Neurosci.* 2000, 12, 4033-7.

[48] Grondin, R., Bedard, P. J., Hadj Tahar, A., Gregoire, L., Mori, A., Kase, H. Antiparkinsonian effect of a new selective adenosine A_{2A} receptor antagonist in MPTP-treated monkeys. *Neurology* 1999, 52, 1673-77.

[49] Kanda, T., Tashiro, T., Kuwana, Y., Jenner, P. Adenosine A_{2A} receptors modify motor function in MPTP-treated common marmosets. *Neuroreport* 1998, 9, 2857-60.

[50] Pinna, A., Pontis, S., Morelli, M. Adenosine A_{2A} receptor antagonists improve deficits in initiation of movement and sensory motor integration in the unilateral 6-hydroxydopamine rat model of Parkinson's disease. *Synapse* 2007, 61, 606-14.

[51] Hauser, R. A., Cantillon, M., Pourcher, E., Micheli, F., Mok, V., Onofrj, M., Huyck, S., Wolski, K. Preladenant in patients with Parkinson's disease and motor fluctuations: a phase 2, double-blind, randomised trial. *Lancet Neurol.* 2011, 10, 221-9.

[52] Hauser, R. A., Olanow, C. W., Kieburtz, K. D., Pourcher, E., Docu-Axelerad, A., Lew, M., Kozyolkin, O., Neale, A., Resburg, C., Meya, U., Kenney, C., Bandak, S. Tozadenant (SYN115) in patients with Parkinson's disease who have motor fluctuations on levodopa: a phase 2b, double-blind, randomised trial. *Lancet Neurol.* 2014, 13, 767-76.

[53] Pourcher, E., Fernandez, H. H., Stacy, M., Mori, A., Ballerini, R., Chaikin, P., Istradefylline for Parkinson's disease patients experiencing motor fluctuations: results of the KW-6002-US-018 study. *Parkinsonism Relat. Disord.* 2012, 18, 178-84.

[54] Van Os, J., Kapur, S. Schizophrenia. *Lancet* 2009, 374, 635-45.

[55] Seeman, P., Chau-Wong, M., Tedesco, J., Wong, K. Brain receptors for antipsychotic drugs and dopamine: direct binding assays. *Proc. Natl. Acad. Sci. U.S.A.* 1975, 72, 4376-080.

[56] Seeman, P. All roads to schizophrenia lead to dopamine supersensitivity and elevated dopamine D_2 receptors. *CNS Neurosci. Ther.* 2011, 17, 118-32.

[57] Ferré, S., O'Connor, W. T., Snaprud, P., Ungerstedt, U., Fuxe, K. Antagonistic interaction between adenosine A_{2A} receptors and dopamine D_2 receptors in the ventral striopallidal system. Implications for the treatment of schizophrenia. *Neuroscience* 1994, 63, 765-73.

[58] Kafka, S. H., Corbett, R. Selective adenosine A_{2A} receptor/dopamine D_2 receptor interactions in animal models of schizophrenia. *Eur. J. Pharmacol.* 1996, 295, 147-54.

[59] Ferré, S. Adenosine-dopamine interactions in the ventral striatum. Implications for the treatment of schizophrenia. *Psychopharmacol.* (Berl) 1997, 133, 107-20.

[60] Cunha, R. A., Ferré, S., Vaugeois, J. M., Chen, J. F. Potential therapeutic interest of adenosine A_{2A} receptors in psychiatric disorders. *Curr. Pharm. Des.* 2008, 14, 1512-24.

[61] Boison, D., Singer, P., Shen, H. Y., Feldon, J., Yee, B. K. Adenosine hypothesis of schizophrenia-opportunities for pharmacotherapy. *Neuropharmacol.* 2012, 62, 1527-43.

[62] Rial, D., Lara, D. R., Cunha, R. A. The adenosine neuromodulation system in schizophrenia. *Int. Rev. Neurobiol.* 2014, 119, 395-449.

[63] Fuxe, K., Borroto-Escuela, D. O., Tarakanov, A. O., Romero-Fernandez, W., Ferraro, L., Tanganelli, S., Perez-Alea, M., Di Palma, M., Agnati, L. F. Dopamine D_2 heteroreceptor complexes and their receptor-receptor interactions in ventral striatum: novel targets for antipsychotic drugs. *Prog. Brain Res.* 2014, 211, 113-39.

[64] Harper, P., *Huntington's Disease,* London, W. B. Saunders Co., 1991.

[65] Mitchell, I. J., Cooper, A. J., Griffiths, M. R. The selective vulnerability of striatopallidal neurons. *Prog. Neurobiol.* 1999, 59, 691-719.

[66] Blum, D, Hourez, R., Galas, M. C., Popoli, P., Schiffmann, S. N. Adenosine receptors and Huntington's disease: implications for pathogenesis and therapeutics. *Lancet Neurol.* 2003, 2, 366-74.

[67] Popoli, P., Blum, D., Martire, A., Ledent, C., Ceruti, S., Abbracchio, M. P. Functions, dysfunctions and possible therapeutic relevance of adenosine A_{2A} receptors in Huntington's disease. *Prog. Neurobiol.* 2007, 81, 331-48.

[68] Popoli, P., Blum, D., Domenici, M. R., Burnouf, S., Chern, Y. A critical evaluation of adenosine A_{2A} receptors as potentially 'druggable' targets in Huntington's disease. *Curr. Pharm. Des.* 2008, 14, 1500-11.
[69] Mangiarini, L., Sathasivam, K., Seller, M., Cozens, B., Harper, A., Hetherington, C., Lawton, M., Trottier, Y., Lehrach, H., Davies, S. W., Bates, G. P. Exon 1 of the HD gene with an expanded CAG repeat is sufficient to cause a progressive neurological phenotype in transgenic mice. *Cell* 1996, 87, 493-506.
[70] Popoli, P., Pèzzola, A., Reggio, R., Caporali, M. G., Scotti de Carolis, A. CGS 21680 antagonizes motor hyperactivity in a rat model of Huntington's disease. *Eur. J. Pharmacol.* 1994, 257, R5-6.
[71] Nasir, J., Frima, N., Pickard, B., Malloy, M. P., Zhan, L., Grünewald, R. Unbalanced whole arm translocation resulting in loss of 18p in dystonia. *Mov. Disord.* 2006, 21, 859-63.
[72] Fahn, S., Bressman, S. B., Marsden, C. D. Classification of dystonia. *Adv. Neurol.* 1998, 78, 1-10.
[73] Augood, S. J., Hollingsworth, Z., Albers, D. S., Yang, L., Leung, J. C., Muller, B., Klein, C., Breakefield, X. O., Standaert, D. G. Dopamine transmission in DYT1 dystonia: a biochemical and autoradiographical study. *Neurology* 2002, 59, 445-8.
[74] Augood, S. J., Hollingsworth, Z., Albers, D. S., Yang, L., Leung, J., Breakefield, X. O., Standaert, D. G. Dopamine transmission in DYT1 dystonia. *Adv. Neurol.* 2004, 94, 53-60.
[75] Perlmutter, J. S., Mink, J. W. Dysfunction of dopaminergic pathways in dystonia. *Adv. Neurol.* 2004, 94, 163-70.
[76] Perlmutter, J. S., Tempel, L. W., Black, K. J., Parkinson, D., Todd, R. D. MPTP induces dystonia and parkinsonism. Clues to the pathophysiology of dystonia. *Neurology* 1997, 49, 1432-8.
[77] Napolitano, F., Pasqualetti, M., Usiello, A., Santini, E., Pacini, G., Sciamanna, G., Errico, F., Tassone, A., Di Dato, V., Martella, G., Cuomo, D., Fisone, G., Bernardi, G., Mandolesi, G., Mercuri, N. B., Standaert, D. G., Pisani, A. Dopamine D_2 receptor dysfunction is rescued by adenosine A_{2A} receptor antagonism in a model of DYT1 dystonia. *Neurobiol. Dis.* 2010, 38, 434-45.
[78] Calabresi, P., Pisani, A., Mercuri, N. B., Bernardi, G. The corticostriatal projection: from synaptic plasticity to dysfunctions of the basal ganglia. *Trends Neurosci.* 1996, 19, 19-24.
[79] Gubellini, P., Saulle, E., Centonze, D., Bonsi, P., Pisani, A., Bernardi, G., Conquet, F., Calabresi, P. Selective involvement of mGlu1 receptors in corticostriatal LTD. *Neuropharmacol.* 2001, 40, 839-46.
[80] Calabresi, P., Centonze, D., Pisani, A., Bernardi, G. Metabotropic glutamate receptors and cell-type-specific vulnerability in the striatum: implications for ischemia and Huntington's disease. *Exp. Neurol.* 1999, 158, 97-108.
[81] Bruno, V., Battaglia, G., Copani, A., Cespédes, V. M., Galindo, M. F., Ceña, V., Sánchez-Prieto, J., Gasparini, F., Kuhn, R., Flor, P. J., Nicoletti, F. An activity-dependent switch from facilitation to inhibition in the control of excitotoxicity by group I metabotropic glutamate receptors. *Eur. J. Neurosci.* 2001, 13, 1469-78.

[82] Testa, C. M., Standaert, D. G., Landwehrmeyer, G. B., Penney, J. B. Jr., Young, A. B. Differential expression of $mGluR_5$ metabotropic glutamate receptor mRNA by rat striatal neurons. *J. Comp. Neurol.* 1995, 354, 241-52.

[83] Cunha, R. A. Neuroprotection by adenosine in the brain: from A_1 receptor activation to A_{2A} receptor blockade. *Purinergic Signalling* 2005, 1, 111-34.

[84] Anwyl, R. Metabotropic glutamate receptors: electrophysiological properties and role in plasticity. *Brain Res. Rev.* 1999, 29, 83-120.

[85] Balschun, D., Manahan-Vaughan, D., Wagner, T., Behnisch, T., Reymann, K. G., Wetzel, W. A specific role for group I mGluRs in hippocampal LTP and hippocampus-dependent spatial learning. *Learn. Mem.* 1999, 6, 138-52.

[86] Sacaan, A. I., Schoepp, D. D. Activation of hippocampal metabotropic excitatory amino acid receptors leads to seizures and neuronal damage. *Neurosci. Lett.* 1992,139, 77-82.

[87] Attucci, S., Clodfelter, G.V., Thibault, O., Staton, J., Moroni, F., Landfield, P. W., Porter, N. M. Group I metabotropic glutamate receptor inhibition selectively blocks a prolonged $Ca(^{2+})$ elevation associated with age-dependent excitotoxicity. *Neuroscience* 2002, 112, 183-94.

[88] Romano, C., Sesma, M. A., McDonald, C. T., O'Malley, K., Van den Pol, A. N., Olney, J. W. Distribution of metabotropic glutamate receptor $mGluR_5$ immunoreactivity in rat brain. *J. Comp. Neurol.* 1995, 355, 455-69.

[89] Rodrigues, R. J., Alfaro, T. M., Rebola, N., Oliveira, C. R., Cunha, R. A. Co-localization and functional interaction between adenosine A(2A) and metabotropic group 5 receptors in glutamatergic nerve terminals of the rat striatum. *J. Neurochem.* 2005, 92, 433-41.

[90] Tebano, M. T., Martire, A., Rebola, N., Pepponi, R., Domenici, M. R., Grò, M. C., Schwarzschild, M. A., Chen, J. F., Cunha, R. A., Popoli, P. Adenosine A_{2A} receptors and metabotropic glutamate 5 receptors are co-localized and functionally interact in the hippocampus: a possible key mechanism in the modulation of N-methyl-D-aspartate effects. *J. Neurochem.* 2005, 95, 1188-200.

[91] Kniazeff, J., Prézeau, L., Rondard, P., Pin, J. P., Goudet, C. Dimers and beyond: The functional puzzles of class C GPCRs. *Pharmacol. Ther.* 2011,130, 9-25.

[92] Ferré, S., Karcz-Kubicha, M., Hope, B. T., Popoli, P., Burgueño, J., Gutiérrez, M. A., Casadó, V., Fuxe, K., Goldberg, S. R., Lluis, C., Franco, R., Ciruela, F. Synergistic interaction between adenosine A_{2A} and glutamate $mGlu_5$ receptors: implications for striatal neuronal function. *Proc. Natl. Acad. Sci. U.S.A.* 2002, 99, 11940-5.

[93] Berke, J. D., Hyman, S. E. Addiction, dopamine, and the molecular mechanisms of memory. *Neuron* 2000, 25, 515-32.

[94] Valjent, E., Caboche, J., Vanhoutte, P. Mitogen-activated protein kinase/extracellular signal-regulated kinase induced gene regulation in brain: a molecular substrate for learning and memory? *Mol. Neurobiol.* 2001, 23, 83-99.

[95] Sweatt, J. D. The neuronal MAP kinase cascade: a biochemical signal integration system subserving synaptic plasticity and memory. *J. Neurochem.* 2001, 76, 1-10.

[96] D'Alcantara, P., Ledent, C., Swillens, S., Schiffmann, S.N. Inactivation of adenosine A_{2A} receptor impairs long term potentiation in the accumbens nucleus without altering basal synaptic transmission. *Neuroscience* 2001, 107, 455-64.

[97] Sung, K. W., Choi, S., Lovinger, D. M. Activation of group I mGluRs is necessary for induction of long-term depression at striatal synapses. *J. Neurophysiol.* 2001, 86, 2405-12.

[98] Uslaner, J., Badiani, A., Norton, C. S., Day, H. E., Watson, S. J., Akil, H., Robinson, T. E. Amphetamine and cocaine induce different patterns of c-fos mRNA expression in the striatum and subthalamic nucleus depending on environmental context. *Eur. J. Neurosci.* 2001, 13, 1977-83.

[99] Domenici, M. R., Pepponi, R., Martire, A., Tebano, M. T., Potenza, R. L., Popoli, P. Permissive role of adenosine A_{2A} receptors on metabotropic glutamate receptor 5 ($mGluR_5$)-mediated effects in the striatum. *J. Neurochem.* 2004, 90, 1276-9.

[100] Nishi, A., Liu, F., Matsuyama, S., Hamada, M., Higashi, H., Nairn, A. C., Greengard, P. Metabotropic $mGlu_5$ receptors regulate adenosine A_{2A} receptor signalling. *Proc. Natl. Acad. Sci. U.S.A.* 2003, 100, 1322-7.

[101] Popoli, P., Betto, P., Reggio, R., Ricciarello, G. Adenosine A_{2A} receptor stimulation enhances striatal extracellular glutamate levels in rats. *Eur. J. Pharmacol.* 1995, 287, 215-7.

[102] Pintor, A., Pèzzola, A., Reggio, R., Quarta, D., Popoli, P. The $mGlu_5$ receptor agonist CHPG stimulates striatal glutamate release: possible involvement of A_{2A} receptors. *Neuroreport* 2000, 11, 3611-4.

[103] Lopes, L. V., Cunha, R. A., Kull, B., Fredholm, B. B., Ribeiro, J. A. Adenosine A(2A) receptor facilitation of hippocampal synaptic transmission is dependent on tonic A(1) receptor inhibition. *Neuroscience* 2002, 112, 319-29.

[104] Popoli, P., Minghetti, L., Tebano, M. T., Pintor, A., Domenici, M. R., Massotti, M. Adenosine A_{2A} receptor antagonism and neuroprotection: mechanisms, lights, and shadows. *Crit. Rev. Neurobiol* 2004, 16, 99-106.

[105] Stone, T. W., Ceruti, S., Abbracchio, M. P. Adenosine receptors and neurological disease: neuroprotection and neurodegeneration. *Handb. Exp. Pharmacol.* 2009, 193, 535-87.

[106] Wei, C. J., Li, W., Chen, J. F. Normal and abnormal functions of adenosine receptors in the central nervous system revealed by genetic knockout studies. *Biochim. Biophys. Acta* 2011, 1808, 1358-79.

[107] Corsi, C., Melani, A., Bianchi, L., Pepeu, G., Pedata, F. Striatal A_{2A} adenosine receptors differentially regulate spontaneous and K^+-evoked glutamate release *in vivo* in young and aged rats. *Neuroreport* 1999, 10, 687-91.

[108] Alagarsamy, S., Marino, M. J., Rouse, S. T., Gereau, R. W., Heinemann, S. F., Conn, P. J. Activation of NMDA receptors reverses desensitization of $mGlu_5R$ in native and recombinant systems. *Nat. Neurosci.* 1999, 2, 234-40.

[109] Bortolotto, Z. A., Fitzjohn, S. M., Collingridge, G. L. Roles of metabotropic glutamate receptors in LTP and LTD in the hippocampus. *Curr. Opin. Neurobiol.* 1999, 9, 299-304.

[110] Orlando, L. R., Alsdorf, S. A., Penney, J. B., Young, A. B. The role of group I and group II metabotropic glutamate receptors in modulation of striatal NMDA and quinolinic acid toxicity. *Exp. Neurol.* 2001, 167, 196-204.

[111] Beal, M. F., Kowall, N. W., Ellison, D. W., Mazurek, M. F., Swartz, K. J., Martin, J. B. Replication of the neurochemical characteristics of Huntington's disease by quinolinic acid. *Nature* 1986, 321, 168-71.

[112] Orlando, L. R., Standaert, D. G., Penney, J. B. Jr., Young, A. B. Metabotropic receptors in excitotoxicity: (S)-4-carboxy-3-hydroxyphenylglycine ((S)-4C3HPG) protects against rat striatal quinolinic acid lesions. *Neurosci Lett.* 1995, 202, 109-12.

[113] Popoli, P., Pintor, A., Domenici, M. R., Frank, C., Tebano, M. T., Pèzzola, A., Scarchilli, L., Quarta, D., Reggio, R., Malchiodi-Albedi, F., Falchi, M., Massotti, M. Blockade of striatal adenosine A_{2A} receptor reduces, through a presynaptic mechanism, quinolinic acid-induced excitotoxicity: possible relevance to neuroprotective interventions in neurodegenerative diseases of the striatum. *J. Neurosci.* 2002, 22, 1967-75.

[114] Coyle, J. The glutamatergic dysfunction hypothesis for schizophrenia. Harv. *Rev. Psychiatry* 1996, 3, 241-53.

[115] Tamminga, C. A. Schizophrenia and glutamatergic transmission. *Crit. Rev. Neurobiol.* 1998, 12, 21-36.

[116] Harrison, P. J., Owen, M. J. Genes for schizophrenia? Recent findings and their pathophysiological implications. *Lancet* 2003, 361, 417-9.

[117] Moghaddam, B. Bringing order to the glutamate chaos in schizophrenia. *Neuron* 2003, 40, 881-4.

[118] Luby, E., Cohen, B., Rosenbaum, G., Gottlieb, J., Kelley, R. Study of a new schizophrenomimetic drug-sernyl. *Am. Med. Assoc. Arch. Neuro.l Psychiatry* 1959, 81, 363-9.

[119] Lahti, A. C., Koffel, B., LaPorte, D., Tamminga, C.A. Subanesthetic doses of ketamine stimulate psychosis in schizophrenia. *Neuropsychopharmacol.* 1995, 13, 9-19.

[120] Malhotra, A. K., Pinals, D. A., Adler, C. M., Elman, I., Clifton, A., Pickar, D., Breier, A. Ketamine-induced exacerbation of psychotic symptoms and cognitive impairment in neuroleptic-free schizophrenics. *Neuropsychopharmacol.* 1997, 17, 141-50.

[121] Javitt, D. C., Zukin, S. R. Recent advances in the phencyclidine model of schizophrenia. *Am. J. Psychiatry* 1991, 148, 1301-8.

[122] Krystal, J. H., Karper, L. P., Seibyl, J. P., Freeman, G. K., Delaney, R., Bremner, J. D., Heninger, G. R., Bowers, M. B. Jr., Charney, D. S. Subanesthetic effects of the noncompetitive NMDA antagonist, ketamine, in humans. Psychotomimetic, perceptual, cognitive, and neuroendocrine responses. *Arch. Gen. Psychiatry* 1994, 51, 199-214.

[123] Olney, J., Farber, N. Glutamate receptor dysfunction and schizophrenia. *Arch. Gen. Psychiatry* 1995, 52, 998-1007.

[124] Javitt, D. C., Zylberman, I., Zukin, S. R., Heresco-Levy, U., Lindenmayer, J.P. Amelioration of negative symptoms in schizophrenia by glycine. *Am. J. Psychiatry* 1994, 151, 1234-6.

[125] Coyle, J. T., Tsai, G., Goff, D. C. Ionotropic glutamate receptors as therapeutic targets in schizophrenia. *Curr. Drug Targets* 2002, 1, 183-9.

[126] Moghaddam, B. Targeting metabotropic glutamate receptors for treatment of the cognitive symptoms of schizophrenia. *Psychopharmacol.* 2004, 174, 39-44.

[127] Pietraszek, M., Nagel, J., Gravius, A., Schäfer, D., Danysz, W. The role of group I metabotropic glutamate receptors in schizophrenia. *Amino Acids* 2007, 32, 173-8.

[128] Goff, D. C., Coyle, J. T. The emerging role of glutamate in the pathophysiology and treatment of schizophrenia. *Am. J. Psychiatry* 2001, 158, 1367-77.

[129] Krystal, J. H., D'Souza, D. C., Mathalon, D., Perry, E., Belger, A., Hoffman, R. NMDA receptor antagonist effects, cortical glutamatergic function, and schizophrenia: toward a paradigm shift in medication development. *Psychopharmacol.* 2003, 169, 215-33.

[130] Ferré, S., Ciruela, F., Woods, A. S. Glutamate $mGlu_5$-adenosine A_{2A}-dopamine D_2 receptor interactions in the striatum. Implications for drug therapy in neuro-psychiatric disorders and drug abuse. *Curr. Med. Chem.* 2003, 3, 1-26.

[131] Schwarzschild, M. A., Chen, J. F., Ascherio, A. Caffeinated clues and the promise of adenosine A_{2A} antagonists in PD. *Neurology* 2002, 58, 1154–60.

[132] Spooren, W. P., Gasparini, F., Bergmann, R., Kuhn, R. Effects of the protypical mGlu(5) receptor antagonist 2-methyl-6-(phenylethynyl)-pyridine on rotarod, locomotor activity and rotational responses in unilateral 6-OHDA-lesioned rats. *Eur. J. Pharmacol.* 2000, 406, 403-10.

[133] Ossowska, K., Konieczny, J., Wolfarth, S., Wieronska, J., Pilc, A. Blockade of the metabotropic glutamate receptor subtype 5 ($mGluR_5$) produces antiparkinsonian-like effects in rats. *Neuropharmacol.* 2001, 41, 413-20.

[134] Breysse, N., Baunez, C., Spooren, W., Gasparini, F., Amalric, M. Chronic but not acute treatment with a metabotropic glutamate 5 receptor antagonist reverses the akinetic deficits in a rat model of parkinsonism, *J. Neurosci.* 2002, 22, 5669-78.

[135] Breysse, N., Amalric, M., Salin, P. Metabotropic glutamate 5 receptor blockade alleviates akinesia by normalizing activity of selective basal-ganglia structures in Parkinsonian rats. *J. Neurosci.* 2003, 23, 8302-9.

[136] Morin, N., Grégoire, L., Morissette, M., Desrayaud, S., Gomez-Mancilla, B., Gasparini, F., Di Paolo, T. MPEP, an $mGlu_5$ receptor antagonist, reduces the development of L-DOPA-induced motor complications in de novo parkinsonian monkeys: biochemical correlates. *Neuropharmacol.* 2013, 66, 355-64.

[137] Morin, N., Morissette, M., Grégoire, L., Gomez-Mancilla, B., Gasparini, F., Di Paolo, T. Chronic treatment with MPEP, an $mGlu_5$ receptor antagonist, normalizes basal ganglia glutamate neurotransmission in L-DOPA-treated parkinsonian monkeys. *Neuropharmacol.* 2013, 73,216-31.

[138] Morin, N., Jourdain, V.A., Morissette, M., Grégoire, L., Di Paolo, T. Long-term treatment with l-DOPA and an $mGlu_5$ receptor antagonist prevents changes in brain basal ganglia dopamine receptors, their associated signaling proteins and neuropeptides in parkinsonian monkeys. *Neuropharmacol.* 2014, 79, 688-706.

[139] Morelli, M., Di Paolo,T., Wardas, J., Calon, F., Xiao, D., Schwarzschild, M. A. Role of adenosine A_{2A} receptors in parkinsonian motor impairment and l-DOPA-induced motor complications. *Prog. Neurobiol.* 2007, 83, 293-309.

[140] Popoli, P., Pèzzola, A., Torvinen, M., Reggio, R., Pintor, A., Scarchilli, L., Fuxe, K., Ferré, S. The selective mGlu(5) receptor agonist CHPG inhibits quinpirole-induced turning in 6-hydroxydopamine-lesioned rats and modulates the binding characteristics of dopamine D(2) receptors in the rat striatum: interactions with adenosine A(2a) receptors. *Neuropsychopharmacol.* 2001, 25, 505-13.

[141] Coccurello, R., Breysse, N., Amalric, M. Simultaneous blockade of adenosine A_{2A} and metabotropic glutamate $mGlu_5$ receptors increase their efficacy in reversing Parkinsonian deficits in rats. *Neuropsychopharmacol.* 2004, 29, 1451-61.

[142] Obeso, J. A., Rodriguez-Oroz, M. C., Rodriguez, M., Lanciego, J. L., Artieda, J., Gonzalo, N., Olanow, C. W. Pathophysiology of the basal ganglia in Parkinson's disease. *Trends Neurosci.* 2000, 23, S8-S19.

[143] Garber, K. B., Visootsak, J., Warren, S. T. Fragile X syndrome. *Eur. J. Hum. Genet.* 2008, 16, 666-72.

[144] Li, Z., Zhang, Y., Ku, L., Wilkinson, K. D., Warren, S. T., Feng, Y. The fragile X mental retardation protein inhibits translation via interacting with mRNA. *Nucleic Acids Res.* 2001, 29, 2276-83.

[145] Antar, L. N., Bassell, G. J. Sunrise at the synapse: the FMRP mRNP shaping the synaptic interface. *Neuron* 2003, 37, 555-8.

[146] Bear, M. F., Huber, K. M., Warren, S. T. The mGluR theory of fragile X mental retardation. *Trends Neurosci.* 2004, 27, 370-7.

[147] Herkenham, M., Lynn, A. B., Little, M. D., Johnson, M. R., Melvin, L. S., de Costa, B. R., Rice, K. C. Cannabinoid receptor localization in brain. *Proc. Natl. Acad. Sci. U.S.A.* 1990, 87, 1932-6.

[148] Herkenham, M., Lynn, A. B., Johnson, M. R., Melvin, L. S., de Costa, B. R., Rice, K. C. Characterization and localization of cannabinoid receptors in rat brain: a quantitative *in vitro* autoradiographic study. *J. Neurosci.* 1991, 11, 563-83.

[149] Freund, T. F., Katona, I., Piomelli, D. Role of endogenous cannabinoids in synaptic signaling. *Physiol. Rev.* 2003, 83, 1017-66.

[150] Piomelli D. The molecular logic of endocannabinoid signalling. *Nat. Rev. Neurosci.* 2003, 4, 873-84.

[151] Mátyás, F., Yanovsky, Y., Mackie, K., Kelsch, W., Misgeld, U., Freund, T. F. Subcellular localization of type 1 cannabinoid receptors in the rat basal ganglia. *Neuroscience* 2006, 137, 337-61.

[152] Uchigashima, M., Narushima, M., Fukaya, M., Katona, I., Kano, M., Watanabe, M. Subcellular arrangement of molecules for 2-arachidonoyl-glycerol-mediated retrograde signaling and its physiological contribution to synaptic modulation in the striatum. *J. Neurosci.* 2007, 27, 3663-76.

[153] Köfalvi, A., Rodrigues, R. J., Ledent, C., Mackie, K., Vizi, E. S., Cunha, R. A., Sperlágh B. Involvement of cannabinoid receptors in the regulation of neurotransmitter release in the rodent striatum: a combined immunochemical and pharmacological analysis. *J. Neurosci.* 2005, 25, 2874-84.

[154] Hoffman, A. F., Laaris, N., Kawamura, M., Masino, S. A., Lupica, C. R. Control of cannabinoid CB_1 receptor function on glutamate axon terminals by endogenous adenosine acting at A_1 receptors. *J. Neurosci.* 2010, 30, 545-55.

[155] Sousa, V. C., Assaife-Lopes, N., Ribeiro, J. A., Pratt, J. A., Brett, R. R., Sebastião, A. M. Regulation of hippocampal cannabinoid CB_1 receptor actions by adenosine A_1 receptors and chronic caffeine administration: implications for the effects of Δ9-tetrahydrocannabinol on spatial memory. *Neuropsychopharmacol.* 2011, 36, 472-87.

[156] Normile, H. J., Barraco, R. A. N^6-cyclopentyladenosine impairs passive avoidance retention by selective action at A_1 receptors. *Brain Res. Bull.* 1991, 27, 101-4.

[157] Suzuki, F., Shimada, J., Shiozaki, S., Ichikawa, S., Ishii, A., Nakamura, J., Nonaka, N. Adenosine A_1 antagonists. 3. Structure-activity relationships on amelioration against scopolamine- or N6-((R)- phenylisopropyl) adenosine induced cognitive disturbance. *J. Med. Chem.* 1993, 36, 2508-18.

[158] Pereira, G. S., Mello e Souza, T., Vinadé, E. R., Choi, H., Rodrigues, C., Battastini, A. M., Izquierdo, I., Sarkis, J. J., Bonan C. D. Blockade of adenosine A_1 receptors in the posterior cingulate cortex facilitates memory in rats. *Eur. J. Pharmacol.* 2002, 437, 151-4.
[159] Von Lubitz, D. K., Paul, I. A., Bartus, R. T., Jacobson K. A. Effects of chronic administration of adenosine A_1 receptor agonist and antagonist on spatial learning and memory. *Eur. J. Pharmacol.* 1993, 249, 271-80.
[160] Terranova, J. P., Storme, J. J., Lafon, N., Pério, A., Rinaldi-Carmona, M., Le Fur, G., Soubrié, P. Improvement of memory in rodents by the selective CB_1 cannabinoid receptor antagonist, SR 141716. *Psychopharmacol.* (Berl) 1996, 126, 165-72.
[161] Lichtman A. H. SR 141716A enhances spatial memory as assessed in a radial-arm maze task in rats. *Eur. J. Pharmacol.* 2000, 404, 175-9.
[162] Takahashi, R. N., Pamplona, F. A., Fernandes, M. S. The cannabinoid antagonist SR141716A facilitates memory acquisition and consolidation in the mouse elevated T-maze. *Neurosci. Lett.* 2005, 380, 270-5.
[163] Wolff, M.C., Leander, J. D. SR141716A, a cannabinoid CB_1 receptor antagonist, improves memory in a delayed radial maze task. *Eur. J. Pharmacol.* 2003, 477, 213-7.
[164] Assini, F. L., Nakamura, C. A., Piermartiri, T. C., Tasca, C. I., Takahashi, R. N. Coadministration of cannabinoid CB_1-receptor and adenosine A_1-receptor antagonists improves the acquisition of spatial memory in mice: participation of glutamatergic neurotransmission. *Behav. Pharmacol.* 2012, 23, 292-301.
[165] Andersson, M., Usiello, A., Borgkvist, A., Pozzi, L., Dominguez, C., Fienberg, A. A., Svenningsson, P., Fredholm, B. B., Borrelli, E., Greengard, P., Fisone, G. Cannabinoid action depends on phosphorylation of dopamine- and cAMP-regulated phosphoprotein of 32 kDa at the protein kinase A site in striatal projection neurons. *J. Neurosci.* 2005, 25, 8432-8.
[166] Soria, G., Castané, A., Berrendero, F., Ledent, C., Parmentier, M., Maldonado, R., Valverde, O. Adenosine A_{2A} receptors are involved in physical dependence and place conditioning induced by THC. *Eur. J. Neurosci.* 2004, 20, 2203-13.
[167] Carriba, P., Ortiz, O., Patkar, K., Justinova, Z., Stroik, J., Themann, A., Müller, C., Woods, A. S., Hope, B. T., Ciruela, F., Casadó, V., Canela, E. I., Lluis, C., Goldberg, S. R., Moratalla, R., Franco, R., Ferré, S. Striatal adenosine A_{2A} and cannabinoid CB_1 receptors form functional heteromeric complexes that mediate the motor effects of cannabinoids. *Neuropsychopharmacol.* 2007, 32, 2249-59.
[168] Tebano, M. T., Martire, A., Chiodi, V., Pepponi, R., Ferrante, A., Domenici, M. R., Frank, C., Chen, J. F., Ledent, C., Popoli, P. Adenosine A_{2A} receptors enable the synaptic effects of cannabinoid CB_1 receptors in the rodent striatum. *J. Neurochem.* 2009, 110, 1921-30.
[169] Rossi, S., De Chiara, V., Musella, A., Mataluni, G., Sacchetti, L., Siracusano, A., Bernardi, G., Usiello, A., Centonze, D. Caffeine drinking potentiates cannabinoid transmission in the striatum: interaction with stress effects. *Neuropharmacol.* 2009, 56, 590-7.
[170] Lerner, T. N., Horne, E. A., Stella, N., Kreitzer, A. C. Endocannabinoid signaling mediates psychomotor activation by adenosine A_{2A} antagonists. *J. Neurosci.* 2010, 30, 2160-4.

[171] Lerner, T. N., Kreitzer, A. C. RGS_4 is required for dopaminergic control of striatal LTD and susceptibility to Parkinsonian motor deficits. *Neuron* 2012, 73, 347-59.

[172] Martire, A., Tebano, M. T., Chiodi, V., Ferreira, S. G., Cunha, R. A., Köfalvi, A., Popoli P. Pre-synaptic adenosine A_{2A} receptors control cannabinoid CB_1 receptor-mediated inhibition of striatal glutamatergic neurotransmission. *J. Neurochem.* 2011, 116, 273 80.

[173] Ferreira, S. G., Gonçalves, F. Q., Marques, J. M., Tomé, A. R., Rodrigues, R. J., Nunes-Correia, I., Ledent, C., Harkany, T., Venance, L., Cunha, R. A., Köfalvi, A. Presynaptic A_{2A} adenosine receptors dampen CB_1 cannabinoid receptor-mediated inhibition of corticostriatal glutamatergic transmission. *Br. J. Pharmacol.* 20014, doi: 10.1111/bph. 12970.

[174] Yao, L., McFarland, K., Fan, P., Jiang, Z., Ueda, T., Diamond I. Adenosine A_{2A} blockade prevents synergy between mu-opiate and cannabinoid CB_1 receptors and eliminates heroin-seeking behavior in addicted rats. *Proc. Natl. Acad. Sci. U.S.A.* 2006, 103, 7877-82.

[175] Justinová, Z., Ferré, S., Redhi, G. H., Mascia, P., Stroik, J., Quarta, D., Yasar, S., Müller, C. E., Franco, R., Goldberg, S. R. Reinforcing and neurochemical effects of cannabinoid CB_1 receptor agonists, but not cocaine, are altered by an adenosine A_{2A} receptor antagonist. *Addict. Biol.* 2011, 16, 405-15.

[176] Justinová, Z., Redhi, G. H., Goldberg, S. R., Ferré, S. Differential effects of presynaptic versus postsynaptic adenosine A_{2A} receptor blockade on Δ9-tetrahydrocannabinol (THC) self-administration in squirrel monkeys. *J. Neurosci.* 2014, 34, 6480-4.

[177] Bonaventura, J., Rico, A. J., Moreno, E., Sierra, S., Sánchez, M., Luquin, N., Farré, D., Müller, C. E., Martínez-Pinilla, E., Cortés, A., Mallol, J., Armentero, M. T., Pinna, A., Canela, E. I., Lluís, C., McCormick, P. J., Lanciego, J. L., Casadó, V., Franco, R. L-DOPA-treatment in primates disrupts the expression of A(2A) adenosine-CB(1) cannabinoid-D(2) dopamine receptor heteromers in the caudate nucleus. *Neuropharmacol.* 2014, 79, 90-100.

[178] Pinna, A., Bonaventura, J., Farré, D., Sánchez, M., Simola, N., Mallol, J., Lluís, C., Costa, G., Baqi, Y., Müller, C.E., Cortés, A., McCormick, P., Canela, E. I., Martínez-Pinilla E., Lanciego, J. L., Casadó V., Armentero, M.T., Franco, R. L-DOPA disrupts adenosine A(2A)-cannabinoid CB(1)-dopamine D(2) receptor heteromer cross-talk in the striatum of hemiparkinsonian rats: biochemical and behavioral studies. *Exp. Neurol.* 2014, 253, 180-91.

[179] Cerri, S., Levandis, G., Ambrosi, G., Montepeloso, E., Antoninetti, G.F., Franco, R., Lanciego, J. L., Baqi, Y., Müller, C. E., Pinna, A., Blandini, F., Armentero, M. T. Neuroprotective potential of adenosine A_{2A} and cannabinoid CB_1 receptor antagonists in an animal model of Parkinson disease. *J. Neuropathol. Exp. Neurol.* 2014, 73, 414-24.

[180] Tebano, M. T., Martire, A., Popoli, P. Adenosine A(2A)-cannabinoid CB(1) receptor interaction: An integrative mechanism in striatal glutamatergic neurotransmission. *Brain Res.* 2012, 1476, 108-18.

[181] Fredholm, B. B., Irenius, E., Kull, B., Schulte, G. Comparison of the potency of adenosine as an agonist at human adenosine receptors expressed in Chinese hamster ovary cells. *Biochem. Pharmacol.* 2001, 61, 443-8.

[182] Correia-de-Sá, P., Ribeiro, J. A. Presynaptic A_1 inhibitory/A_{2A} facilitatory activation balance depends on motor nerve stimulation paradigm at the rat hemidiaphragm. *J. Neurophysiol.* 1996, 76, 3910-9.

[183] Rebola, N., Rodrigues, R. J., Lopes, L. V., Richardson, P. J., Oliveira, C. R., Cunha, R. A. Adenosine A_1 and A_{2A} receptors are co-expressed in pyramidal neurons and colocalized in glutamatergic nerve terminals of the rat hippocampus. *Neuroscience* 2005, 133, 79-83.

[184] Ciruela, F., Casado, V., Rodrigues, R.J., Lujan, R., Burgueno, J., Canals, M., Borycz, J., Rebola, N., Goldberg, S. R., Mallol, J., Cortes, A., Canela, E. I., Lopez-Gimenez, J. F., Milligan, G., Lluis, C., Cunha, R. A., Ferré, S., Franco, R. Presynaptic control of striatal glutamatergic neurotransmission by adenosine A_1-A_{2A} receptor heteromers. *J. Neurosci.* 2006, 26, 2080-7.

[185] Lopes, L. V., Cunha, R. A., Ribeiro, J. A. Cross talk between A(1) and A(2A) adenosine receptors in the hippocampus and cortex of young adult and old rats. *J. Neurophysiol.* 1999, 82, 3196-203.

[186] Cunha, R. A., Johansson, B., van der Ploeg, I., Sebastião, A. M., Ribeiro, J. A., Fredholm, B. B. Evidence for functionally important adenosine A_{2A} receptors in the rat hippocampus. *Brain Res.* 1994, 649, 208-16.

[187] O'Kane, E. M., Stone, T. W. Interaction between adenosine A_1 and A2 receptor-mediated responses in the rat hippocampus *in vitro*. *Eur. J. Pharmacol.* 1998, 362, 17-25.

[188] Lopes, L. V., Cunha, R. A., Ribeiro, J. A. ZM 241385, an adenosine A(2A) receptor antagonist, inhibits hippocampal A(1) receptor responses. *Eur. J. Pharmacol.* 1999, 383, 395-8.

[189] Ferré, S., Borycz, J., Goldberg, S. R., Hope, B. T., Morales, M., Lluis, C., Franco, R., Ciruela, F., Cunha, R. Role of adenosine in the control of homosynaptic plasticity in striatal excitatory synapses. *J. Integr. Neurosci.* 2005, 4, 445-64.

[190] Sebastião, A. M., Ribeiro, J. A. Adenosine receptors and the central nervous system. *Handb. Exp. Pharmacol.* 2009, 193, 471-534.

[191] Quarta, D., Borycz, J., Solinas, M., Patkar, K., Hockemeyer, J., Ciruela, F., Lluis, C., Franco, R., Woods, A. S., Goldberg, S. R., Ferré, S. Adenosine receptor-mediated modulation of dopamine release in the nucleus accumbens depends on glutamate neurotransmission and N-methyl-D-aspartate receptor stimulation. *J. Neurochem.* 2004, 91,873-80.

[192] Ciruela, F., Ferré, S., Casado, V., Cortes, A., Cunha, R. A., Lluis, C., Franco, R. Heterodimeric adenosine receptors: a device to regulate neurotransmitter release. *Cell. Mol. Life Sci.* 2006, 63, 2427-31.

[193] Ferré, S., Ciruela, F., Woods, A.S., Lluis, C., Franco, R. Functional relevance of neurotransmitter receptor heteromers in the central nervous system. *Trends Neurosci.* 2007, 30, 440-6.

[194] Ferré, S., Ciruela, F., Quiroz, C., Luján, R., Popoli, P., Cunha, R. A., Agnati, L. F., Fuxe, K., Woods, A. S., Lluis, C., Franco, R. Adenosine receptor heteromers and their integrative role in striatal function. *Scientific World Journal* 2007, 7, 74-85.

[195] Cristovao-Ferreira, S., Navarro, G., Brugarolas, M., Pérez-Capote, K., Vaz, S. H., Fattorini, G., Conti, F., Lluis, C., Ribeiro, J. A., McCormick, P. J., Casadó, V., Franco,

R., Sebastião, A. M. $A_1RA_{2A}R$ heteromers coupled to Gs and Gi/0 proteins modulate GABA transport into astrocytes. *Purinergic Signalling* 2013, 9, 433-49.

[196] Lipton, S. A., Rosenberg, P. A. Excitatory amino acids as a final common pathway for neurologic disorders. *N. Engl. J. Med.* 1994, 330, 613-22.

[197] Gomes, C. V., Kaster, M. P., Tomé, A. R., Agostinho, P. M., Cunha, R.A. Adenosine receptors and brain diseases: neuroprotection and neurodegeneration. *Biochim. Biophys. Acta* 2011, 1808, 1380-99.

[198] Cunha, R. A. Adenosine as a neuromodulator and as a homeostatic regulator in the nervous system: different roles, different sources and different receptors. *Neurochem. Int.* 2001, 38, 107-125.

[199] Chen, J. F., Eltzschig, H. K., Fredholm, B. B. Adenosine receptors as drug targets--what are the challenges? *Nat. Rev. Drug Discov.* 2013, 12, 265-86.

[200] Jacobson, K. A., Gao, Z. G. Adenosine receptors as therapeutic targets. *Nat. Rev. Drug Discov.* 2006, 5, 247-61.

[201] Latini, S., Pedata, F. Adenosine in the central nervous system: release mechanisms and extracellular concentrations. *J. Neurochem.* 2001, 79, 463-84.

[202] Cunha, R. A., Vizi, E. S., Ribeiro, J. A., Sebastião, A. M. Preferential release of ATP and its extracellular catabolism as a source of adenosine upon high- but not low-frequency stimulation of rat hippocampal slices. *J. Neurochem.* 1996, 67, 2180-7.

[203] Lovatt, D., Xu, Q., Liu, W., Takano, T., Smith, N. A., Schnermann, J., Tieu, K., Nedergaard, M. Neuronal adenosine release, and not astrocytic ATP release, mediates feedback inhibition of excitatory activity. *Proc. Natl. Acad. Sci. U.S.A.* 2012, 109, 6265-70.

[204] Napieralski, R., Kempkes, B., Gutensohn, W. Evidence for coordinated induction and repression of ecto-50-nucleotidase (CD73) and the A_{2A} adenosine receptor in a human B cell line. *Biol. Chem.* 2003, 384, 483-7.

[205] Boeck, C. R., Kroth, E. H., Bronzatto, M. J., Vendite, D. Effect of the L- or D-aspartate on ecto-5 nucleotidase activity and on cellular viability in cultured neurons: participation of the adenosine A_{2A} receptors. *Amino Acids* 2007, 33, 439-44.

[206] Langer, D., Hammer, K., Koszalka, P., Schrader, J., Robson, S., Zimmermann, H. Distribution of ectonucleotidases in the rodent brain revisited. *Cell Tissue Res.* 2008, 334, 199-217.

[207] Rebola, N., Lujan, R., Cunha, R. A., Mulle, C. Adenosine A_{2A} receptors are essential for long-term potentiation of NMDA-EPSCs at hippocampal mossy fiber synapses. *Neuron* 2008, 57, 121-34.

[208] Augusto, E., Matos, M., Sevigny, J., El-Tayeb, A., Bynoe, M. S., Muller, C. E., Cunha, R. A., Chen, J. F. Ecto-5 -Nucleotidase (CD73)-mediated Formation of Adenosine Is Critical for the Striatal Adenosine A_{2A} Receptor Functions. *J. Neurosci.* 2013, 33, 11390-9.

[209] Ballarín, M., Fredholm, B. B., Ambrosio, S., Mahy, N. Extracellular levels of adenosine and its metabolites in the striatum of awake rats: inhibition of uptake and metabolism. *Acta Physiol. Scand.* 1991, 142, 97-103.

[210] Pedata, F., Corsi, C., Melani, A., Bordoni, F., Latini, S. Adenosine extracellular brain concentrations and role of A_{2A} receptors in ischemia. *Ann. N. Y. Acad. Sci.* 2001, 939, 74-84.

[211] Stone, T. W. Purines and neuroprotection. *Adv. Exp. Med. Biol.* 2002, 513, 249-80.

[212] Dunwiddie, T. V., Masino, S. A. The role and regulation of adenosine in the central nervous system. *Annu. Rev. Neurosci.* 2001, 24, 31-55.

[213] Chen, J. F., Sonsalla, P. K., Pedata, F., Melani, A., Domenici, M. R., Popoli, P., Geiger, J., Lopes, L. V., de Mendonça, A. Adenosine A_{2A} receptors and brain injury: broad spectrum of neuroprotection, multifaceted actions and 'fine tuning' modulation. *Prog. Neurobiol.* 2007, 83, 310-31.

[214] Dai, S. S., Zhou, Y. G., Li, W., An, J. H., Li, P., Yang, N., Chen, X. Y., Xiong, R. P., Liu, P., Zhao, Y., Shen, H. Y., Zhu, P. F., Chen, J. F. Local glutamate level dictates adenosine A_{2A} receptor regulation of neuroinflammation and traumatic brain injury. *J. Neurosci.* 2010, 30, 5802-10.

[215] Melani A., Corti, F., Cellai, L., Vannucchi, M. G., Pedata, F. Low doses of the selective adenosine A_{2A} receptor agonist CGS21680 are protective in a rat model of transient cerebral ischemia. *Brain Res.* 2014, 1551, 59-72.

[216] Pedata, F., Pugliese, A. M., Coppi, E., Dettori, I., Maraula, G., Cellai, L., Melani A. Adenosine A_{2A} receptors modulate acute injury and neuroinflammation in brain ischemia. *Mediators Inflamm.* 2014, - doi: 10.1155/2014/805198.

[217] Jones, P. A., Smith, R. A., Stone, T. W. Protection against intra hippocampal kainite excitotoxicity by intracerebral administration of an adenosine A_{2A} receptor antagonist. *Brain Res.* 1998, 800, 328-35.

[218] Jones, P. A., Smith, R. A., Stone, T. W. Protection against kainate-induced excitotoxicity by adenosine A_{2A} receptor agonists and antagonists. *Neuroscience* 1998, 85, 229-37.

[219] During, M. J., Spencer, D. D. Adenosine: a potential mediator of seizure arrest and postictal refractoriness. *Ann. Neurol.* 1992, 32, 618-24.

[220] Young, D., Dragunow, M. Status epilepticus may be caused by loss of adenosine anticonvulsant mechanisms. *Neuroscience* 1994, 58, 245-61.

[221] Dunwiddie, T. V. Adenosine and suppression of seizures. *Adv. Neurol.* 1999, 79, 1001-10.

[222] Boison, D. Adenosine and epilepsy: from therapeutic rationale to new therapeutic strategies. *Neuroscientist* 2005, 11, 25-36.

[223] Ochiishi, T., Takita, M., Ikemoto, M., Nakata, H., Suzuki, S.S. Immunohistochemical analysis on the role of adenosine A_1 receptors in epilepsy. *Neuroreport* 1999, 10, 3535-41.

[224] Ekonomou, A., Sperk, G., Kostopoulos, G., Angelatou, F. Reduction of A_1 adenosine receptors in rat hippocampus after kainic acid-induced limbic seizures. *Neurosci. Lett.* 2000, 284, 49-52.

[225] Rebola, N., Coelho, J. E., Costenla, A. R., Lopes, L. V., Parada, A., Oliveira, C. R., Soares-da-Silva, P., de Mendonca, A., Cunha, R.A. Decrease of adenosine A_1 receptor density and of adenosine neuromodulation in the hippocampus of kindled rats. *Eur. J. Neurosci.* 2003, 18, 820-8.

[226] Glass, M., Faull, R. L., Bullock, J. Y., Jansen, K., Mee, E. W., Walker, E. B., Synek, B. J., Dragunow, M. Loss of A_1 adenosine receptors in human temporal lobe epilepsy. *Brain Res.* 1996, 710, 56-68.

[227] Rebola, N., Porciúncula, L. O., Lopes, L. V., Oliveira, C. R., Soares-da-Silva, P., Cunha, R. A. Long-term effect of convulsive behavior on the density of adenosine A_1 and A_{2A} receptors in the rat cerebral cortex. *Epilepsia* 2005, 46,159-65.

[228] Huicong, K., Zheng, X., Furong, W., Zhouping, T., Feng, X., Qi, H., Xiaoyan, L., Xiaojiang, H., Na, Z., Ke, X., Zheng, Z., Suiqiang, Z. The imbalanced expression of adenosine receptors in an epilepsy model corrected using targeted mesenchymal stem cell transplantation. *Mol. Neurobiol.* 2013, 48, 921-30.

[229] Bonan, C. D., Walz, R., Pereira, G. S., Worm, P. V., Battastini, A. M., Cavalheiro, E. A., Izquierdo, I., Sarkis, J. J. Changes in synaptosomal ectonucleotidase activities in two rat models of temporal lobe epilepsy. *Epilepsy Res.* 2000, 39, 229-38.

[230] Fedele, D. E., Li, T., Lan, J. Q., Fredholm, B. B., Boison, D. Adenosine A_1 receptors are crucial in keeping an epileptic focus localized. *Exp. Neurol.* 2006, 200, 184-90.

[231] Kochanek, P. M., Vagni, V. A., Janesko, K. L. W., Clark, R. S. B., Homanics, G. E., Dixon, C. E., Schnermann, J., Jackson, E. K. Adenosine A_1 receptor knockout mice develop lethal status epilepticus after experimental traumatic brain injury. *J. Cereb. Blood Flow Metab.* 2006, 26, 565-75.

[232] Boison, D. The adenosine kinase hypothesis of epileptogenesis. *Prog. Neurobiol.* 2008, 84, 249-62.

In: Adenosine Signaling Mechanisms
Editors: V. Ramkumar and R. Paes de Carvalho
ISBN: 978-1-63483-186-4

Chapter 14

EMERGING ROLES FOR ADENOSINE IN NERVOUS SYSTEM PLASTICITY

Henrique Rocha-Mendonça[1], Vanessa G. Goulart[1], Michele B. de Carvalho[1], Felipe Cabral-Miranda[2], Ana L. Tavares-Gomes[1] and Paula Campello-Costa[1*]

[1]Programa de Neurociências, Departamento de Neurobiologia, Instituto de Biologia, Universidade Federal Fluminense, Niterói, RJ, Brazil
[2]Instituto de Biofísica, Universidade Federal do Rio de Janeiro, Rio de Janeiro, Cidade Universitária, RJ, Brazil

ABSTRACT

Adenosine is an important modulator of the central nervous system (CNS) that acts mainly by A1 (A1AR) and A2a (A2aAR) receptors, which are found in neurons and glial cells, and are coupled to different intracellular signaling pathways. These receptors modulate neurotransmitter release and regulate different aspects of neuroplasticity, an important process to ensure circuit development and remodeling after lesion. Neural plasticity involves pre- and postsynaptic changes in circuit strength. In this context, adenosine maintains the relation of neuronal excitability within the network, allowing plasticity to occur. Adenosine receptors are expressed throughout lifespan, but their expression and function varies accordingly to the stage of development in different structures, suggesting that they can contribute differently to the formation, maintenance and modification of circuits. Usually, A1AR activation has been associated with decrease in neurotransmitter release or excitatory postsynaptic current and long-term depression (LTD), whereas A2aAR activation has opposite effects. Therefore, A1AR activation is important to synaptic elimination of weaker connections, while A2aAR activity is related to the establishment of appropriated stronger contacts, indicating that adenosine produces a dynamic adjustment in the synaptic strength. This adjustment is also necessary during disorder-induced plasticity, such as observed in different kinds of injury and in neurodegenerative and neuropsychiatric disorders. In Parkinson´s disease (PD),

* Author to whom correspondence should be addressed: E-mail: pcamp@vm.uff.br or pcampello.lopes@gmail.com, Phone: +5521-26292277, Fax: +5521-26292268.

adenosine plays a key role in the fine-tuning of striatal function by coordinating responses to dopamine thereby serves as promising non-dopaminergic target for PD treatment. Besides, adenosine receptors have also been associated to anxiogenic or anxiolytic effects. At cellular level it can leads to hippocampal LTD and LTP, suggesting that it could be related to establishment of aversive or non-aversive memories. Thus, adenosine plays an active role in regulating both natural and the plasticity subsequent to disorders in the CNS. Hence, adenosinergic receptors emerge as candidates for clinical intervention regarding different pathologies in the central nervous system.

Introduction

Adenosine is an important modulator of the nervous system that is synthetized both in the intracellular and extracellular fluid. In the cytosol, adenosine is derived mainly from AMP (adenosine monophosphate) through the action of a family of enzymes known as 5'-nucleotidases. In the extracellular medium, it is derived from the breakdown of other adenine nucleotides, such as ATP (adenosine triphosphate) by ecto-nucleotidases [1]. Adenosine is metabolized mainly by its phosphorylation to AMP by adenosine kinase (ADK) or by its degradation to inosine by adenosine deaminase [2, 3]. There are two bidirectional transporters families that play a role in the control of intra- and extracellular levels of adenosine: (1) Equilibratives transporters (ENT) carry purines and pyrimidines following their concentration gradient. These transporters may be sensitive or insensitive to nanomolar concentrations of the transporter inhibitor, nitrobenzylthioinosine (NBMPR); (2) concentrative transporters (CNT) activity depends on transmembrane sodium gradient that mediate adenosine influx against its concentration gradient (Figure 1) [2, 4].

The actions of adenosine are mediated by four different G protein-coupled receptors (GPCRs) called: A1, A2a, A2b and A3 (Figure 1) [5, 6]. These receptors are widely distributed in the nervous system, where the high affinity A1 (A1AR) and A2a (A2aAR) receptors account for the fine tuning of neurotransmitter release and modulation of other signaling molecules involved in neuronal plasticity [7]. Therefore, in this chapter we will discuss particularly the biochemical, physiological and pathophysiological aspects of these two subtypes of adenosine receptors and its relation to development and plasticity of central nervous system.

A1A Receptors

It has been reported that different signaling cascades are activated by A1A receptors. It can be coupled to Gi/Go proteins [8], which inhibit adenylyl cyclase (AC) activity thereby decreasing cyclic AMP levels, protein kinase A (PKA) activity and cyclic AMP response element binding protein (CREB) phosphorylation [8, 9]. A1ARs can also couple to other G proteins, activating phospholipase C (PLC), phospholipase A2 [10] and mitogen-activated protein kinase (MAPK) pathways [8, 11]. Alternatively, they can directly modulate K^+ channels and voltage sensitive Ca^{2+} channels (types Q, N and P) [12-14].

A1A receptors are widely distributed in the central nervous system, being found in different regions such as cerebral cortex, hippocampus, basal ganglia, thalamus, superior colliculus, retina, among others [9, 15, 16]. At cellular level, A1AR are expressed in neurons [17] and glial cells (Figure 1) [18, 19]. In astrocytes, A1AR activity has been associated with inhibition of proliferation in culture [19] and protects them from damage and cell death [18, 20]. In oligodendroglia, A1AR activation stimulates their migration [21] and in microglia, it reduces their activation upon inflammation [18, 22]. In neurons, A1ARs are highly localized to synaptic regions, where they mediate inhibitory effects through postsynaptic hyperpolarization due to activation of K^+ channels [17, 23] and mainly via presynaptic inhibition of neurotransmitter release [23].

Usually, a long-term exposure of cells to agonist leads to desensitization or down regulation of the receptor. It was demonstrated that under these conditions postsynaptic A1AR undergo a faster desensitization compared to presynaptic ones [24]. Therefore, this may explain the mainly action of A1A receptor at presynaptic terminal. In this context, it has been shown that A1AR inhibits glutamate and serotonin release in the hippocampus [25-27], acetylcholine in neuromuscular junction, dopamine in the striatum, and gamma-aminobutyric acid (GABA) in the cerebral cortex [12, 28-30].

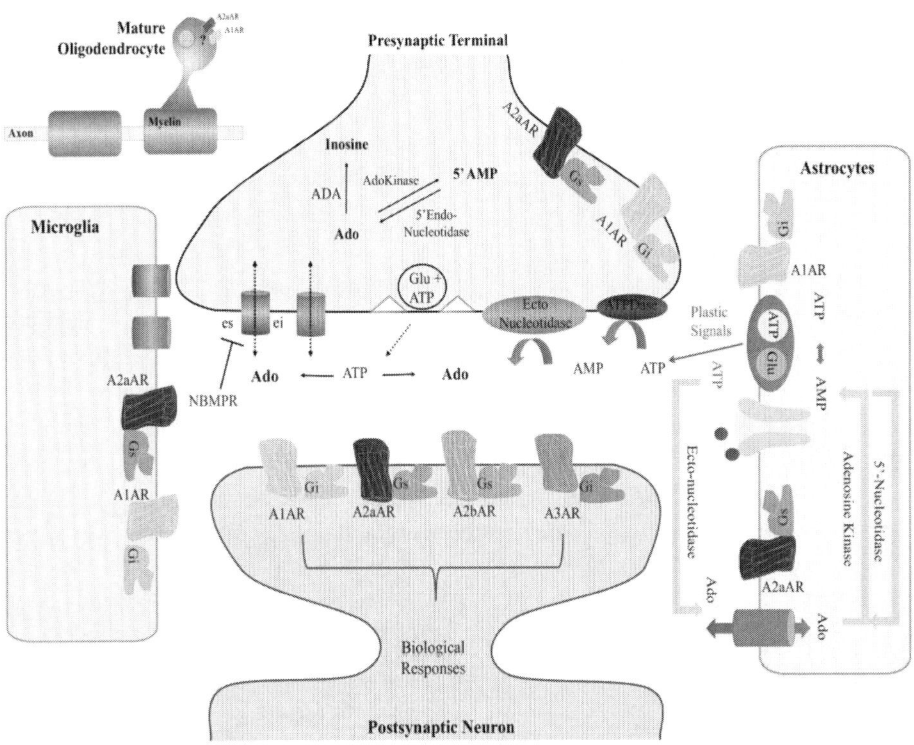

Figure 1. Adenosinergic system modulates synaptic transmission in the context of the quadripartite synapse. Pathways of adenosine synthesis, metabolism and transport are present in presynaptic neurons and astrocytes. In the extracellular space, adenosine activates different G protein coupled receptors present in pre- and postsynaptic neurons, astrocytes, microglia and oligodendrocytes. Thus, adenosine may elicit different biological responses depending on the cell type. For instance, in the context of quadripartite synapse, adenosine can modulate: (1) neurotransmitter release from presynaptic neuron, (2) postsynaptic current through regulate ion channels; gliotransmitter release from astrocytes and microglia to control synaptic transmission; (3) oligodendrocyte activity to ensure the success of nerve impulse conduction.

A2aA Receptors

A2aA receptors actions are commonly coupled to GS [8, 31], but can also be linked to Golf and G15/16 proteins [8]. These activated G protein interacts and regulates many effectors such as AC, which increase cAMP production, PKA activity and CREB phosphorylation. A2aAR activation can also lead to increase protein kinase C (PKC) activity [8]. The A2aA receptors are responsible for the excitatory actions of adenosine in different regions of the central nervous system (CNS), such as the striatum, cerebral cortex, hippocampus and superior colliculus [32, for review see 33]. At cellular levels, A2aAR was identified both in neurons and glial cells [18, 19, 34]. In glial cells, their expression is specially enhanced after lesion, but it has been also shown that adenosine acting on glial cells regulates the nervous system development. In culture system, A2aAR inhibits differentiation of oligodendrocyte progenitor cells [35]. On the other hand, it was demonstrated that an unspecific adenosine receptor agonist, inhibits oligodendrocyte progenitor cells proliferation, stimulate differentiation, and promote the formation of myelin, suggesting that functional adenosine receptors are important for oligodendrocyte development and myelination in response to neuronal activity [34]. A2aARs are also expressed in microglia, a cell related to different physiopathological conditions [36]. In the healthy brain, microglia functions as a sensor to detect changes in the microenvironment. In this context, its activity has been recognized as important in modulating synaptic transmission in response to increased neuronal activity. Thus, microglia interacts with neurons and astrocytes at quad-partite synapses, which is important for synapse formation and maturation (Figure 1) [37]. Astrocytes, the main population of glia in the brain, express all elements involved in metabolism of adenosine, which regulate the astrogliosis and the synaptic transmission through gliotransmitters release (Figure 1). Several lines of evidence point to a critical role for astrocytic source of adenosine in modulation of synaptic plasticity [18, 19, 38]. Indeed, activation of astrocytic A2aAR control glutamate uptake [39] and astrocytic-derived adenosine directly modulate neurotransmitter release through activating A2a adenosine receptors in the presynaptic neuron [40]. Some studies have shown that the A2aAR excitatory effects occur through an increase in neurotransmitter release at presynaptic terminal [41] mediated by regulation of Ca^{2+} channels induced by activation of PKA (channel P-type) or PKC (the N-type channel) [42]. However, it is not clear yet if the A2a adenosine receptor directly controls the release of neurotransmitters or whether it acts indirectly through the attenuation of A1AR -mediated inhibition [43, 44]. The release of acetylcholine induced by A2aAR involves the activation of PKA and is independent of A1AR [45, 46] while facilitation of glutamate release induced by A2aAR results from activation of PKC [47] and desensitization of inhibitory action of A1AR on synaptic transmission [44]. On the other hand, in slices of hippocampus, the expression of this receptor was demonstrated in dendritic spines that receive glutamatergic afferents, suggesting that some of the excitatory effects of A2aARs could be mediated by post-synaptic mechanisms [48]. Also, in supraoptical nucleus, A2aAR enhances neuronal excitability, increasing membrane depolarisations and transmitter release from glutamatergic afferents projecting to the nucleus, through astrocytes [49].

FUNCTIONAL INTERACTIONS OF ADENOSINE RECEPTORS

Endogenous adenosine activates both A1AR and A2aAR, which may coexist on the same nerve terminal [50, 51]. Indeed, activation of presynaptic A1AR tonically inhibits glutamate release, whereas activation of A2aAR enhances synaptic transmission [52].

The pattern of electrical activity, the extracellular source of adenosine, as well as its concentration determine which one will be preferentially activated [53]. Thus, high frequency stimulation provides high ATP release, which is converted into elevated amount of adenosine in the extracellular space, leading to A2aAR activation. On the other hand, low frequency stimulation favors the direct and low release of adenosine through transporters and A1AR activation [54-56]. Also, it was demonstrated by Lopes and co-workers that A2aAR has the ability to inhibit A1AR function [43].

As observed for other GPCRs, adenosine receptors also undergo desensitization, an important process to regulate neuronal excitability and function [15]. This process seems to be faster in A2aAR than in A1AR [57], which in turn could also account for the effect of endogenous adenosine to the fine-tuning in synaptic circuits.

Besides this indirect functional interaction, both receptors can form physical links to each other, forming functional homodimers or even with different kinds of GPCRs, forming working heterodimers [58, 59]. Usually, the new receptor complex presents different biochemical and functional characteristics and the stimulation of one of the receptor interfere with the activation of the other by its agonist [60].

Cortical A1AR can form homodimers that are regulated by endogenous adenosine [61]. Moreover, A1AR could be arranged in dimeric complex with others GPCRs. A functional interaction between A1AR and metabotropic glutamate 1 receptor (mGluR1) has been shown in cortical neuronal cultures [62]. In hippocampal slices, A1AR activation leads to a decrease in the sensitization of mGluR1 to agonist, suggesting a novel mechanism to change synaptic function [63]. Another important partner of A1AR is D1 dopamine receptor (D1R) [64-67]. This interaction has been described in striatal neurons and will be discussed later in the context of Parkinson's disease. Yoshioka and co-workers reported oligomeric association between A1AR and P2Y1R in cells following activation of both receptors [68, 69]. A1ARs are also functionally linked to P2Y2R [70, 71]. Heteromerization of A1AR and A2aAR have been observed in human embryonic kidney (HEK) cells, where A2aAR activation diminishes A1AR affinity by its agonist. Furthermore, in striatal synaptosomes A2aAR activation abrogates the inhibitory action of A1AR in glutamate release, suggesting that these heteromers can be important to the fine-tuning of synaptic transmission [72]. These heteromers are also identified in primary cultures of astrocytes, where they can modulate GABA uptake [73]. These data suggest that adenosine signaling in astrocytes contribute to neurotransmission at the tripartite synapses.

Studies show that the A2aAR aggregate as functional homodimers and trimeric oligomers at the cell membrane [74, 75]. Another classic partner of A2aAR is D2 dopamine receptor (D2R) [66], and A2aAR-D2R heterodimeric complexes are found extensively [76, 77]. They seem to exert an important modulatory role in striatal neurons in physiological and pathological conditions that will be discussed later in this chapter. Others relevant functional interactions in the striatum are mediated by assembly between A2aAR and metabotropic glutamate 5 receptor (mGluR5) [78]. Furthermore, more complex trimeric oligomers formed by mGluR5/D2R/A2aAR assembly [79] and D2R/A2aAR associated with cannabinoid (CB)-1 receptors [80] have also been identified, although their functions have not yet been established. These data collectively suggest that a broader range of actions for purinergic signaling in the nervous system could be mediated by oligomers. These oligomers emerge as possible targets for the treatment of neurodegenerative and neuropsychiatric disorders.

Therefore, activation of multiple forms of adenosine receptors may account for the different actions of adenosine in the central nervous system, such as in sleep homeostasis, learning and memory consolidation, promoting anti-nociceptive and anti-inflammatory effects, neuroprotection, anxiety modulation and synaptic plasticity [81-83].

ADENOSINE AND NEUROPLASTICITY

The precise connectivity patterns of CNS are tuned during development through several mechanisms that rely on genetic fate counterbalanced by experience [84-87]. The ability to undergo neural reorganization based on experience is termed neuroplasticity. This phenomenon is a dynamic process, which includes modulation of conduction velocity and intrinsic axonal excitability, changes in the synaptic strength, and generation or elimination of synapses [88-90]. In addition, homeostatic plasticity plays a role in maintaining neuron excitability constant relative to the network, providing stable output in the form of action potentials, and preventing hyperexcitability [91]. Adenosine has been shown to modulate most of these processes, playing a pivotal role in neuroplasticity.

The modulation of conduction velocity is mainly achieved by alterations in the thickness of myelin sheath relative to axon caliber [92]. It has been shown in the premotor cortex that neuronal activity induces OPC proliferation and differentiation into mature oligodendrocytes, which in turns add new myelin layers to active axons, increasing the conduction velocity [93]. Impulse activity promotes non-synaptic adenosine release at the nodes of Ranvier. Studies using organotypic cultures of dorsal root ganglion and cerebellar slices showed that adenosine acts on OPC, reducing proliferation and inducing differentiation to oligodendrocyte marker 4 (O4) expressing immature oligodendrocytes and then in myelinating (myelin basic protein positive - MBP+) oligodendrocytes with multiple wraps of myelin separated by interperiod lines, which is indicative of compact myelin (Figure 2) [34, 94-96]. In contrast, it has been shown that A2aAR activation inhibits OPC differentiation [35]. Thus, adenosine may induce changes in the conduction velocity of the action potential by modulating temporal integration on the postsynaptic density.

Intrinsic excitability refers to the propensity of a neuron to fire action potentials when exposed to an input signal. Its properties rely on the ion channels expressed at the axon initial segment and their biochemical properties and kinetics. Since they generate neuronal output, alterations in the properties, spatial distribution or abundance of these ion channels result in plasticity for neurons and their networks [97]. It has been shown in cultured hippocampal neurons that adenosine signals through A1AR, enhancing the K^+ efflux from the G protein-activated potassium rectifying channel (GIRK) in the soma and dendrites, limiting dendritic spines excitability (Figure 2). This signaling is required for synaptic depolarization, thus determining if the synapse sustain long-term potentiation (LTP) shortly after induction [98]. Accordingly, increased expression of ADK increases neuronal excitability, whereas decreases in ADK render the brain resistant to seizures [99]. Besides, A1AR tonic activation restrains the action potential waveform on CA3 axons. The blockade of A1AR promotes an increase in the action potential width, which correlates with robust calcium influx in the presynaptic bouton and facilitates synaptic transmission to postsynaptic neurons (Figure 2) [100]. Additionally, CA1-CA3 long-term depression of kainate receptors requires A2aAR activation. When active, kainate receptors activate non-canonical G-protein, blocking the slow afterhyperpolarization (sAHP). Kainate receptor long-term depression (LTD) relieves inhibition of the sAHP, rendering the neurons less excitable (Figure 2) [101]. Therefore, adenosine can reduce excitability both through A1AR and A2aAR, distancing membrane potential from the action potential threshold and reducing its width.

Synaptic plasticity is the best-comprehended form of neuroplasticity, which involves pre- and postsynaptic changes in synaptic strength. Long-term potentiation and depression underlies key steps in learning and memory [102]. Regarding LTP, there is extensive literature showing reduction with A1AR activation and enhancement with A2aAR activation [for review, see ref. 103].

Hippocampal LTP was negatively regulated by inhibition of the adenosine transporter or A1AR blockade, arguing for adenosine tonus activating preferentially A1AR [104-106]. Also, A1AR null mice showed impaired paired pulse facilitation at CA1, demonstrating a role of A1AR in presynaptic plasticity (Figure 2). Also, in the CA3 region, deletion of A1AR reduced the LTP inhibition mediated by adenosine [107]. Though, A1AR activation in the somatosensory cortex stimulates the membrane expression of N-methyl D-aspartate (NMDA) receptor subunit 2B (GluN2B) containing NMDA receptors through its phosphorylation by SRC kinase (Figure 2) [108]. Conversely, A1AR deletion neither modulates LTP nor spatial learning. Nonetheless, emotional context habituation behavior was impaired with A1AR deletion [109].

In terms of synaptic depression, A1AR knockout mice present impaired cortico- striatal LTD induced by NMDA or adenosine application in the striatum [110]. Alternatively, A1AR activation impairs cerebellar LTD, possibly through mGluR1/A1AR heterodimerization (Figure 2) [63]. Similarly, A1AR agonist or antagonist application does not modulate LTD in the striatum [111]. So, A1AR modulation of synaptic plasticity might depend on the induction protocol and the area investigated.

Although A2aARs do play a role in LTD [101], LTP is their main plasticity outcome. Lopes and coworkers [44] have shown that A2aAR activation not only facilitates CA1 LTP directly, but also through reduction of A1A receptor-mediated inhibition. Moreover, A2aAR activation resulted in an increase in paired-pulse ratio, suggesting an augmentation of presynaptic neurotransmitter release. Postsynaptically, it has been shown that A2aAR induces PKA activation, which in turns phosphorylates α-amino-3-hydroxy-5-methyl-4-isoxazolepropionic acid (AMPA) receptor subunit 1 (GluA1) subunit of the AMPA receptor, facilitating AMPA-evoked excitatory currents (Figure 2) [44, 112]. In the CA3 region, postsynaptic A2aAR, NMDAR and mGluR5 co-activation results in LTP of NMDA receptor currents in a SRC kinase activity-dependent manner [113]. Alternatively, the effects of the A2aAR activity on LTP might rely on its stimulation to BDNF production (Figure 2). A2aAR null mice present less hippocampal brain-derived neurotrophic factor (BDNF) than control animals, which increase BDNF production under A2aAR activation [114-116]. On the other hand, A2aAR null mice present enhancement in working memory, whereas A2aAR overexpression leads to memory deficits [117, 118]. Thus, A2aAR activation preferentially stimulates LTP, although its correlation with behavior modulation needs further study.

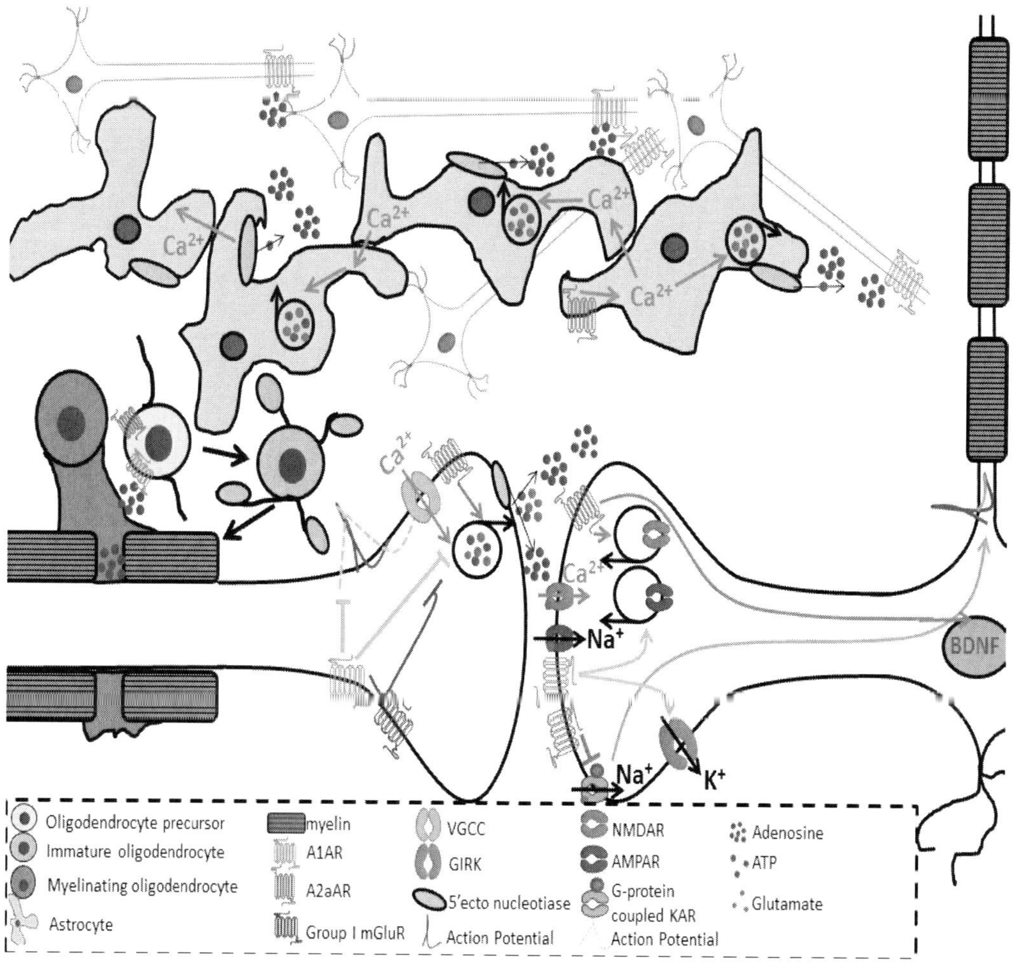

Figure 2. Mechanisms of adenosine action on neuroplasticity. Neuronal activity induces ATP/adenosine release from axons, which in turns regulate OPC differentiation and myelination. A1AR activation on the presynaptic neuron reduces the action potential width and transmitter release, whereas co-activation with group 1 metabotropic glutamate receptor blocks the A1AR mediated inhibition of transmitter release. A2aAR activation stimulates transmitter release. On the postsynaptic terminal, A1AR activation stimulates potassium efflux through GIRK channels and stimulates AMPA receptor membrane insertion. A2aAR activation induces LTD of kainate receptors and reduces the sAHP kainate-induced inhibition. Besides, A2aAR stimulates NMDA receptor membrane insertion and BDNF production. Extracellular adenosine activates A2aAR on astrocytes, inducing calcium waves on the astrocytic network that induces the release of gliotransmitters. Astrocyte-derived adenosine activates A1R on the neuronal net, reducing excitability.

Adenosine plays a pivotal role in homeostatic mechanisms [119]. It is released from the postsynaptic terminal after excessive activation of NMDAR and in turn reduces glutamate release through A1AR presynaptic activation [120]. Moreover, tonic adenosine levels provided by astrocytes tunes down the hippocampal synaptic transmission, expanding the

LTP available range [106]. In the presynaptic side, adenosine is released from highly active terminals, where it binds to presynaptic A1AR, which reduces transmitter release, protecting against neurotransmitter depletion [121]. Additionally, ATP derived from highly active synapses is converted to adenosine in the extracellular space, where it induces metaplasticity. Adenosine activates A2 adenosine receptors in the astrocytic network, inducing calcium transients that stimulate gliotransmitter release. Among these astrocyte-derived factors, adenosine is hypothesized to activate neuronal A1AR, reducing the network activity (Figure 2) [122, 123]. Hence, adenosine maintains the relation of neuronal excitability within the network, making further plastic changes possible.

Adenosine and Central Nervous System Development: A Pivotal Role upon Natural Plasticity

The CNS development involves progressive events that include the generation of different cell types and, in the case of neurons, the growth of axonal processes to the target, as well as regressive events mediated by neuronal cell death and refinement of inappropriate connections [124]. During the period when neuronal circuits are constructed, referred to as the 'critical period', the nervous system is highly plastic, being highly susceptible to changes in its environment. Afterwards, accurate neural circuits allow the correct processing of sensory, motor and cognitive abilities [125-127].

Adenosine may influence plastic changes in the nervous system during development. In the chick retina, extracellular adenosine arises as an important signaling molecule for the embryonic development and its presence regulates the survival and differentiation of retinal neurons [128, for review 9]. The neuroprotective role of adenosine has been extensively shown under different kinds of injury and it has long been suggested that it is mediated by the activation of A1AR [129]. Indeed, it was demonstrated that an A1AR up-regulation leads to the attenuation in neuronal damage evoked by NMDA-induced excitotoxicity [130]. In a pilocarpine model of temporal lobe epilepsy, it was evaluated the effect of A1AR agonist and A2aAR antagonist on seizures. This work showed that neuroprotection induced by combined A1AR activation and A2aAR inhibition leads to decline in seizure occurrence indicated by a decrease in neuronal degeneration [131]. Therefore, it is reasonable to propose that during normal development, adenosine participates as a signaling molecule to modulate the number of cells and the number of cellular contacts.

In accordance with a role played by adenosine in distinct processes during the formation and maintenance of CNS circuits, it has been reported that the expression of both A1AR and A2aAR changes during the development (Figure 3). In the superior colliculus (SC), both receptors are expressed throughout postnatal life; A1AR is higher until the fourth postnatal week, while A2aAR are more evident after this period [16]. These expression patterns are consistent with the functional activity of the receptors [16]. A similar development pattern of A1AR and A2aAR expression was demonstrated in cortex and hippocampus [132, 133].

Besides the decrease in the A1AR content in aged hippocampus, there is also a reduction in the ability of the selective A1AR agonist, N(6)-cyclopentyladenosine, to inhibit synaptic transmission in those circuits [133]. The functional relevance of A1AR in the early stages of

SC development were evaluated after a short-term blockade of adenosine deaminase, which leads to an increase in adenosine levels in the cerebrospinal fluid [134]. Indeed, during the second postnatal week, preferential activation of A1AR induces a disruption of the retinotectal topography. In the goldfish optic tectum, the application of adenosine reduces field potential by decreasing neurotransmitter release induced by A1AR in association with metabotropic glutamate receptor type 2 [135]. In hippocampus, A1AR activation and downstream signaling pathway is involved in CA3-CA1 LTD, which is correlated to synaptic elimination [136]. Moreover, Lo and Mize demonstrated that long-term depression is correlated to refinement of the retinocollicular pathway [137]. Likewise, in primary cultures of cortical and hippocampal neurons it was demonstrated that A1AR activation is able to inhibit nerve growth factor (NGF)-induced neurite growth and induce stress fiber formation, in a Rho kinase dependent-way [138], suggesting that A1AR participate in the neurite pruning process. Therefore, these data suggest that A1AR, through decrease neurotransmitter release or excitatory postsynaptic current, weakens afferents, and point to A1AR as an important regulator of sensory topographic map formation (Figure 3) [16].

On the other hand, the classic facilitatory effects achieved by A2aAR on neurotransmission have also been reported in retinotectal pathway [134, 139, 140]. Studies performed in guinea-pig superior colliculus have demonstrated that the adenosine excitatory effect involves PKA signaling pathway [141]. The literature clearly shows the cAMP is directly linked to the maintenance of synaptic strength and stabilization [142, 143]. On the other hand, the reduction of extracellular adenosine by treatment with L-homocysteine-thiolactone attenuated the effects of adenosine on synaptic potentiation [139]. These findings are consistent with that reported above, where the main A2aAR expression occurs after the end of the critical period of retinotectal development. Moreover, A2aAR activation in other systems has also been correlated to cellular mechanisms that are capable of increasing the synaptic strength [144-146]. Studies demonstrated that the main facilitatory effect of A2aAR on the neurotransmission in aged hippocampus can be due not only to the highest expression of receptors but also by the modification of transducing system coupled to A2aAR [43, 145]. This was shown in the analysis of excitatory postsynaptic potential by an A2aAR agonist, which was prevented by a protein kinase C inhibitor in hippocampus slices of young rats and blocked by a PKA inhibitor on aged rats [145]. Interestingly, the modification of signaling pathway coupled to A2aAR during development results in a different action of A2aAR. While in hippocampus of young animals the A2aAR is involved with the control of A1AR-induced synaptic inhibition [43], in aged animals its facilitatory action occurs directly on synaptic transmission, by production of cAMP [145]. Alternatively, the preferential A2aAR activation observed in hippocampus of older animals could also be associated with high levels of adenosine in synaptic cleft provided by the decrease of bidirectional transporter activity followed by lower adenosine uptake (Figure 3). Additionally, although a decrease in both adenosine evoked release and ATP extracellular catabolism is found in aged rats, the ecto-5'-nucleotidases activity is five-fold higher [147].

Collectively, the data show that A1AR activation is important to synaptic elimination of weaker connections, while A2aAR activity is related to the establishment of appropriated stronger contacts and indicates that adenosine produces a dynamic adjustment in the synaptic strength (Figure 3).

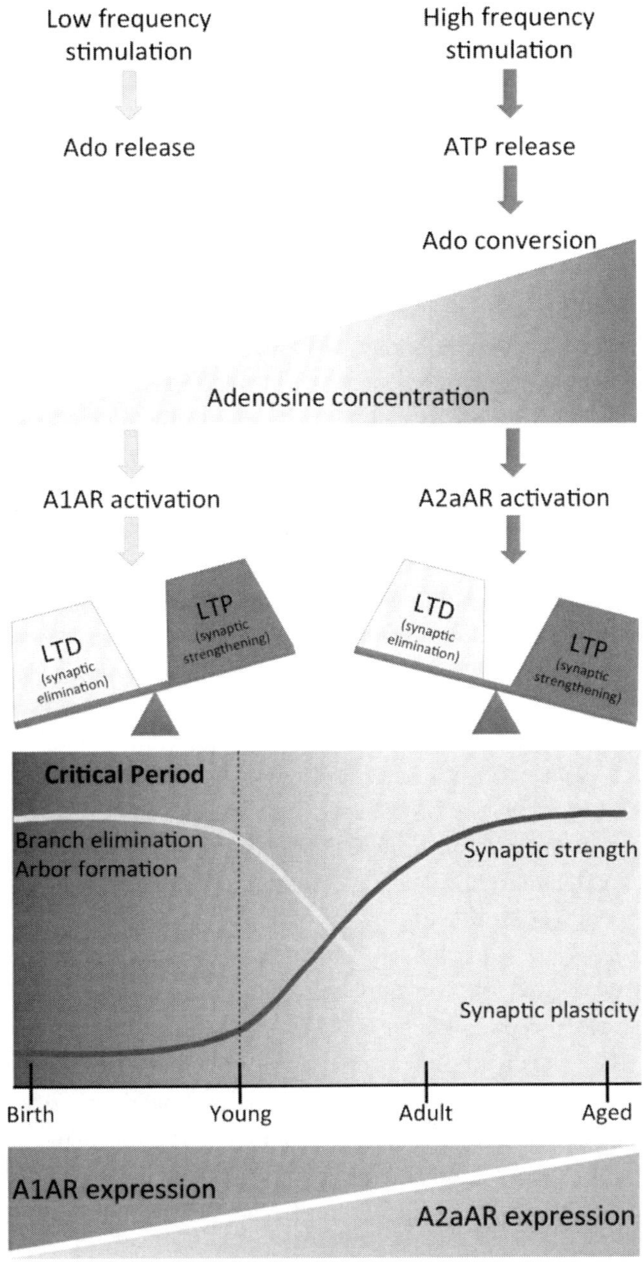

Figure 3. Adenosine in the central nervous system development: involvement with synaptic elimination and strengthening. The adenosine level changes during development. Low frequency stimulation leads to a smaller adenosine concentration, favoring the A1AR activation and LTD process. High frequency stimulation releases ATP which produces a higher adenosine concentration and A2aAR activation, triggering synaptic strengthening (LTP) (Top). Although A1AR and A2aAR are expressed throughout lifespan, there is a prevalence of A1AR in the early stages of development, during the critical period, when the synaptic plasticity is intense. During this period, selective stabilization of synapses occurs through mechanisms that involve the formation of correct arbors and elimination of inappropriate branches. On the other hand, A2aAR expression is higher after the critical period, when the synaptic plasticity reduces and the synaptic strength increase (Bottom).

In this context, it is necessary to include glial cells as an important active member of the synaptic modulation, composing the synapse *tripartite* scenario [148, 149]. Nowadays, it is known that glial cells are able to release gliotransmitters [150] and express all the molecular machinery required for vesicular release [151]. One of the first gliotransmitters described was ATP and studies have shown that, under physiological conditions, vesicular ATP release by astrocytes seems to be the main source of synaptic adenosine [106 119]. Also, astrocytes uptakes adenosine from synaptic cleft, phosphorylates it to AMP by adenosine kinase [19] and presents all adenosine receptors [for review 83]. These receptors are involved with physiological aspects of glial cells such as control of ion channel expression, transient increase in intracellular calcium concentration, regulation of proliferation, migration and apoptosis, and above all, modulation of synaptic transmission and plasticity [152-158].

Astrocytes are also known to participate in the response to neuronal activity and that adenosine is involved in this process. An elegant research demonstrated, through a dominant-negative SNARE domain selectively expressed in astrocytes, that interruption on gliotransmitter release prevents adenosine-mediated heterosynaptic depression via A1AR in the hippocampus [106]. Following this work, using the same animal model, Fellin and coworkers observed a reduction on NMDAR currents [159]. Later on, they showed that dominant- negative SNARE (soluble N-ethylmaleimide-sensitive factor attachment protein receptor) in astrocytes exhibited a reduction of surface NMDA receptor 2 subunits content on postsynaptic neuron [108]. This response was triggered by absence of A1AR activation found in dominant-negative SNARE mice that is consistent with similar data obtained by administration of A1AR antagonist in wild type slices and rescued in transgenic mice by the incubation with A1AR agonist [108]. In view of these results, it was suggested that astrocytes-derived adenosine activates A1AR, which is closely associated with the control of synaptic plasticity, modulating the synaptic strength, and directly involved with the neuron-glia cross-talk, [for review see 160].

In addition to A1AR, the astrocyte also modulates synaptic transmission through A2aAR activation. Recently, it was shown that astrocytes could detect synaptic activity by metabotropic glutamate receptor, which induces an increase in the intracellular calcium concentration that culminates with purines release and presynaptic A2aAR activation, leading to an enhancement in the efficiency of hippocampus basal synaptic transmission [40]. In turn, the astrocytic A2aAR can control the synaptic plasticity. Indeed, both *in vivo* and *in vitro* models, the exposure to A2aAR agonist blocked the glutamate uptake and led to a decrease in glutamate transporters, GLT-I and GLAST [161]. The administration of A2aAR antagonist prevents the inhibition of glutamate uptake and this effect was abolished in astrocytes cultures of A2aAR knockout mice [161]. Thus, the astrocytic A2aAR can play a critical role in synaptic plasticity by increasing the extracellular glutamate levels and, in the same way, it can be involved with the neurodegeneration process induced by excitotoxicity.

In summary, glial cells can modulate the synaptic efficacy and natural plasticity not only regulating adenosine levels at synapses, but also changing gliotransmitters and neurotransmitters levels in response to adenosine receptors activation.

ADENOSINE AND LESION INDUCED PLASTICITY

It has been known that lesions in the central and peripheral nervous system, which permanently remove inputs to specific areas, are followed by the rearrangement of projections that connect the lesioned area to its targets [162-165]. This phenomenon has been reported in different sites of projections, such as neocortex [166]; lateral geniculate nucleus (LGN) [167] and the superior colliculus [168-170], raising the question whether subcortical alteration may also be crucial for cortical reorganization followed by lesions [171, 172].

A classic experiment performed in the 70's by Hubel and Wiesel shows that plastic alterations induced by visual deprivation by suturing the lids of cats in different ages may induce rearrangements in the ocular dominance columns in a time dependent manner [173]. These experiments show that younger cats present a robust plastic response after visual deprivation compared to older cats that does not. Additionally, Serfaty and co-workers showed that retinal lesions lead to a rapid rearrangement of retinal afferents into superior colliculus in young animals, while a slower response was detected in older animals [168].

Lesion-induced plasticity relies on several signaling molecules that orchestrate circuit changes [86, 169, 170, 174, 175]. Not surprisingly, purinergic modulation account as one of these molecules in a range of systems [119]. In an *in vitro* model of glucose and oxygen deprivation in the hippocampus, adenosine signaling plays a critical role in the electrophysiology of projections to CA1 region, where neurons are more susceptible to ischemic lesions [176]. Additionally, *in vivo* data showed that A2aAR knockout mice present a greater extent of lesions in white matter following cerebral hypoperfusion [177], sustaining that purinergic modulation is directly related to plastic alterations following lesions in the central nervous system. Indeed, during optic nerve regeneration, adenosine decreases basal neurotransmission but enhances the field potential after LTP-inducing protocol [178].

Regarding neuroregeneration after lesions, the process of re-innervation of degenerated sites is guided by nuances in synaptic modulations in the peripheral regions of the lesion. This process seems to be related to purinergic modulation by its interaction with growth factors and the regulation of neurogenesis in the adult brain [179]. Recently, it was reviewed by Santiago and colleagues that in retinal and brain neurodegenerative diseases, activated microglia up-regulates A2aAR, which, in turn, may regulate microglial proliferation and levels of anti- and proinflammatory cytokines [36]. Neuroinflammation has been associated with neuroplasticity. Actually, it was demonstrated that intravitreous treatment with either interleukin 2 or zymosan disrupts retinocollicular topography [180]. Also, lens injury induces axonal extension after optic nerve crush through macrophage-derived factors [181, 182]. Therefore it is possible that microglial A2aAR contribute to the structural reorganization of the nervous system observed after lesion.

Besides, it was previously shown that a systemic treatment with caffeine, the most popular psychoactive drug in the society and, at the dose chosen, a non-specific antagonist for A1AR and A2aAR [183] promote a differential modulation on retinal lesion-induced plasticity. In young animals, a short-term treatment with caffeine leads to a dramatic reduction in the sprouting on intact retinal fibers to the ipsilateral superior colliculus that follows a contralateral retinal lesion, compared to the match-control group (Figure 4).

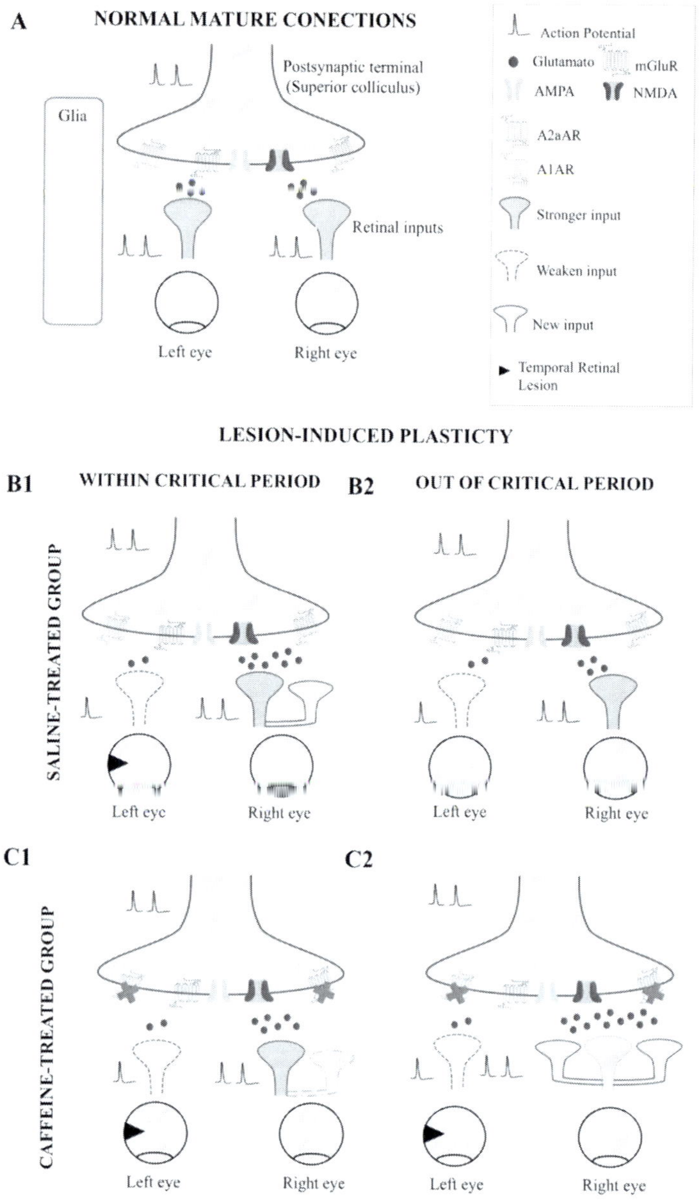

Figure 4. Dual effect of purinergic blockade with caffeine upon lesion-induced plasticity.
(A) In the visual system, retinal afferents with correlated activity converge upon the same postsynaptic target in the superior colliculus arising from a Hebbian synapse. The retinocollicular pathway develops until the third postnatal week, which comprise the critical period of these connections. (B1) A temporal retinal lesion (TRL) performed within the critical period of development leads to a rapid plastic response, characterized by the sprouting of contralateral intact axons to the correspondent denervated area. (B2) TRL out of the critical period the plastic rearrangement is barely detectable within a one-week survival. On the other hand, a short-term treatment with caffeine (30 mg/kg ip during 4 days) reduces the rearrangement of the intact, uncrossed retinal axons induced by contralateral TRL within the critical period (C1) and amplifies the plasticity after the third postnatal week, suggesting opposite effects of adenosine receptors blockade depending on the developmental time window.

On the other hand, animals submitted to retinal lesion after the closure of the critical period of development and treated with caffeine for a week exhibit a greatly enhanced the plasticity of the uncrossed retinotectal pathway compared to vehicles (Figure 4) [184]. These results suggest that caffeine has different effect on neuronal circuit, depending on the stage of development. This dual effect of caffeine may be useful to block undesired reorganization avoiding maladaptive regulation of synaptic transmission or to stimulate plasticity in adulthood.

ADENOSINE AND PARKINSON´S DISEASE

Parkinson´s disease (PD) is characterized by progressive impairment in motor function that is frequently associated with mood and cognitive instabilities [185]. At molecular level, the motor impairments of PD are caused by a gradual loss of dopamine (DA) producing nigral neurons (ventral midbrain) and parallel loss of DA input to forebrain (striatum). The decrease in DA levels leads to dysregulation of striatal function that produces the classic PD symptoms like resting tremor, muscular rigidity, and bradykinesia [186, 187]. All basal ganglia-related nuclei are connected through well-established neurochemical circuits [185]. Briefly, the two functional basal ganglia circuits, the direct and indirect pathways, originate from distinct populations of striatal medium spiny neurons (MSNs) that project to different output structures and have opposite effects on movement. Specifically, the activity of direct pathway MSNs, enriched with D1 dopamine receptor (D1R), promote movement, whereas the activation of indirect pathway, MSNs that preferentially express D2 dopamine receptor (D2R), stops the movement [188]. These two GPCRs have distinctive linkages to intracellular signaling cascades and targets, leading to fundamentally different cellular responses to extracellular DA [189]. In the context of PD, direct and the indirect pathways are differentially disturbed by the loss of dopamine [187].

Adenosine is a fundamental neuromodulator that coordinates responses to dopamine and other neurotransmitters in areas of the brain that are responsible for motor function, mood, and learning and memory [186]. As previously mentioned, A1AR and A2aAR are physiologically important in striatal regulation of behavior. Presynaptically, A1AR and A2aAR are found in glutamatergic nerve terminals, where they jointly modulate glutamate release. Postsynaptically, dimers of A1AR with D1R are present on neurons of direct pathways, whereas A2aAR with D2R are on neurons of indirect pathways (Figure 5). These dimers act through opposing intracellular signaling to modulate neuronal output and, eventually, the amount of locomotor activity [190].

Most of the A1AR in striatum are located presynaptically and act to modulate the release of glutamate, dopamine and also adenosine itself, acting like an autoreceptor [29, 191, 192]. Blockade and stimulation of A1AR respectively augments and reduces glutamate release in striatum, respectively. A1AR suppress dopamine release directly and through an indirect effect on glutamate-induced dopamine release and thereby modulate the extracellular dopamine concentration and D1R- and D2R-mediated dopaminergic neurotransmission [193].

In striatum, presynaptic A2aAR modulate glutamate release [191]. Using bioluminescence resonance energy transfer (BRET) and time-resolved fluorescence resonance energy transfer (TR-FRET) techniques, Ciruela and colleagues demonstrated the formation of

A1AR and A2aAR heteromers in striatal glutamatergic nerve terminals [72]. In this heteromeric complex form, A2aAR activation can reduce the affinity of the A1AR for agonists, providing a switch mechanism by which low and high concentrations of adenosine inhibit and stimulate, respectively, glutamatergic neurotransmission [72]. However, striatal A2aAR is predominantly postsynaptic [191]. They are coupled to excitatory G protein (G_{olf}) and act to promote neuronal activity. Selective A2aAR agonist promotes the phosphorylation of postsynaptic targets through AC–cAMP-PKA in striatal slices in a time- and dose-dependent manner, whereas presynaptic PKA substrates are unresponsive, consistent with the postsynaptic localization of A2aAR [194].

Postsynaptically, A2aAR form several heteromeric complex and modulate different aspects of MSNs activity [195, 196]. As heteromers with D2R, A2aAR decrease the affinity of D2R by its endogenous agonist, leading to the opposite effect of dopamine, *i.e.* increase motor activity [191, 196]. Oligomerization between A2aAR and D2R with mGluR5 [79] or CB1R [197] as well as homodimerization [198] have also been described (Figure 5). These heteromers provide new indications regarding potential mechanisms to control PD symptoms. Therefore, based on affinity of different antagonists for pre- or postsynaptic adenosine receptors and on differential pharmacological activity observed in heteromers, different compounds can be used to obtain more effective antidyskinetic and antiparkinsonian effects [196].

Epidemiological studies have raised the possibility of caffeine serving as a neuroprotective agent in PD [199]. Caffeine blocks adenosine receptors, at both pre- and postsynaptic sites, producing a relative strengthening of dopaminergic neurotransmission [190]. By blocking the effects of endogenous adenosine at these receptors, caffeine promotes vigilance, arousal, and attention, which in rodents are reflected by a significant increase in locomotion [190].

The A2aAR is a leading non-dopaminergic therapeutic target in PD and consistent demonstrations indicates that A2aAR antagonists afford neuroprotection against MPTP (1-methyl 4-phenyl 1,2,3,6-tetrahydropyridine) or 6-OHDA (6-hydroxydopamine)-induced dopaminergic neurotoxicity. In this context, the involvement of the glial A2aAR becomes an attractive possibility, since glial function and neuroinflammation is commonly associated with diverse pathological conditions such PD and the expression of A2aARs in glial cells is induced following brain insults [19]. The induction of A2aARs in glial cells by brain insults and inflammatory signals, coupled with a local increase in adenosine and pro-inflammatory cytokines (such as interleukin-1β, which further induces A2aAR expression), may serve as part of an important feed-forward mechanism to locally control neuroinflammatory responses in the brain. Moreover, it has been shown that adenosine, acting on A2aAR, can increase extracellular levels of glutamate, both by reducing glutamate uptake via GLT-1 and by direct release [200, 201].

It has been reported in PD patients and in 6-OHDA-lesioned rats, that post-synaptic D2 receptors as well as mRNA and protein and functionality of A2aAR are increased in the striatum [202]. In this context, A2aAR up-regulation further impair motor symptoms and A2aAR antagonists have been used to normalize motor functions by facilitating D2R signaling. Moreover, an abnormal increase in A2aAR signaling in the striatum of 6-OHDA lesioned rats and in MPTP treated primates, as well as in PD patients chronically treated with levodopa (L-DOPA) produces an intense tone of A2aAR. Therefore, blockade of the A2aAR inhibitory tone on D2R could be one of the factors underlying the positive effects produced by A2aAR antagonists in PD [203, 186]. Several studies also show that A2aAR may regulate the conductance of NMDAR [204]. Therefore, this must have important implications for dyskinesia, since NMDAR play a major role in neuroplasticity phenomena, including those which take place in motor circuits, and may underlie abnormal motor responses to dopamine-replacement therapy [205]. In the 6-OHDA rat, striatal synaptic plasticity has been shown to be impaired, although chronic treatment with levodopa was able to restore it [206].

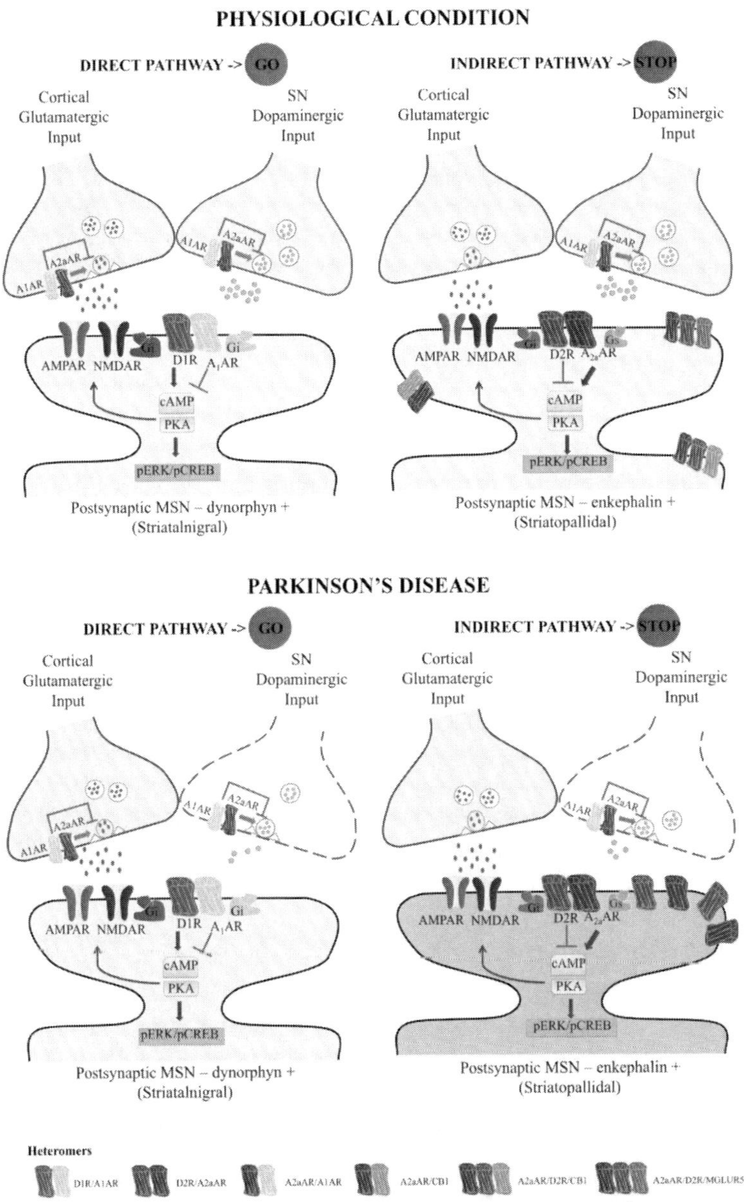

Figure 5. Differential expression of adenosine receptors in the striatum and their contribution to coordinate movement. (A) Under physiological conditions, medium spiny striatal neurons (MSN) receive dopaminergic inputs from the substantia nigra (SN) and glutamatergic input from cortex and give rise to direct and indirect pathways that instructs to start (go) or stop movement, respectively. Adenosine receptors differentially modulate both pathways, mainly by forming heteromers.

Presynaptically, A1AR and A2aAR heteromers control both glutamate and dopamine release to dynorphyn positive MSNs (DYN-MSN) and dopamine to enkephalin positive MSNs (ENK-MSN). Postsynaptically, A1AR forms heteromers with dopamine 1 receptor (D1R) in DYN- MSN (direct pathway – striatalnigral) while A2aAR forms heteromers with D2R, CB1 and with D2R/CB1 and with D2R/mGluR5 in ENK-MSN (indirect pathway – striato-pallidal). While dopamine (DA) activates neurons belonging to direct pathway through D1Rs and depresses neurons from indirect pathway through D2Rs, adenosine does the opposite, inhibiting the direct pathway by activation of A1AR and activating the indirect pathway through the A2aAR. A balanced level of activity of the direct and indirect pathways is the basis to the proper movement execution. (B) In Parkinson's disease, the degeneration of the dopaminergic afferent, decrease dopamine in the striatum. This results in a reduced activation of the direct pathway and in a disinhibition of the indirect pathway, which is associated with the elevation of A2aAR transmission. Such unbalanced activity of the striatal output pathways is the basis of the motor impairment observed in Parkinson's disease. Also, the differential activation of intracellular signaling pathway by dopamine and adenosine receptors can contribute to long-term plasticity observed both in normal and pathological conditions.

As previously mentioned, changes in pre- and postsynaptic activity are important to control the efficacy of synaptic transmission. Bidirectional synaptic plasticity regulates learning and adaptative behavior in an experience-dependent manner. In this context, striatal neurons exhibit both LTP and LTD when appropriated stimulated by glutamatergic or dopaminergic presynaptic afferents [207-209]. Physiologically, phasic dopamine signaling modifies medium spiny neuron excitability in the striatonigral and striatopallidal pathways. In Parkinson's disease, dopamine depletion induces inappropriate potentiation in the inhibitory striato-pallidal pathway that gradually deteriorates motor functioning (Figure 5) [210]. Again, A2aAR has been implicated in the synaptic modifications of striatal neurons acting both at pre- or postsynaptic levels in the indirect pathway [211]. Shen and coworkers showed that AC activation by G_{olf} under A2aAR activation induces LTP on striato-pallidal circuits [212]. Moreover, co-activation of postsynaptic FGF and A2aA receptors facilitates the induction of LTP at corticostriatal synapses of striato-pallidal MSNs [213].

Together, these studies postulate that adenosine plays a key role in the fine-tuning of striatal function, coordinating responses to dopamine and thus emerge like a non-dopaminergic target promise to PD treatment.

Adenosine and Anxiety

Anxiety is a primal behavior that helps animals to survive when challenged in aversive environment. However, anxiety disorders are highly prevalent [214]. This behavior is associated with others cognitive processes like learning and memory, and involves a great degree of plasticity [215, 216]. There are different structures and neurotransmitters related to anxiety [217-220].

The hypothalamic-pituitary-adrenal (HPA) axis coordinates the primary physiological response to acute stress, which is generally considered adaptive for re-establishing homeostasis. Typically the stress response has been identified as a "fight or flight" reaction, but may also include an increased state of vigilance, which is often accompanied by increased anxiety. There is now substantial evidence that repeated exposure to stress induces persistent changes in HPA axis functionality that may produce maladaptive outcomes [221]. The main action of HPA axis occurs through the secretion of glucocorticoids (Figure 6). Indeed, chronic stress stimulates changes in HPA axis functionality that may produce an increase or decrease glucocorticoid levels, leading to different psychopathologies in animals and humans [221-228]. Chronic stress induces a reduction on the dendritic length and ramification, reducing synaptic inputs on hippocampal or medial prefrontal cortex neurons, in a process dependent on glucocorticoids [215, 221]. On the other hand, it also induces the opposite effects on orbitofrontal cortex and amygdala [215, 229]. Besides, human hippocampus has been shown to undergo shrinkage associated with post-traumatic stress disorder, whereas the amygdala is more active in anxiety disorders patients [215].

Several systems of neurotransmission / neuromodulation are involved in the activation of the defense motivational system. Data from clinical studies have implicated adenosine in different types of anxiety disorders: panic attacks, phobic or generalized anxiety states and post-traumatic stress disorder [230-232].

A1AR are extensively distributed in the central nervous system, with high levels of expression in glucocorticoid receptor-rich areas like hippocampus, pituitary gland, hypothalamus and cerebral and cerebellar cortex [233-236]. A2aAR have a more restricted distribution and are mostly found in striatum. These striatal neurons also show elevated levels of immunoreactivity for glucocorticoid receptors [237]. These data suggest that glucocorticoids may regulate adenosine receptors in the brain. In fact, there is some evidence that stressful stimuli, which are known to affect glucocorticoid levels, can influence A1AR [238, 239] in the central nervous system. Svenningsson and Fredholm [240] showed that one week after adrenalectomy, the expression of mRNA for A1AR was significantly decreased, as was the number of binding sites for DPCPX (8-cyclopentyl-1,3-dipropylxanthine, a potent and selective A1AR antagonist). However, did not significantly affect the number of CGS 21680 (2-[*p*-(2-carbonylethyl)phenylethylamino]-59-*N*-ethylcarboxami-doadenosine, a selective A2aAR agonist) binding sites in striatum or the mRNA encoding adenosine A2aAR.

On the other hand, adenosine can also control the release of corticotrophin and cortisol or corticosterone, the latter being major adrenal cortex hormones in humans and rodents [241]. The neurohormonal analysis parameters of the A2aAR KO mice may support this hypothesis about the relationship between the A2aAR and the HPA axis. A2aAR gene disruption induced a significant increase in alpha-MSH (alpha-melanocyte-stimulating hormone) content in the amygdala and cerebral cortex. Moreover, in the anterior lobe of the pituitary, a substantial increase in ACTH was observed, and plasma corticosterone levels were also significantly higher in the A2aAR KO mice, revealing hyperactivity of their pituitary-adrenocortical axis [241]. Furthermore, some hormonal parameters that correlate with anxiety may be modified by caffeine. Administration of caffeine to healthy male volunteers has been shown to increase the activity of the hypothalamus–pituitary–adrenal (HPA) axis, as evidenced by elevations in the adrenocorticotropic hormone (ACTH) and cortisol [242]. In contrast, some reports indicate that caffeine can have beneficial effects in anxiety-prone patients [243]. Also, another A2aAR blocker restores the behavior, eletrophysiological and morphological impairments induced by maternal separation in early stages of development [244].

Adenosine receptors in the hippocampus are important modulators of synaptic transmission and neuronal excitability [103]. Glutamatergic synaptic transmission in physiological conditions is controlled negatively by the dominant A1 adenosine receptors and positively to a lesser degree by A2aAR [245]. Interestingly, this pattern appears to be modified in the aged hippocampus, with an increase in the number of A2aAR and a decrease in the number of A1AR, leading to unbalanced adenosine receptors [132]. These changes facilitate the release of glutamate by A2aAR [145]. This is also observed in situations of acute stress, which suggests a deleterious contribution of A2aAR to these conditions [246]. Sub-chronic stress did not cause hippocampal neurodegeneration but induced loss of synaptophysin, a synaptic marker, in rat hippocampal nerve terminals. The blockade of A2aAR by a the selective antagonist SCH 58261 (7-(2-phenylethyl)-5-amino-2-(2-furyl)-pyrazolo-[4,3-e]-1,2,4-triazolo[1,5-c]pyrimidine) attenuated the loss of synaptophysin immunoreactivity in these rats [246]. Therefore, the effects of A2aAR antagonists may depend, at least in part, on the direct or indirect control of the HPA axis (Figure 6).

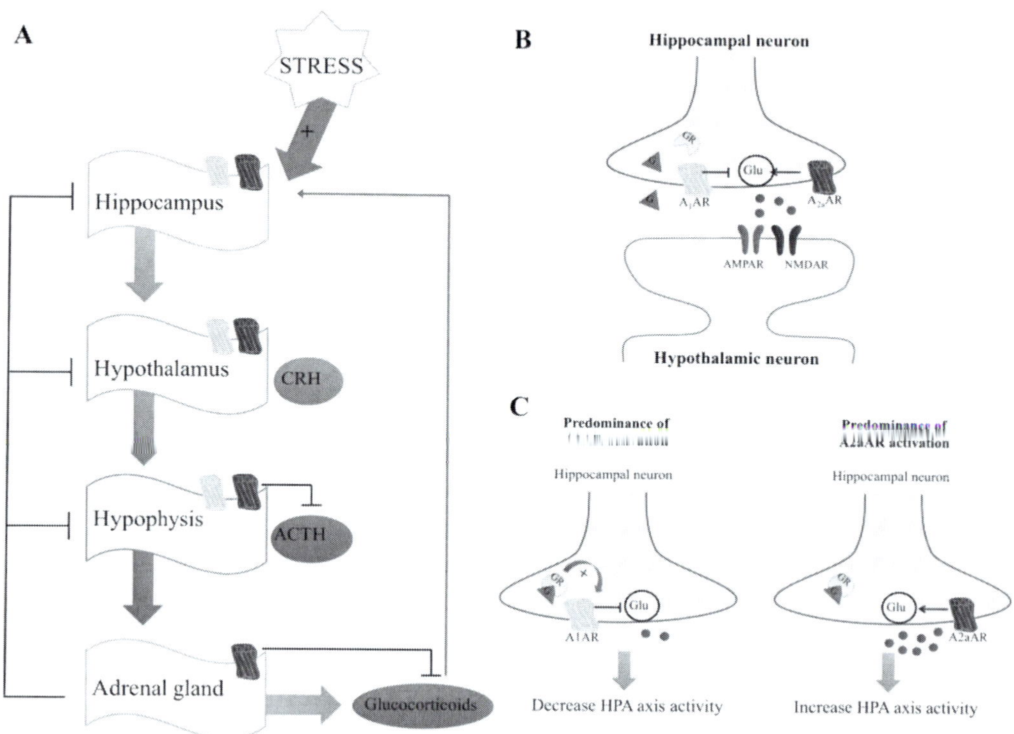

Figure 6. Modulation of HPA axis by adenosine receptors. (A) Chronic stress stimulates changes in HPA axis functionality that may produce an increase or decrease glucocorticoid (G) levels depending on the psychopathology. A1 and A2a adenosine receptors are present in the hippocampus and throughout HPA axis and may regulate G levels in the brain. A2aAR decrease ACTH levels in the anterior lobe of the pituitary and plasma corticosterone levels, decreasing pituitary-adrenocortical axis activity. (B) Hippocampal neurons express glucocorticoids receptors as well as A1AR and A2aAR that control glutamate release to hypothalamic cells. (C) The increase in G level is detected by glucocorticoid receptor (GR) on hippocampal neurons. GR stimulation increase A1AR activation, which in turns decrease glutamate release and HPA axis activity (left panel). On the other hand, the A2aAR activation in hippocampus leads to an increase in glutamate release and HPA axis activity (right panel).

Despite the growing body of evidence linking adenosine receptors to anxiety-like behavior, the data still seem contradictory. Mice lacking A1AR display increased anxiety-like behavior in the plus maze and dark-light box [247, 248]. Additionally, selective activation of

A1AR decreases anxiety-like behavior in naïve mice [249], and also reduces ethanol withdrawal-associated anxiety [250].

The data obtained from drugs of selective A1AR antagonists show some contradictory results. For example, chronic administration of DCPCX, resulted in a reduction on the time spent in the light zone of the dark/light test in mice [251] and in the open arm of a plus maze [252], suggesting an anxiogenic effect. However, in other study, DCPCX reduced exploratory behavior in APPswe/PS1dE9 (APdE9) double transgenic mice assessed by the open field test, being suggestive of reduced anxiety [253]. Therefore, in different studies the same drug has anxiogenic effects or anxiolytic properties. In accordance, at eletrophysiological level, DPCPX not only induces LTD but also increase theta burst-induced hippocampal LTP [254]. These results could be related to establishment of memories for aversive or non-aversive tasks and for anxiety-like behavior.

The anxiolytic effect of the selective A1AR agonist 2-chloro-N6- cyclopentyladenosine (CCPA) was investigated in CD1 mice by the elevated plus-maze and the light/dark test [255]. Similarly, CCPA reduced the anxiogenic-like response of ethanol hangover that was reversed by pretreatment with the selective A1AR antagonist, DPCPX [250]. Studies involving CPA (a selective A1AR agonist) had marked anxiolytic-like activity with no effect on locomotion [249].

A2aAR knock-out mice exhibited reduced exploratory activity on the open-field, spent more time in the closed protected arms of an elevated plus-maze and also in the less threatening dark area of a light/dark box, possibly indicating an increased level of anxiety [256]. However, acute or chronic treatment with selective A2aAR antagonists (DMPX, ZM241385 and SCH58261) has no effect on these anxiety indices [257, 258]. The A2aAR agonist, DPMA (N6-[2-(3,5-dimethoxyphenyl)-2(2-methylphenyl) ethyl] adenosine) had no effect on anxiety behavior [249]. On the other hand, the A2aAR selective agonist CGS 21680 (CGS 21680-2-p-(2-carboxyethyl) phenethylamino-5' Nethylcarboxamidoadenosine hydrochloride) has anxiolytic effects on mice in a plus maze [257]. CGS 21680 reduced the anxiogenic effects of theophylline, but did not reverse the anxiogenic effect of A1AR antagonist DPCPX [251].

Further studies evaluating the correlation between adenosine receptors and anxiety-related plasticity could increase the knowledge of pathophysiology of neuropsychiatric disorders and indicate that purinergic agonists or antagonists may emerge as promising candidates for clinical intervention.

Conclusion

In conclusion, adenosine signaling plays an active role in regulating neuronal plasticity, a critical event for normal development and also for a broad range of disorders in the central nervous system. A better understanding of these processes would shed light into the molecular tuning that drives different stages of development and also target purinergic receptors as candidates for clinical intervention regarding different pathologies in the central nervous system.

REFERENCES

[1] Dunwiddie, T.V.; Diao, L.; Proctor, W.R. Adenine nucleotides undergo rapid, quantitative conversion to adenosine in the extracellular space in rat hippocampus. *J Neurosci.* 1997, 17(20), 7673-7682.

[2] Latini, S.; Pedata, F. Adenosine in the central nervous system: release mechanisms and extracellular concentrations. *J Neurochem.*2001, 79(3), 463-484.

[3] Matz, H.; Hertz, L. Adenosine metabolism in neurons and astrocytes in primary cultures. *J Neurosci Res.*1989, 24(2), 260-267.

[4] Podgorska, M.; Kocbuch, K.; Pawelczyk, T. Recent advances in studies on biochemical and structural properties of equilibrative and concentrative nucleoside transporters. *Acta Biochim Pol.*2005, 52(4), 749-758.

[5] Fredholm, B.B.; IJzerman, A.P.; Jacobson, K.A.; Linden, J.; Muller, C.E. International Union of Basic and Clinical Pharmacology. LXXXI. Nomenclature and classification of adenosine receptors-an update. *Pharmacol Rev.*63(1), 1-34.

[6] Fredholm, B.B.; Chen, J.F.; Masino, S.A.; Vaugeois, J.M. Actions of adenosine at its receptors in the CNS: insights from knockouts and drugs. *Annu Rev Pharmacol Toxicol.*2005, 45, 385-412.

[7] Sebastiao, A.M.; Ribeiro, J.A. Tuning and fine-tuning of synapses with adenosine. *CurrNeuropharmacol.*2009, 7(3),180-194.

[8] Fredholm, B.B.; IJzerman, A.P.; Jacobson, K.A.; Klotz, K.N.; Linden, J. International Union of Pharmacology. XXV. Nomenclature and classification of adenosine receptors. *Pharmacol Rev.*2001, 53(4),527-552.

[9] Paes-De-Carvalho, R. Adenosine as a signaling molecule in the retina: biochemical and developmental aspects. *An Acad Bras Cienc.*2002, 74(3), 437-451.

[10] Akbar, M.; Okajima, F.; Tomura, H.; Shimegi, S.; Kondo, Y. A single species of A1 adenosine receptor expressed in Chinese hamster ovary cells not only inhibits cAMP accumulation but also stimulates phospholipase C and arachidonate release. *Mol Pharmacol.*1994, 45(5), 1036-1042.

[11] Schulte, G.; Fredholm, B.B. Signalling from adenosine receptors to mitogen-activated protein kinases. *Cell Signal.*2003, 15(9), 813-827.

[12] Cunha, R.A. Adenosine as a neuromodulator and as a homeostatic regulator in the nervous system: different roles, different sources and different receptors. *Neurochem Int.*2001, 38(2), 107-125.

[13] Ponzio, T.A.; Hatton, G.I. Adenosine postsynaptically modulates supraoptic neuronal excitability. *J Neurophysiol.*2005, 93(1), 535-547.

[14] Clark, B.D.; Kurth-Nelson, Z.L.; Newman, E.A. Adenosine-evoked hyperpolarization of retinal ganglion cells is mediated by G-protein-coupled inwardly rectifying K^+ and small conductance Ca^{2+}-activated K^+ channel activation. *J Neurosci.*2009, 29(36), 11237-11245.

[15] Sheth, S.; Brito, R.; Mukherjea, D.; Rybak, L.P., Ramkumar, V. Adenosine receptors: expression, function and regulation. *Int J Mol Sci.*2014, 15(2), 2024-2052.

[16] Tavares Gomes, A.L.; Maia, F.B.; Oliveira-Silva, P.; Marques Ventura, A.L.; Paes-De-Carvalho, R.; Serfaty, C.A.; Campello Costa, P. Purinergic modulation in the

development of the rat uncrossed retinotectal pathway. *Neuroscience.*2009, 163(4), 1061-1068.

[17] Dunwiddie, T.V.; Fredholm, B.B. Adenosine A1 receptors inhibit adenylate cyclase activity and neurotransmitter release and hyperpolarize pyramidal neurons in rat hippocampus. *J Pharmacol Exp Ther.*1989, 249(1), 31-37.

[18] Dare, E.; Schulte, G.; Karovic, O.; Hammarberg, C.; Fredholm, B.B. Modulation of glial cell functions by adenosine receptors. *Physiol Behav.*2007, 92(1-2), 15-20.

[19] Boison, D.; Chen, J.F.; Fredholm, B.B. Adenosine signaling and function in glial cells. *Cell Death Differ.*2010, 17(7), 1071-1082.

[20] Ciccarelli, R.; D'Alimonte, I.; Ballerini, P.; D'Auro, M.; Nargi, E.; Buccella, S.; Di Iorio, P.; Bruno, V.; Nicoletti, F.; Caciagli, F. Molecular signalling mediating the protective effect of A1 adenosine and mGlu3 metabotropic glutamate receptor activation against apoptosis by oxygen/glucose deprivation in cultured astrocytes. *Mol Pharmacol.*2007, 71(5), 1369-1380.

[21] Othman, T.; Yan, H.; Rivkees, S.A. Oligodendrocytes express functional A1 adenosine receptors that stimulate cellular migration. *Glia.*2003, 44(2), 166-172.

[22] Luongo, L.; Guida, F.; Imperatore, R.; Napolitano, F.; Gatta, L.; Cristino, L.; Giordano, C.; Siniscalco, D.; Di Marzo, V.; Bellini, G.; Petrelli, R.; Cappellacci, L.; Usiello, A.; de Novellis, V.; Rossi, F.; Maione, S. The A1 adenosine receptor as a new player in microglia physiology. *Glia.*2014, 62(1),122-132.

[23] Thompson, S.M.; Haas, H.L.; Gahwiler, B.H. Comparison of the actions of adenosine at pre- and postsynaptic receptors in the rat hippocampus in vitro. *J Physiol.*1992, 451, 347-363.

[24] Wetherington, J.P.; Lambert, N.A. Differential desensitization of responses mediated by presynaptic and postsynaptic A1 adenosine receptors. *J Neurosci.*2002, 22(4), 1248-1255.

[25] Barrie, A.P.; Nicholls, D.G. Adenosine A1 receptor inhibition of glutamate exocytosis and protein kinase C-mediated decoupling. *J Neurochem.*1993, 60(3), 1081-1086.

[26] Wu, L.G.; Saggau, P. Adenosine inhibits evoked synaptic transmission primarily by reducing presynaptic calcium influx in area CA1 of hippocampus. *Neuron.*1994, 12(5), 1139- 1148.

[27] Feuerstein, T.J.; Bar, K.I.; Lucking, C.H. Activation of A1 adenosine receptors decreases the release of serotonin in the rabbit hippocampus, but not in the caudate nucleus. *Naunyn Schmiedebergs Arch Pharmacol.*1988, 338(6), 664-670.

[28] Garcia, N.; Priego, M.; Obis, T.; Santafe, M.M.; Tomas, M.; Besalduch, N.; Lanuza, M.A.; Tomas, J. Adenosine A(1) and A(2)A receptor-mediated modulation of acetylcholine release in the mice neuromuscular junction. *Eur J Neurosci.*2013, 38(2), 2229-2241.

[29] Borycz, J.; Pereira, M.F.; Melani, A.; Rodrigues, R.J.; Kofalvi, A.; Panlilio, L.; Pedata, F.; Goldberg, S.R.; Cunha, R.A.; Ferre, S. Differential glutamate-dependent and glutamate- independent adenosine A1 receptor-mediated modulation of dopamine release in different striatal compartments. *J Neurochem.*2007, 101(2), 355-363.

[30] Li, Y.; Fan, S.; Yan, J.; Li, B.; Chen, F.; Xia, J.; Yu, Z.; Hu, Z. Adenosine modulates the excitability of layer II stellate neurons in entorhinal cortex through A1 receptors. *Hippocampus.*2011, 21(3), 265-280.

[31] Olah, M.E.; Stiles, G.L. The role of receptor structure in determining adenosine receptor activity. *Pharmacol Ther.*2000, 85(2), 55-75.

[32] Kobayashi, K.; Okada, Y. Excitatory effects of adenosine are not mediated by inhibition of GABAergic system in slices of superior colliculus and hippocampus from guinea pig. *Neurosci Lett.*1999, 264(1-3), 37-40.

[33] Sebastiao, A.M.; Ribeiro, J.A. Adenosine A2 receptor-mediated excitatory actions on the nervous system. *Prog Neurobiol.*1996, 48(3), 167-189.

[34] Stevens, B.; Porta, S.; Haak, L.L.; Gallo, V.; Fields, R.D. Adenosine: a neuron-glial transmitter promoting myelination in the CNS in response to action potentials. *Neuron.*2002, 36(5), 855-868.

[35] Coppi, E.; Cellai, L.; Maraula, G.; Pugliese, A.M.; Pedata, F. Adenosine A(2)A receptors inhibit delayed rectifier potassium currents and cell differentiation in primary purified oligodendrocyte cultures. *Neuropharmacology.*2013, 73, 301-310.

[36] Santiago, A.R.; Baptista, F.I.; Santos, P.F.; Cristovao, G.; Ambrosio, A.F.; Cunha, R.A.; Gomes, C.A. Role of microglia adenosine A(2A) receptors in retinal and brain neurodegenerative diseases. *Mediators Inflamm.*2014, 2014, 465694.

[37] Schafer, D.P.; Lehrman, E.K.; Stevens, B. The "quad-partite" synapse: microglia-synapse interactions in the developing and mature CNS. *Glia.*2013, 61(1), 24-36.

[38] Florian, C.; Vecsey, C.G.; Halassa, M.M.; Haydon, P.G.; Abel, T. Astrocyte-derived adenosine and A1 receptor activity contribute to sleep loss-induced deficits in hippocampal synaptic plasticity and memory in mice. *J Neurosci.*2011, 31(19), 6956-6962.

[39] Matos, M.; Augusto, E.; Agostinho, P.; Cunha, R.A.; Chen, J.F. Antagonistic interaction between adenosine A2A receptors and Na^+/K^+-ATPase-alpha2 controlling glutamate uptake in astrocytes. *J Neurosci.*2013, 33(47), 18492-18502.

[40] Panatier, A.; Vallee, J.; Haber, M.; Murai, K.K.; Lacaille, J.C.; Robitaille, R. Astrocytes are endogenous regulators of basal transmission at central synapses. *Cell.*2011, 146(5), 785-798.

[41] Li, H.; Henry, J.L. Adenosine A2 receptor mediation of pre- and postsynaptic excitatory effects of adenosine in rat hippocampus in vitro. *Eur J Pharmacol.*1998, 347(2-3), 173-182.

[42] Gubitz, A.K.; Widdowson, L.; Kurokawa, M.; Kirkpatrick, K.A.; Richardson, P.J. Dual signalling by the adenosine A2a receptor involves activation of both N- and P-type calcium channels by different G proteins and protein kinases in the same striatal nerve terminals. *J Neurochem.*1996, 67(1), 374-381.

[43] Lopes, L.V.; Cunha, R.A.; Ribeiro, J.A. Cross talk between A(1) and A(2A) adenosine receptors in the hippocampus and cortex of young adult and old rats. *J Neurophysiol.*1999, 82(6), 3196-3203.

[44] Lopes, L.V.; Cunha, R.A.; Kull, B.; Fredholm, B.B.; Ribeiro, J.A. Adenosine A(2A) receptor facilitation of hippocampal synaptic transmission is dependent on tonic A(1) receptor inhibition. *Neuroscience.*2002, 112(2), 319-329.

[45] Rebola, N.; Oliveira, C.R.; Cunha, R.A. Transducing system operated by adenosine A(2A) receptors to facilitate acetylcholine release in the rat hippocampus. *Eur J Pharmacol.*2002, 454(1), 31-38.

[46] Cunha, R.A.; Johansson, B.; Fredholm, B.B.; Ribeiro, J.A.; Sebastiao, A.M. Adenosine A2A receptors stimulate acetylcholine release from nerve terminals of the rat hippocampus. *Neurosci Lett.*1995, 196(1-2), 41-44.

[47] Cunha, R.A.; Ribeiro, J.A. Adenosine A2A receptor facilitation of synaptic transmission in the CA1 area of the rat hippocampus requires protein kinase C but not protein kinase A activation. *Neurosci Lett.*2000, 289(2), 127-130.

[48] Fredholm, B.B. Adenosine receptors as targets for drug development. *Drug News Perspect.*2003, 16(5), 283-289.

[49] Ponzio, T.A.; Wang, Y.F.; Hatton, G.I. Activation of adenosine A2A receptors alters postsynaptic currents and depolarizes neurons of the supraoptic nucleus. *Am J Physiol Regul Integr Comp Physiol.*2006, 291(2), R359-366.

[50] Rebola, N.; Rodrigues, R.J.; Lopes, L.V.; Richardson, P.J.; Oliveira, C.R.; Cunha. R.A. Adenosine A1 and A2A receptors are co-expressed in pyramidal neurons and co-localized in glutamatergic nerve terminals of the rat hippocampus. *Neuroscience.*2005, 133(1), 79-83.

[51] Correia-de-Sa, P.; Sebastiao, A.M.; Ribeiro, J.A. Inhibitory and excitatory effects of adenosine receptor agonists on evoked transmitter release from phrenic nerve ending of the rat. *Br J Pharmacol.*1991, 103(2), 1614-1620.

[52] Marchi, M.; Raiteri, L.; Risso, F.; Vallarino, A.; Bonfanti, A.; Monopoli, A.; Ongini, E.; Raiteri, M. Effects of adenosine A1 and A2A receptor activation on the evoked release of glutamate from rat cerebrocortical synaptosomes. *Br J Pharmacol.*2002, 136(3), 434-440.

[53] Cunha, R.A. Different cellular sources and different roles of adenosine: A1 receptor-mediated inhibition through astrocytic-driven volume transmission and synapse-restricted A2A receptor-mediated facilitation of plasticity. *Neurochem Int.*2008, 52(1-2), 65-72.

[54] Cunha-Reis, D.; Fontinha, B.M.; Ribeiro, J.A.; Sebastiao, A.M. Tonic adenosine A1 and A2A receptor activation is required for the excitatory action of VIP on synaptic transmission in the CA1 area of the hippocampus. *Neuropharmacology.*2007, 52(2), 313-320.

[55] Correia-de-Sa, P.; Ribeiro, J.A. Adenosine uptake and deamination regulate tonic A2a receptor facilitation of evoked [3H]acetylcholine release from the rat motor nerve terminals. *Neuroscience.*1996, 73(1), 85-92.

[56] Bannon, N.M.; Zhang, P.; Ilin, V.; Chistiakova, M.; Volgushev, M. Modulation of synaptic transmission by adenosine in layer 2/3 of the rat visual cortex in vitro. *Neuroscience.*2014, 260, 171-184.

[57] Klaasse, E.C.; Ijzerman, A.P.; de Grip, W.J.; Beukers, M.W. Internalization and desensitization of adenosine receptors. *Purinergic Signal.*2008, 4(1), 21-37.

[58] Carriba, P.; Navarro, G.; Ciruela, F.; Ferre, S.; Casado, V.; Agnati, L.; Cortes, A.; Mallol, J.; Fuxe, K.; Canela, E.I.; Lluís, C.; Franco, R. Detection of heteromerization of more than two proteins by sequential BRET-FRET. *Nat Methods.*2008, 5(8), 727-733.

[59] George, S.R.; O'Dowd, B.F.; Lee, S.P. G-protein-coupled receptor oligomerization and its potential for drug discovery. *Nat Rev Drug Discov.*2002, 1(10), 808-820.

[60] Ferre, S.; Baler, R.; Bouvier, M.; Caron, M.G.; Devi, L.A.; Durroux, T.; Fuxe, K.; George, S.R.; Javitch, J.A.; Lohse, M.J.; Mackie, K.; Milligan, G.; Pfleger, K.D.; Pin,

J.P.; Volkow, N.D.; Waldhoer, M.; Woods, A.S.; Franco, R. Building a new conceptual framework for receptor heteromers. *Nat Chem Biol.*2009, 5(3), 131-134.

[61] Ciruela, F.; Casado, V.; Mallol, J.; Canela, E.I.; Lluis, C.; Franco, R. Immunological identification of A1 adenosine receptors in brain cortex. *J Neurosci Res.*1995, 42(6), 818-828.

[62] Ciruela, F.; Escriche, M., Burgueno, J.; Angulo, E.; Casado, V.; Soloviev, M.M.; Canela, E.I.; Mallol, J.; Chan, W.Y.; Lluis, C.; McIlhinney, R.A.; Franco, R. Metabotropic glutamate 1alpha and adenosine A1 receptors assemble into functionally interacting complexes. *J Biol Chem.*2001, 276(21), 18345-18351.

[63] Kamikubo, Y.; Shimomura, T.; Fujita, Y.; Tabata, T.; Kashiyama, T.; Sakurai, T.; Fukurotani, K.; Kano, M. Functional cooperation of metabotropic adenosine and glutamate receptors regulates postsynaptic plasticity in the cerebellum. *J Neurosci.*2013, 33(47), 18661-18671.

[64] Gines, S.; Hillion, J.; Torvinen, M.; Le Crom, S.; Casado, V.; Canela, E.I.; Rondin, S.; Lew, J.Y.; Watson, S.; Zoli, M.; Agnati, L.F.; Verniera, P.; Lluis, C.; Ferré, S.; Fuxe, K.; Franco, R. Dopamine D1 and adenosine A1 receptors form functionally interacting heteromeric complexes. Proc Natl Acad Sci U S A.2000, 97(15), 8606-8611.

[65] Franco, R.; Lluis, C.; Canela, E.I.; Mallol, J.; Agnati, L.; Casado, V.; Ciruela, F.; Ferre, S.; Fuxe, K. Receptor-receptor interactions involving adenosine A1 or dopamine D1 receptors and accessory proteins. *J Neural Transm.*2007, 114(1), 93-104.

[66] Franco, R.; Ferre, S.; Agnati, L.; Torvinen, M.; Gines, S.; Hillion, J.; Casado, V.; Lledo, P.; Zoli, M.; Lluis, C.; Fuxe, K. Evidence for adenosine/dopamine receptor interactions: indications for heteromerization. *Neuropsychopharmacology.*2000, 23(4 Suppl), S50-59.

[67] Agnati, L.F.; Ferre, S.; Lluis, C.; Franco, R.; Fuxe, K. Molecular mechanisms and therapeutical implications of intramembrane receptor/receptor interactions among heptahelical receptors with examples from the striatopallidal GABA neurons. *Pharmacol Rev.*2003, 55(3), 509-550.

[68] Yoshioka, K.; Saitoh, O.; Nakata, H. Heteromeric association creates a P2Y-like adenosine receptor. *Proc Natl Acad Sci U S A.*2001, 98(13), 7617-7622.

[69] Yoshioka, K.; Saitoh, O.; Nakata, H. Agonist-promoted heteromeric oligomerization between adenosine A(1) and P2Y(1) receptors in living cells. *FEBS Lett.*2002, 523(1-3), 147- 151.

[70] Suzuki, T.; Namba, K.; Tsuga, H.; Nakata, H. Regulation of pharmacology by hetero-oligomerization between A1 adenosine receptor and P2Y2 receptor. *Biochem Biophys Res Commun.*2006, 351(2), 559-565.

[71] Nakata, H.; Yoshioka, K.; Kamiya, T.; Tsuga, H.; Oyanagi, K. Functions of heteromeric association between adenosine and P2Y receptors. *J Mol Neurosci.*2005, 26(2-3), 233-238.

[72] Ciruela, F.; Casado, V.; Rodrigues, R.J.; Lujan, R.; Burgueno, J.; Canals, M.; Borycz, J.; Rebola, N.; Goldberg, S.R.; Mallol, J.; Cortés, A.; Canela, E.I.; López-Giménez, J.F.; Milligan, G.; Lluis, C.; Cunha, R.A.; Ferré, S.; Franco, R. Presynaptic control of striatal glutamatergic neurotransmission by adenosine A1-A2A receptor heteromers. *J Neurosci.*2006, 26(7), 2080- 2087.

[73] Cristovao-Ferreira, S.; Navarro, G.; Brugarolas, M.; Perez-Capote, K.; Vaz, S.H.; Fattorini, G.; Conti, F.; Lluis, C.; Ribeiro, J.A.; McCormick, P.J.; Casadó, V.; Franco,

R.; Sebastião, A.M. A1R- A2AR heteromers coupled to Gs and G i/0 proteins modulate GABA transport into astrocytes. *Purinergic Signal.*2013, 9(3), 433-449.

[74] Canals, M.; Burgueno, J.; Marcellino, D.; Cabello, N.; Canela, E.I.; Mallol, J.; Agnati, L.; Ferre, S.; Bouvier, M.; Fuxe, K.; Ciruela, F.; Lluis, C.; Franco, R. Homodimerization of adenosine A2A receptors: qualitative and quantitative assessment by fluorescence and bioluminescence energy transfer. J Neurochem.2004, 88(3), 726-734.

[75] Vidi, P.A.; Chen, J.; Irudayaraj, J.M.; Watts, V.J. Adenosine A(2A) receptors assemble into higher-order oligomers at the plasma membrane. FEBS Lett.2008, 582(29), 3985-3990.

[76] Canals, M.; Marcellino, D.; Fanelli, F.; Ciruela, F.; de Benedetti, P.; Goldberg, S.R.; Neve, K.; Fuxe, K.; Agnati, L.F.; Woods, A.S.; Ferré, S.; Lluis, C.; Bouvier, M.; Franco, R. Adenosine A2A- dopamine D2 receptor-receptor heteromerization: qualitative and quantitative assessment by fluorescence and bioluminescence energy transfer. J Biol Chem.2003, 278(47), 46741-46749.

[77] Kamiya, T.; Saitoh, O.; Yoshioka, K.; Nakata, H. Oligomerization of adenosine A2A and dopamine D2 receptors in living cells. Biochem Biophys Res Commun.2003, 306(2), 544- 549.

[78] Ferre, S.; Karcz-Kubicha, M.; Hope, B.T.; Popoli, P.; Burgueno, J.; Gutierrez, M.A.; Casado, V.; Fuxe, K.; Goldberg, S.R.; Lluis, C.; Franco, R.; Ciruela F. Synergistic interaction between adenosine A2A and glutamate mGlu5 receptors: implications for striatal neuronal function. *Proc Natl Acad Sci U S A.*2002, 99(18), 11940-11945.

[79] Cabello, N.; Gandia, J.; Bertarelli, D.C.; Watanabe, M.; Lluis, C.; Franco, R.; Ferre, S.; Lujan, R.; Ciruela, F. Metabotropic glutamate type 5, dopamine D2 and adenosine A2a receptors form higher-order oligomers in living cells. *J Neurochem.*2009, 109(5), 1497-1507.

[80] Ferre, S.; Goldberg, S.R.; Lluis, C.; Franco, R. Looking for the role of cannabinoid receptor heteromers in striatal function. *Neuropharmacology.*2009, 56 Suppl 1, 226-234.

[81] Chen, J.F.; Eltzschig, H.K.; Fredholm, B.B. Adenosine receptors as drug targets--what are the challenges? *Nat Rev Drug Discov.*2013, 12(4), 265-286.

[82] Chen, J.F. Adenosine receptor control of cognition in normal and disease. *Int Rev Neurobiol.*2014, 119, 257-307.

[83] Chen, J.F.; Lee, C.F.; Chern, Y. Adenosine receptor neurobiology: overview. *Int Rev Neurobiol.*2014, 119, 1-49.

[84] Allen, N.J.; Barres, B.A. Neuroscience: Glia - more than just brain glue. *Nature.*2009, 457(7230), 675-677.

[85] Cohen-Cory, S. The developing synapse: construction and modulation of synaptic structures and circuits. *Science.*2002, 298(5594), 770-776.

[86] de Velasco, P.C.; Mendonca, H.R.; Borba, J.M.; Andrade da Costa, B.L.; Guedes, R.C.; Navarro, D.M.; Santos, G.K.; Faria-Melibeu, A.C.; Campello Costa, P.; Serfaty, C.A. Nutritional restriction of omega-3 fatty acids alters topographical fine tuning and leads to a delay in the critical period in the rodent visual system. *Exp Neurol.*2012, 234(1), 220-229.

[87] Tschetter, W.W.; Alam, N.M.; Yee, C.W.; Gorz, M.; Douglas, R.M.; Sagdullaev, B.; Prusky, G.T. Experience-enabled enhancement of adult visual cortex function. *J Neurosci.*2013, 33(12), 5362-5366.

[88] Harkavy, K.L. Mortality rates ignore population segment. *Am J Obstet Gynecol.*1990, 163(1 Pt 1), 269-270.

[89] Wang, S.; Young, K.M. White matter plasticity in adulthood. *Neuroscience.*2014, 276, 148-160.

[90] Masino, S.A.; Kawamura, M. Jr.; Ruskin, D.N.; Geiger, J.D.; Boison, D. Purines and neuronal excitability: links to the ketogenic diet. *Epilepsy Res.*2012, 100(3), 229-238.

[91] Davis, G.W. Homeostatic signaling and the stabilization of neural function. *Neuron.*2013, 80(3), 718-728.

[92] Smith, R.S.; Koles, Z.J. Myelinated nerve fibers: computed effect of myelin thickness on conduction velocity. *Am J Physiol.*1970, 219(5), 1256-1258.

[93] Gibson, E.M.; Purger, D.; Mount, C.W.; Goldstein, A.K.; Lin, G.L.; Wood, L.S.; Inema, I.; Miller, S.E.; Bieri, G.; Zuchero, J.; Barres, B.A.; Woo, P.J.; Vogel, H.; Monje, M. Neuronal activity promotes oligodendrogenesis and adaptive myelination in the mammalian brain. *Science.*2014, 344(6183), 1252304.

[94] Stevens, B.; Fields, R.D. Response of Schwann cells to action potentials in development. *Science.*2000, 287(5461), 2267-2271.

[95] Maire, J.C.; Medilanski, J.; Straub, R.W. Release of adenosine, inosine and hypoxanthine from rabbit non-myelinated nerve fibres at rest and during activity. *J Physiol.*1984, 357, 67- 77.

[96] Kuperman, A.S.; Volpert, W.A.; Okamoto, M. Release of Adenine Nucleotide from Nerve Axons. *Nature.*1964, 204, 1000-1001.

[97] Schulz, D.J. Plasticity and stability in neuronal output via changes in intrinsic excitability: it's what's inside that counts. *J Exp Biol.*2006, 209 (Pt 24), 4821-4827.

[98] Chung, H.J.; Ge, W.P.; Qian, X.; Wiser, O.; Jan, Y.N.; Jan, L.Y. G protein-activated inwardly rectifying potassium channels mediate depotentiation of long-term potentiation. *Proc Natl Acad Sci U S A.*2009, 106(2), 635-640.

[99] Boison, D. Adenosine dysfunction in epilepsy. *Glia.*2012, 60(8), 1234-1243.

[100] Sasaki, T.; Matsuki, N.; Ikegaya, Y. Action-potential modulation during axonal conduction. *Science.*2011, 331(6017), 599-601.

[101] Chamberlain, S.E.; Sadowski, J.H.; Teles-Grilo Ruivo, L.M.; Atherton, L.A.; Mellor, J.R. Long-term depression of synaptic kainate receptors reduces excitability by relieving inhibition of the slow afterhyperpolarization. *J Neurosci.*2013, 33(22), 9536-9545.

[102] Cossenza, M.; Socodato, R.; Portugal, C.C.; Domith, I.C.; Gladulich, L.F.; Encarnacao, T.G.; Calaza, K.C.; Mendonca, H.R.; Campello-Costa, P.; Paes-de-Carvalho, R. Nitric oxide in the nervous system: biochemical, developmental, and neurobiological aspects. *Vitam Horm.*2014, 96, 79-125.

[103] Duster, R.; Prickaerts, J.; Blokland, A. Purinergic signaling and hippocampal long-term potentiation. *Curr Neuropharmacol.*2014, 12(1), 37-43.

[104] Forghani, R.; Krnjevic, K. Adenosine antagonists have differential effects on induction of long-term potentiation in hippocampal slices. *Hippocampus.*1995, 5(1), 71-77.

[105] de Mendonca, A.; Ribeiro, J.A. Endogenous adenosine modulates long-term potentiation in the hippocampus. *Neuroscience.*1994, 62(2), 385-390.

[106] Pascual, O.; Casper, K.B.; Kubera, C.; Zhang, J.; Revilla-Sanchez, R.; Sul, J.Y.; Takano, H.; Moss, S.J.; McCarthy, K.; Haydon, P.G. Astrocytic purinergic signaling coordinates synaptic networks. *Science.*2005, 310(5745), 113-116.
[107] Scammell, T.E.; Arrigoni, E.; Thompson, M.A.; Ronan, P.J.; Saper, C.B.; Greene, R.W. Focal deletion of the adenosine A1 receptor in adult mice using an adeno-associated viral vector. *J Neurosci.*2003, 23(13), 5762-5770.
[108] Deng, Q.; Terunuma, M.; Fellin, T.; Moss, S.J.; Haydon, P.G. Astrocytic activation of A1 receptors regulates the surface expression of NMDA receptors through a Src kinase dependent pathway. *Glia.*2011, 59(7), 1084-1093.
[109] Gimenez-Llort, L.; Masino, S.A.; Diao, L.; Fernandez-Teruel, A.; Tobena, A.; Halldner, L.; Fredholm, B.B. Mice lacking the adenosine A1 receptor have normal spatial learning and plasticity in the CA1 region of the hippocampus, but they habituate more slowly. *Synapse.*2005, 57(1), 8-16.
[110] Schotanus, S.M.; Fredholm, B.B.; Chergui, K. NMDA depresses glutamatergic synaptic transmission in the striatum through the activation of adenosine A1 receptors: evidence from knockout mice. *Neuropharmacology.*2006, 51(2), 272-282.
[111] Lovinger, D.M.; Choi, S. Activation of adenosine A1 receptors initiates short-term synaptic depression in rat striatum. *Neurosci Lett.*1995, 199(1), 9-12.
[112] Dias, R.B.; Ribeiro, J.A.; Sebastiao, A.M. Enhancement of AMPA currents and GluR1 membrane expression through PKA-coupled adenosine A(2A) receptors. *Hippocampus.*2012, 22(2), 276-291.
[113] Rebola, N.; Lujan, R.; Cunha, R.A.; Mulle, C. Adenosine A2A receptors are essential for long-term potentiation of NMDA-EPSCs at hippocampal mossy fiber synapses. *Neuron.*2008, 57(1), 121-134.
[114] Fontinha, B.M.; Diogenes, M.J.; Ribeiro, J.A.; Sebastiao, A.M. Enhancement of long-term potentiation by brain-derived neurotrophic factor requires adenosine A2A receptor activation by endogenous adenosine. *Neuropharmacology.*2008, 54(6), 924-933.
[115] Tebano, M.T.; Martire, A.; Potenza, R.L.; Gro, C.; Pepponi, R.; Armida, M.; Domenici, M.R.; Schwarzschild, M.A.; Chen, J.F.; Popoli, P. Adenosine A(2A) receptors are required for normal BDNF levels and BDNF-induced potentiation of synaptic transmission in the mouse hippocampus. *J Neurochem.*2008, 104(1), 279-286.
[116] Jeon, S.J.; Rhee, S.Y.; Ryu, J.H.; Cheong, J.H.; Kwon, K.; Yang, S.I.; Park, S.H; Lee, J.; Kim, H.Y.; Han, S.H.; Ko, K.H.; Shin, C.Y. Activation of adenosine A2A receptor up-regulates BDNF expression in rat primary cortical neurons. *Neurochem Res.*2011, 36(12), 2259-2269.
[117] Zhou, S.J.; Zhu, M.E.; Shu, D.; Du, X.P.; Song, X.H.; Wang, X.T.; Zheng, R.Y.; Cai, X.H.; Chen, J.F.; Hc, J.C. Preferential enhancement of working memory in mice lacking adenosine A(2A) receptors. *Brain Res.*2009, 1303, 74-83.
[118] Gimenez-Llort, L.; Schiffmann, S.N.; Shmidt, T.; Canela, L.; Camon, L.; Wassholm, M.; Canals, M.; Terasmaa, A.; Fernandez-Teruel, A.; Tobena, A.; Popova, E.; Ferré, S.; Agnati, L.; Ciruela, F.; Martínez, E.; Scheel-Kruger, J.; Lluis, C.; Franco, R.; Fuxe, K.; Bader, M. Working memory deficits in transgenic rats overexpressing human adenosine A2A receptors in the brain. *Neurobiol Learn Mem.*2007, 87(1), 42-56.
[119] Dias, R.B.; Rombo, D.M.; Ribeiro, J.A.; Henley, J.M.; Sebastiao, A.M. Adenosine: setting the stage for plasticity. *Trends Neurosci.*2013, 36(4), 248-257.

[120] Lovatt, D.; Xu, Q.; Liu, W.; Takano, T.; Smith, N.A.; Schnermann, J.; Tieu, K.; Nedergaard, M. Neuronal adenosine release, and not astrocytic ATP release, mediates feedback inhibition of excitatory activity. *Proc Natl Acad Sci U S A.*2012, 109(16), 6265-6270.

[121] Klyuch, B.P.; Dale, N.; Wall, M.J. Receptor-mediated modulation of activity-dependent adenosine release in rat cerebellum. *Neuropharmacology.*2012, 62(2), 815-824.

[122] Hulme, S.R.; Jones, O.D.; Raymond, C.R.; Sah, P.; Abraham, W.C. Mechanisms of heterosynaptic metaplasticity. *Philos Trans R Soc Lond B Biol Sci.*2014, 369(1633), 20130148.

[123] Jones, O.D.; Hulme, S.R.; Abraham, W.C. Purinergic receptor- and gap junction-mediated intercellular signalling as a mechanism of heterosynaptic metaplasticity. *Neurobiol Learn Mem.*2013, 105, 31-39.

[124] Vanderhaeghen, P.; Cheng, H.J. Guidance molecules in axon pruning and cell death. *Cold Spring Harb Perspect Biol.*2010, 2(6), a001859.

[125] Berardi, N.; Pizzorusso, T.; Maffei, L. Critical periods during sensory development. *Curr Opin Neurobiol.* 2000, 10(1), 138-145.

[126] Hensch, T.K. Critical period plasticity in local cortical circuits. *Nat Rev Neurosci.*2005, 6(11), 877-888.

[127] Takesian, A.E.; Hensch, T.K. Balancing plasticity/stability across brain development. *Prog Brain Res.*2013, 207, 3-34.

[128] Socodato, R.; Brito, R.; Calaza, K.C.; Paes-de-Carvalho, R. Developmental regulation of neuronal survival by adenosine in the in vitro and in vivo avian retina depends on a shift of signaling pathways leading to CREB phosphorylation or dephosphorylation. *J Neurochem.*2011, 116(2), 227-239.

[129] Sperlagh, B.; Vizi, E.S. The role of extracellular adenosine in chemical neurotransmission in the hippocampus and Basal Ganglia: pharmacological and clinical aspects. *Curr Top Med Chem.*2011, 11(8), 1034-1046.

[130] Serchov, T.; Atas, H.C.; Normann, C.; van Calker, D.; Biber, K. Genetically controlled upregulation of adenosine A(1) receptor expression enhances the survival of primary cortical neurons. *Mol Neurobiol.*2012, 46(2), 535-544.

[131] Rosim, F.E; Persike, D.S.; Nehlig, A.; Amorim, R.P.; de Oliveira, D.M.; Fernandes, M.J. Differential neuroprotection by A(1) receptor activation and A(2A) receptor inhibition following pilocarpine-induced status epilepticus. *Epilepsy Behav.*2011, 22(2), 207-213.

[132] Cunha, R.A.; Constantino, M.C.; Sebastiao, A.M.; Ribeiro, J.A. Modification of A1 and A2a adenosine receptor binding in aged striatum, hippocampus and cortex of the rat. *Neuroreport.*1995, 6(11), 1583-1588.

[133] Sebastiao, A.M.; Cunha, R.A.; de Mendonca, A.; Ribeiro, J.A. Modification of adenosine modulation of synaptic transmission in the hippocampus of aged rats. *Br J Pharmacol.*2000, 131(8), 1629-1634.

[134] Ishikawa, S.; Saijoh, K.; Okada, Y. Endogenous adenosine facilitates neurotransmission via A2A adenosine receptors in the rat superior colliculus in vivo. *Brain Res* 1997, 757(2), 268-275.

[135] Zhang, C.; Schmidt, J.T. Adenosine A1 and class II metabotropic glutamate receptors mediate shared presynaptic inhibition of retinotectal transmission. *J Neurophysiol.*1999, 82(6), 2947-2955.

[136] Santschi, L.A.; Zhang, X.L.; Stanton, P.K. Activation of receptors negatively coupled to adenylate cyclase is required for induction of long-term synaptic depression at Schaffer collateral-CA1 synapses. *J Neurobiol.*2006, 66(3), 205-219.

[137] Lo, F.S.; Mize, R.R. Synaptic regulation of L-type $Ca^{(2+)}$ channel activity and long-term depression during refinement of the retinocollicular pathway in developing rodent superior colliculus. *J Neurosci.*2000, 20(3), RC58.

[138] Thevananther, S.; Rivera, A.; Rivkees, S.A. A1 adenosine receptor activation inhibits neurite process formation by Rho kinase-mediated pathways. *Neuroreport.*2001, 12(14), 3057-3063.

[139] Hirai, H.; Okada, Y. Adenosine facilitates in vivo neurotransmission in the superior colliculus of the rat. *J Neurophysiol.*1995, 74(3), 950-960.

[140] Hirai, H.; Okada, Y. Adenosine facilitates glutamate release in a protein kinase-dependent manner in superior colliculus slices. *Eur J Pharmacol.*1994, 256(1), 65-71.

[141] Hirai, H.; Ikeuchi, Y.; Okada, Y. The contribution of PKA to the excitatory mechanism of adenosine in guinea pig superior colliculus slices. *Neurosci Lett.*1994, 182(1), 33-36.

[142] Chern, Y. Regulation of adenylyl cyclase in the central nervous system. *Cell Signal.*2000, 12(4), 195-204.

[143] Lessmann, V.; Heumann, R. Cyclic AMP endogenously enhances synaptic strength of developing glutamatergic synapses in serum-free microcultures of rat hippocampal neurons. *Brain Res.*1997, 763(1), 111-122.

[144] d'Alcantara, P.; Ledent, C.; Swillens, S.; Schiffmann, S.N. Inactivation of adenosine A2A receptor impairs long term potentiation in the accumbens nucleus without altering basal synaptic transmission. *Neuroscience.*2001, 107(3), 455-464.

[145] Rebola, N.; Sebastiao, A.M.; de Mendonca, A.; Oliveira, C.R.; Ribeiro, J.A.; Cunha, R.A. Enhanced adenosine A2A receptor facilitation of synaptic transmission in the hippocampus of aged rats. *J Neurophysiol.*2003, 90(2), 1295-1303.

[146] Schiffmann, S.N.; Dassesse, D.; d'Alcantara, P.; Ledent, C.; Swillens, S.; Zoli, M. A2A receptor and striatal cellular functions: regulation of gene expression, currents, and synaptic transmission. *Neurology.*2003, 61(11 Suppl 6), S24-29.

[147] Cunha, R.A.; Almeida, T.; Ribeiro, J.A. Parallel modification of adenosine extracellular metabolism and modulatory action in the hippocampus of aged rats. *J Neurochem.*2001, 76(2), 372-382.

[148] Araque, A.; Parpura, V.; Sanzgiri, R.P.; Haydon, P.G. Tripartite synapses: glia, the unacknowledged partner. *Trends Neurosci.*1999, 22(5), 208-215.

[149] Halassa, M.M.; Fellin, T.; Haydon, P.G. Tripartite synapses: Roles for astrocytic purines in the control of synaptic physiology and behavior. *Neuropharmacology.*2009, 57 (4), 343-346.

[150] Halassa, M.M.; Fellin, T.; Haydon, P.G. The tripartite synapse: roles for gliotransmission in health and disease. *Trends Mol Med.*2007, 13(2), 54-63.

[151] Zhang, Q.; Fukuda, M.; Van Bockstaele, E.; Pascual, O.; Haydon, P.G. Synaptotagmin IV regulates glial glutamate release. *Proc Natl Acad Sci USA.*2004, 101(25), 9441-9446.

[152] Hindley, S.; Herman, M.A.; Rathbone, M.P. Stimulation of reactive astrogliosis in vivo by extracellular adenosine diphosphate or an adenosine A2 receptor agonist. *J Neurosci Res.*1994, 38(4), 399-406.

[153] Inoue, K. Microglial activation by purines and pyrimidines. *Glia.*2002, 40(2), 156-163.

[154] Kust, B.M.; Biber, K.; van Calker, D.; Gebicke-Haerter, P.J. Regulation of K^+ channel mRNA expression by stimulation of adenosine A2a-receptors in cultured rat microglia. *Glia.*1999, 25(2), 120-130.

[155] Ogata, T.; Schubert, P. Programmed cell death in rat microglia is controlled by extracellular adenosine. *Neurosci Lett.*1996, 218(2), 91-94.

[156] Newman, E.A. New roles for astrocytes: regulation of synaptic transmission. *Trends Neurosci.*2003, 26(10), 536-542.

[157] Volterra, A.; Meldolesi, J. Astrocytes, from brain glue to communication elements: the revolution continues. *Nat Rev Neurosci.*2005, 6(8), 626-640.

[158] Haydon, P.G. GLIA: listening and talking to the synapse. *Nat Rev Neurosci.*2001, 2(3), 185-193.

[159] Fellin, T.; Halassa, M.M.; Terunuma, M.; Succol, F.; Takano, H.; Frank, M.; Moss, S.J.; Haydon, P.G. Endogenous nonneuronal modulators of synaptic transmission control cortical slow oscillations in vivo. *Proc Natl Acad Sci U S A.*2009, 106(35), 15037-15042.

[160] Araque, A.; Carmignoto, G.; Haydon, P.G.; Oliet, S.H.; Robitaille, R.; Volterra, A. Gliotransmitters travel in time and space. *Neuron.*2014, 81(4), 728-739.

[161] Matos, M.; Augusto, E.; Santos-Rodrigues, A.D.; Schwarzschild, M.A.; Chen, J.F.; Cunha, R.A.; Agostinho, P. Adenosine A2A receptors modulate glutamate uptake in cultured astrocytes and gliosomes. *Glia.*2012, 60(5), 702-716.

[162] Rasmusson, D.D. Reorganization of raccoon somatosensory cortex following removal of the fifth digit. *J Comp Neurol.*1982, 205(4), 313-326.

[163] Kaas, J.H.; Krubitzer, L.A.; Chino, Y.M.; Langston, A.L.; Polley, E.H.; Blair, N. Reorganization of retinotopic cortical maps in adult mammals after lesions of the retina. *Science.*1990, 248(4952), 229-231.

[164] Devor, M.; Wall, P.D. Reorganisation of spinal cord sensory map after peripheral nerve injury. *Nature.*1978, 276(5683), 75-76.

[165] Mayer de Oliveira, A.; Cabral Miranda, F. Are the mechanisms driving somatosensory reorganization cortical or subcortical? *Front Neuroanat.*2014, 8, 62.

[166] Darian-Smith, C.; Gilbert, C.D. Topographic reorganization in the striate cortex of the adult cat and monkey is cortically mediated. *J Neurosci.*1995, 15(3 Pt 1), 1631-1647.

[167] Eysel, U.T.; Gonzalez-Aguilar, F.; Mayer, U. Time-dependent decrease in the extent of visual deafferentation in the lateral geniculate nucleus of adult cats with small retinal lesions. *Exp Brain Res.*1981, 41(3-4), 256-263.

[168] Serfaty, C.A.; Campello-Costa, P.; Linden, R. Rapid and long-term plasticity in the neonatal and adult retinotectal pathways following a retinal lesion. *Brain Res Bull.*2005, 66(2), 128-134.

[169] Campello-Costa, P.; Fosse-Junior, A.M.; Oliveira-Silva, P.; Serfaty, C.A. Blockade of arachidonic acid pathway induces sprouting in the adult but not in the neonatal uncrossed retinotectal projection. *Neuroscience.*2006, 139(3), 979-989

[170] Oliveira-Silva, P.; Jurgilas, P.B.; Trindade, P.; Campello-Costa, P.; Perales, J.; Savino, W.; Serfaty, C.A. Matrix metalloproteinase-9 is involved in the development and

plasticity of retinotectal projections in rats. *Neuroimmunomodulation.*2007, 14(3-4), 144-149.

[171] Kaas, J.H.; Florence, S.L.; Jain, N. Subcortical contributions to massive cortical reorganizations. *Neuron.*1999, 22(4), 657-660.

[172] Kambi, N.; Halder, P.; Rajan, R.; Arora, V.; Chand, P.; Arora, M.; Jain, N. Large-scale reorganization of the somatosensory cortex following spinal cord injuries is due to brainstem plasticity. *Nat Commun.*2014, 5, 3602.

[173] Hubel, D.H.; Wiesel, T.N. The period of susceptibility to the physiological effects of unilateral eye closure in kittens. *J Physiol.*1970, 206(2), 419-436.

[174] Campello-Costa, P.; Fosse, A.M. Jr.; Ribeiro, J.C.; Paes-De-Carvalho, R.; Serfaty, C.A. Acute blockade of nitric oxide synthesis induces disorganization and amplifies lesion-induced plasticity in the rat retinotectal projection. *J Neurobiol.*2000, 44(4), 371-381.

[175] Mendonca, H.R.; Araujo, S.E.; Gomes, A.L.; Sholl-Franco, A.; da Cunha Faria Melibeu, A.; Serfaty, C.A.; Campello-Costa, P. Expression of GAP-43 during development and after monocular enucleation in the rat superior colliculus. *Neurosci Lett.*2010, 477(1), 23-27.

[176] Dennis, S.H.; Jaafari, N.; Cimarosti, H.; Hanley, J.G.; Henley, J.M.; Mellor, J.R. Oxygen/glucose deprivation induces a reduction in synaptic AMPA receptors on hippocampal CA3 neurons mediated by mGluR1 and adenosine A3 receptors. *J Neurosci.*2011, 31(33), 11941-11952.

[177] Duan, W.; Gui, L.; Zhou, Z.; Liu, Y.; Tian, H.; Chen, J.F.; Zheng, J. Adenosine A2A receptor deficiency exacerbates white matter lesions and cognitive deficits induced by chronic cerebral hypoperfusion in mice. *J Neurol Sci.*2009, 285(1-2), 39-45.

[178] Zhang, C.; Schmidt, J.T. Adenosine A1 receptors mediate retinotectal presynaptic inhibition: uncoupling by C-kinase and role in LTP during regeneration. *J Neurophysiol.*1998, 79(2), 501-510.

[179] Gomes, C.V.; Kaster, M.P.; Tome, A.R.; Agostinho, P.M.; Cunha, R.A. Adenosine receptors and brain diseases: neuroprotection and neurodegeneration. *Biochim Biophys Acta.*2011, 1808(5), 1380-1399.

[180] Espirito-Santo, S.; Mendonca, H.R.; Menezes, G.D.; Goulart, V.G.; Gomes, A.L.; Marra, C.; Melibeu, A.C.; Serfaty, C.A.; Sholl-Franco, A.; Campello-Costa, P. Intravitreous interleukin-2 treatment and inflammation modulates glial cells activation and uncrossed retinotectal development. *Neuroscience.*2012, 200, 223-236.

[181] Leon, S.; Yin, Y.; Nguyen, J.; Irwin, N.; Benowitz, L.I. Lens injury stimulates axon regeneration in the mature rat optic nerve. *J Neurosci.*2000, 20(12), 4615-4626.

[182] Yin, Y.; Cui, Q.; Li, Y.; Irwin, N.; Fischer, D.; Harvey, A.R.; Benowitz, L.I. Macrophage- derived factors stimulate optic nerve regeneration. *J Neurosci.*2003, 23(6), 2284-2293.

[183] Fredholm, B.B.; Battig, K.; Holmen, J.; Nehlig, A.; Zvartau, E.E. Actions of caffeine in the brain with special reference to factors that contribute to its widespread use. *Pharmacol Rev.*1999, 51(1), 83-133.

[184] Cabral-Miranda, F.; Serfaty, C.A.; Campello-Costa, P. A time-dependent effect of caffeine upon lesion-induced plasticity. *Neurosci Res.*2011, 71(1), 99-102.

[185] Dauer, W.; Przedborski, S. Parkinson's disease: mechanisms and models. *Neuron.*2003, 39(6), 889-909.

[186] Shook, B.C.; Jackson, P.F. Adenosine A(2A) Receptor Antagonists and Parkinson's Disease. *ACS Chem Neurosci.*2011, 2(10), 555-567.

[187] Benazzouz, A.; Mamad, O.; Abedi, P.; Bouali-Benazzouz, R.; Chetrit, J. Involvement of dopamine loss in extrastriatal basal ganglia nuclei in the pathophysiology of Parkinson's disease. *Front Aging Neurosci.*2014, 6, 87.

[188] Calabresi, P.; Picconi, B.; Tozzi, A.; Ghiglieri, V.; Di Filippo, M. Direct and indirect pathways of basal ganglia: a critical reappraisal. *Nat Neurosci.*2014, 17(8), 1022-1030.

[189] Gerfen, C.R.; Surmeier, D.J. Modulation of striatal projection systems by dopamine. *Annu Rev Neurosci.*2011, 34, 441-466.

[190] Xie, X.; Ramkumar, V.; Toth, L.A. Adenosine and dopamine receptor interactions in striatum and caffeine-induced behavioral activation. *Comp Med.*2007, 57(6), 538-545.

[191] Ferre, S.; Quiroz, C.; Orru, M.; Guitart, X.; Navarro, G.; Cortes, A.; Casado, V.; Canela, E.I.; Lluis, C.; Franco, R. Adenosine A(2A) Receptors and A(2A) Receptor Heteromers as Key Players in Striatal Function. *Front Neuroanat.*2011, 5, 36.

[192] Cechova, S.; Elsobky, A.M.; Venton, B.J. A1 receptors self-regulate adenosine release in the striatum: evidence of autoreceptor characteristics. *Neuroscience.*2010, 171(4), 1006-1015.

[193] Nagel, J.; Schladebach, H.; Koch, M.; Schwienbacher, I.; Muller, C.E.; Hauber, W. Effects of an adenosine A2A receptor blockade in the nucleus accumbens on locomotion, feeding, and prepulse inhibition in rats. *Synapse.*2003, 49(4), 279-286.

[194] Okada, M.; Mizuno, K.; Kaneko, S. Adenosine A1 and A2 receptors modulate extracellular dopamine levels in rat striatum. *Neurosci Lett.*1996, 212(1), 53-56.

[195] Armentero, M.T.; Pinna, A.; Ferre, S.; Lanciego, J.L.; Muller, C.E.; Franco, R. Past, present and future of A(2A) adenosine receptor antagonists in the therapy of Parkinson's disease. *Pharmacol Ther.*2011, 132(3), 280-299.

[196] Brugarolas, M.; Navarro, G.; Martinez-Pinilla, E.; Angelats, E.; Casado, V.; Lanciego, J.L.; Franco, R. G-protein-coupled receptor heteromers as key players in the molecular architecture of the central nervous system. *CNS Neurosci Ther.*2014, 20(8), 703-709.

[197] Navarro, G.; Carriba, P.; Gandia, J.; Ciruela, F.; Casado, V.; Cortes, A.; Mallol, J.; Canela, E.I.; Lluis, C.; Franco, R. Detection of heteromers formed by cannabinoid CB1, dopamine D2, and adenosine A2A G-protein-coupled receptors by combining bimolecular fluorescence complementation and bioluminescence energy transfer. *ScientificWorldJournal.*2008, 8, 1088-1097.

[198] Ciruela, F.; Ferre, S.; Casado, V.; Cortes, A.; Cunha, R.A.; Lluis, C.; Franco, R. Heterodimeric adenosine receptors: a device to regulate neurotransmitter release. *Cell Mol Life Sci.*2006, 63(21), 2427-2431.

[199] Xu, K.; Xu, Y.H.; Chen, J.F.; Schwarzschild, M.A. Neuroprotection by caffeine: time course and role of its metabolites in the MPTP model of Parkinson's disease. *Neuroscience.*2010, 167(2), 475-481.

[200] Li, X.X.; Nomura, T.; Aihara, H.; Nishizaki, T. Adenosine enhances glial glutamate efflux via A2a adenosine receptors. *Life Sci.*2001, 68(12), 1343-1350.

[201] Nishizaki, T.; Nagai, K.; Nomura, T.; Tada, H.; Kanno, T.; Tozaki, H.; Li, X.X.; Kondoh, T.; Kodama, N.; Takahashi, E.; Sakai, N.; Tanaka, K.; Saito, N. A new neuromodulatory pathway with a glial contribution mediated via A(2a) adenosine receptors. *Glia.*2002, 39(2), 133-147.

[202] Varani, K.; Vincenzi, F.; Tosi, A.; Gessi, S.; Casetta, I.; Granieri, G.; Fazio, P.; Leung, E.; MacLennan, S.; Granieri, E.; Borea, P.A. A2A adenosine receptor overexpression and functionality, as well as TNF-alpha levels, correlate with motor symptoms in Parkinson's disease. *FASEB J.*2010, 24(2), 587-598.

[203] Morelli, M.; Blandini, F.; Simola, N.; Hauser, R.A. A(2A) Receptor Antagonism and Dyskinesia in Parkinson's Disease. *Parkinsons Dis.*2012, 2012, 489853.

[204] Wirkner, K.; Gerevich, Z.; Krause, T.; Gunther, A.; Koles, L.; Schneider, D.; Norenberg, W.; Illes, P. Adenosine A2A receptor-induced inhibition of NMDA and GABAA receptor- mediated synaptic currents in a subpopulation of rat striatal neurons. *Neuropharmacology.*2004, 46(7), 994-1007.

[205] Calabresi, P.; Galletti, F.; Saggese, E.; Ghiglieri, V.; Picconi, B. Neuronal networks and synaptic plasticity in Parkinson's disease: beyond motor deficits. *Parkinsonism Relat Disord.*2007, 13 Suppl 3, S259-262.

[206] Pisani, A.; Centonze, D.; Bernardi, G.; Calabresi, P. Striatal synaptic plasticity: implications for motor learning and Parkinson's disease. *Mov Disord.*2005, 20(4), 395-402.

[207] Lovinger, D.M. Neurotransmitter roles in synaptic modulation, plasticity and learning in the dorsal striatum. *Neuropharmacology.*2010, 58(7), 951-961.

[208] Augustin, S.M.; Beeler, J.A.; McGehee, D.S.; Zhuang, X. Cyclic AMP and afferent activity govern bidirectional synaptic plasticity in striatopallidal neurons. *J Neurosci.*2014, 34(19), 6692-6699.

[209] Hawes, S.L.; Gillani, F.; Evans, R.C.; Benkert, E.A.; Blackwell, K.T. Sensitivity to theta-burst timing permits LTP in dorsal striatal adult brain slice. *J Neurophysiol.*2013, 110(9), 2027- 2036.

[210] Zhuang, X.; Mazzoni, P.; Kang, U.J. The role of neuroplasticity in dopaminergic therapy for Parkinson disease. *Nat Rev Neurol.*2013, 9(5), 248-256.

[211] Schiffmann, S.N.; Fisone, G.; Moresco, R.; Cunha, R.A.; Ferre, S. Adenosine A2A receptors and basal ganglia physiology. *Prog Neurobiol.*2007, 83(5), 277-292.

[212] Shen, W.; Flajolet, M.; Greengard, P.; Surmeier, D.J. Dichotomous dopaminergic control of striatal synaptic plasticity. *Science.*2008, 321(5890), 848-851.

[213] Flajolet, M.; Wang, Z.; Futter, M.; Shen, W.; Nuangchamnong, N.; Bendor, J.; Wallach, I.; Nairn, A.C.; Surmeier, D.J.; Greengard, P. FGF acts as a co-transmitter through adenosine A(2A) receptor to regulate synaptic plasticity. *Nat Neurosci.*2008, 11(12), 1402-1409.

[214] Kessler, R.C.; Chiu, W.T.; Demler, O.; Merikangas, K.R.; Walters, E.E. Prevalence, severity, and comorbidity of 12-month DSM-IV disorders in the National Comorbidity Survey Replication. *Arch Gen Psychiatry.*2005, 62(6), 617-627.

[215] Hunter, R.G.; McEwen, B.S. Stress and anxiety across the lifespan: structural plasticity and epigenetic regulation. *Epigenomics.*2013, 5(2), 177-194.

[216] McEwen, B.S.; Eiland, L.; Hunter, R.G.; Miller, M.M. Stress and anxiety: structural plasticity and epigenetic regulation as a consequence of stress. *Neuropharmacology.*2012, 62(1), 3-12.

[217] Adhikari, A. Distributed circuits underlying anxiety. *Front Behav Neurosci.*2014, 8, 112.

[218] Gadek-Michalska, A.; Tadeusz, J.; Rachwalska, P.; Bugajski, J. Cytokines, prostaglandins and nitric oxide in the regulation of stress-response systems. *Pharmacol Rep.*2013, 65(6), 1655-1662.

[219] Farb, D.H.; Ratner, M.H. Targeting the Modulation of Neural Circuitry for the Treatment of Anxiety Disorders. *Pharmacol Rev.*2014, 66(4), 1002-1032.

[220] Maniam, J.; Antoniadis, C.; Morris, M.J. Early-Life Stress, HPA Axis Adaptation, and Mechanisms Contributing to Later Health Outcomes. *Front Endocrinol (Lausanne).*2014, 5, 73.

[221] McEwen, B.S. Physiology and neurobiology of stress and adaptation: central role of the brain. *Physiol Rev.*2007, 87(3), 873-904.

[222] Gillespie, C.F.; Nemeroff, C.B. Hypercortisolemia and depression. *Psychosom Med.*2005, 67 Suppl 1, S26-28.

[223] Daban, C.; Vieta, E.; Mackin, P.; Young, A.H. Hypothalamic-pituitary-adrenal axis and bipolar disorder. *Psychiatr Clin North Am.*2005, 28(2), 469-480.

[224] Yehuda, R. Advances in understanding neuroendocrine alterations in PTSD and their therapeutic implications. *Ann N Y Acad Sci.*2006, 1071, 137-166.

[225] Abelson, J.L.; Khan, S.; Liberzon, I.; Young, E.A. HPA axis activity in patients with panic disorder: review and synthesis of four studies. *Depress Anxiety.*2007, 24(1), 66-76.

[226] Gustafsson, P.E.; Gustafsson, P.A.; Ivarsson, T.; Nelson, N. Diurnal cortisol levels and cortisol response in youths with obsessive-compulsive disorder. *Neuropsychobiology.*2008, 57(1-2), 14-21.

[227] Van Houdenhove, B.; Van Den Eede, F.; Luyten, P. Does hypothalamic-pituitary-adrenal axis hypofunction in chronic fatigue syndrome reflect a 'crash' in the stress system? *Med Hypotheses.*2009, 72(6), 701-705.

[228] Lenze, E.J.; Mantella, R.C.; Shi, P.; Goate, A.M.; Nowotny, P.; Butters, M.A.; Andreescu, C.; Thompson, P.A.; Rollman, B.L. Elevated cortisol in older adults with generalized anxiety disorder is reduced by treatment: a placebo-controlled evaluation of escitalopram. *Am J Geriatr Psychiatry.*2011, 19(5), 482-490.

[229] Vyas, A.; Mitra, R.; Shankaranarayana Rao, B.S.; Chattarji, S. Chronic stress induces contrasting patterns of dendritic remodeling in hippocampal and amygdaloid neurons. *J Neurosci.*2002, 22(15), 6810-6818.

[230] Hohoff, C.; Mullings, E.L.; Heatherley, S.V.; Freitag, C.M.; Neumann, L.C.; Domschke, K.; Krakowitzky, P.; Rothermundt, M.; Keck, M.E.; Erhardt, A.; Unschuld, P.G.; Jacob, C.; Fritze, J.; Bandelow, B.; Maier, W.; Holsboer, F.; Rogers, P.J.; Deckert, J. Adenosine A(2A) receptor gene: evidence for association of risk variants with panic disorder and anxious personality. *J Psychiatr Res.*2010, 44(14), 930-937.

[231] Huang, Z.L.; Liu, R.; Bai, X.Y.; Zhao, G.; Song, J.K.; Wu, S.; Du, G.H. Protective effects of the novel adenosine derivative WS0701 in a mouse model of posttraumatic stress disorder. *Acta Pharmacol Sin.*2014, 35(1), 24-32.

[232] Correa, M.; Font, L. Is there a major role for adenosine A2A receptors in anxiety? *Front Biosci.*2008, 13, 4058-4070.

[233] Goodman, R.R.; Synder, S.H. Autoradiographic localization of adenosine receptors in rat brain using [3H]cyclohexyladenosine. *J Neurosci.*1982, 2(9), 1230-1241.

[234] Weaver, D.R. A2a adenosine receptor gene expression in developing rat brain. *Brain Res Mol Brain Res.*1993, 20(4), 313-327.

[235] Zhang, J.; Yin, D.; Wu, F.; Zhang, G.; Jiang, C.; Li, Z.; Wang, L.; Wang, K. Microinjection of adenosine into the hypothalamic ventrolateral preoptic area enhances wakefulness via the A1 receptor in rats. *Neurochem Res.*2013, 38(8), 1616-1623.

[236] Fastbom, J.; Pazos, A.; Palacios, J.M. The distribution of adenosine A1 receptors and 5'- nucleotidase in the brain of some commonly used experimental animals. *Neuroscience.*1987, 22(3), 813-826.

[237] Cintra, A.; Fuxe, K.; Solfrini, V.; Agnati, L.F.; Tinner, B.; Wikstrom, A.C; Staines, W.; Okret, S.; Gustafsson, J.A. Central peptidergic neurons as targets for glucocorticoid action. Evidence for the presence of glucocorticoid receptor immunoreactivity in various types of classes of peptidergic neurons. *J Steroid Biochem Mol Biol.*1991, 40(1-3), 93-103.

[238] Boulenger, J.P.; Uhde, T.W.; Wolff, E.A.3rd.; Post, R.M. Increased sensitivity to caffeine in patients with panic disorders. Preliminary evidence. *Arch Gen Psychiatry.*1984, 41(11), 1067- 1071.

[239] Boulenger, J.P.; Marangos, P.J.; Zander, K.J.; Hanson, J. Stress and caffeine: effects on central adenosine receptors. *Clin Neuropharmacol.*1986, 9(1), 79-83.

[240] Svenningsson, P.; Fredholm, B.B. Glucocorticoids regulate the expression of adenosine A1 but not A(2A) receptors in rat brain. *J Pharmacol Exp Ther.*1997, 280(2), 1094-1101.

[241] Jegou, S.; El Yacoubi, M.; Mounien, L.; Ledent, C,; Parmentier, M.; Costentin, J.; Vaugeois, J.M.; Vaudry, H. Adenosine A2A receptor gene disruption provokes marked changes in melanocortin content and pro-opiomelanocortin gene expression. *J Neuroendocrinol.*2003, 15(12), 1171-1177.

[242] Lin, A.S.; Uhde, T.W.; Slate, S.O.; McCann, U.D. Effects of intravenous caffeine administered to healthy males during sleep. *Depress Anxiety.*1997, 5(1), 21-28.

[243] Koran, L.M.; Aboujaoude, E.; Gamel, N.N. Double-blind study of dextroamphetamine versus caffeine augmentation for treatment-resistant obsessive-compulsive disorder. *J Clin Psychiatry.*2009, 70(11), 1530-1535.

[244] Batalha, V.L.; Pego, J.M.; Fontinha, B.M.; Costenla, A.R.; Valadas, J.S.; Baqi, Y.; Radjainia, H.; Muller, C.E.; Sebastiao, A.M.; Lopes, L.V. Adenosine A(2A) receptor blockade reverts hippocampal stress-induced deficits and restores corticosterone circadian oscillation. *Mol Psychiatry.*2013, 18(3), 320-331.

[245] Cunha, R.A.; Johansson, B.; van der Ploeg, I.; Sebastiao, A.M.; Ribeiro, J.A.; Fredholm, B.B. Evidence for functionally important adenosine A2a receptors in the rat hippocampus. *Brain Res.*1994, 649(1-2), 208-216.

[246] Cunha, G.M.; Canas, P.M.; Oliveira, C.R.; Cunha, R.A. Increased density and synapto-protective effect of adenosine A2A receptors upon sub-chronic restraint stress. *Neuroscience.*2006, 141(4), 1775-1781.

[247] Johansson, B.; Halldner, L.; Dunwiddie, T.V.; Masino, S.A.; Poelchen, W.; Gimenez-Llort, L.; Escorihuela, R.M.; Fernandez-Teruel, A.; Wiesenfeld-Hallin, Z.; Xu, X.J.; Hårdemark, A.; Betsholtz, C.; Herlenius, E.; Fredholm, B.B. Hyperalgesia, anxiety, and decreased hypoxic neuroprotection in mice lacking the adenosine A1 receptor. *Proc Natl Acad Sci U S A.*2001, 98(16), 9407- 9412.

[248] Gimenez-Llort, L.; Fernandez-Teruel, A.; Escorihuela, R.M.; Fredholm, B.B.; Tobena, A.; Pekny, M.; Johansson, B. Mice lacking the adenosine A1 receptor are anxious and

aggressive, but are normal learners with reduced muscle strength and survival rate. *Eur J Neurosci.*2002, 16(3), 547-550.

[249] Jain, N.; Kemp, N.; Adeyemo, O.; Buchanan, P.; Stone, T.W. Anxiolytic activity of adenosine receptor activation in mice. *Br J Pharmacol.*1995, 116(3), 2127-2133.

[250] Prediger, R.D.; da Silva, G.E.; Batista, L.C.; Bittencourt, A.L.; Takahashi, R.N. Activation of adenosine A1 receptors reduces anxiety-like behavior during acute ethanol withdrawal (hangover) in mice. *Neuropsychopharmacology.*2006, 31(10), 2210-2220.

[251] Imaizumi, M.; Miyazaki, S.; Onodera, K. Effects of xanthine derivatives in a light/dark test in mice and the contribution of adenosine receptors. *Methods Find Exp Clin Pharmacol.*1994, 16(9), 639-644.

[252] Prediger, R.D.; Batista, L.C.; Takahashi, R.N. Adenosine A1 receptors modulate the anxiolytic-like effect of ethanol in the elevated plus-maze in mice. *Eur J Pharmacol.*2004, 499(1-2), 147-154.

[253] Vollert, C.; Forkuo, G.S.; Bond, R.A.; Eriksen, J.L. Chronic treatment with DCPCX, an adenosine A(1) antagonist, worsens long-term memory. *Neurosci Lett.*2013, 548, 296-300.

[254] de Mendonca, A.; Ribeiro, J.A. Long-term potentiation observed upon blockade of adenosine A1 receptors in rat hippocampus is N-methyl-D-aspartate receptor-dependent. *Neurosci Lett.*2000, 291(2), 81-84.

[255] Florio, C.; Prezioso, A.; Papaioannou, A.; Vertua, R. Adenosine A1 receptors modulate anxiety in CD1 mice. *Psychopharmacology (Berl).*1998, 136(4), 311-319.

[256] Deckert, J. The adenosine A(2A) receptor knockout mouse: a model for anxiety? *Int J Neuropsychopharmacol.*1998, 1(2), 187-190.

[257] El Yacoubi, M.; Ledent, C.; Parmentier, M.; Costentin, J.; Vaugeois, J.M. The anxiogenic- like effect of caffeine in two experimental procedures measuring anxiety in the mouse is not shared by selective A(2A) adenosine receptor antagonists. *Psychopharmacology (Berl).*2000, 148(2), 153-163.

[258] Thorsell, A.; Johnson, J.; Heilig, M. Effect of the adenosine A2a receptor antagonist 3,7- dimethyl-propargylxanthine on anxiety-like and depression-like behavior and alcohol consumption in Wistar Rats. *Alcohol Clin Exp Res.*2007, 31(8), 1302-1307.

In: Adenosine Signaling Mechanisms
Editors: V. Ramkumar and R. Paes de Carvalho

ISBN: 978-1-63483-186-4

Chapter 15

ADENOSINE RECEPTORS: AN OPPORTUNITY FOR NANOMEDICINE

José Luis Albasanz*[*] *and Mairena Martín

Departamento de Química Inorgánica, Orgánica y Bioquímica,
Facultad de Ciencias y Tecnologías Químicas /
Facultad de Medicina de Ciudad Real, Centro Regional de Investigaciones Biomédicas,
Universidad de Castilla-La Mancha, Ciudad Real; Spain

ABSTRACT

Nanotechnology is transforming medicine, manufacturing, energy production, and other fundamental features of everyday life in the 21st Century. Nanomedicine involves the use of nanoparticles for therapeutic and diagnostic purposes (nanotheranostics). Among the diseases where the use of nanoparticles has been considered are malignant tumors- namely breast and prostate cancer, prion disease, myocardial infarction and heart failure, inflammatory atherosclerosis, glaucoma and neurological and neurodegenerative disorders. In many of these diseases adenosine receptors have been implicated, mainly in cancer and neurodegenerative diseases. This chapter summarizes recent findings in relation to adenosine receptors and its signalling obtained using different nanoparticles such as dendrimers, fullerenes, gold nanoparticles and quantum dots, among others approaches,. Several aspects related to their use in biomedical applications, as structure, functionalization and safety, are also discussed. Data presented in this chapter clearly evidence nanoparticles as promising pharmacological tools for diagnostic and therapeutic purposes targeting adenosine receptors. We conclude that adenosine receptors represent an opportunity and an emerging prospect for effective Nanomedicine.

[*] Correspondence to: José Luis Albasanz, Facultad de CC y Tecnologías Químicas. Avda. Camilo José Cela 10, 13071 Ciudad Real, Spain. E-mail: jose.albasanz@uclm.es

INTRODUCTION

Nanotechnology is the science of the ultra-small. It is the creation of functional materials, devices and systems through control of matter at the scale of 1 to 100 nanometers (nm), and the exploitation of novel properties and phenomena at the same scale. For a better perspective, a simple water molecule represent 0.01 nm while 100,000,000 nanometers corresponding to the size of a tennis ball. One nanometer equals one-billionth of a meter, which means it would take 100,000 nm lined up side by side to equal the diameter of a human hair. In 1959 Richard Feynman, a Quantum theorist and Nobel Laureate wrote the book entitled "*There's Plenty of Room at the Bottom.*" In 1981, Eric Drexler published the first journal article on molecular nanotechnology [1]. In 1986, he published the book "*Engines of Creation: The Coming Era of Nanotechnology.*"

Nanotechnology is transforming medicine, manufacturing, energy production, and other fundamental features of everyday life in the 21st Century [2]. The application of nanotechnology to medicine is termed "nanomedicine." Therefore, nanomedicine encompasses diagnosing, treating, and preventing disease and traumatic injury, relieving pain, and preserving and improving human health, by taking advantage of molecular tools at nano scale and current molecular knowledge of the human body [3, 4]. Thus, nanomedicine involves the use of nanoparticles for therapeutic and diagnostic purposes. The word "theranostics" refers to the simultaneous integration of diagnosis and therapy [5]. Accordingly, "nanotheranostics" involves the application and further development of nanomedicine strategies for advanced theranostics.

It is important to evaluate the safety of nanoparticles in order to achieve biocompatibility and desired activity.[6] Nanocarriers such as polymer conjugations, dendrimers, micelles, liposomes, metal and inorganic nanoparticles, carbon nanotubes, nanoparticles of biodegradable polymers for sustained, controlled and targeted co-delivery of diagnostic and therapeutic agents should be therefore fully investigated for better theranostic effects and fewer side effects. Thus, these applications of nanomedicine harbinger a revolution in modern medicine comparable to that of discovery of antibiotics and invention of vaccines.

The central objectives of nanomedicine is to diagnose and treat the diseases at their earliest stage, when the diseases are most likely curable or at least treatable.[7] Among the diseases where nanomedicine has been considered are malignant tumors [8-9]- breast cancer [10], prostate cancer [11], prion disease [12], myocardial infarction and heart failure [13], inflammatory atherosclerosis [14], glaucoma [15] and neurological [16] and neurodegenerative disorders [17], especially in Alzheimer's disease [18-20] and Parkinson's disease [21]. These studies demonstrate the new frontiers that novel theranostics applications are making strides [22].

In many of these different diseases adenosine receptors have been proposed to serve as effective therapeutic targets. Adenosine (P1) and nucleotide (P2) receptors are widely expressed in the brain on both neurons and glial cells. There is increasing interest in the involvement of purinergic signalling in the pathophysiology of the CNS, including trauma, ischaemia, epilepsy, neurodegenerative diseases, neuropsychiatric and mood disorders, and cancers, including gliomas [23].

Adenosine is a nucleoside widely distributed in both central and peripheral nervous system where acts as neuromodulator and homeostatic regulator mainly through four types of receptors: A_1 and A_3 receptors inhibit adenylyl cyclase activity decreasing cAMP level, while A_{2A} and A_{2B} receptors stimulate adenylyl cyclase activity increasing cAMP levels. A_1 and A_{2A} receptors are highly expressed in brain where adenosine acting through these receptors participates in learning and memory processes by modulating neurotransmitter release as glutamate [23]. The neuroprotective action of adenosine is mainly mediated through A_1 receptors because of their inhibition of glutamate release when they are activated. Conversely, A_{2A} overactivation is generally associated to neurodegeneration. Therefore, adenosine receptors have been involved in neurodegenerative diseases as Alzheimer, Parkinson and other dementias [24].

Tomalia (2009) has extensively and thoroughly reviewed the different classes of the nano-sized building blocks and proposed a classification roadmap for nanomaterials [2]. Accordingly, all nanomatter could be classified into category I: discrete, well-defined and category II: statistical, undefined nanoparticles. Among category I are hard and soft nanoparticles. Hard nanoparticles have been further categorized as nanoclusters, nanocrystals (quantum dots), amorphous nanoparticles (silica nanoparticles), rigid carbon allotropes (fullerenes, carbon nanotubes). Soft nanoparticles have been categorized as synthetic (dendrimers/dendrons, nanolatexes, polymeric nanoparticles) and biological (protein, viruses, RNA/DNA) [2].

Some of these nanoparticles have been employed in research about GPCRs, i.e., adenosine receptors, as indicated below.

Dendrimers

Dendrimers are nano-sized, radially symmetric molecules with well-defined, homogeneous, and monodisperse structure consisting of tree-like arms or branches which have several medicinal and practical applications [25].They allow the precise control of size, shape and placement of functional groups and combine typical surface characteristics of small organic molecules and polymers that result in special physical and chemical properties, giving to these dendritic polymers advantage in biomedical applications through easy functionalization. Poly(amido)amine (PAMAM) dendrimers represent a new class of such polymers [26-27]. The manufacturing process involves a series of repetitive steps starting with a central initiator core. Amino-terminated PAMAM dendrimers are considered full-generation, whereas ester-terminated PAMAM dendrimers are termed half-generation. Then onwards each subsequent growth step represents a new 'generation' (G) of polymer with a larger molecular diameter and twice the number of reactive surface sites. With each new generation from G0 to G10, the molecular weights and the numbers of active surface units of PAMAM dendrimers increases exponentially, while their diameters increases linearly [28]. A main interest in PAMAM compounds stems from their biomedical effectiveness as drug and as drug carriers [28].

Safe and effective methods to facilitate delivery of drugs into the brain are needed for the treatment of central nervous system disorders. However, these treatments are limited by the extremely low permeability of the blood brain barrier (BBB) [29].This barrier limits the entry of molecules into the brain through two main strategies. First, tight junctions (TJs) that seal the endothelial cells regulate permeability and induce a high intercellular electrical resistance which leads to an extremely low permeability to blood-borne molecules [30]. Second, there are few trans-cellular transport pathways but high expression levels of active efflux transport proteins such as P-glycoprotein on brain capillary endothelial cells [31]. Recently, it has been reported that adenosine receptor signalling modulates BBB permeability *in vivo*. Adenosine A_1 and A_{2A} receptor activation facilitated the entry of intravenously administered macromolecules, including large dextrans and antibodies to β-amyloid, into murine brains. Interestingly, these changes in BBB permeability were found to be dose-dependent and temporally discrete [32]. Although the complete mechanism has not been clearly elucidated, this effect was used to develop a series of "nanoagonists" (based on PAMAM G5 dendrimers) that not only up-regulated brain delivery of macromolecular model drugs by specifically signalling A_{2A}R but also demonstrated the capability to temporarily tune the BBB opening time-window under *in vivo* conditions using imprinting control region (ICR) mice and athymic nude mice [33]. Therefore, compared to small molecules, nanoparticles demonstrate advantages in pharmaceutical applications including their tunable circulation lifetimes and adjustable receptor binding affinities [34].

The route of administration (e.g., intranasal, intravenous, intratracheal, or intraperitoneal), dosage (low or high dose), and duration of treatment (acute, subchronic, chronic, or intermittent) may also influence the toxicological and pharmacological effects of dendrimers *in vivo* [35]. In line with this, a recent study showed by confocal microscopy that dendrimers are able to cross the cell membrane of primary neurons and are also able to diffuse to the CNS tissues after intraparenchymal or intraventricular injections [36]. To determine whether PAMAM dendrimers cause toxicity in an animal model, male BALB/c mice (7 weeks old) were used to assess the neurotoxic effects of the single intranasal administration of PAMAM dendrimers on neurological biomarkers in the mouse brain. No remarkable changes were observed in serum biochemical markers and brain inflammatory biomarkers. Moreover, the administration of an acute single high dose of dendrimer caused mild effects on neurological gene expressions in the olfactory bulb, hippocampus and cerebral cortex [35].

The first example of dendrimer application to induce the multivalent intracellular signal transduction across GPCRs as adenosine receptors was reported by Kim et al., (2008) [37]. These authors selected the carboxylic acid derivative CGS21680 as adenosine A_{2A} receptor agonist for the attachment to the peripheral amino groups of a third generation (G3) PAMAM dendrimer with an ethylenediamine core. Here, the PAMAM dendrimer linked to 2-[4-(2-carboxyethyl)phenylethylamino]-5'-N-ethylcarboxamidoadenosine (CGS21680) through its aliphatic amino end group can be considered a part of the distal modification under the functionalized congener concept to modulate the biological activity of the original ligand [38]. Interestingly, to aid in the microscopic visualization in biological assays, they prepared a fluorescent dendrimer derivative based on Alexa Fluor 488 fluorophore, which has similar absorption/emission maxima to those of fluorescein, but is reported to have a higher photostability, pH insensitivity, and good water solubility [39]. The average molecular weights and the degree of ligand attachments for a couple of dendrimers constructions were determined by ^{1}H NMR integration and MALDI, indicating the formation of nearly fully loaded dendrimers. These constructs were able to activate A_{2A} receptors as in preliminary *in vitro* functional assays they demonstrated to have a potent anti-aggregation effect on the human platelet in response to ADP application. In addition, the fluorescent dendrimer construct was found to be internalized into the platelet.

Although potential steric complications makes polyethylene glycol (PEG) a not very useful tool for covalent modification of GPCR ligands, it was employed to modify these PAMAM dendrimers and increase its aqueous solubility. The PEGylation of CGS21680 containing PAMAMs increased the bioavailability of dendritic carriers for GPCR ligands and showed that some of the conjugates were modestly enhanced in potency compared with the monomeric agonist. These antithrombotic PEGylated GPCR ligand–dendrimer (GLiDe) conjugates can be considered to be unimolecular micelles in which the outer hydrophilic layer determines the interface for interaction with the medium, membrane surfaces, and biopolymers [40].

Xanthine amine congener (XAC), a potent adenosine receptor antagonist, was also appended to G4 PAMAM dendrimers to form triazoles. These conjugates also contained triazole-linked PEG groups for increasing water-solubility and optionally prosthetic groups for spectroscopic characterization and affinity labelling. Interestingly, human adenosine receptor binding affinity increased progressively with the degree of xanthine substitution to reach K_i values in the nanomolar range. Thus, this substitution confers sensitivity enabling detection of qualitative difference in responses to a potent, multivalent xanthine conjugate and its corresponding monomer mediated A_1R antagonism. This was the first systematic probe of a potent adenosine receptor antagonist tethered on a dendrimer and its activity as a function of loading variability [41].

The A_1R agonist N^6-[4-[[[4-[[[(2-aminoethyl)amino]carbonyl]methyl] anilino]-carbonyl]methyl]phenyl]adenosine (ADAC) was chosen for conjugation with G2.5 and G3 PAMAM dendrimers intended for selectivity for A_1R. Surprisingly, a high selectivity in binding at the A_3R in comparison to the monomeric nucleosides was achieved by G2.5 dendrimer-ligand conjugates, which depended on the nature of the linker moiety (i.e., a secondary amine linkage resulted in greater than 100-fold A_3R selectivity in both radioligand binding and cAMP assays). This selectivity was not seen with the G3 dendrimer-ligand conjugates, which were approximately equipotent at the A_1 and A_3R in both radioligand binding and cAMP assays [42]. This decrease in affinity at the A_1R and increase in affinity at A_3R was also seen when using $A_{2A}R$-directed G3 dendrimer-ligand conjugates [43]. Thus, the modulation and even enhancement of the pharmacological profile of a GPCR ligand through its conjugation to a nanocarrier is possible. A G5.5 PAMAM dendrimeric conjugate, a larger variation in molecular weight of G2.5 PAMAM structure, was able to protect HL-1 mouse cardiomyocytes exposed to H_2O_2 for the induction of apoptosis acting through A_3R. This protective effect was found to have significantly greater potency (IC_{50} 35 nM) with the dendrimeric conjugate than the corresponding monomeric nucleosides, which protected in the μmolar range [44].

Conjugation of MRS5233, an A_3R-specific agonist, to a PAMAM dendrimer [45] greatly improved its *in vitro* affinity and potency compared to the monomer form. In comparing the activities of the conjugated MRS5246 and the monomer MRS5233 compounds, MRS5246 possessed a much higher affinity (98.5-fold) to the mouse A_3R compared to the monomer. The conjugated MRS5246 was also 200-fold more potent A_3R agonist than the monomer, MRS5233, in an *in vitro* cAMP assay. Other GLiDe conjugates showed significantly improved affinity and potency over the unconjugated monomeric agonists [46-47]. Different chemical linking strategies by click chemistry through functionalized chains attached at the adenine C_2 position [48] or through the N^6 position of methanocarba nucleoside ligands [49] achieved higher affinities than coupling by amide formation. Therefore, a careful design of such conjugates would help improve potency or specificity of a desired ligand.

It has been determined that extravasation of PAMAM dendrimers across the microvascular endothelium is charge, size and molecular weight-dependent. Thus, with each size-increasing generation of PAMAM dendrimers, up to the G4, there was a weight- and size-dependent decrease in the extravasation of the dendrimers [50]. It can be argued that *in vitro* assays (e.g., radioligand binding and cAMP assays) for functional analysis of these dendrimer conjugates are performed in the absence of biological barriers such as would be encountered *in vivo*. However, *in vivo*, the GLiDe conjugates must reach the myocardial targets by extravasation from the coronary microvasculature. This issue was confirmed by the cardioprotective effect detected in an isolated mouse heart model of ischemia/reperfusion injury treated with the conjugated MRS5246 and the monomer MRS5233 compounds. They were both found to be effective in a concentration dependent manner (0.1, 1 and 10 nM) [45]. Three dendrimer derivatives of A_3R agonists (MRS5216, MRS5246 and MRS5539), similarly to a monomeric agonist of the A_3R, were cardioprotective as they demonstrated anti-hypoxic effects on neonatal rat cardiomyocyte cultures and isolated hearts. However, the dendrimer conjugates had significantly greater potency (10 - 100 pM) than an A_3R-selective monomeric nucleoside Cl-IB-MECA, which was effective in the 100 nM range. It should be noted that protection or greater injury of the neonatal rat cultured cardiomyocytes was observed depending on dendrimer concentration. Thus, while nM and pM concentrations provided protection to the cultured cells against hypoxia, μM concentrations were injurious. The mechanism of this injury was not explored, but it might be a function of greater internalization [51]. These effects, however, suggest another important conclusion, the envisaged potential use in *in vivo* systems of these dendrimer conjugates, providing appropriated precautions.

Oligomerization among G protein-coupled receptors has been clearly established from experimental, particularly *in vitro*, studies. Moreover, homo and heterodimerization provide new unexpected clues for explaining the molecular mechanisms underlying some diseases in which GPCRs signalling might be affected. Therefore, a promising therapeutic approach would be to develop bivalent ligands that emerge as very powerful molecules interacting with receptor homomers or heteromers [52]. The theoretical possibility of bivalent binding of a dendrimer, covalently appended with multiple copies of a small ligand, to a homodimer of a GPCR was investigated through docking experiments with a molecular model of a third generation poly(amidoamine) dendrimer condensed with multiple copies of the potent $A_{2A}R$ agonist CGS21680 [53]. The PAMAM-CGS21680 dendrimer was bound to an $A_{2A}R$ homodimer and it was shown that both subunits of the dimer can be occupied simultaneously by CGS21680 moieties of the PAMAM-CGS21680 dendrimer. Moreover, the binding mode of CGS21680 moieties linked to the PAMAM dendrimer and docked to the $A_{2A}R$ was found to be similar to the binding mode of a monomeric CGS21680 ligand. Therefore, a rational design of novel multivalent ligands for adenosine receptors and other GPCR homo- and heterodimers is envisaged and it would be interesting to develop multivalent ligands for GPCRs to create a dendrimer containing a set of heterogeneous agonists/antagonists [36] in order to induce combined synergistic biological effects via binding to different receptor subtypes simultaneously on a cellular membrane [54-55]. Synergistic effects of activating multiple GPCRs with a 'smart drug' as a single dendrimer conjugate might be useful in disease treatment [55].

In line with this, a bifunctional conjugate (MRS5259) containing an A_3R agonist (MRS5221) linked through a click reaction of its 2-alkynyl functionalized chain and an amide-linked uridine-5'-diphosphoglucuronic acid (UDPGA), potently activated both A_3 and $P2Y_{14}$ receptors with selectivity in comparison to other adenosine receptors and P2Y receptors. However, while a functional assay of adenylate cyclase inhibition carried out in C6 cells expressing the human $P2Y_{14}$ receptor demonstrated that potency at this receptor was enhanced in the dendrimer conjugate 8 (MRS5259) over that of the monomer. The Ki value of MRS5259 in binding to the A_3R (39.5 nM) indicated that it was considerably less potent than the conjugates which contained only the A_3R agonist ligand [48]. This was the first example of targeting two different GPCRs with the same dendrimer conjugate, which is intended for activation of heteromeric GPCR aggregates.

FULLERENES

Buckminsterfullerene (C_{60}) is a carbon allotrope discovered in 1985 [56]. Since then, fullerenes have attracted much attention and been widely investigated toward their applications in different physicochemical and pharmacological fields, and new breakthroughs and uses are constantly being discovered [57]. After the main problem of poor solubility was solved by chemical or supramolecular approaches (usually a chemical modification on the surface of the sphere), various functionalized fullerenes were synthesized to achieve promising results [58]. Thus, fullerene C_{60} and derivatives can be used effectively in the photo-acoustic imaging of cancer and tumor cells [59], orthopaedic research [60], photodynamic therapy for cancer [61-63] and rheumatoid arthritis [64-65].

Pristine C_{60} has no acute or sub-acute toxicity in a wide range of living organisms as bacteria, fungal, drosophila, mice, rats, guinea pigs and humans. In contrast, some C_{60} derivatives can be highly toxic. For instance, under light exposure, C_{60} is an efficient singlet oxygen sensitizer. Therefore, if pristine C_{60} is absolutely nontoxic under dark conditions, this is not the case under UV-Visible irradiation and in the presence of O_2 where fullerene solutions can be highly toxic through $1O_2$ formation [66].

The toxicity information of C_{60} on the lungs as a hazard assessment was reviewed elsewhere [67]. The no-observed-adverse-effect level (NOAEL) of C_{60} (geometric mean of 96 nm and geometric standard deviation of 2.0) on rat lung toxicity was estimated to be 3.1 mg/m^3 based on the results of an intratracheal instillation test, and the period limited acceptable exposure level (AEL (PL)) values of this C_{60} were estimated to be 0.39 mg/m^3 for healthy workers and $1.4x10^{-2}$ mg/m^3 for the general human population. Because this is the NOAEL for subchronic toxicity, an AEL (PL) for humans was proposed, which assumed 15 years of exposure and modification within the next 10 years since more knowledge will be gained in the future [67].

Neuroprotection is an active field of research where C_{60} fullerenes have been also considered [68]. Fullerene derivative compounds provide effective neuroprotection against excitotoxic cell death, apoptosis initiated by different types of triggers (growth factor deprivation, NMDA receptor blockade in immature neurons) and metabolic insults (oxygen glucose deprivation) in cultured cells [69]. Water-soluble fullerols have potential applications to Parkinson's disease therapy [70] and the therapeutical utility of fullerene derivatives for the treatment of Alzheimer disease has also been considered [18]. Accordingly, an aqueous molecule-colloidal solution of C_{60} might be a good source for anti-amyloid drug therapy due to its antioxidant and anti-aggregative properties [71]. Moreover, carboxyl-functionalized fullerenes efficiently cross the blood-brain barrier [72], which in an interesting feature in the development of new C_{60} pharmacological tools.

Excitotoxicity due to imbalanced glutamate metabolism and consequent increased extracellular glutamate concentration is one important mechanism by which neurodegeneration occurs in multiple sclerosis [73], but also in other neurodegenerative diseases as Alzheimer and Parkinson [74].

In line with this, oxidative stress may mediate excitotoxic neuronal damage, resulting in an excessive stimulation of, for example, glutamate receptors [75] as well as the apoptotic death of neurons [76]. Apoptosis can contribute to the loss of neurons in pathological conditions such as cerebral ischemia, Huntington's disease or Alzheimer's disease [77-78].

Fullerenes are considered to be effective scavengers of free radicals based on the large numbers of conjugated double π bonds with low energy unoccupied molecular orbitals (LUMO), which can easily take up an electron, facilitating reactions with radical species [79]. The most commonly used free radical scavengers are water-soluble C_{60} derivatives such as PEG-C_{60}, PVP-C_{60}, and $C_{60}(OH)n$ [80]. Fullerenols as $C_{60}(OH)_{24}$ can be protective in both radiotherapy and chemotherapy due to its antioxidant properties; even they could have a potential role in the treatment of neurodegenerative diseases [81].

Reversal of axonal loss and disability in a mouse model of progressive multiple sclerosis has been reported by using fullerene ABS-75, a water-soluble adamantyl-oligoethyleneglycol-fullerene hybrid, in which NMDA receptor-targeting (antagonist) adamantyl groups are connected to an antioxidant carboxyfullerene moiety via oligoethyleneglycol bridges [82]. The combination of antioxidant and anti-excitotoxic properties in the same molecule was critical for the effect of ABS-75 in reducing neurodegeneration and disease progression in this model. In fact, treatment with the unsubstituted fullerene (fullerene core without the attached NMDA receptor antagonist) or with memantine alone had no effect on disease progression.

Six different carboxyfullerene superoxide dismutase (SOD) mimetics with distinct electronic and biophysical characteristics were tested in an NMDA-induced neuronal death model. Two of them were neuroprotective probably due to the SOD activity of these cell-permeable compounds. The rescue from neuronal death by carboxyfullerenes is not due to NMDA receptor blockade, as molecular docking calculations revealed a sterically prohibited and thermodynamically highly unfavourable binding to this receptor [83].

Hydrophilic fullerene derivatives, as $C_{60}(OH)_{12}$ and $C_{60}(OH)_{18-20}O_{3-7}$, decreased excitotoxicity induced neuronal death of mouse neocortical neurons up to 80% and were antiapoptotic [84]. Fullerenols exerted a neuroprotective effect on neuronal cultures *in vitro* by blocking glutamate receptors and reducing the concentration of intracellular calcium [85]. Fullerenols can also modulate the organophosphate-induced acetylcholinesterase inhibition in SH-SY5Y cells [86] and were protective towards NMDA, AMPA, GABA, and KA receptors under conditions of increased oxidative stress [84-85].

The oxidative stress consequent to low oxygen availability (hypoxia) has been related to cell death processes, generally apoptotic, upon the formation of reactive oxygen species (ROS) and subsequently oxidative modifications of lipid, DNA, cell membrane proteins and other target molecules. The oxidative stress is also related to the development of some neurodegenerative diseases as Alzheimer's and Parkinson's [87]. Thus a dramatic increase in ROS has been related to cell death induced by oxidative stress and to neurodegenerative diseases as Alzheimer's [88] and it has been proposed that the first step of Alzheimer's disease is the oxidative damage originated by ROS [89-90]. As mentioned above, fullerene and their derivatives are effective redox-active compounds towards ROS such as superoxide anion radical (O_2^-), hydroperoxide (ROOH), and hydroxyl radical ($^{\bullet}OH$) produced during oxidative stress and responsible for different type of cell damages [91]. It was demonstrated that the bis-functionalization of [60]fullerene by means of azomethineylide 1,3 dipolar cycloaddition leading to bispyrrolidine derivatives, confers a certain electronic stability to the fullerene carbon cage without depriving its quenching properties towards radical species [92]. In that sense, a series of highly hydrosoluble [60]fullerene bis-adducts were synthesized and encouraging results on assays in biological substrates have been obtained [93]. One of them was the trans-3 isomer of [60]fullerene bis-adduct (t3ss). This fullerene derivative protected SH-SY5Y cells subjected to hypoxic conditions and the involvement of adenosine and metabotropic glutamate receptors in this t3ss-mediated effect was suggested [94]. MTT assays revealed that the significant cell death detected after 6 and 24 h of hypoxic insult (5% O_2) was strongly recovered, and in a concentration dependent manner, when t3ss was present during hypoxia. The possible role of t3ss on gene expression modulation of receptors involved in neurodegenerative processes and diseases, such as adenosine (A_1, A_{2A}, A_{2B}) and metabotropic glutamate receptors ($mGlu_1$, $mGlu_5$) was also analysed. Interestingly, while no t3ss effect on gene expression was detected in normoxic conditions, a significant reduction in the hypoxia-induced increase in A_{2A} expression was achieved. Moreover, t3ss also increased A_1 and $mGlu_1$ expression after 24 h in hypoxia, being this effect maintained at 48 and 72 h. However, t3ss did not modify A_{2B} and $mGlu_5$ gene expression during hypoxia. These results show that the protective effect of [60]fullerene derivative against hypoxia, with increased effectiveness at higher concentrations, could be attributed, at least in part, to its modulatory effect on the expression of these receptor genes. Thus, the observed increase in A_1R gene expression by t3ss derivative could be related to the described protective role exerted by these receptors during hypoxic conditions [95]. A similar A_1 / A_{2A} receptor imbalance was reported in rat C6 glioma cells, where hypoxia (5% O_2) caused a significant decrease in A_1 receptors while A_{2A} receptors were significantly increased through a mechanism in which endogenous adenosine and tonic A_1 receptor activation is involved [96]. The use of antagonists of adenosine A_{2A} receptors during oxidative stress and related toxicity may be considered a valid therapeutic approach [97-99]. Therefore, although the exact mechanism by which t3ss is able to modulate gene expression of these receptors is not known, this effect represents one potential application for such fullerene derivatives.

Glutamate, the most abundant and important amino acid acting as an excitatory neurotransmitter in the mammalian central nervous system, plays a relevant role in many physiological processes as developmental synaptogenesis, plasticity, long-term potentiation, and learning and memory[100-101] but also in pathological processes as neurodegeneration [102]. Glutamate excitotoxic effect was detected in rat cortical neurons exposed to 100 μM L-glutamate for 2 or 6 hours, a highly toxic concentration for this cellular model with short periods of exposure [102-103]. Interestingly, presence of 5 μM t3ss clearly showed a protective effect, recovering cell viability loss induced by L-glutamate at 2 (almost completely) and 6 hours. Moreover, gene expression analysis at different time (24, 48, 72h) and dose of glutamate (0.1, 1 μM) and/or t3ss (1, 5 μM) revealed that a frequent effect of L-glutamate treatment was the decreased expression of A_{2A} and $mGlu_1$ receptors and the increased levels of A_{2B} mRNA. Concerning t3ss, a general effect was the increase in the gene expression of A_{2B}, A_1 and $mGlu_5$ receptors while the gene expression of A_{2A} and $mGlu_1$ results decreased [104]. Therefore, the modulation by t3ss of these receptors analyzed was different depending on the concentration employed, the time of treatment and, more interesting, the type of receptor. This modulation of adenosine and glutamate receptors gene expression could be, at least in part, responsible for the t3ss fullerene derivative protective effect against glutamate toxicity in cortical neurons.

The modulation of adenosine receptors gene expression and their corresponding protein levels by C_{60} bis-adduct hydrosoluble derivative t3ss was first reported in SK-N-MC cells derived from human neuroepithelioma [105]. Treatment of SK-N-MC cells using a single nontoxic concentration of t3ss (5 μM) led to an increase in A_{2A} and A_{2B} receptor mRNA expression and higher A_1 and A_{2A} protein levels. In this *in vitro* model, no significant cell viability loss was detected at doses equal or less than 10 μM t3ss; even at 1 mM, which largely exceeds any therapeutic dose, viability loss was 40% ($p < 0.001$). The fullerene derivative t3ss was unable to act as an A_1 receptor antagonist in radioligand binding assays, even at a range of high concentrations, including around 10 mM, i.e., t3ss cannot substitute N^6-cyclopentyladenosine (CPA) in the unspecific binding determination. This result excludes that t3ss may occupy the specific binding site of A_1, suggesting another way to regulate these receptors [105]. In line with this, other reported effect for C_{60} fullerene such as a reduced contractile response for acetylcholine, histamine, and 5-HT in muscle of guinea pig and rat may be due to changes at postreceptor level [106]. Low levels of fullerene C_{60} nanocrystals (nano C_{60}) enhanced learning and memory of rats through persistent Ca^{2+}/calmodulin-dependent protein kinase II (CaMKII) activation. These nano C_{60} enhanced the autonomous activity and T286 phosphorylation level of CaMKII in rat hippocampus, leading to enhancement of ERK and CREB phosphorylation, and facilitation of AMPA/NMDA ratio and CaMKII translocation to dendritic spines, and, finally, enhanced learning and memory *in vivo* [107]. The long-lasting activation of CaMKII by nano C_{60}, confirmed by using the FRET reporter Camuia (YFPCaMKIIa-CFP) expressed in cultured neurons, is coupled with CaMKII translocation to spines of neurons [108]. Considering that inhibitors of CaMKII prevented NMDA-evoked adenosine and inosine release and decreased nucleoside transporter function in neurons [109] and that stimulated CaMKII acts to promote the adenosine activated A_{2A}-mediated potentiation of cAMP production in astrocytes, being relatively specific for those responses induced by A_{2A} coupled to Gs over β-adrenoceptors coupled to Gs or $mGlu_{2/3}$ coupled to Gi [110], it would be possible to speculate that CaMKII could somehow contribute to C_{60} effect on adenosine receptors.

Research on predictive computational studies performed on 1207 protein entries (drug target database, PDTD) has shown that [60]fullerene can interact with hundreds of human proteins, thus potentially leading to a series of unpredictable ''off-target'' side effects [111]. The possible effects deriving from freely-diffusing [60]fullerene-based drugs were undertaken by confining two [60]fullerene derivatives bearing a C-unsubstituted pyrrolidine linkage and a ferrocenyl side group on a self-assembled monolayer (SAM) deposited on Au(111)-coated mica substrates. Scanning electron microscopy (SEM) images showed no evident morphological differences between undifferentiated neuroblastoma (SH-SY5Y) cells grown on a Petri dish, used as control, and those grown on the [60]fullerene-containing SAMs. Unprecedentedly, both [60]fullerene-based SAMs provided near-complete protection of SH-SY5Y cells exposed to excitotoxic levels (100 μM) of L-Glutamate for 6 hours [112]. In addition, SAMs revealed to be fully compatible materials for the retinoic acid-induced differentiation of SH-SY5Y cells into neurons, thus representing novel nanomaterials for neuronal regeneration in damaged tissues as in the case of ischemia. Therefore, surface immobilization of redox-active [60] fullerene derivatives and the consequent neuroprotective effects toward L-glutamate induced excitotoxicity in undifferentiated SH-SY5Y cells [112] could suggest an extracellular effect of C_{60} derivatives, discarding an intracellular role after its endocytosis. In this sense, fullerene-substituted poly-lysine derivative of a peptide with a FITC label (Baa-Lys(FITC)-$(Lys)_8$-OH) endocytosis in human HEK cells is mediated by caveolae/lipid raft. Moreover, both F-actin and microtubules regulated endocytosis. This ATP-dependent endocytosis most likely occurred through a scavenger receptor specific pathway being more complex than a simple endocytic pathway [113].

Non-viral vector-mediated gene delivery has been facilitated by a variety of polycations such as polyethylenimine (PEI), polyL-lysine (PLL) or polyallylamine (PAA). These polycations can condense DNA and bind to cells through electrostatic interaction between the net positive surface charges of complexes and negatively charges of cellular membrane. Accidentally, it was discovered that oligonucleotide $C_{10}A_{20}$ could mediate PAA-HIS/DNA complexes to exhibit high gene expression in both *in vitro* and *in vivo* systems [114-115]. The '*receptor*' on the cell surface for oligonucleotide-mediated gene delivery is an adenosine receptor, as pretreatment of HeLa cells with 8-SPT, an adenosine receptor antagonist, significantly reduced the transfection efficiency of PAAHIS/DNA/$C_{10}A_{20}$ complexes by 80%. Upon receptor binding, oligonucleotide-coated PAA-HIS/DNA complexes are internalized via caveolae-mediated pathway to result in effective intracellular processing for gene expression, showing a feasible approach to further develop new anionic nanoparticles for gene delivery [115].

Gold Nanoparticles

Gold nanoparticles (AuNPs) have received tremendous attention in colorimetric biosensors for possessing size- and distance dependent optical properties. The colour of the colloidal gold solution changes from red to purple/blue during AuNP aggregation or from purple/blue to red during redispersion of AuNP aggregates due to varying interparticle plasmon coupling and the resulting surface plasmon band shift [116-117].

The combination of AuNPs with aptamers (short ssDNAs or RNAs selected with a combinatorial technique called Systematic Evolution of Ligands by EXponential enrichment 7or SELEX) have been developed a homogeneous multiplex aptasensor by using more than one class of aptamers to stabilize AuNPs to measure, among others, adenosine, with a detection limit of 500 ng/mL [118].

An aptamer tethered to the surface of carboxylated polystyrene nanoparticles was able to detect adenosine at levels as low as 500 nM [119], approximately one order of magnitude lower concentration than what was previously reported by the selective recognition of adenosine with an adenosine-sensitive oligonucleotide, studied via the electrochemical oxidation of guanine bases of the aptamer in solution [120]. Another aptamer-based sensor microbeads hybridized with AuNPs–HRP conjugates could respond to 0.1 pM of adenosine

[121]. Similar adenosine aptasensor based on platinum nanoparticles can detect 1-750 nM adenosine [122]. In general, there are a wide diversity of nanosized devices developed to detect adenosine, ATP and adenosine deaminase, which may reflect the importance of purinergic signalling in biomedical research [117, 123-126].

Application of gold nanoparticles in biomedical and drug delivery has been reviewed elsewhere [127]. Among these, one concerning adenosine receptors is the conjugation of small molecular ligands for these receptors to AuNPs with retention of biological activity [128]. Due to the widespread expression of a given receptor there is a need for more selective and targeted ligands (agonist/antagonist) with the aim of avoid, or at least reduce, undesirable side effects [129]. In line with this, different agonist and antagonist ligands for A_{2A} and A_3Rs where attached to AuNPs in such a way that their selectivity for receptor was retained, even in some cases increased. As the ligands were covalently tethered to AuNPs, these constructions could serve both targeting and therapeutic purposes [128]. This methodology is promising as it could potentially be applied to other GPCR ligands.

QUANTUM DOTS

Colloidal semiconductor quantum dots (QDs) are single crystals a few nanometers in diameter whose size and shape can be precisely controlled by the duration, temperature, and ligand molecules used in the synthesis. QDs absorption and emission spectra are composition and size-dependent. The QDs fluorescent light is inversely proportional to its size, i.e., the smaller the size, the bluer the emission and the larger the size, the redder the emission. They are typically composed of atoms from groups II-VI (e.g., CdSe, CdS, CdTe, ZnSe), III-V (InP and InAs) and IV-VI (PbSe) [130].

QDs modified with PEG and primary antibodies have been synthesized to be used as fluorescent probes for a cell-based HTS system. In particular, GFP-tagged A_3 adenosine receptor (A_3R) have been used to detect molecules with agonist activity on these receptors [131].

The nucleoside-based agonist CGS21680 has been also thetered to CdSe/ZnS QDs. In this case, it was additionally used PAMAM D5 dendrons as multivalent spacer groups, grafted on the QD surface through a thioctic acid moiety, which suitably increased water solubility and maintained the ability of the QD conjugate to bind to the GPCR [132-133].

An interesting application of QDs related to adenosine receptors is the single particle tracking of QD-labeled receptors upon agonist binding. This technology served to observe differences in diffusivity and diffusion mode of two forms of the A_{2A} receptors, the wild type and a mutant engineered to have a palmitoylation site, in human embryonic kidney (HEK) 293 cells [134]. Later, it was also employed in hippocampal neurons to compare the agonist-induced switch to slow mobility of the wild type (full length) A_{2A} and a C-terminally truncated version of the receptor [135].

Although not yet directly applied to adenosine receptors, ~~or~~ its endogenous ligand, adenosine, has been utilized as a stabilizer of fluorescent sensor based on adenosine capped CdSe/ZnS quantum dots for highly sensitive detection of dopamine in human urine samples [136].

POLYMERIC NANOPARTICLES

It has been reported an attempt to encapsulate cyclopentyl adenosine (CPA), an A_1 receptor agonist, in poly(lactic acid) nanospheres which was not fruitful. However, the prodrug 5'-octanoyl CPA, converted into CPA only by plasma esterases, allowed its encapsulation to obtain the prodrug sustained release and the decrease of its hydrolysis rate in human whole blood. Moreover, the presence of nanospheres does not interfere with the CPA interaction to its action site, i.e., the adenosine A_1 receptor [137].

The identification of focal adenosine deficiency as major cause for seizure generation provides a neurochemical rationale for focal adenosine augmentation therapies (AATs). Focal AATs can be accomplished by implanting adenosine releasing devices into the vicinity of an epileptogenic focus [138]. To this end, different silk-based microspheres containing adenosine have been developed with therapeutic efficacy in the widely used rat kindling model [139].

SILICA NANOPARTICLES

The design of two-dyed fluorescent silica nanoparticles for ATP detection and measurement have been reported [140]. The indicator dye possesses a dipicolylamine (DPA) unit complexed with Zn(II) as a receptor function for ATP while a rhodamine derivative is used as the reference dye.

By combining different nanoparticles, i.e., Rhodamine B doped fluorescent silica nanoparticles (SiNPs), gold nanoparticles (AuNPs) and single-strand DNA/RNA oligonucleotides (aptamers), a novel device to measure adenosine with a detection limit as low as 45 nM has been developed. Adenosine causes the change of the aptamer conformation and weaken the FRET between the fluorescent SiNPs and AuNPs leading to the recovery of the fluorescence of SiNPs quenched by AuNPs [141].

CONCLUSION

In this chapter we have reviewed the state of art of using nanoparticles in several aspects concerning to adenosine receptors (Figure 1). Available data show that dendrimers and fullerene derivatives have been explored, among others, as nanoparticles able to bind and even modulate adenosine receptors. A_{2A} and A_3 are the mainly analysed receptors to date, probably because of their important role in cancer and neurodegeneration. From a therapeutic point of view, a careful design of such nanoparticles would permit some degree of potency or selectivity improvement for a desired ligand for a given receptor, or even oligomeric cluster of receptors (heteromers), which could increase the specificity in the site of action and, therefore, decrease undesirable side effects. Moreover, the toxicity of such compounds can be controlled by this design, achieving the low toxicity levels needed. As adenosine receptors have been involved in pathologies as cancer and neurodegenerative diseases, an important field of research is now open for the use of nanoparticles in diagnosis by imaging techniques and treatment of these diseases.

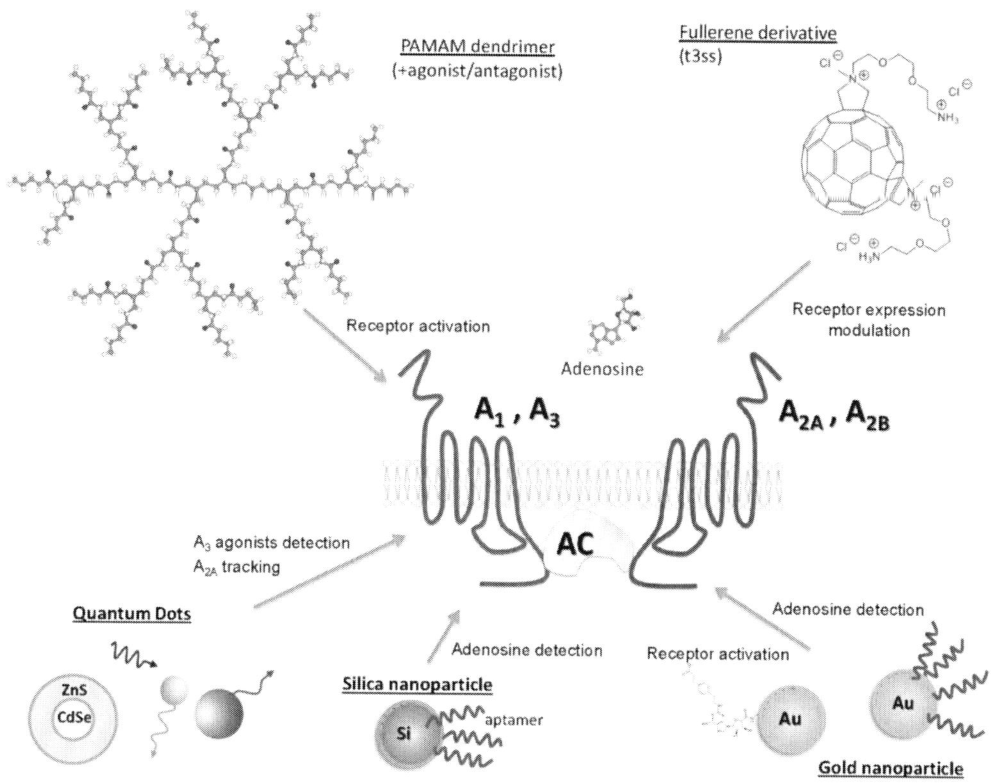

Figure 1. *Nanoparticles used in adenosine receptors research.* Different applications related to adenosine receptors have been reported ranging from receptor activation, tracking and modulation to endogenous ligand (adenosine) detection. For details see text.

REFERENCES

[1] Drexler, K.E. Molecular engineering: An approach to the development of general capabilities for molecular manipulation. *Proc. Natl. Acad. Sci. U S A.* 1981, 78, 5275-5278.

[2] Tomalia, D.A. In quest of a systematic framework for unifying and defining nanoscience. *J. Nanopart. Res.* 2009, 11, 1251-1310.

[3] Venkatraman, S. Has nanomedicine lived up to its promise? *Nanotechnology* 2014, *25*, 372501.

[4] Freitas, R. A., Jr. What is nanomedicine? *Dis Mon. 2005, 51*, 325-341.

[5] Sumer, B; Gao, J. (2008) Theranostic nanomedicine for cancer. *Nanomedicine (Lond)* 2008, 3, 137-140.

[6] Wolfram, J.; Zhu, M.; Yang, Y.; Shen, J.; Gentile, E.; Paolino, D.; Fresta, M.; Nie, G.; Chen, C.; Shen, H.; Ferrari, M.; Zhao, Y. Safety of Nanoparticles in Medicine. *Curr Drug Targets* 2014, epub ahead of print.

[7] Muthu, MS.; Leong, D.T.; Mei, L.; Feng, S.S. (2014) Nanotheranostics - application and further development of nanomedicine strategies for advanced theranostics. *Theranostics* 2014, 4, 660-677.

[8] Seigneuric, R.; Markey, L.; Nuyten, D.S.; Dubernet, C.; Evelo, C.T.; Finot, E.; Garrido, C. From nanotechnology to nanomedicine: applications to cancer research. *Curr Mol Med.* 2010, 10, 640-652.

[9] Agostinelli, E.; Vianello, F.; Magliulo, G.; Thomas, T.; Thomas, T.J. Nanoparticle strategies for cancer therapeutics: Nucleic acids, polyamines, bovine serum amine oxidase and iron oxide nanoparticles (Review), *Int J Oncol.* 2014, 46, 5-16.

[10] Saadeh, Y.; Leung, T.; Vyas, A.; Chaturvedi, L.S.; Perumal, O.; Vyas, D. Applications of nanomedicine in breast cancer detection, imaging, and therapy. *J. Nanosci. Nanotechnol.* 2014, 14, 913-923.

[11] Ouvinha de Oliveira, R.; de Santa Maria, L.C.; Barratt, G. Nanomedicine and its applications to the treatment of prostate cancer. *Ann. Pharm. Fr.* 2014, 72, 303-316.

[12] McCarthy, J.M.; Appelhans, D.; Tatzelt, J.; Rogers, M.S. Nanomedicine for prion disease treatment: new insights into the role of dendrimers. *Prion* 2013 7, 198-202.

[13] Ferreira, M. P.; Balasubramanian, V; Hirvonen, J; Ruskoaho, H.;Santos, H. A. Advanced Nanomedicines for the Treatment and Diagnosis of Myocardial Infarction and Heart Failure. *Curr Drug Targets,* 2014, epub ahead of print.

[14] Mulder, W.J.; Jaffer, F.A.; Fayad, Z.A.; Nahrendorf, M. Imaging and nanomedicine in inflammatory atherosclerosis. *Sci. Transl. Med.* 2014, 6, 239sr231.

[15] Kim, N.J.; Harris, A.; Gerber, A.; Tobe, L.A.; Amireskandari, A.; Huck, A.; Siesky, B. Nanotechnology and glaucoma: a review of the potential implications of glaucoma nanomedicine. *Br. J. Ophthalmol.* 2014, 98, 427-431.

[16] Sriramoju, B.; Kanwar, R.K.; Kanwar, J.R. Nanomedicine Based Nanoparticles for Neurological Disorders. *Curr. Med. Chem.* 2014, 21, 4154-4168.

[17] Re, F.; Gregori, M.; Masserini, M. Nanotechnology for neurodegenerative disorders, *Maturitas.* 2012, 73, 45-51.

[18] Brambilla, D.; Le Droumaguet, B.; Nicolas, J.; Hashemi, S.H.; Wu, L.P.; Moghimi, S.M.; Couvreur, P.; Andrieux, K. Nanotechnologies for Alzheimer's disease: diagnosis, therapy, and safety issues. *Nanomedicine.* 2011, 7, 521-540.

[19] Andrieux, K.; Couvreur, P. Nanomedicine as a promising approach for the treatment and diagnosis of brain diseases: the example of Alzheimer's disease. *Ann. Pharm. Fr.* 2013, 71, 225-233.

[20] Ruozi, B.; Belletti, D.; Pederzoli, F.; Veratti, P.; Forni, F.; Vandelli, M. A.; Tosi, G. Nanotechnology and Alzheimer's Disease: What Has Been Done and What to Do. *Curr Med Chem.* 2014 epub ahead of print.

[21] Martinez-Fong, D.; Bannon, M.J.; Trudeau, L.E.; Gonzalez-Barrios, J.A.; Arango-Rodriguez, M.L.; Hernandez-Chan, N.G.; Reyes-Corona, D.; Armendariz-Borunda, J.; Navarro-Quiroga, I. NTS-Polyplcx: a potential nanocarrier for neurotrophic therapy of Parkinson's disease. *Nanomedicine* 2012, 8, 1052-1069.

[22] Garbayo, E.; Estella-Hermoso de Mendoza, A.; Blanco-Prieto, M.J. Diagnostic and Therapeutic Uses of Nanomaterials in the Brain. *Curr Med Chem.* 2014. epub ahead of print.

[23] Burnstock, G. Introduction to purinergic signalling in the brain. *Adv Exp Med Biol.* 2013, 986, 1-12.

[24] Gomes, C.V.; Kaster, M.P.; Tome, A.R.; Agostinho, P.M.; Cunha, R.A. Adenosine receptors and brain diseases: neuroprotection and neurodegeneration, *Biochim Biophys Acta.* 2011, 1808, 1380-1399.

[25] Abbasi, E.; Aval, S.F.; Akbarzadeh, A.; Milani, M.; Nasrabadi, H.T.; Joo, S.W.; Hanifehpour, Y.; Nejati-Koshki, K.; Pashaei-Asl, R. Dendrimers: synthesis, applications, and properties, *Nanoscale Res. Lett.* 2014, *9*, 247.

[26] Duncan, R.; Izzo, L. Dendrimer biocompatibility and toxicity. *Adv. Drug Deliv. Rev.* 2005, 57, 2215-2237.

[27] Yellepeddi, V.K., Kumar, A., Palakurthi, S. Surface modified poly(amido)amine dendrimers as diverse nanomolecules for biomedical applications, *Expert Opin. Drug Deliv.* 2009, 6, 835-850.

[28] Jedrych, M.; Borowska, K.; Galus, R.; Jodlowska-Jedrych, B. The evaluation of the biomedical effectiveness of poly(amido)amine dendrimers generation 4.0 as a drug and as drug carriers: a systematic review and meta-analysis, *Int J Pharm.* 2014, 462, 38-43.

[29] Zlokovic, B.V. The blood-brain barrier in health and chronic neurodegenerative disorders. *Neuron* 2008, 57, 178-201.

[30] Huber, J.D.; Egleton, R.D.; Davis, T.P. (2001) Molecular physiology and pathophysiology of tight junctions in the blood-brain barrier. *Trends Neurosci.* 2001, 24, 719-725.

[31] Daneman, R. The blood-brain barrier in health and disease, *Ann Neurol.* 2012 72, 648-672.

[32] Carman, A.J.; Mills, J.H.; Krenz, A.; Kim, D.G.; Bynoe, M.S. Adenosine receptor signaling modulates permeability of the blood-brain barrier, *J. Neurosci.* 2011, 31, 13272-13280.

[33] Gao, X.; Qian, J.; Zheng, S.; Changyi, Y.; Zhang, J.; Ju, S.; Zhu, J.; Li, C. Overcoming the blood brain barrier for delivering drugs into the brain by using adenosine receptor nanoagonist, *ACS Nano.* 2014, 8, 3678-3689.

[34] Albanese, A.; Tang, P. S.; Chan, W. C. The effect of nanoparticle size, shape, and surface chemistry on biological systems, *Annu Rev Biomed Eng.* 2012, 14, 1-16.

[35] Win-Shwe, T. T.; Sone, H.; Kurokawa, Y.; Zeng, Y.; Zeng, Q.; Nitta, H.; Hirano, S. Effects of PAMAM dendrimers in the mouse brain after a single intranasal instillation, *Toxicol. Lett.* 2014, 228, 207-215.

[36] Albertazzi, L.; Gherardini, L.; Brondi, M.; Sulis Sato, S.; Bifone, A.; Pizzorusso, T.; Ratto, G. M.; Bardi, G. In vivo distribution and toxicity of PAMAM dendrimers in the central nervous system depend on their surface chemistry. *Mol. Pharm.* 2013, 10, 249-260.

[37] Kim, Y.; Hechler, B.; Klutz, A.M.; Gachet, C.; Jacobson, K.A. Toward multivalent signaling across G protein-coupled receptors from poly(amidoamine) dendrimers, *Bioconjug. Chem.* 2008, 19, 406-411.

[38] Jacobson, K. A. Functionalized congener approach to the design of ligands for G protein-coupled receptors (GPCRs), *Bioconjug Chem.* 2009, 20, 1816-1835.

[39] Panchuk-Voloshina, N.; Haugland, R. P.; Bishop-Stewart, J.; Bhalgat, M. K.; Millard, P. J.; Mao, F.; Leung, W. Y.; Haugland, R. P. Alexa dyes, a series of new fluorescent dyes that yield exceptionally bright, photostable conjugates, *J Histochem Cytochem.* 1999, 47, 1179-1188.

[40] Kim, Y.; Hechler, B.; Gao, Z. G.; Gachet, C.; Jacobson, K.A. PEGylated dendritic unimolecular micelles as versatile carriers for ligands of G protein-coupled receptors *Bioconjug. Chem.* 2009, 20, 1888-1898.

[41] Kecskes, A.; Tosh, D. K.; Wei, Q.; Gao, Z. G.; Jacobson, K.A. GPCR ligand dendrimer (GLiDe) conjugates: adenosine receptor interactions of a series of multivalent xanthine antagonists, *Bioconjug. Chem.* 2011, 22, 1115-1127.

[42] Klutz, A. M.; Gao, Z. G.; Lloyd, J.; Shainberg, A.; Jacobson, K. A. Enhanced A3 adenosine receptor selectivity of multivalent nucleoside-dendrimer conjugates, *J Nanobiotechnology.* 2008, 6, 12.

[43] Kim, Y.; Klutz, A. M.; Hechler, B.; Gao, Z. G.; Gachet, C.; Jacobson, K. A. Application of the functionalized congener approach to dendrimer-based signaling agents acting through A(2A) adenosine receptors, *Purinergic Signal.* 2009 5, 39-50.

[44] Keene, A. M.; Balasubramanian, R.; Lloyd, J.; Shainberg, A.; Jacobson, K. A. Multivalent dendrimeric and monomeric adenosine agonists attenuate cell death in HL-1 mouse cardiomyocytes expressing the A(3) receptor, *Biochem Pharmacol.* 2010, 80, 188-196.

[45] Wan, T. C.; Tosh, D. K.; Du, L.; Gizewski, E. T.; Jacobson, K. A.; Auchampach, J. A. Polyamidoamine (PAMAM) dendrimer conjugate specifically activates the A3 adenosine receptor to improve post-ischemic/reperfusion function in isolated mouse hearts, *BMC Pharmacol.* 2011, 11.

[46] Tosh, D. K,; Chinn, M,; Ivanov, A. A,; Klutz, A. M,; Gao, Z. G,; Jacobson, K. A. Functionalized congeners of A3 adenosine receptor-selective nucleosides containing a bicyclo[3.1.0]hexane ring system, *J Med Chem.* 2009, 52, 7580-7592.

[47] Tosh, D. K,; Chinn, M,; Yoo, L. S,; Kang, D. W,; Luecke, H,; Gao, Z. G,; Jacobson, K. A. 2-Dialkynyl derivatives of (N)-methanocarba nucleosides: 'Clickable' A(3) adenosine receptor-selective agonists, *Bioorg Med Chem.* 2010, 18, 508-517.

[48] Tosh, D. K,; Yoo, L. S,; Chinn, M,; Hong, K,; Kilbey, S. M,; 2nd, Barrett, M. O,; Fricks, I. P,; Harden, T. K,; Gao, Z. G,; Jacobson, K. A. Polyamidoamine (PAMAM) dendrimer conjugates of "clickable" agonists of the A3 adenosine receptor and coactivation of the P2Y14 receptor by a tethered nucleotide, *Bioconjug Chem.* 2010, 21, 372-384.

[49] Tosh, D. K.; Phan, K.; Deflorian, F.; Wei, Q.; Yoo, L. S.; Gao, Z. G.; and Jacobson, K. A. Click modification in the N6 region of A3 adenosine receptor-selective carbocyclic nucleosides for dendrimeric tethering that preserves pharmacophore recognition, *Bioconjug Chem.* 2012, 23, 232-247.

[50] El-Sayed, M.; Kiani, M. F.; Naimark, M. D.; Hikal, A. H.; Ghandehari, H. Extravasation of poly(amidoamine) (PAMAM) dendrimers across microvascular network endothelium, *Pharm Res.* 2001, 18, 23-28.

[51] Chanyshev, B.; Shainberg, A.; Isak, A.; Litinsky, A.; Chepurko, Y.; Tosh, D. K.; Phan, K.; Gao, Z. G.; Hochhauser, E.; Jacobson, K. A. Anti-ischemic effects of multivalent dendrimeric A(3) adenosine receptor agonists in cultured cardiomyocytes and in the isolated rat heart. *Pharmacol Res.* 2012, 65, 338-346.

[52] Tena-Campos, M.; Ramon, E.; Rivera, D.; Borroto-Escuela, D.O.; Romero-Fernandez, W.; Fuxe, K.; Garriga, P. G-protein-coupled receptors oligomerization: emerging signaling units and new opportunities for drug design. *Curr. Protein Pept. Sci.* 2014, 15, 648-658.

[53] Ivanov, A.A.; Jacobson, K.A. Molecular modeling of a PAMAM-CGS21680 dendrimer bound to an A2A adenosine receptor homodimer. *Bioorg. Med. Chem. Lett.* 2008, 18, 4312-4315.

[54] Jacobson, K.A.; Xie, R.; Young, L.; Chang, L.; Liang, B.T. A novel pharmacological approach to treating cardiac ischemia. Binary conjugates of A1 and A3 adenosine receptor agonists. *J. Biol. Chem.* 2000, 275, 30272-30279.

[55] Jacobson, K.A. GPCR ligand-dendrimer (GLiDe) conjugates: future smart drugs? *Trends Pharmacol. Sci.* 31, 575-579.

[56] Kroto, H.W., Heath, J.R., O'Brien, S.C.; Curl, R.F.; Smalley, R.E. C60: Buckminsterfullerene. *Nature* 1985, 318, 162 - 163.

[57] Yang, X.; Ebrahimi, A.; Li, J.; Cui, Q. Fullerene-biomolecule conjugates and their biomedicinal applications. *Int. J. Nanomedicine* 2014, *9*, 77-92.

[58] Bosi, S.; Da Ros, T.; Spalluto, G.; Prato, M. Fullerene derivatives: an attractive tool for biological applications. *Eur. J. Med. Chem. 2003,* 38, 913-923.

[59] Chen, Z.; Ma, L.; Liu, Y.; Chen, C. Applications of functionalized fullerenes in tumor theranostics. *Theranostics* 2012, *2*, 238-250.

[60] Liu, Q.; Cui, Q.; Li, X. J.; Jin, L. The applications of buckminsterfullerene C60 and derivatives in orthopaedic research. *Connect Tissue Res.* 2014, *55*, 71-79.

[61] Li, Z.; Pan, L.L.; Zhang, F.L.; Zhu, X.L.; Liu, Y.; Zhang, Z.Z. 5-Aminolevulinic acid-loaded fullerene nanoparticles for in vitro and in vivo photodynamic therapy, *Photochem Photobiol.* 2014, *90*, 1144-1149.

[62] Li, Z.; Pan, L.L.; Zhang, F.L.; Wang, Z.; Shen, Y.Y.; Zhang, Z.Z. Preparation and characterization of fullerene (C60) amino acid nanoparticles for liver cancer cell treatment. *J. Nanosci. Nanotechnol.* 2014, *14*, 4513-4518.

[63] Hu, Z.; Zhang, C.; Huang, Y.; Sun, S.; Guan, W.; Yao, Y. Photodynamic anticancer activities of water-soluble C(60) derivatives and their biological consequences in a HeLa cell line. *Chem Biol Interact 195*, 86-94.

[64] Mamontova, T. V.; Mykytiuk, M.V.; Bobrova, N.O.; Kutsenko, L.O.; Vesnina, L.E.; Kaidashev, I.P. [The anti-inflammatory effect of fullerene C60 on adjuvant arthritis in rats]. *Fiziol. Zh.* 2013, *59*, 102-110.

[65] Henderson, C.S.; Madison, A.C.; Shah, A. Size matters - nanotechnology and therapeutics in rheumatology and immunology. *Curr. Rheumatol. Rev. 2014,* 10, 11-21.

[66] Kolosnjaj, J.; Szwarc, H.; Moussa, F. Toxicity studies of fullerenes and derivatives, *Adv. Exp. Med. Biol.* 2007, 620, 168-180.

[67] Shinohara, N.; Gamo, M.; Nakanishi, J. Fullerene c60: inhalation hazard assessment and derivation of a period-limited acceptable exposure level. *Toxicol. Sci. 2011, 123*, 576-589.

[68] Dugan, L.L.; Tian, L.; Quick, K.L.; Hardt, J.I.; Karimi, M.; Brown, C.; Loftin, S.; Flores, H.; Moerlein, S.M.; Polich, J.; Tabbal, S.D.; Mink, J.W.; Perlmutter, J.S. Carboxyfullerene neuroprotection postinjury in Parkinsonian nonhuman primates, *Ann. Neurol.* 2014, 76, 393-402.

[69] Dugan, L.L.; Lovett, E.G.; Quick, K.L.; Lotharius, J.; Lin, T.T.; O'Malley, K.L. Fullerene-based antioxidants and neurodegenerative disorders. *Parkinsonism Relat. Disord., 2001,* 7, 243-246.

[70] Linazasoro, G. Potential applications of nanotechnologies to Parkinson's disease therapy, *Parkinsonism Relat. Disord.* 2008, *14*, 383-392.

[71] Makarova, E.G.; Gordon, R.Y.; Podolski, I.Y. Fullerene C60 prevents neurotoxicity induced by intrahippocampal microinjection of amyloid-beta peptide. *J. Nanosci. Nanotechnol.* 2012, *12*, 119-126.

[72] Yamago, S.; Tokuyama, H.; Nakamura, E.; Kikuchi, K.; Kananishi, S.; Sueki, K.; Nakahara, H.; Enomoto, S.; Ambe, F. In vivo biological behavior of a water-miscible fullerene: 14C labeling, absorption, distribution, excretion and acute toxicity. *Chem. Biol.* 1995, 2, 385-389.

[73] Groom, A.J.; Smith, T.; Turski, L. Multiple sclerosis and glutamate. *Ann N Y Acad Sci. 2003, 993*, 229-275; discussion 287-228.

[74] Koutsilieri, E.; and Riederer, P. Excitotoxicity and new antiglutamatergic strategies in Parkinson's disease and Alzheimer's disease. *Parkinsonism Relat. Disord.* 2007, 13 Suppl 3, S329-331.

[75] Dugan, L.L.; Bruno, V.M.; Amagasu, S.M.; Giffard, R.G. Glia modulate the response of murine cortical neurons to excitotoxicity: glia exacerbate AMPA neurotoxicity. J. Neurosci. 1995, 15, 4545-4555.

[76] Greenlund, L.J.; Deckwerth, T.L.; Johnson, E.M. Jr. Superoxide dismutase delays neuronal apoptosis: a role for reactive oxygen species in programmed neuronal death. *Neuron* 1995, 14, 303-315.

[77] Dragunow, M.; Faull, R.L.; Lawlor, P.; Beilharz, E.J.; Singleton, K.; Walker, E.B.; Mee, E. In situ evidence for DNA fragmentation in Huntington's disease striatum and Alzheimer's disease temporal lobes. *Neuroreport* 1995, 6, 1053-1057.

[78] Calissano, P.; Matrone, C.; Amadoro, G. Apoptosis and in vitro Alzheimer disease neuronal models. *Commun. Integr. Biol.2009, 2*, 163-169.

[79] Krusic, P.J.; Wasserman, E.; Keizer, P.N.; Morton, J.R.; Preston, K.F. Radical reactions of c60. *Science* 1991, 254, 1183-1185.

[80] Krokosz, A. Fullerenes in biology. *Postepy. Biochem.* 2007, 53, 91-96.

[81] Grebowski, J.; Kazmierska, P.; Krokosz, A. Fullerenols as a new therapeutic approach in nanomedicine. *Biomed. Res. Int. 2013*, 751913.

[82] Basso, A.S.; Frenkel, D.; Quintana, F.J.; Costa-Pinto, F.A.; Petrovic-Stojkovic, S.; Puckett, L.; Monsonego, A.; Bar-Shir, A.; Engel, Y.; Gozin, M.; Weiner, H.L. Reversal of axonal loss and disability in a mouse model of progressive multiple sclerosis. *J. Clin. Invest.* 2008, 118, 1532-1543.

[83] Ali, S.S.; Hardt, J.I.; Dugan, L.L. SOD activity of carboxyfullerenes predicts their neuroprotective efficacy: a structure-activity study. *Nanomedicine* 2008, 4, 283-294.

[84] Dugan, L.L.; Gabrielsen, J.K.; Yu, S.P.; Lin, T.S.; Choi, D.W. Buckminsterfullerenol free radical scavengers reduce excitotoxic and apoptotic death of cultured cortical neurons. *Neurobiol. Dis.* 1996, 3, 129-135.

[85] Jin, H.; Chen, W.Q.; Tang, X.W.; Chiang, L.Y.; Yang, C.Y.; Schloss, J.V.; Wu, J.Y. Polyhydroxylated C(60), fullerenols, as glutamate receptor antagonists and neuroprotective agents. *J. Neurosci. Res.* 2000, 62, 600-607.

[86] Ehrich, M.; Van Tassell, R.; Li, Y.; Zhou, Z.; Kepley, C.L. Fullerene antioxidants decrease organophosphate-induced acetylcholinesterase inhibition in vitro. *Toxicol. In Vitro 2011, 25*, 301-307.

[87] Barnham, K.J.; Masters, C.L.; Bush, A.I. Neurodegenerative diseases and oxidative stress. *Nat. Rev. Drug Discov.* 2004, 3, 205-214.

[88] De Leo, M.E.; Borrello, S.; Passantino, M.; Palazzotti, B.; Mordente, A.; Daniele, A.; Filippini, V.; Galeotti, T.; Masullo, C. Oxidative stress and overexpression of manganese superoxide dismutase in patients with Alzheimer's disease. *Neurosci. Lett.* 1998, 250, 173-176.

[89] Nunomura, A.; Perry, G.; Aliev, G.; Hirai, K.; Takeda, A.; Balraj, E.K.; Jones, P.K.; Ghanbari, H.; Wataya, T.; Shimohama, S.; Chiba, S.; Atwood, C.S.; Petersen, R.B.; Smith, M.A. Oxidative damage is the earliest event in Alzheimer disease, *J Neuropathol Exp. Neurol.* 2001, *60*, 759-767.

[90] Wirz, K.T.; Keitel, S.; Swaab, D.F.; Verhaagen, J.; Bossers, K. Early molecular changes in Alzheimer disease: can we catch the disease in its presymptomatic phase? *J. Alzheimers Dis.* 2014, *38*, 719-740.

[91] Rajendran, P.; Nandakumar, N.; Rengarajan, T.; Palaniswami, R.; Gnanadhas, E.N.; Lakshminarasaiah, U.; Gopas, J.; Nishigaki, I. Antioxidants and human diseases. *Clin. Chim. Acta* 2014, 436, 332-347.

[92] Guldi, DM; Prato, M. Excited-state properties of C(60) fullerene derivatives, *Acc. Chem. Res.* 2000, 33, 695-703.

[93] Bosi, S.; Feruglio, L.; Milic, D.; Prato, M. Synthesis and water solubility of novel fullerene bisadduct derivatives. *Eur. J. Org. Chem.* 2003, 24, 4741–4747.

[94] Giust, D.; Da Ros, T.; Martin, M.; Albasanz, J. Fullerene derivative modulates adenosine and metabotropic glutamate receptors gene expression: a possible protective effect against hypoxia. *J. Nanobiotechnology* 2014, *12*, 27.

[95] Arrigoni, E.; Crocker, A.J.; Saper, C.B.; Greene, R.W.; Scammell, T.E. Deletion of presynaptic adenosine A1 receptors impairs the recovery of synaptic transmission after hypoxia. *Neuroscience* 2005, 132, 575-580.

[96] Castillo, C.A.; Leon, D.; Ruiz, M.A.; Albasanz, J.L.; Martin, M. Modulation of adenosine A1 and A2A receptors in C6 glioma cells during hypoxia: involvement of endogenous adenosine. *J. Neurochem.* 2008, *105*, 2315-2329.

[97] Cieslak, M.; Komoszynski, M.; Wojtczak, A. Adenosine A(2A) receptors in Parkinson's disease treatment. *Purinergic Signal.* 2008, *4*, 305-312.

[98] Leite, M.R.; Wilhelm, E.A.; Jesse, C.R.; Brandao, R.; Nogueira, C.W. Protective effect of caffeine and a selective A2A receptor antagonist on impairment of memory and oxidative stress of aged rats. *Exp. Gerontol.* 2011, 46, 309-315.

[99] Rivera-Oliver, M.; Diaz-Rios, M. Using caffeine and other adenosine receptor antagonists and agonists as therapeutic tools against neurodegenerative diseases: a review. *Life Sci.* 2014, *101*, 1-9.

[100] Riedel, G.; Platt, B.; Micheau, J. Glutamate receptor function in learning and memory, *Behav. Brain Res.* 2003, 140, 1-47.

[101] Meldrum, B.S. (2000) Glutamate as a neurotransmitter in the brain: review of physiology and pathology. *J. Nutr.* 2000, 130, 1007S-1015S.

[102] Dong, X.X.; Wang, Y.; Qin, Z.H. Molecular mechanisms of excitotoxicity and their relevance to pathogenesis of neurodegenerative diseases. *Acta Pharmacol. Sin.* 2009, 30, 379-387.

[103] Nicoletti, F.; Bockaert, J.; Collingridge, G.L.; Conn, P.J.; Ferraguti, F.; Schoepp, D.D.; Wroblewski, J.T.; Pin, J.P. Metabotropic glutamate receptors: from the workbench to the bedside. *Neuropharmacology* 2011, 60, 1017-1041.

[104] Giust, D.; Da Ros, T.; Martin, M.; Albasanz, J.L. Modulation of gene expression of adenosine and metabotropic glutamate receptors in rat's neuronal cells exposed to L-glutamate and [60]fullerene. *J. Biomed. Nanotechnol.* 2014, 10, 1610 1619.

[105] Giust, D.; Leon, D.; Ballesteros-Yanez, I.; Da Ros, T.; Albasanz, J.L.; Martin, M. Modulation of adenosine receptors by [60]fullerene hydrosoluble derivative in SK-N-MC cells. *ACS Chem. Neurosci.* 2011, 2, 363-369.

[106] Satoh, M.; Matsuo, K.; Takanashi, Y.; Takayanagi, I. Effects of acute and short-term repeated application of fullerene C60 on agonist-induced responses in various tissues of guinea pig and rat. *Gen. Pharmacol.* 1995, 26, 1533-1538.

[107] Chen, L.; Miao, Y.; Chen, L.; Xu, J.; Wang, X.; Zhao, H.; Shen, Y.; Hu, Y.; Bian, Y.; Shen, Y.; Chen, J.; Zha, Y.; Wen, L. P.; Wang, M. The role of low levels of fullerene C60 nanocrystals on enhanced learning and memory of rats through persistent CaMKII activation. *Biomaterials* 2014, 35, 9269-9279.

[108] Miao, Y.; Xu, J.; Shen, Y.; Chen, L.; Bian, Y.; Hu, Y.; Zhou, W.; Zheng, F.; Man, N.; Shen, Y.; Zhang, Y.; Wang, M.; Wen, L. Nanoparticle as signaling protein mimic: robust structural and functional modulation of CaMKII upon specific binding to fullerene C60 nanocrystals, *ACS Nano.* 2014, 8, 6131-6144.

[109] Zamzow, C. R.; Bose, R.; Parkinson, F.E. N-methyl-D-aspartate-evoked adenosine and inosine release from neurons requires extracellular calcium. *Can. J. Physiol. Pharmacol.* 2009, 87, 850-858.

[110] Moldrich, R.X.; Aprico, K.; Diwakarla, S.; O'Shea, R.D.; Beart, P.M. Astrocyte mGlu(2/3)-mediated cAMP potentiation is calcium sensitive: studies in murine neuronal and astrocyte cultures, *Neuropharmacology* 2002, 43, 189-203.

[111] Calvaresi, M.; Zerbetto, F. Baiting proteins with C60. *ACS Nano.* 2010, 4, 2283-2299.

[112] Giust, D.; Albasanz, J.L.; Martin, M.; Marega, R.; Delforge, A.; Bonifazi, D. [60]Fullerene-based monolayers as neuroprotective biocompatible hybrid materials. *Chem Commun (Camb).* 2011, 47, 10617-10619.

[113] Zhang, L. W.; Yang, J.; Barron, A.R.; Monteiro-Riviere, N.A. Endocytic mechanisms and toxicity of a functionalized fullerene in human cells, *Toxicol. Lett.* 2009, 191, 149-157.

[114] Chung, Y.C.; Hsieh, W.Y.; Young, T.H. Polycation/DNA complexes coated with oligonucleotides for gene delivery. *Biomaterials* 2011, 31, 4194-4203.

[115] Chung, Y.C.; Cheng, T.Y.; Young, T.H. The role of adenosine receptor and caveolae-mediated endocytosis in oligonucleotide-mediated gene transfer, *Biomaterials* 2010, 32, 4471-4480.

[116] Ghosh, S.K.; Pal, T. Interparticle coupling effect on the surface plasmon resonance of gold nanoparticles: from theory to applications. *Chem. Rev.* 2007, 107, 4797-4862.

[117] Liu, D.; Wang, Z.; Jiang, X. Gold nanoparticles for the colorimetric and fluorescent detection of ions and small organic molecules. *Nanoscale* 2011, 3, 1421-1433.

[118] Niu, S.; Lv, Z.; Liu, J.; Bai, W.; Yang, S.; Chen, A. Colorimetric aptasensor using unmodified gold nanoparticles for homogeneous multiplex detection, *PLoS One* 2014, 9, e109263.

[119] Alsager, O.A.; Kumar, S.; Willmott, G.R.; McNatty, K.P.; Hodgkiss, J.M. Small molecule detection in solution via the size contraction response of aptamer functionalized nanoparticles. *Biosens. Bioelectron.* 2014, 57, 262-268.

[120] Kim, J.; Kim, I.Y.; Choi, M.S.; Wu, Q. Label-free electrochemical detection of adenosine based on electron transfer from guanine bases in an adenosine-sensitive aptamer. *Chem. Commun. (Camb)*, 2009, 4747-4749.

[121] Zhang, H,; Hu, X,; Fu, X. Aptamer-based microfluidic beads array sensor for simultaneous detection of multiple analytes employing multienzyme-linked nanoparticle amplification and quantum dots labels. *Biosens. Bioelectron.* 2014, *57*, 22-29.

[122] Shahdost-fard, F,; Salimi, A,; Khezrian, S. Highly selective and sensitive adenosine aptasensor based on platinum nanoparticles as catalytical label for amplified detection of biorecognition events through H2O2 reduction. *Biosens. Bioelectron* 2014, 53, 355-362.

[123] Sun, W,; Wang, X,; Sun, X,; Deng, Y,; Liu, J,; Lei, B,; Sun, Z. Simultaneous electrochemical determination of guanosine and adenosine with graphene-ZrO2 nanocomposite modified carbon ionic liquid electrode. *Biosens. Bioelectron* 2013, 44, 146-151.

[124] Zhang, M,; Guo, S. M.; Li, Y. R.; Zuo, P.; Ye, B.C. A label-free fluorescent molecular beacon based on DNA-templated silver nanoclusters for detection of adenosine and adenosine deaminase. *Chem. Commun. (Camb)* 2012, 48, 5488-5490.

[125] Zhang, L.; Zhao, J.; Jiang, J.; Yu, R. Enzyme-regulated unmodified gold nanoparticle aggregation: a label free colorimetric assay for rapid and sensitive detection of adenosine deaminase activity and inhibition *Chem. Commun. (Camb)* 2012, 48, 10996-10998.

[126] Hao, S.; Chen, G.; Yang, C. Sensing using rare-earth-doped upconversion nanoparticles. *Theranostics* 2013, *3*, 331-345.

[127] Daraee, H.; Eatemadi, A.; Abbasi, E.; Fekri Aval, S.; Kouhi, M.; Akbarzadeh, A. Application of gold nanoparticles in biomedical and drug delivery. *Artif. Cells Nanomed. Biotechnol*. 2014, 1-13.

[128] Jayasekara, P.S.; Phan, K.; Tosh, D.K.; Kumar, T.S.; Moss, S.M.; Zhang, G.; Barchi, J.J. Jr; Gao, Z.G.; Jacobson, K.A. Modulation of G protein-coupled adenosine receptors by strategically functionalized agonists and antagonists immobilized on gold nanoparticles, *Purinergic Signal.* 2013, 9, 183-198.

[129] Smith, N.J.; Bennett, K.A.; Milligan, G. When simple agonism is not enough: emerging modalities of GPCR ligands. *Mol. Cell. Endocrinol.* 2011, 331, 241-247.

[130] Michalet, X.; Pinaud, F.F.; Bentolila, L.A.; Tsay, J.M.; Doose, S.; Li, J.J.; Sundaresan, G.; Wu, A.M.; Gambhir, S.S.; Weiss, S. Quantum dots for live cells, in vivo imaging, and diagnostics, *Science* 2005, 307, 538-544.

[131] Lee, J.; Kwon, Y. J.; Choi, Y.; Kim, H. C.; Kim, K.; Kim, J.; Park, S.; Song, R. Quantum dot-based screening system for discovery of G protein-coupled receptor agonists. *Chembiochem.* 2012, 13, 1503-1508.

[132] Das, A.; Sanjayan, G.J.; Kecskes, M.; Yoo, L.; Gao, Z.G.; Jacobson, K.A. Nucleoside conjugates of quantum dots for characterization of G protein-coupled receptors: strategies for immobilizing A2A adenosine receptor agonists. *J. Nanobiotechnology* 2010, 8, 11.

[133] Jacobson, K.A. Structure-based approaches to ligands for G-protein-coupled adenosine and P2Y receptors, from small molecules to nanoconjugates. *J. Med. Chem.* 2013, 56, 3749-3767.

[134] Keuerleber, S.; Thurner, P.; Gruber, C.W.; Zezula, J.; Freissmuth, M. Reengineering the collision coupling and diffusion mode of the A2A-adenosine receptor: palmitoylation in helix 8 relieves confinement, *J. Biol. Chem.* 2012, 287, 42104-42118.

[135] Thurner, P.; Gsandtner, I.; Kudlacek, O.; Choquet, D.; Nanoff, C.; Freissmuth, M.; Zezula, J. A two-state model for the diffusion of the A2A adenosine receptor in hippocampal neurons: agonist-induced switch to slow mobility is modified by synapse-associated protein 102 (SAP102), *J. Biol. Chem.* 2014, 289, 9263-9274.

[136] Mu, Q.; Xu, H.; Li, Y.; Ma, S.; Zhong, X. Adenosine capped QDs based fluorescent sensor for detection of dopamine with high selectivity and sensitivity. *Analyst* 2014, 139, 93-98.

[137] Dalpiaz, A.; Leo, E.; Vitali, F.; Pavan, B.; Scatturin, A.; Bortolotti, F.; Manfredini, S.; Durini, E.; Forni, F.; Brina, B.; Vandelli, M.A. Development and characterization of biodegradable nanospheres as delivery systems of anti-ischemic adenosine derivatives. *Biomaterials* 2005, 26, 1299-1306.

[138] Boison, D. Adenosine-based cell therapy approaches for pharmacoresistant epilepsies. *Neurodegener. Dis.* 2007, 4, 28-33.

[139] Wilz, A.; Pritchard, E.M.; Li, T.; Lan, J.Q.; Kaplan, D.L.; Boison, D. Silk polymer-based adenosine release: therapeutic potential for epilepsy. *Biomaterials* 2008, *29*, 3609-3616.

[140] Moro, A.J.; Schmidt, J.; Doussineau, T.; Lapresta-Fernandez, A.; Wegener, J.; Mohr, G.J. Surface-functionalized fluorescent silica nanoparticles for the detection of ATP, *Chem. Commun. (Camb)* 2011, 47, 6066-6068.

[141] Qiang, W.; Liu, H.; Li, W.; Chen, X.; Xu, D. Label-free detection of adenosine based on fluorescence resonance energy transfer between fluorescent silica nanoparticles and unmodified gold nanoparticles. *Anal. Chim. Acta* 2014, 828, 92-98.

INDEX

#

A

B

C

D

E

F

G

H

I

K

L

M

N

O

P

T